Biotechnology: Prospects and Applications

R.K. Salar • S.K. Gahlawat
P. Siwach • J.S. Duhan
Editors

Biotechnology: Prospects and Applications

Springer

Editors
R.K. Salar
Department of Biotechnology
Chaudhary Devi Lal University
Sirsa, Haryana, India

S.K. Gahlawat
Department of Biotechnology
Chaudhary Devi Lal University
Sirsa, Haryana, India

P. Siwach
Department of Biotechnology
Chaudhary Devi Lal University
Sirsa, Haryana, India

J.S. Duhan
Department of Biotechnology
Chaudhary Devi Lal University
Sirsa, Haryana, India

ISBN 978-81-322-1682-7 ISBN 978-81-322-1683-4 (eBook)
DOI 10.1007/978-81-322-1683-4
Springer New Delhi Heidelberg New York Dordrecht London

Library of Congress Control Number: 2013958395

Printed on acid-free paper

Springer is part of Springer Science+Business Media (www.springer.com)

Dedicated to our Teachers

Preface

The past few decades have witnessed a more exciting time in our knowledge of biotechnology and continue to fascinate many scientists throughout the world. Having said that, it is also quite mesmerizing how little this has helped in generating new biotechnology products. Incredible advances in this field are occurring at a dizzying pace, and biotechnology has made an impact on many aspects of our everyday lives. This excitement about the new advances is well justified; however, we have a long way to go.

Based on diverse gamut of biotechnology, several distinct subdisciplines of biotechnology have emerged. An edited book is the collaborative result of hard work from many dedicated scientists and researchers. *Biotechnology: Prospects and Applications* is such a treatise that is divided into four parts, namely, Agriculture and Food Biotechnology, Microbial Biotechnology, Environmental Biotechnology, and Animal Biotechnology. The advances in these and allied areas of biotechnology have created unprecedented opportunities for cross talk among the scientific communities.

The upsurge in research has justified the inclusion of a diverse range of topics in this book. To improve production and productivity of crops, to conserve biodiversity, to generate cheap alternative sources of fuel, to design drugs for dreaded diseases, to protect the environment in ever-growing industrial age, and to make the life easier, there is no option but to harness the potential of biotechnology in various areas of research and development. Keeping this perspective in view, *Biotechnology: Prospects and Applications* provides a comprehensive and authoritative resource on these and allied biological subdisciplines.

We hope that the outlook in this book is distinctly global. An international authorship was, therefore, inevitable, and we as editors are grateful to all contributing authors for providing necessary expertise and commitment. Without their contribution, this book would have not been possible. Our sincere thanks go to the University Grants Commission, New Delhi, for sponsoring *International Conference on Biotechnology: Emerging Trends* which was organized by Department of Biotechnology, Chaudhary Devi Lal University, Sirsa,

India, during September 2012 and attended by most contributing authors. Last but not the least, we also thank our family and friends for their support and encouragement while we spent endless hours on this project.

Sirsa, India

R.K. Salar
S.K. Gahlawat
P. Siwach
J.S. Duhan

Contents

Part I

Agriculture and Food Biotechnology

1 Emerging Dynamics of Brassinosteroids Research

Isha Sharma, Navdeep Kaur, Shivani Saini, and Pratap Kumar Pati

Abstract

Brassinosteroids are a class of naturally occurring steroidal compounds that play vital role in plant growth and development. Extensive research on BRs biosynthesis, distribution, signal perception, and transduction has broadened our understanding about this important phytohormone. The mode of action of brassinosteroids involves the perception of its signal by the cell surface receptor and the subsequent activation of downstream transcription factors and genes. Development of modern tools, availability of biological resources, and use of genetic, biochemical, and proteomic approaches have greatly advanced our understanding on the key regulatory elements in the BRs signaling networks. At present, brassinosteroids are implicated in various important functions in plants such as growth and development as well as stress amelioration. However, further advancement of knowledge in the area will enable scientists to precisely answer some key fundamental questions related to its versatile roles in plants.

1.1 Introduction

Brassinosteroids (BRs) are a class of plant hormones that are ubiquitously present in plant kingdom and play an important role in multiple developmental and physiological processes, including seed germination, stem elongation, leaf expansion, and xylem differentiation, and in biotic and abiotic stress resistance (Gudesblat and Russinova 2011; Fridman and Savaldi-Goldstein 2013). BRs, being steroidal in nature, share a high degree of similarity with their animal counterparts which play an important role in embryonic development and adult homeostasis in animals (Clouse 2011). Owing to this analogy with animal system, as well as their ability to generate the physiological response at very low concentrations with concomitant ability to alter gene expression, these compounds are named as hormones (Davies 1995). It happened a little over 40 years ago, when this c-28 steroid with an unusual lactone B-ring structure was isolated from *Brassica* pollen extracts and was found to have unique growth-promoting activity. Till date, nearly 70 different analogs of BRs have been added to the family of BRs (Clouse 2011). Brassinolide (BL), however, shows the highest

I. Sharma • N. Kaur • S. Saini • P.K. Pati (✉)
Department of Biotechnology, Guru Nanak Dev University, Amritsar 143005, Punjab, India
e-mail: pkpati@yahoo.com

R.K. Salar et al. (eds.), *Biotechnology: Prospects and Applications*,
DOI 10.1007/978-81-322-1683-4_1, © Springer India 2013

biological activity among various analogs (Mandava 1988). Castesterone (CS), an immediate precursor to BL, shows only 10 % of the activity of BL, while the rest of the analogs are the intermediates of the BL biosynthetic pathway or inactivated by-products of various BR catabolic reactions (Zhao and Li 2012). Though several studies confirmed the physiological and developmental responses induced by brassinosteroids, their indispensability for the plant growth and development was not established until mid-1990s when BR-insensitive and BR-deficient mutants were identified independently by four groups (Clouse et al. 1996; Kauschmann et al. 1996; Li et al. 1996; Szekeres et al. 1996). It was observed that the mutants of various genes involved in BRs signaling and biosynthesis exhibited characteristic phenotype and their analysis provided convincing genetic and biochemical evidence for the essential role of BRs in normal plant growth and development. These findings led to a growing interest of many researchers to work in the field of brassinosteroids. For the past one decade, there has been a significant progress in our understanding of BR biology involving BR biosynthesis, their signal perception and transduction mechanism, as well as their versatile role in governing growth and developmental processes in plants. The present chapter highlights the latest developments in the field of BR research with more emphasis to new key players identified in the BR biosynthesis, regulation of their homeostasis as well as their signaling network, and the growing implications of BRs research in crop improvement.

1.2 Brassinosteroids Biosynthesis and Regulation of Homeostasis

Various biochemical studies were conducted for the elucidation of the BRs biosynthetic pathway. However, the clear understanding of the BRs biosynthesis was obtained by feeding various isotope-labeled intermediates to *Catharanthus roseus* suspension culture seedlings and subsequent analysis by gas chromatography-mass spectrometric (GC-MS) approaches to identify various metabolic products and their reaction orders. The study has revealed that BL biosynthesis occurs from campesterol which is first converted to campestanol and then bifurcates into two parallel-branched pathways, namely, the early and late C-6 oxidation pathways following series of reductions, hydroxylations, epimerizations, and oxidations, and then ultimately converging at castesterone which leads to the formation of BL (Fujioka and Sakurai 1997; Sakurai and Fujioka 1997; Noguchi et al. 2000). However, studies suggest that a certain degree of cross talk occurs between the parallel pathways revealing the complex networking of brassinosteroid synthesis (Fujioka and Yokota 2003). Moreover, the enzymes involved in catalyzing each of the reactions from campesterol to BL have been dealt with in great details in different plant species via mutant analyses (Noguchi et al. 2000). Besides, the classical early and late C6-oxidation pathway, another novel shortcut route to the BL biosynthesis, has been proposed recently which has been thought to be the predominant pathway of BRs synthesis (Ohnishi et al. 2012). This study suggested that plants may prefer to use 8-step conversion involving campestanol-independent route and a late C-6 oxidation pathway instead of 10-step campestanol-dependent pathway. For better understanding of the BRs biosynthetic pathway, several key genes involved in their biosynthesis have been characterized, viz., De-etiolated-2 (DET2), which is a 5α-reductase; constitutive photomorphogenesis and dwarfism (CPD), a true C-3 oxidase; DWARF4, a C-22 hydroxylase; and ROT3 (CYP90C1) and CYP90D1, C-23 hydroxylases. Various genetic and biochemical evidences suggest their important role in the biosynthesis of BRs as well as indicate towards a great extent of functional redundancy among them (Wang et al. 2012a).

Since BRs do not undergo a long-distance transport, the regulation of BR homeostasis at the tissue or even at the cellular level is extremely important for normal plant growth and development. Experimental analysis has shown that most of the genes involved in BR biosynthesis and inactivation are feedback regulated (Tanaka et al.

2005; Sun et al. 2010; Yu et al. 2011). Studies have shown that BR-specific biosynthesis genes (DET2, DWF4, CPD, BR6ox1, and ROT3) are upregulated in BR-depleted wild-type plants grown under a BR biosynthesis inhibitor, brassinazole, while they were downregulated on the application of exogenous brassinolide. On the other hand, a BR inactivation gene (BAS1) was upregulated on application of brassinolide (Tanaka et al. 2005; Goda et al. 2002). Also, plants defective in BR signaling (*bri1* mutants) had considerably higher levels of accumulated BRs revealing that feedback regulation of endogenous BRs is achieved through intact perception and signaling pathway of BRs. Further studies have made it clear that two transcription factors BZR1 and BES1 are involved in control of BR biosynthesis (Mora-Garcia et al. 2004; Sun et al. 2010). Moreover, several novel transcription factors (BRX, CESTA, RAVL1, TCP1) have been lately identified to regulate the expression of key BR biosynthetic genes such as *CPD* and *DWF4* (Mouchel et al. 2006; Poppenberger et al. 2011; Guo et al. 2010; Je et al. 2010). Some bHLH-like transcription factors like CESTA and TCP1 are involved in positive regulation of BRs signaling (Wang et al. 2012a, b). CES acts as an activator of BR biosynthetic genes by binding to G-box motifs residing in promoters of the BR biosynthesis gene CPD (Poppenberger et al. 2011). Another transcription factor of bHLH family, TCP1, also positively regulates BR synthetic gene DWF4 by interaction with its promoter, while its own expression appears to be regulated by BRs (Guo et al. 2010). Another transcription factor, RAVL1 (RELATED TO ABI3/VP1, ABA INSENSITIVE 3/VIVAPARIOUS 1), plays a significant role in maintaining BRs homeostasis by positively regulating E-box cis-elements in the promoters of various biosynthetic (D2, D11, and BRD1) and receptor (*OsBRI1*) genes in rice (Je et al. 2010). In addition to the array of transcription factors regulating BRs level, BRs catabolism/metabolism also plays a crucial role in maintaining the optimum levels of bioactive BRs in the cell. A study indicated that BAS1, a C-26 hydroxylase, converts active CS and BL to their inactive C-26-hydroxylated derivatives (Turk et al. 2005). Recently, two research groups have independently found novel genes belonging to the family of BAHD acyltransferases, *BIA1* (*BRASSINOSTEROID INACTIVATOR1*) (Roh et al. 2012) and *ABS1 abnormal shoot* (*abs-1*) (Wang et al. 2012a, b), which are involved in BRs homeostasis. Overexpression of BIA1 was observed to inhibit the biosynthesis of some BR compounds and alter the expression of genes encoding BR-biosynthetic enzymes, while overexpression of ABS1 resulted in plants with BR-deficient and BR signaling mutant-like phenotype along with the activation of BR biosynthetic genes. Similarly, other genes, BEN1 and BNST3, with differential specificities to CS and BL inactivate active BRs by various mechanism involving reduction and sulfonation (Yuan et al. 2007; Marsolais et al. 2007). Brassinosteroid biosynthesis is also regulated by its cross talk with other hormones. BRX (BREVIS RADIX) is a gene that controls the expression of BR biosynthetic enzyme, CPD, and promotes root and shoot growth, while its own expression is regulated by auxin (Mouchel et al. 2006; Ye et al. 2011). Another study has shown that auxin signaling induces the expression of *DWF4* by reducing BZR1/BES1 binding to *DWF4* (Chung et al. 2011). Moreover, BRs biosynthesis is also shown to be regulated by external stimuli like salt and temperature stress (Sharma et al. 2013a; Maharajan and Choe 2011).

To maintain receptor homeostasis and signaling, BRI1 (Brassinosteroids insensitive 1) undergoes a constitutive recycling between plasma membrane and endosomes (Irani et al. 2012). Recently, a suppressor of bri1, sbi1, has been shown to cause selective accumulation of BR-activated BRI1. SBI1 encodes a leucine carboxyl methyltransferase (LCMT) which methylates PP2A at its catalytic subunit leading to enrichment of methylated PP2A at membranes. It brings the methylated PP2A at close proximity with ligand-activated BRI1 and results in ultimately dephosphorylating BRI1 and heading it for subsequent degradation. Exogenous application of BRs increases the expression level of *SBI1*, thus SBI1 and PP2A hypothesized to work in a feedback mechanism that helps in BRI1 homeostasis after BR activation of the signaling pathway (Wu et al. 2011). To further support the process of endocytosis as a major factor regulating BRs signaling and receptor degradation, Irani et al. (2012) developed a bioactive, fluorescent BR analog, Alexa Fluor 647-castasterone (AFCS), and visualized the endocytosis of BRI1-AFCS complexes in living *A. thaliana* cells for the first time in plants. Impairment of endocytosis was found to enhance BR signaling by retaining active BRI1-ligand complexes at the plasma membrane. Moreover, BRI1 can deactivate itself by autophosphorylation of serine-891 in the ATP-binding domain thereby inhibiting BRI1 activity and BR signaling in vivo (Oh et al. 2012b). It was also observed that calmodulin binds to the cytoplasmic domain of the recombinant BRI1 in vitro in a Ca^{2+}-dependent manner and co-expression of CaM with BRI1 in *Escherichia coli* specifically inhibited both the autophosphorylation of BRI1 and BRI1-mediated transphosphorylation of *E. coli* proteins hinting that CaM/Ca^{2+} may attenuate the kinase activity of BRI1 (Oh et al. 2012a).

1.3 Brassinosteroid Transport

Plants exhibit widespread distribution of BRs among various tissues with greatest accumulation in reproductive organs and young actively growing parts of the plants (Shimada et al. 2003). In contrast, bioactive level of BRs is much reduced in vegetative tissues, with roots possessing the least content of BRs. The finding draws significance due to the role of BRs in promoting cell division and elongation (Symons and Reid 2004). BRs synthesis is associated with internal cell membrane associated with endoplasmic reticulum. Indeed, many BRs biosynthetic enzymes which are predominately belonging to the class of cytochrome P450 proteins are membrane-bound, and the synthesis of BRs on or near such membranes is thought to be necessary because of the relative hydrophobicity of the steroidal BRs (Bishop et al. 1996; Nomura and Bishop 2006). At the same time, it is known that BRs are perceived at the exterior cell surface which implies that BRs are either passively or actively transported to the external cell surface (Clouse 2011).

An array of evidences have suggested that BRs undergo short-distance transport from the site of their synthesis to the external surface for perception by the same cell or the neighboring cells, though they exert a long-distance effect by their cross talk with other hormones like auxins (Symons et al. 2008). The short-distance transport of BRs might be involving the carrier mechanism (BR conjugates formed by binding of BRs to fatty acids or glucose) or through specific proteinaceous transporters. However, the only candidate proteins for binding and delivering brassinosteroids to the cell the membrane are those belonging to the class of pathogenesis-related (PR) 10 family of proteins. Betv1, a well-studied PR-10, can bind to a range of hydrophobic molecules, including brassinosteroids, and has been speculated as a potential transporter for BRs into the cytoplasm (Markovic-Housley et al. 2003; Kerr et al. 2011). Since membrane-bound, ABC (ATP-binding cassette) transporters are involved in steroid transport in animal cells, they have also been suggested as putative BR transporters (Symons et al. 2008).

1.4 Brassinosteroids Perception and Signal Transduction

Molecular genetic studies of BRs response mutants in *Arabidopsis* not only led to the identification of enzymes of the BRs biosynthetic pathway but also the BR receptor and downstream signaling component. Forward genetic screening led to the identification of multiple alleles of two loci causing insensitivity, bri1 and bin2, which led to the identification of BRI1 as the BR receptor and BIN2 as a negative downstream regulatory element (Clouse 2011). Mutations in the BR homologs have led to the identification of BR-insensitive mutants in other species also including pea, rice, barley, and tomato (Wang et al. 2012a, b). *bri1* mutant was first identified by its ability to cause root elongation in the presence of 24-epibrassinolide in a genetic screen of 70,000 EMS-mutagenized seedlings (Clouse et al. 1996). Later, other groups also found other BR-insensitive mutants, a vast majority of them turned out be allelic to a single locus, *BRI1* (Kauschmann et al. 1996; Li and Chory 1997; Noguchi et al. 1999). The *BRI1* gene was first isolated by positional cloning and verified by sequencing numerous mutant alleles (Li and Chory 1997). The gene encodes a 1,196-bp amino acid protein, containing the three major domains present in both animal and plant receptor kinases: the extracellular ligand-binding domain, the transmembrane domain, and the cytoplasmic kinase domain (Clouse 2011). The extracellular domain of BRI1 contains 25-leucine-rich repeats (LRRs) and an island domain between 21st and 22nd repeat followed by a single transmembrane domain (amino acids 792–814), a juxtamembrane region (amino acids 815–882), a kinase domain, and a C-terminal regulatory region (amino acids 1,156–1,196) (Li and Chory 1997; Ye et al. 2011). Several groups independently conformed BRI1 as the receptor for BR by various lines of experimental evidence lizing chimeric BRI1-Xa21 receptor kinase, binding of immunoprecipitated BRI1 with radio-labeled brassinolide, and binding assays with recombinant BRI1 protein (He et al. 2000; Wang et al. 2001; Kinoshita et al. 2005). Moreover, three additional homologs of BRI1 have been identified, two of which have been shown to have ability to bind to BRs and play important role in vascular development and show partial redundancy in functioning with BRI1 (Zhou et al. 2004; Cano-Delgado et al. 2004).

Cellular localization of BRI1 has been established to be the plasma membrane and early endosomes (Friedrichsen et al. 2000; Ye et al. 2011). Two independent laboratories have recently elucidated the three-dimensional structure of the extracellular domain of BRI1 using X-ray diffraction of crystallized BRI1 extracellular domain derived from recombinant protein expressed in insect cells (Hothorn et al. 2011; She et al. 2011). They have revealed the structure of the ligand-binding domain of *Arabidopsis* BRI1 in its free form and bound to the steroidal ligand brassinolide. As against the expected regular horseshoe-shaped structure of LRR-proteins, BRI1 forms the right-handed superhelix composed of extracellular 25 LRRs and a large insertion comprising of 70-residue segment corresponding to the island domain of BRI1. It forms a small domain that folds back into the interior of the superhelix, forming a binding pocket for one molecule of BL per BRI1 monomer. These studies reveal that the LRR superhelix as well as the island domain both extensively contributes to formation of the hormone-binding site. The A–D rings of the BL bind to a hydrophobic surface provided by LRRs 23–25 that maps to the inner side of the BRI1 superhelix, while its alkyl chain fits into a small pocket formed by residues from LRRs 21 and 22. The three-dimensional structure of BRI1 has, however, suggested that the superhelical shape of the BRI1 ectodomain is incompatible with homodimerization and that the isolated ectodomain behaves as a monomer even in the presence of steroid which leads to a hypothesis that some unknown protein factor could bind to the interaction platform in BRI1 and provides a dimerization interface for two BRI1 molecules. Moreover, the overall structure of BRI1 remains largely unchanged upon BL binding though significant structural rearrangement in the interdomain loops around the brassinolide-binding site forms a region referred to as the brassinolide-created surface. Interestingly, BAK1 has five LRRs and that the new brassinolide-created surface is also located about five LRRs from the membrane surface. Thus, it is assumed that the brassinolide-created surface could possibly be involved in interaction with BAK1, allowing formation of more stable BRI1–BAK1 receptor complexes which could be a reflection that BRs initiate signaling by promoting protein-protein interactions like other hormones (Hothorn et al. 2011; She et al. 2011; Che et al. 2010).

BRI1 can remain in the homodimer state independently of the ligand binding. However, ligand binding stabilizes and activates the homodimer (Fig. 1.1). However, for full activation of BR-induced signal transduction, heteroligomerization of BRI1 with members of the SOMATIC EMBROYGENESIS RECEPTOR KINASE (SERK) subfamily of LRR-RLKs is absolutely

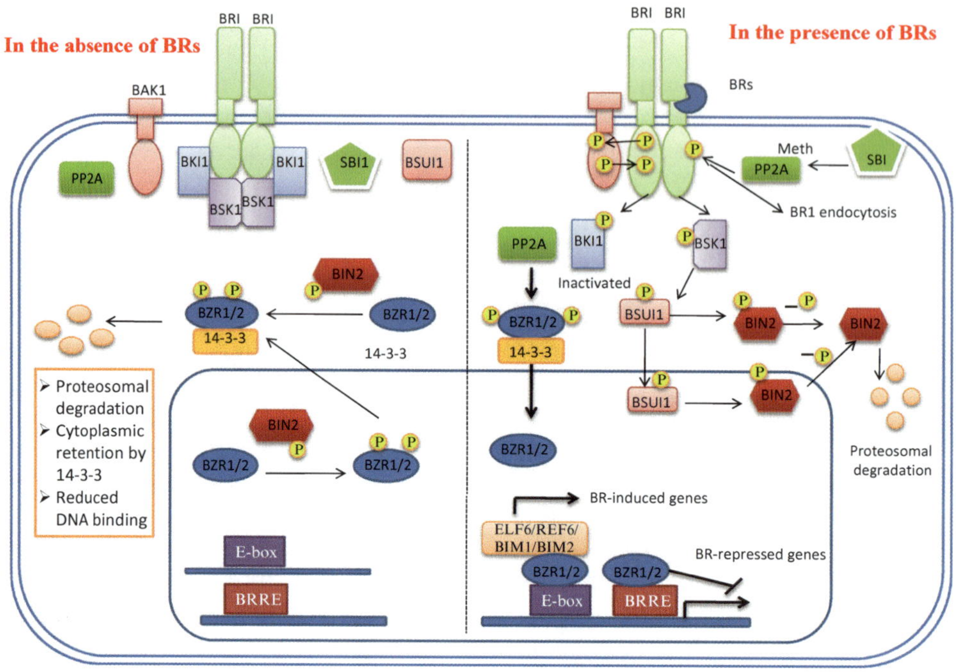

Fig. 1.1 Current model of BRs signal transduction in the presence or absence of brassinosteroids. The *small circles* containing letter P denotes phosphate residues. In the absence of BRs, BRI1 is kept in the inactive form by BKI1 by inhibiting its kinase functions and does not let it to heterodimerize with its co-receptor BAK1. In the cytosol, a GSK3-like kinase BIN2 (BRASSINOSTEROID INSENSITIVE 2), which is a negative regulator of BR signaling, phosphorylates BZR1 (BRASSINAZOLE RESISTANT 1) and BES1 (BRI1-EMS SUPRESSOR 1) family transcription factors which results in the inhibition of their function by one of the several methods including proteosomal degradation, reduced DNA binding, or cytoplasmic retention by 14-3-3 proteins. When BRs bind to the extracellular domain of BRI1, it undergoes autophosphorylation promoting the basal BRI1 kinase activity which leads to the phosphorylation of the negative regulator BKI1 and hence its dissociation from the plasma membrane. BRI1 associates with BAK1 and undergoes a series of transphosphorylation events to fully activate BRI1 kinase. BRI1 then phosphorylates BSK1 (BR-SIGNALING KINASES) to activate it which then associates with BSU1 (BRI1 SUPPRESSOR 1) phosphatase. It dephosphorylates BIN2 to inactivate it and thus releases it from 14-3-3 protein resulting in the accumulation of dephosphorylated BZR1 and BES1 with the help of PP2A (Protein Phosphatase 2A). Dephosphorylated BZR1 and BES1 undergo nuclear localization where they may interact with other accessory transcriptions factors BIM1, ELF6/REF6 to regulate the expression of BR target genes. Besides dephosphorylating BZR1/2, PP2A also regulates BRI homeostasis by phosphorylating BRI1 and heading it for endocytosis under the effect of its methylation by SBI1 (SUPPRESSOR OF BRI1)

necessary (Clouse 2011). BRI1-ASSOCIATED RECEPTOR KINASE1 (BAK1), also known as SERK3, is a co-receptor of BRI1 identified by activation tagging screen of *bri1–5* suppressor and yeast two-hybrid screen (Li et al. 2002; Nam and Li 2002). It has small extracellular domain with 5 LRRs and interacts with BRI1 in a BR-dependent manner and promotes the phosphorylation of both BRI1 and BAK1 (Russinova et al. 2004; Wang et al. 2005, 2008b). Besides its role in BRs signaling, BAK1 is involved in multiple signaling pathways by interaction with other LRR-RLKs and binding to different ligands. Genetic analysis has revealed that BAK1 has a redundant role to play with another member of SERK family, BKK1 in suppressing cell death independently of the BR signaling. BAK1 has been recently shown to be linking brassinosteroid signaling to plant immunity. Upon treatment with flagellin, BAK1 acts as a co-receptor to the flagellin receptor, FLS2 (an LRR-RLK protein) (Gao et al. 2009; Li 2010; Belkhadir et al. 2011). Both BRI1-BAK1 and

BAK1-FLS2 complex coexist in the cell, though the differential phosphorylation of BAK1 by FLS2 and BRI1 indicates signaling specificity (Oh et al. 2010). However, a link has been established between BRI1 signaling and immune response as has recently been shown through the genetic analysis that BRI1-activated BAK1 enhances FLS2 signaling through direct regulation of BR-induced FLS2 signaling in a BAK1-independent manner (Albrecht et al. 2012).

Since the last decade, a great deal of advancement has been made in understanding the BRs signal transduction using various proteomics and functional studies (Kim and Wang 2010; Ye et al. 2011). BRI1 and BAK1 are found to have both serine/threonine (S/T) and tyrosine (Y) phosphorylation activities, which play pivotal role in kinase activation, substrate modification, and specific BR responses (Oh et al. 2010, 2011; Jaillais et al. 2011). LC-MS/MS analysis has identified various in vitro and in vivo phosphorylation sites for BRI1 and BAK1, and it has been shown that the specific residues Th-455 and Th-1049 in BAK1 and BRI1 activation loop respectively are absolutely indispensable for the activation of BRs signaling (Wang et al. 2005). In the absence of BRs, BKI1 associates with the C-terminal tail of the BRI1 and inhibits its activity. The C-terminal domain of BRI1 can also function to inhibit BRI1 kinase activity (Wang and Chory 2006). Concurrently, active BIN2 (BRASSINOSTEROID INSENSITIVE 2), a GSK3/Shaggy-like kinase phosphorylates BES1/BZR2 family transcription factors and inhibits their activity by protein degradation, reduced DNA binding and/or cytoplasmic retention by 14–3–3 proteins (He et al. 2002; Yin et al. 2002; Ye et al. 2011; Hao et al. 2013). Binding of BRs to BRI1 activates BRI1 which phosphorylates BKI1 and results in dissociation of BKI1 from plasma membrane and association of BRI1 with co-receptor BAK1. It is followed by a series of phosphorylation and transphosphorylation events which fully activates the receptor complex (Wang et al. 2008b). Activated BRI1 then initiates a series of phosphorylation events to its downstream substrates such as receptor-like cytoplasmic kinases (RLCKs), BR SIGNALING KINASES (BSKs) and CONSTITUTIVE DIFFERENTIAL GROWTH 1 (CDG1), TRIP-1 (TGF-β RECEPTOR-INTERACTING PROTEIN-1) and TTL (TRANSTHYRETIN-LIKE PROTEIN), and potentially a proton-ATPase (P-ATPase) to transduce the BR signal to downstream targets (Tang et al. 2008; Kim et al. 2011), which then activate a phosphatase, BRI1-SUPPRESSOR 1 (BSU1). While the exact functioning of TRIP-1 and TTL in BR signaling is not known, BRI1 phosphorylation of BSK1 leads to its association with BSU1 phosphatase (BSU1 (BRI1-SUPPRESSOR 1)) which inhibits BIN2 kinase. Recent study by Kim et al. 2011 has shown that CDG1 functions much like BSKs. Both CGD1 and BSK1 get phosphorylated by BRI1 at Ser-234 and Ser-230 respectively and then activate BSU1 (Kim et al. 2011). BZR1 and BZR2/BES1 are then dephosphorylated by protein phosphatase 2A (PP2A) and released from 14-3-3 proteins (Tang et al. 2011), leading to their nuclear localization, and bind to the genomic DNA to regulate target gene expression (Ye et al. 2011; Wang et al. 2012a, b).

1.5 BRs-Mediated Regulation of BES1 and BZR1

BZR1 and BZR2/BES1 belong to a class of plant-specific transcription factors that share a significant 88 % overall amino acid sequence identity and 97 % identity in the DNA-binding domain. They possess a bHLH DNA-binding domain (DBD), a BIN2 phosphorylation domain containing 22 putative BIN2 phosphorylation sites, a PEST motif (Pro-, Glu-, Ser-, and Thr-rich) involved in protein degradation, and a 14–3–3 binding motif interacting with 14–3–3 when BES1 and BZR1 are phosphorylated (Hao et al. 2013). The C-terminal domain, which acts as a transcription activation domain, is highly conserved between two proteins, and thus it is predicted that like BZR1, C-terminal domain of BES1 is also involved in binding to BIN2 allowing to phosphorylate BES1 and BZR1 (Yin et al. 2002; Ryu et al. 2008, 2010; Peng et al. 2010). Inspite of the various similar biochemical and genetically redundant functions of BZR1 and BZR2/BES1, the mutant analysis (*bzr1-1D* and *bes1-D*) has revealed distinct phenotypes suggesting that a difference exist in the two proteins either at the level of expression pattern or interaction with other protein partners (Yin et al. 2002; Sun et al. 2010).

Though most of the major components of BRs signaling pathway were elucidated earlier, the major gap in our understanding remained to unravel the mechanism by which phosphorylated BZR1 and BES1 located in cytoplasm are dephosphorylated and transported into the nucleus to regulate gene expression. BSU1 was thought to perform this function, but later studies by Kim et al. (2009) revealed that they directly bind to BIN2 to regulate BRI1 signaling. However, a year later, another study suggested that BSU1 interacts with cytoplasmic BES1 and BZR1, while BIN2 phosphorylate BES1 and BZR1 in the nucleus to activate its cytoplasmic localization (Ryu et al. 2010). A recent proteomics study in tandem with a range of biochemical and genetic approaches has shown that a cytoplasmic protein phosphatase PP2A dephosphorylates BES1 and BZR1 thereby activating these transcription factors and promoting BR signaling (Tang et al. 2011). PP2A serine/threonine phosphatase is composed of a scaffolding subunit A, catalytic subunit C, and a regulatory substrate-binding B subunit (Janssens et al. 2008). B subunit of PP2A can directly bind with the Pro-, Glu-, Ser-, and Thr-rich (PEST) domain of BZR1 and lead to dephosphorylation of BZR1 sites phosphorylated by BIN2 (Tang et al. 2011). It was further confirmed through various in vitro and in vivo experiments that at least 4 PP2A isoforms of B subunit were able to interact with BZR1 through PEST domain. The loss-of-function PP2A mutant showed reduced BRs signaling by accumulating more phosphorylated BZR1 leading to a typical BR dwarf phenotype. On the other hand, the overexpression BZR1 having PEST domain removed resulted in accumulation of phosphorylated BZR1 and gave a typical BR mutant phenotype suggesting that phosphorylated BZR1 has a negative effect on BR signal transduction. It also points towards the negative role of PEST in protein degradation and positive

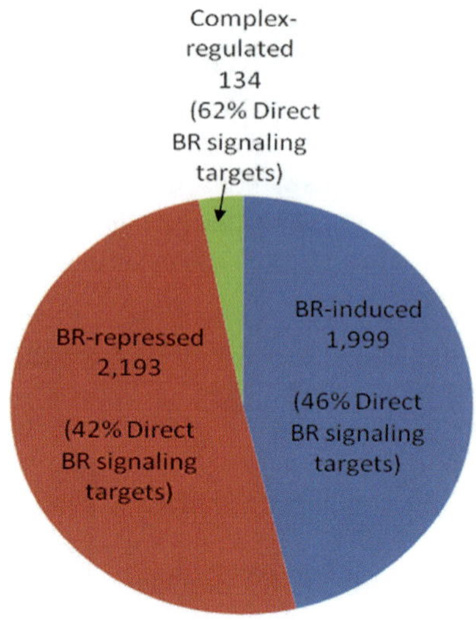

Fig. 1.2 A graphical representation of the number of genes altered under the effect of Brassinosteroids. Out of total of 4,326 altered in expression by exogenous BR treatment or by mutation with *bri1* and *bzr1-1D*, 1,999 are BR-induced genes, 2,193 are BR-repressed genes, and another 134 genes show complex expression under different experimental conditions. Among these 46 % of the 1,999 BR-induced genes, 42 % of the 2,193 BR-repressed genes, 62 % of the 134 complex-regulated genes are the direct targets of brassinosteroids signaling pathway

function in recruiting PP2A. Yeast two-hybrid screens have revealed that PP2A can interact with BES1 and thereby implying that PP2A can also regulate BES1 (Wu et al. 2011).

Studies using chromatin immunoprecipitation microarray (ChIP-chip) experiments have vastly enhanced our knowledge about the genes directly targeted by BES1 and BZR1 unraveling a massive gene regulatory networks that modulate various BR-regulated responses (Sun et al. 2010; Yu et al. 2011). Brassinosteroids participate in an array of growth responses primarily by inducing BR-mediated changes in the expression of genes (Fig. 1.2). A total of 4,326 genes have been identified which undergo alteration in the expression either by exogenous BR treatment or by mutation with *bri1* and *bzr1-1D* (Sun et al. 2010) which can further be trifurcated into those including 1,999 BR-induced genes, 2,193 BR-repressed genes, and 134 genes that show complex expression in different experiments. Of the total 4,326 genes, 1,927 are the direct targets of BR signaling pathway that include 46 % of the 1,999 BR-induced genes, 42 % of the 2,193 BR-repressed genes, and 62 % of the 134 complex-regulated genes (Wang et al. 2012a, b). For the better part of the last decade, BZR1 was chiefly believed to suppress transcription by binding to the BR response element (BRRE) sequence CGTG(T/C)G; on the other hand, BES1 was largely considered transcriptional activator binding to the E-box element (CANNTG) in the form of a homodimer or after heterodimerization with other transcription factors (He et al. 2005; Yin et al. 2005). However, combining chromatin immunoprecipitation coupled with *Arabidopsis* tiling arrays (ChIP-chip) and gene expression studies, it has become quite evident that both BZR1 and BES1 can either activate or repress transcription, depending on the specific target gene promoter, and that BZR1 and BES1 can recognize both the BRRE and E-box domains. Microarray data has shown that though BZR1 targets about 953 genes downstream from the BRI1-mediated signaling pathway while BES1 controls 250 genes out of which 120 genes show an overlap with the BZR1 target genes indicating that both the transcription factors can bind to same individual gene (Sun et al. 2010). It is through the regulation of these target genes that BRs exercise their effect on various growth and developmental process.

1.6 Implications of the Advancement of Brassinosteroid Research

At present, the focus of majority of the research occurring in the field of plant biology is to boost the yield of crop plants either by improving their agronomical traits like plant growth, flowering time, photosynthesis, and plant architecture or by enhancing their ability to survive changing environmental conditions. BRs have a direct role to play in plants growth and development thus the manipulation of their levels or signal transduction pathway have profound effect on the yield and sustainability of plants under various environmental challenges (Vriet et al. 2012). The growing insights into the BRs biology have extensively been exploited to study various BRs-regulated processes like cell elongation, cell differentiation, epigenetic regulation of gene expression, as well as various reproductive processes that are directly implicated in yield enhancement (Clouse 2011; Vriet et al. 2012). Identification of key components of BRs signaling has provided a huge potential of modulating the expression of BRs-regulated genes to improve the agricultural output.

Various genetic and molecular studies have shown that BRs promote cell elongation largely by regulating the expression of genes involved in cell-wall modifications, ion and water transport, and cytoskeleton rearrangements (Clouse and Sasse 1998; Clouse 2011) and many of these genes have now been found to be the direct targets of BES1 and BZR1 (Sun et al. 2010). Recent studies have shown various other novel genes involved in cell expansion like DEVELOPMENTALLY REGULATED PLASMAMEMBRANE POLYPEPTIDE (DREPP) gene which is a direct target of BR-induced BZR1 (Sun et al. 2010). Moreover, several cell cycle genes have been identified as the direct targets of both the BRZ1and BES1 which imply a transcriptional control of the cell cycle by BRs (Sun et al. 2010; Gudesblat and Russinova 2011; Yu et al. 2011). Microarray data has shown that BRs also regulate *Arabidopsis EXPANSIN A5* (*AtEXPA5*), a member of expansins family that are cell-wall proteins having the ability to induce cell-wall expansion and thus regulate various events in the growth and development (Mussig et al. 2002). *AtEXPA5* was enhanced by exogenously applied brassinosteroids in wild-type plants, and further analysis provided a convincing evidence that *AtEXPA5* expression is regulated by BRs signaling downstream of the BZR1 (Park et al. 2010). BRs are also required for cell differentiation as well as determining the time of differentiation in the root (Gonzalez-Garcia et al. 2011). Analysis of cell division, expansion, and differentiation in *Arabidopsis* leaves of the BR-deficient mutant constitutive photomorphogenesis and dwarfism (*cpd*) showed that BR production and BR receptor-dependent signaling differentially control the balance between cell division and expansion in the leaf. Thus, BRs are required to maintain normal cell cycle activity and cell expansion to ensure the coherent gradient of cell progression, from the apical to the basal meristem.

Methylation at histone and lysine plays a crucial role in epigenetic regulation of gene expression in plants. In rice, SDG725 encodes a H3K36 methyltransferase that plays a key role in plant growth and development, while the downregulation of *SDG725* is known to cause various growth and developmental defects including dwarfism, shortened internodes, erect leaves, and small seeds which are reminiscent of brassinosteroid-knockdown mutants. SDG725 is known to bind directly to chromatin to methylate H3K36 and eventually activate expression of the BR-related genes D11, BRI1, and BU1 to modulate several important traits of growth (Sui et al. 2012).

BRs control various other factors of plant growth that have direct implication on yield enhancement. Leaf bending is one such architectural feature in rice which affects grain yield and is under the control of BRs (Wang et al. 2008a). A key transcription factor LIC (LEAF AND TILLER ANGLE INCREASED CONTROLLER) has been identified as a negative regulator of OsBZR1, and both work antagonistically and repress each other during transcription in a BR-dependent manner. LIC antagonizes BZR1 to halt BR-mediated leaf bending in rice (Zhang et al. 2012). The newly found antagonistic role of LIC and BZR1 in mediating leaf bending may play a vital role in designing ideal plant architecture for improving photosynthesis efficiency during rice development to breed rice plants for high yield. BR signaling also has a role in determination of plant architecture by spatial regulation of its activity during organ boundary formation which is critical for coordination of organogenesis and meristem maintenance in cells. A tightly regulated reduction in brassinosteroid activity is critical for specification of organ boundaries (Gendron et al. 2012; Bell et al. 2012).

Besides growth, BRs are also known to regulate various phenomenon of reproduction in plants which greatly determine the yield of the crop. BR mutants have reduced expression of key genes involved in microspore mother cell development (SPL/NZZ), microspore development (TDF1, AMS, and AyMYB103), and those involved in tapetal development and pollen wall formation (MS1/MS2) resulting in defective anther and pollen development including reduced filament length, fewer numbers of pollen grains, defects in tapetal development, pollen wall formation, and pollen release (Ye et al. 2010). Brassinosteroids are also involved in ovule outer integument and gynoecial medial domain development (Nole-Wilson et al. 2010). The Arabidopsis SEUSS (SEU) gene encodes a transcriptional adaptor protein essential for gynoecium, ovule and embryo development. Genetic screens identified a BR biosynthetic enzyme as a suppressor of mutations in the *Arabidopsis* SEUSS gene, suggesting a novel role for brassinolide synthesis in gynoecial and ovule development. Moreover, analysis of BRs-deficient and BR-insensitive mutants has shown delayed flowering implying the role of BRs in regulating the time of flowering (Li and Chory 1997; Azpiroz et al. 1998). Later studies revealed that BRs signaling promote flowering by repressing a potent floral repressor FLOWERING LOCUS C (FLC) (Domagalska et al. 2007).

Along with their role in plant growth and development, BRs are one of the major players in providing tolerance against a number of abiotic and biotic stresses in plants. Exogenous application of BRs can directly provide tolerance against water stress (Vardhini et al. 2011), salt stress (Sharma et al. 2013a), heavy metal stress (Hasan et al. 2008; Choudhary et al. 2012), high- and low-temperature stress (Kagale et al. 2007), and various biotic stresses (Clouse and Sasse 1998). Work done in our lab also shows that in response to various abiotic stresses, BRs elicit the antioxidative defense system in plants to promote expression and activity of various antioxidative enzymes like superoxide dismutase, catalase, and ascorbate peroxidase as well enhance the levels of various metabolites like proline to fight stress condition (Sharma et al. 2012, 2013a, b; Bajguz and Hayat 2009). Under biotic stresses, mainly inflicted by bacterial, fungal, and viral infections, BRs stimulate synthesis of abscisic acid, ethylene, and salicylic acid to generate cellular responses for stress amelioration (Bajguz and Hayat 2009). Though the stress ameliorative property of BRs has been widely known, the molecular mechanism that leads to stress tolerance, however, remains unclear. A recent study has shown that BR signaling mediated through BRI1 is necessary for salt stress tolerance in plants. BRs regulate a ubiquitin-conjugating enzyme UBC32 which is a functional component of the endoplasm reticulum (ER)-associated protein degradation (ERAD) pathway and play a vital role in BR-mediated salt tolerance in plants. In *Arabidopsis*, a positive correlation was found between *ubc32*, a stress-inducible ubiquitin-conjugating enzyme mutant, and protein level of BRI1. *ubc32* mutant has enhanced salt tolerance, and this could possibly be linked to increased accumulation of BRI1 (Cui et al. 2012). Another study using semiquantitative PCR has shown that the transcript level of OsBRI1 increases in response to combined treatment with NaCl and BRs hinting towards an interactive mechanism for salt tolerance (Sharma et al. 2013a). Increasing body of evidence suggests that BR-induced stress tolerance is associated with enhanced accumulation of reactive oxygen species (ROS). BR triggers a periodic increase in the level of H_2O_2, a potent ROS that can function as a signaling molecule in response to various stimuli both in plant and animal cells (Neill et al. 2002). BR signal perception may result in activation of NADPH oxidase to produce ROS, which initiates a protein phosphorylation cascade through MAPKs to activate transcription factors to target specific genes participating in cellular protection (Xia et al. 2009). Nitric oxide is also induced by BR in a H_2O_2-dependent manner to play a critical role in BR-induced stress tolerance and causes the upregulation of the expression and the activities of antioxidant enzymes in BR-treated cucumber plants (Zhang et al. 2010).

Brassinosteroids alone or in cross talk with other hormones can confer tolerance against biotic stress in plants. Upon BR treatment, enhanced activity of NADPH oxidase elevates the ROS production and level of pathogen-related 1 (PR1) proteins under biotic stress. This increased ROS level causes respiratory burst and leads to programmed cell death which is a protective phenomenon (Hofiusa et al. 2007). It also leads to the activation of MAPK cascade that may result in increased activity of antioxidant enzymes and other protective proteins which can confer tolerance (Xia et al. 2009). Flagellin 22 (flg 22), a microbial-associated molecular patterns (MAMP) molecule that elicit defense responses, is recognized by leucine-rich repeat (LRR)-receptor-like kinases (LRR-RLKs) that binds to them to activate the innate immune response (Chinchilla et al. 2007; Heese et al. 2007; Schwessinger et al. 2011). BAK1 besides being a co-receptor to BRI1 rapidly associates with FLS2 to induce immune responses (Albrecht et al. 2012; Belkhadir et al. 2011).

1.7 Future Perspectives

For the past one decade, scientists working in the field of BRs have considerably deciphered the brassinosteroids signaling pathway, and relatively, a holistic view of BRs perception and signal transduction is emerging. With each new finding, we are heading one step further to fully elucidate the complex network of BRs signaling events to relate cell surface BR receptor to plant growth and development, and thus, so far, BR pathway symbolizes the first and the only best known pathway among rest of over 400 receptor kinases. However, with the identification of hundreds of direct BRs responsive transcription factors, extensive genetic and biochemical analysis of the targets with undefined functions is required which may further unravel the unknown physiological processes affected by BR signaling. Owing to the importance of BRs in the crop improvement, further investigations are needed to identify the key regulatory elements in BR signaling pathway and the underlying mechanism of BR-modulated growth and developmental responses in major crop plants to design optimal strategies to enhance crop yield and improve their performance under stress conditions.

Acknowledgements Isha Sharma was supported by grants received from University Grants Commission (UGC), New Delhi. Navdeep Kaur is a recipient of DBT-JRF fellowship from the Department of Biotechnology, New Delhi, and Shivani Saini was supported by Innovation in Science Pursued for Inspired Research (INSPIRE) Program and Department of Science and Technology, New Delhi.

References

Albrecht C, Boutrot F, Segonzac C, Schwessinger B, Gimenez-Ibanez S, Chinchilla D, Rathjen JP, de Vries SC, Zipfel C (2012) Brassinosteroids inhibit pathogen-associated molecular pattern-triggered immune signaling independent of the receptor kinase BAK1. Proc Natl Acad Sci U S A 109:303–308

Azpiroz R, Wu Y, LoCascio JC, Feldmann KA (1998) An *Arabidopsis* brassinosteroid- dependent mutant is blocked in cell elongation. Plant Cell 10:219–230

Bajguz A, Hayat S (2009) Effects of brassinosteroids on the plant responses to environmental stresses. Plant Physiol Biochem 47:1–8

Belkhadir Y, Jaillais Y, Epple P, Balsemao–Pires E, Dangl JL, Chory J (2011) Brassinosteroids modulate the efficiency of plant immune responses to microbe- associated molecular patterns. Proc Natl Acad Sci U S A 109:297–302

Bell EM, Lina W, Husbandsa AY, Yua L, Jaganathaa V, Jablonskaa B, Mangeona A, Neffb MM, Girkea T, Springera PS (2012) *Arabidopsis* lateral organ boundaries negatively regulates brassinosteroid accumulation to limit growth in organ boundaries. Proc Natl Acad Sci U S A 109:21146–21151

Bishop GJ, Harrison K, Jones JDG (1996) The tomato Dwarf gene isolated by heterologous transposon tagging encodes the first member of a new cytochrome p450 family. Plant Cell 8:959–969

Cano-Delgado A, Yin Y, Vafeados D, Mora-Garcia S, Cheng J-C, Nam KH, Li J, Chory J (2004) BRL1 and BRL3 are novel brassinosteroid receptors that function in vascular differentiation in *Arabidopsis*. Development 131:5341–5351

Che P, Bussell JD, Zhou W, Estavillo GM, Pogson BJ, Smith SM (2010) Signaling from the endoplasmic reticulum activates brassinosteroid signaling and promotes acclimation to stress in Arabidopsis. Sci Signal 3:ra69

Chinchilla D, Zipfel C, Robatzek S, Kemmerling B, Nurnberger T, Jones JD, Felix G, Boller T (2007) A Flagellin-induced complex of the receptor FLS2 and BAK1 initiates plant defense. Nature 448:497–500

Choudhary SP, Oral HV, Bhardwaj R, Yu J, Tran LP (2012) Interaction of brassinosteroids and polyamine enhances copper stress tolerance in *Raphanus sativus*. J Exp Bot 63:5659–5675

Chung Y, Maharjan PM, Lee O, Fujioka S, Jang S, Kim B, Takatsuto S, Tsujimoto M, Kim H, Cho S, Park T, Cho H, Hwang I, Choe S (2011) Auxin stimulates DWARF4 expression and brassinosteroid biosynthesis in *Arabidopsis*. Plant J 66:564–578

Clouse SD (2011) Brassinosteroid signal transduction: from receptor kinase activation to transcriptional networks regulating plant development. Plant Cell 23:1219–1230

Clouse SD, Sasse JM (1998) Brassinosteroids: essential regulators of plant growth and development. Annu Rev Plant Physiol Plant Mol Biol 49:427–451

Clouse SD, Langford M, McMorris TC (1996) A brassinosteroid-insensitive mutant in *Arabidopsis thaliana* exhibits multiple defects in growth and development. Plant Physiol 111:671–678

Cui F, Liu L, Zhao Q, Zhang Z, Li Q, Lin B, Wu Y, Tang S, Xie Q (2012) *Arabidopsis* ubiquitin conjugase UBC32 is an ERAD component that functions in brassinosteroid- mediated salt stress tolerance. Plant Cell. doi:http://dx.doi.org/10.1105/tpc.111.093062

Davies PJ (1995) The plant hormones: their nature, occurrence and functions. In: Davies PJ (ed) Plant hormones: physiology, biochemistry and molecular biology. Kluwer Academic Publishers, Dordrecht, pp 1–15

Domagalska MA, Schomburg FM, Amasino RM, Vierstra RD, Nagy F, Davis SJ (2007) Attenuation of brassinosteroid signaling enhances FLC expression and delays flowering. Development 134:2841–2850

Fridman Y, Savaldi-Goldstein S (2013) Brassinosteroids in growth control: how, when and where. Plant Sci 209:24–31

Friedrichsen DM, Joazeiro CA, Li J, Hunter T, Chory J (2000) Brassinosteroid-insensitive-1 is a ubiquitously expressed leucine rich repeat receptor serine/threonine kinase. Plant Physiol 123:1247–1256

Fujioka S, Sakurai A (1997) Biosynthesis and metabolism of brassinosteroids. Physiol Plant 100:710–715

Fujioka S, Yokota T (2003) Biosynthesis and metabolism of brassinosteroids. Annu Rev Plant Biol 54:137–164

Gao M, Wang X, Wang D, Xu F, Ding X, Zhang Z, Bi D, Cheng YT, Chen S, Li X, Zhang Y (2009) Regulation of cell death and innate immunity by two receptor-like kinases in *Arabidopsis*. Cell Host Microbe 6:34–44

Gendron GM, Liua J, Fana M, Baia M, Wenkela S, Springerb PS, Bartona MK, Wanga Z (2012) Brassinosteroids regulate organ boundary formation in the shoot apical meristem of *Arabidopsis*. Proc Natl Acad Sci U S A 109:21152–21157

Goda H, Shimada Y, Asami T, Fujioka S, Yoshida S (2002) Microarray analysis of brassinosteroid-regulated genes in Arabidopsis. Plant Physiol 130:1319–1334

Gonzalez-Garcia MP, Vilarrasa-Blasi J, Zhiponova M, Divol F, Mora-Garcia S, Russinova E, Cano-Delgado AI (2011) Brassinosteroids control meristem size by promoting cell cycle progression in Arabidopsis roots. Development 138:849–859

Gudesblat GE, Russinova E (2011) Plants grow on brassinosteroids. Curr Opin Plant Biol 14:530–537

Guo Z, Fujioka S, Blancaflor EB, Miao S, Gou X, Li J (2010) TCP1 modulates brassinosteroid biosynthesis by regulating the expression of the key biosynthetic

gene DWARF4 in *Arabidopsis thaliana*. Plant Cell 22:1161–1173
Hao J, Yin Y, Fei S (2013) Brassinosteroid signaling network: implications on yield and stress tolerance. Plant Cell Rep 32:1017–1030
Hasan SA, Hayat S, Ali B, Ahmad A (2008) 28-homobrassinolide protects chickpea (*Cicer arietinum*) from cadmium toxicity by stimulating antioxidants. Environ Pollut 151:60–66
He Z, Wang ZY, Li J, Zhu Q, Lamb C, Ronald P, Chory J (2000) Perception of brassinosteroids by the receptor kinase BRI1. Nature 474:467–471
He JX, Gendron JM, Yang Y, Li J, Wang ZY (2002) The GSK3-like kinase BIN2 phosphorylates and destabilizes BZR1, a positive regulator of the brassinosteroid signaling pathway in *Arabidopsis*. Proc Natl Acad Sci U S A 99:10185–10190
He JX, Gendron JM, Sun Y, Gampala SS, Gendron N et al (2005) BZR1 is a transcriptional repressor with dual roles in brassinosteroid homeostasis and growth responses. Science 307:1634–1638
Heese A, Hann DR, Gimenez-Ibanez S, Jones AM, He K, Li J, Schroeder JI, Peck SC, Rathjen JP (2007) The receptor-like kinase SERK3/BAK1 is a central regulator of innate immunity in plants. Sci Signal 104:12217
Hofiusa D, Dimitrios I, Tsitsigiannis T, Jonathan DG, Mundy JJ (2007) Inducible cell death in plant immunity. Semin Cancer Biol 17:166–187
Hothorn M, Belkhadir Y, Dreux M, Dabi T, Noel JP, Wilson IA, Chory J (2011) Structural basis of steroid hormone perception by the receptor kinase BRI1. Nature 474:467–472
Irani NG, Rubbo SD, Mylle E, Begin JV, Schneider-Pizon J, Hnilikova J, Sisa M, Buyst D, Vilarrasa-Blasi J, Szatmári A, Damme DV, Mishev K et al (2012) Fluorescent castasterone reveals BRI1 signaling from the plasma membrane. Nat Chem Biol 8:583–589
Jaillais Y, Hothorn M, Belkhadir Y, Dabi T, Nimchuk ZL, Meyerowitz EM, Chory J (2011) Tyrosine phosphorylation controls brassinosteroid receptor activation by triggering membrane release of its kinase inhibitor. Genes Dev 25:232–237
Janssens V, Longin S, Goris J (2008) PP2A holoenzyme assembly: in *Cauda venenum* (the sting is in the tail). Trends Biochem Sci 33:113–121
Je BI, Piao HL, Park SJ, Park SH, Kim CM, Xuan YH, Huang J, Do Choi Y, An G, Wong HL, Fujioka S, Kim MC, Shimamoto K, Han CD (2010) RAV-Like1 maintains brassinosteroid homeostasis via the coordinated activation of BRI1 and biosynthetic genes in rice. Plant Cell 22:1777–1791
Kagale S, Divi UK, Krochko JE, Keller WA, Krishna P (2007) Brassinosteroid confers tolerance in *Arabidopsis thaliana* and *Brassica napus* to a range of abiotic stresses. Planta 225:353–364
Kauschmann A, Jessop A, Koncz C, Szekeres M, Willmitzer L, Altmann T (1996) Genetic evidence for an essential role of brassinosteroids in plant development. Plant J 9:701–713
Kerr I, Carrier D, Twycross J (2011) Hormone transport. In: Murphy AS, Schulz B, Peer W (eds) The plant plasma membrane. Springer, Berlin, pp 379–397
Kim T, Wang Z (2010) Brassinosteroid signal transduction from receptor kinases to transcription factors. Annu Rev Plant Biol 61:681–704
Kim TW, Guan S, Sun Y, Deng Z, Tang W, Shang JX, Sun Y, Burlingame AL, Wang ZY (2009) Brassinosteroid signal transduction from cell-surface receptor kinases to nuclear transcription factors. Nat Cell Biol 11:1254–1260
Kim TW, Guan S, Burlingame AS, Wang ZY (2011) The CDG1 kinase mediates brassinosteroid signal transduction from BRI1 receptor kinase to BSU1 phosphatase and GSK3-like kinase BIN2. Mol Cell 43:561–571
Kinoshita T, Cano-Delgado A, Seto H, Hiranuma S, Fujioka S, Yoshida S, Chory J (2005) Binding of brassinosteroids to the extracellular domain of plant receptor kinase BRI1. Nature 433:167–171
Li J (2010) Multi-tasking of somatic embryogenesis receptor-like protein kinases. Curr Opin Plant Biol 13:509–514
Li J, Chory J (1997) A putative leucine-rich repeat receptor kinase involved in brassinosteroid signal transduction. Cell 90:929–938
Li J, Nagpal P, Vitart V, McMorris TC, Chory J (1996) A role for brassinosteroids in light- dependent development of *Arabidopsis*. Science 272:398–401
Li J, Wen J, Lease KA, Doke JT, Tax FE, Walker JC (2002) BAK1, an *Arabidopsis* LRR receptor-like protein kinase, interacts with BRI1 and modulates brassinosteroid signaling. Cell 110:213–222
Maharajan PM, Choe S (2011) High temperature stimulates DWARF4 (DWF4) expression to increase hypocotyl elongation in Arabidopsis. J Plant Biol 54:425–429
Mandava NB (1988) Plant growth-promoting brassinosteroids. Annu Rev Plant Physiol Plant Mol Biol 39:23–52
Markovic-Housley Z, Degano M, Lamba D, von Roepenack-Lahaye E, Clemens S, Susani M, Ferreira F, Scheiner O, Breiteneder H (2003) Crystal structure of a hypoallergenic isoform of the major birch pollen allergen Bet v 1 and its likely biological function as a plant steroid carrier. Plant Mol Biol 325:123–133
Marsolais F, Boyd J, Paredes Y, Schinas AM, Garcia M, Elzein S, Varin L (2007) Molecular and biochemical characterization of two brassinosteroid sulfotransferases from *Arabidopsis*, AtST4a (At2g14920) and AtST1 (At2g03760). Planta 225:1233–1244
Mora-Garcia S, Vert G, Yin Y, Cano-Delgado A, Cheong H, Chory J (2004) Nuclear protein phosphatases with Kelch-repeat domains modulate the response to brassinosteroids in Arabidopsis. Genes Dev 18:448–460
Mouchel CF, Osmont KS, Hardtke CS (2006) BRX mediates feedback between brassinosteroid levels and auxin signalling in root growth. Nature 443:458–461
Mussig C, Fischer S, Altman T (2002) Brassinosteroid-regulated gene expression. Plant Physiol 129(3): 1241–1251
Nam KH, Li J (2002) BRI1/BAK1, a receptor kinase pair mediating brassinosteroid signaling. Cell 110:203–212

Neill S, Desikan R, Hancock J (2002) Hydrogen peroxide signalling. Curr Opin Plant Biol 5:388–395

Noguchi T, Fujioka S, Choe S, Takatsuto S, Yoshida S, Yuan H, Feldmann KA, Tax FE (1999) Brassinosteroid-insensitive dwarf mutants of *Arabidopsis* accumulate brassinosteroids. Plant Physiol 121:743–752

Noguchi T, Fujioka S, Choe S, Takatsuto S, Tax FE, Yoshida S, Feldmann KA (2000) Biosynthetic pathways of brassinolide in *Arabidopsis*. Plant Physiol 124:201–209

Nole-Wilson S, Rueschhoff EE, Bhatti H, Franks RG (2010) Synergistic disruptions in seuss cyp85A2 double mutants reveal a role for brassinolide synthesis during gynoecium and ovule development. BMC Plant Biol 10:198

Nomura T, Bishop GJ (2006) Cytochrome P450s in plant steroid hormone synthesis and metabolism. Phytochem Rev 5:421–432

Oh MH, Wang X, Wu X, Zhao Y, Clouse SD, Huber SC (2010) Autophosphorylation of Tyr-610 in the receptor kinase BAK1 plays a role in brassinosteroid signaling and basal defense gene expression. Proc Natl Acad Sci U S A 107:17827–17832

Oh MH, Wu X, Clouse SD, Huber SC (2011) Functional importance of BAK1 tyrosine phosphorylation *in vivo*. Plant Signal Behav 6:400–405

Oh MH, Kim HS, Wu X, Clouse SD, Zielinski RE, Huber SC (2012a) Calcium/calmodulin inhibition of the *Arabidopsis* BRASSINOSTEROID-INSENSITIVE 1 receptor kinase provides a possible link between calcium and brassinosteroid signalling. Biochem J 443:515–523

Oh MH, Wang X, Clouse SD, Huber SC (2012b) Deactivation of the *Arabidopsis* BRASSINOSTEROID INSENSITIVE 1 (BRI1) receptor kinase by autophosphorylation within the glycine-rich loop. Proc Natl Acad Sci U S A 109:327–332

Ohnishi ST, Godza B, Watanabe B, Fujioka S, Hategan L, Ide K, Shibata K, Yokota K, Szekeres M, Mizutani M (2012) CYP90A1/CPD, a brassinosteroid biosynthetic cytochrome P450 of *Arabidopsis*, catalyzes C-3 oxidation. J Biol Chem 287:31551–31560

Park CH, Kim TW, Son SH, Hwang JY, Lee SC, Chang SC, Kim SH, Kim SW, Kim SK (2010) Brassinosteroids control *AtEXPA5* gene expression in *Arabidopsis thaliana*. Phytochemistry 71:380–387

Peng P, Zhao J, Zhu Y, Asami T, Li J (2010) A direct docking mechanism for a plant GSK3-like kinase to phosphorylate its substrates. J Biol Chem 285:24646–24653

Poppenberger B, Rozhon W, Khan M, Husar S, Adam G, Luschnig C, Fujioka S, Sieberer T (2011) CESTA, a positive regulator of brassinosteroid biosynthesis. EMBO J 30:1149–1161

Roh H, Jeong CW, Fujioka S, Kim YK, Lee S, Ahn JH, Choi YD, Lee JS (2012) Genetic evidence for the reduction of brassinosteroid levels by a BAHD acyltransferase-like protein in *Arabidopsis*. Plant Physiol 159:696–709

Russinova E, Borst JW, Kwaaitaal M, Cano-Delgado A, Yin Y, Chory J, Vries SC (2004) Heterodimerization and endocytosis of *Arabidopsis* brassinosteroid receptors BRI1 and SERK3 (BAK1). Plant Cell 16:3216–3229

Ryu H, Kim K, Hwang I (2008) Spatial redistribution of key transcriptional regulators in brassinosteroid signaling. Plant Signal Behav 3:278–280

Ryu H, Kim K, Cho H, Hwang I (2010) Predominant actions of cytosolic BSU1 and nuclear BIN2 regulate subcellular localization of BES1 in brassinosteroid signaling. Mol Cells 29:291–296

Sakurai A, Fujioka S (1997) Studies on biosynthesis of brassinosteroids. Biosci Biotechnol Biochem 61:757–762

Schwessinger B, Roux M, Kadota Y, Ntoukakis V, Sklenar J, Jones A, Zipfel C (2011) Phosphorylation-dependent differential regulation of plant growth, cell death, and innate immunity by the regulatory receptor-like kinase BAK1. PLoS Genet 7:e1002046

Sharma I, Bhardwaj R, Pati PK (2012) Mitigation of adverse effects of chlorpyrifos by 24- epibrassinolide and analysis of stress markers in a rice variety Pusa Basmati-1. Ecotoxicol Environ Safe 8:72–81

Sharma I, Bhardwaj R, Pati PK (2013a) Stress modulation response of 24-epibrassinolide against imidacloprid in an elite Indica rice variety Pusa Basmati-1. Pest Biochem Physiol 105:144–153

Sharma I, Ching E, Saini S, Bhardwaj R, Pati PK (2013b) Exogenous application of brassinosteroid offers tolerance to salinity by altering stress responses in rice variety Pusa Basmati-1. Plant Physiol Biochem 69:17–26

She J, Han Z, Kim TW, Wang J, Cheng W, Chang J, Shi S, Yang M, Wang ZY, Chai J (2011) Structural insight into brassinosteroid perception by BRI1. Nature 474:472–476

Shimada Y, Goda H, Nakamura A, Takatsuto S, Fujioka S, Yoshida S (2003) Organ-specific expression of brassinosteroid biosynthetic genes and distribution of endogenous brassinosteroids in *Arabidopsis*. Plant Physiol 131:287–297

Sui P, Jin J, Ye S, Mu C, Gao J, Feng H, Shen WH, Yu Y, Dong A (2012) H3K36 methylation is critical for brassinosteroid-regulated plant growth and development in rice. Plant J 70:340–347

Sun Y, Fan XY, Cao DM, Tang W et al (2010) Integration of brassinosteroid signal transduction with the transcription network for plant growth regulation in *Arabidopsis*. Dev Cell 19:765–777

Symons GM, Reid JB (2004) Brassinosteroids do not undergo long-distance transport in pea. Implications for the regulation of endogenous brassinosteroid levels. Plant Physiol 135:2196–2206

Symons GM, Ross JJ, Jager CE, Reid JB (2008) Brassinosteroid transport. J Exp Bot 59:17–24

Szekeres M, Nemeth K, Koncz-Kalman Z, Mathur J, Kauschmann A et al (1996) Brassinosteroids rescue the deficiency of CYP90, a cytochrome P450, controlling cell elongation and de-etiolation in *Arabidopsis*. Cell 85:171–182

Tanaka K, Asami T, Yoshida S, Nakamura Y, Matsuo T, Okamoto S (2005) Brassinosteroid homeostasis in *Arabidopsis* is ensured by feedback expressions of multiple genes involved in its metabolism. Plant Physiol 138:1117–1125

Tang W, Kim TW, Oses-Prieto JA, Sun Y, Deng Z, Zhu S, Wang R, Burlingame AL, Wang ZY (2008) BSKs mediate signal transduction from the receptor kinase BRI1 in *Arabidopsis*. Science 321:557–560

Tang W, Yuan M, Wang R, Yang Y, Wang C et al (2011) PP2A activates brassinosteroid- responsive gene expression and plant growth by dephosphorylating BZR1. Nat Cell Biol 13:124–131

Turk EM, Fujioka S, Seto H, Shimada Y, Takatsuto S, Yoshida S, Wang H, Torres QI, Ward JM, Murthy G, Zhang J, Walker JC, Neff MM (2005) BAS1 and SOB7 act redundantly to modulate *Arabidopsis* photomorphogenesis via unique brassinosteroid inactivation mechanisms. Plant J 42:23–34

Vardhini BV, Sujatha E, Rao SSR (2011) Brassinosteroids: alleviation of water stress in certain enzymes of sorghum seedlings. J Phytol 3:38–43

Vriet C, Russinova E, Reuzeaua C (2012) Boosting crop yields with plant steroids. Plant Cell 24:842–857

Wang X, Chory J (2006) Brassinosteroids regulate dissociation of BKI1, a negative regulator of BRI1 signaling, from the plasma membrane. Science 313:1118–1122

Wang ZY, Seto H, Fujioka S, Yoshida S, Chory J (2001) BRI1 is a critical component of a plasma-membrane receptor for plant steroids. Nature 410:380–383

Wang X, Goshe MB, Soderblom EJ, Phinney BS, Kuchar JA et al (2005) Identification and functional analysis of in vivo phosphorylation sites of the *Arabidopsis* BRASSINOSTEROID-INSENSITIVE1 receptor kinase. Plant Cell 17:1685–1703

Wang L, Xu Y, Zhang C, Ma Q, Joo SH, Kim SK, Xu Z, Chong K (2008a) OsLIC, a novel CCCH-type zinc finger protein with transcription activation, mediates rice architecture via brassinosteroids signaling. PLoS ONE 3:e3521

Wang X, Kota U, He K, Blackburn K, Li J et al (2008b) Sequential transphosphorylation of the BRI1/BAK1 receptor kinase complex impacts early events in brassinosteroid signaling. Dev Cell 15:220–235

Wang H, Nagegowda DA, Rawat R, Bouvier-Nave P, Guo D, Bach TJ, Chye ML (2012a) Overexpression of *Brassica juncea* wild-type and mutant HMG-CoA synthase 1 in *Arabidopsis* upregulates genes in sterol biosynthesis and enhances sterol production and stress tolerance. Plant Biotechnol J 10:31–42

Wang ZY, Bai MY, Oh E, Zhu JY (2012b) Brassinosteroid signaling network and regulation of photomorphogenesis. Annu Rev Genet 46:701–724

Wu G, Wang X, Li X, Kamiya Y, Otegui MS, Chory J (2011) Methylation of a phosphatase specifies dephosphorylation and degradation of activated brassinosteroid receptors. Sci Signal 4:ra29

Xia XJ, Wang YJ, Zhou YH, Tao Y, Mao WH, Shi K, Asami T, Chen ZX, Yu JQ (2009) Reactive oxygen species are involved in brassinosteroid—induced stress tolerance in cucumber. Plant Physiol 150:801–814

Ye Q, Zhu W, Li L, Zhang S, Yin Y, Ma H, Wang X (2010) Brassinosteroids control male fertility by regulating the expression of key genes involved in *Arabidopsis* anther and pollen development. Proc Natl Acad Sci U S A 107:6100–6105

Ye H, Li L, Yin Y (2011) Recent advances in the regulation of brassinosteroid signaling and biosynthesis pathways. J Int Plant Biol 53:455–468

Yin YH, Wang ZY, Mora-Garcia S, Li JM, Yoshida S, Asami T, Chory J (2002) BES1 accumulates in the nucleus in response to brassinosteroids to regulate gene expression and promote stem elongation. Cell 109:181–191

Yin Y, Vafeados D, Tao Y, Yokoda T, Asami T, Chory J (2005) A new class of transcription factors mediate brassinosteroid-regulated gene expression in *Arabidopsis*. Cell 120:249–259

Yu X, Li L, Zola J, Aluru M, Ye H, Foudree A, Guo H, Anderson S, Aluru S, Liu P, Rodermel S, Yin Y (2011) A brassinosteroid transcriptional network revealed by genome-wide identification of BESI target genes in *Arabidopsis thaliana*. Plant J 65:634–646

Yuan T, Fujioka S, Takatsuto S, Matsumoto S, Gou X, He K, Russell SD, Li J (2007) *BEN1*, a gene encoding a dihydroflavonol 4-reductase (DFR)-like protein, regulates the levels of brassinosteroids in *Arabidopsis thaliana*. Plant J 51:220–233

Zhang J, Li W, Xiang T, Liu Z, Laluk K, Ding X et al (2010) Receptor-like cytoplasmic kinases integrate signaling from multiple plant immune receptors and are targeted by *Pseudomonas syringae* effector. Cell Host Microb 7:290–301

Zhang C, Xu Y, Guo S, Zhu J, Huan Q et al (2012) Dynamics of brassinosteroid response modulated by negative regulator LIC in rice. PLoS Genet 8:e1002686

Zhao B, Li J (2012) Regulation of brassinosteroid biosynthesis and inactivation. J Integr Plant Biol 54: 746–759

Zhou A, Wang H, Walker JC, Li J (2004) BRL1, a leucine-rich repeat receptor-like protein kinase, is functionally redundant with BRI1 in regulating *Arabidopsis* brassinosteroid signaling. Plant J 40:399–409

2 Molecular Basis of Salt Tolerance in Wheat and Other Crop Plants

Veenti Rana, Sewa Ram, and Kiran Nehra

Abstract

Among abiotic stresses, salinity has become a major problem adversely affecting the growth of crop plants grown under saline conditions. In India, an area of about 5.5 mha is already under salinity and 3.6 mha under sodicity problem, and still larger area is coming under potential salinity problem due to injudicious use of water under canal irrigation system. The plant growth in saline conditions is inhibited mainly by the toxic effect of increased sodium accumulation and nonavailability of water. Thus, salt stress is a major agricultural issue that demands an immediate attention. Various mechanisms to avoid salt accumulation inside cells of root and leaf include sodium exclusion, its sequestration in vacuoles, production of compatible solute, and acceleration of reactive oxygen species scavenging systems. Understanding these mechanisms at physiological and molecular levels is needed to identify new genetic sources of salt tolerance and improve the selection process in breeding. More recently, in addition to the conventional mechanisms, several genomic approaches are being used in improving salt tolerance in plants. A significant progress has been made on identification of genes for salt tolerance in wheat, and the studies related to gene expression are underway. This review describes the various tolerance mechanisms adopted by crop plants, especially wheat, under saline conditions.

V. Rana • S. Ram
Directorate of Wheat Research,
Karnal, Haryana, India

K. Nehra (✉)
Department of Biotechnology, DCRUS&T,
Murthal, Sonepat, Haryana, India
e-mail: nehrakiran@gmail.com

2.1 Introduction

With a tremendous increase in population, the demand for food grains has increased in recent years. Wheat is one of the most important world food crops, and its productivity directly affects human survival and quality of life. Increasing production of wheat calls for varieties of wheat

R.K. Salar et al. (eds.), *Biotechnology: Prospects and Applications*,
DOI 10.1007/978-81-322-1683-4_2, © Springer India 2013

with greater tolerance to hostile soil conditions and erratic precipitation (Rengasamy et al. 2003). Various abiotic stresses affect yield of food grains, and among them salinity is the major problem affecting yield because increased soil salinity can disturb plant's ionic homeostasis and create a hyperosmotic state. Salinity also affects crop output and can even cause crop death. Generally, a soil salinity of 0.2–0.5 % negatively affects plant growth. In India about 7 M ha of land is salt or sodicity affected (Hollington 1998). This is becoming worse due to unreasonable use of water, increasing scarcity of fresh water, improper land irrigation, and fertilization practices, resulting into a lot of land becoming secondary salinized. Therefore, salt stress is a primary agricultural concern that requires development of new methods and salt-tolerant varieties to increase crop resistance to saline soil. This in turn, requires a thorough knowledge of the mechanism of salt tolerance. Plants imply basic five mechanisms to overcome salt stress, i.e., ion selectivity, ion accumulation, osmotic adjustment, organic solutes, and water-use efficiency (Shannon 1997). Bread wheat is a sodium excluder and it excludes excess of salt through leaves (Husain et al. 2003). Tester and Davenport (2003) and Munns et al. (2006) suggested that Na^+ exclusion in it is attained by low uptake of Na^+ by the root cortex. Salt tolerance is a complicated trait controlled by multiple genes. Recently, significant progress has been made in marker-assisted selection, and many genes related to salt tolerance in wheat and other crops have been identified and expression studies related to them are being conducted. This review aims at understanding the basic mechanisms of salt tolerance and describes some of the recent approaches being made to improve salt tolerance.

2.2 Soil Salinity

2.2.1 Measurement of Soil Salinity

The concentration of dissolved salt in soil solution at usual field condition is difficult to measure due to sampling problem. So salinity is measured by adopting a simplified procedure. The soil sample is mixed with sufficient water to produce a saturated solution, and then an electric current is passed through the soil solution. The ability of the solution to carry a current is called electrical conductivity (EC). EC is measured in deciSiemens per meter (dS/m), which is the numerical equivalent to the old measure of millimhos per centimeter. The measured conductance is the result of the solution's salt concentration and the electrode geometry (Whipker and Cavins 2000). The effects of electrode geometry are embodied in the cell constant, and this is related to the distance between electrodes divided by their effective cross-sectional area. The cell constant is commonly obtained by calibration with KCl solutions of known concentration. Measuring the electrical conductivity (EC) of a saturation extract has an advantage in that saturation percentage is directly related to field moisture range.

Table 2.1 Classification of salt-affected soil (from NRCS guidelines) (McCauley 2005)

Soil classification	EC value (mmhos/cm^{-1})	ESP	SAR	pH
Saline soil	>4.0	<15	<12	<8.5
Sodic soil	<4.0	>15	>12	>8.5
Saline-sodic soil	>4.0	>15	>12	<8.5

2.2.2 Classification of Salt-Affected Soil

Salt-affected soil can be classified into three categories based on general electrical conductivity (EC), exchangeable sodium percentage (ESP), sodium adsorption ratio (SAR), and pH, namely, saline, sodic, and saline-sodic soil (Table 2.1).

2.2.2.1 Saline Soil

This type of soil contains excessive concentration of soluble carbonate, chloride, and sulfate salts that increase the electrical conductivity of soil above 4 mmhos/cm. At lower concentrations Ca and Mg ions improve soil texture by their tendency to flocculate soil colloids, thus, increasing aggregation and macroporosity. But at higher concentrations, the carbonates of these ions which are relatively insoluble form a white crust on the soil.

2.2.2.2 Sodic Soil

This type of soil has a relatively low EC but having high pH (above 8.5). The high pH of the soil is the result of high Na^+ ion concentration at exchange sites. The Na^+ ions cause soil colloids to disperse in insufficient amount of flocculating cations (Ca^{+2} and Mg^{+2}). Dispersed colloids reduce the soil's ability to transport water and air by clogging soil pores. Sodic soils are also prone to extreme swelling and shrinking during periods of drying and wetting.

2.2.2.3 Saline and Sodic Soil

The saline-sodic soils are soils that have chemical characteristics of both saline soils (EC greater than 4 mmhos/cm and pH less than 8.5) and sodic soils (ESP greater than 15). Therefore, plant growth in saline-sodic soils is affected by both excess salts and excess Na^+. Physical characteristics of saline-sodic soils are intermediate between saline and sodic soils; flocculating salts help moderate the dispersing action of Na^+, and structure is not as poor as in sodic soils (McCauley 2005).

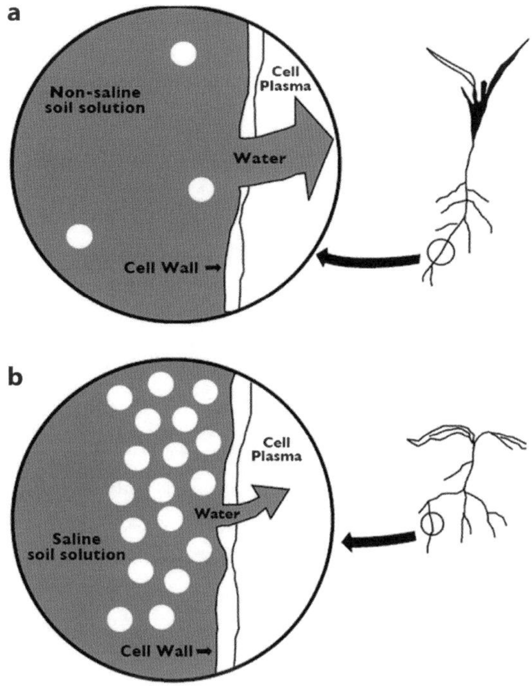

Fig. 2.1 Effect of salt on water uptake in plants. Water uptake by a plant in non-saline soil (**a**) and a saline soil (**b**) (Seelig 2000)

2.2.3 Effect of Salinity on Plant Growth

The growth response of plant to salinity occurs in two phases. The first phase of growth response is a result of salt outside the plant. When excess of salt accumulates near the root zone, it prevents the roots from taking up water from the surrounding soil (Fig. 2.1). This results in a decrease in the amount of water available to the plant regardless of the availability of water in soil and causes the plant to expend more energy to exclude salts and take up pure water. So, the metabolic and cellular stress symptoms that the plant experiences are similar to drought stress. During this phase, there is no accumulation of Na^+ and Cl^- in growing tissues that inhibit growth, as the meristematic tissues formed are fed largely in the phloem resulting in exclusion of salt, and elongating cells with their expanding vacuoles can accumulate the salt that arrives in xylem (Munns 2005). The second phase of growth response is due to build up of sodium ions inside the cell resulting in sodium toxicity. Salt taken up by the plant continues building up through transpiration stream in older leaves. When the concentration of ions exceeds the limit that can be compartmentalized in vacuoles, it causes cell death. The excess ions may either inhibit the enzyme activity or may build up in the cell wall causing cell dehydration.

2.3 Mechanisms of Salt Tolerance

2.3.1 Ion Selectivity

During salt stress there is alteration of ion ratio, resulting from influx of sodium through pathways that function in acquisition of potassium. This occurs due to similarity between the hydrated Na^+ and K^+ ions, which makes it difficult for the transport proteins to discriminate between the two ions. This discrimination problem also forms the basis for Na^+ toxicity, where key biochemical processes in the plant cell are inhibited by the competition by sodium for potassium-binding sites. Plants could use several strategies to maintain a high K^+/Na^+ ratio in the cytosol: diminishing the entry of Na^+ ions into the cells, extrusion of Na^+ ions out of the cell, and vacuolar compartmentation of Na^+ ions. Sodium exclusion trait is correlated with salt tolerance in many species (Munns and James 2003) including rice (Lee et al. 2003) and wheat (Poustini and Siosemardeh 2004). In monocots, there are four control points for Na^+ exclusion: (1) selectivity of uptake by root cells in cortex and stele; (2) loading of xylem by xylem parenchyma cells in root; (3) removal of salt from the xylem in the upper parts of the roots, the stem, or leaf sheaths by xylem parenchyma cells; and (4) loading of the phloem.

2.3.2 Vacuolar Sodium Compartmentation

The compartmentalization of Na^+ into vacuoles averts the deleterious effects of Na^+ in cytosol. Moreover, the compartmentalization of Na^+ (and Cl^-) into the vacuole allows the plants to use NaCl as an osmoticum, maintaining an osmotic potential that drives water into the cells. Earlier biochemical and tonoplast transport analysis led to the model that Na^+/H^+ antiporter localized at tonoplast level in vacuoles is involved in sequestration of sodium in vacuoles. These Na^+/H^+ antiporters mediate Na^+ uptake into vacuoles, which is driven by the vacuolar proton gradient established by the vacuolar (V-type) proton ATPase that acidifies the vacuolar lumen. In addition to NA^+/H^+ exchange in tonoplast, vesicle AtNHX1 also mediates potassium-proton exchange in tomato plant and yeast (Zhang and Blumwald 2001).

2.3.3 Synthesis of Compatible Solutes

The cellular response of salt-tolerant plants to both long-term and short-term salinity stress includes the synthesis and accumulation of compatible solutes which act as osmoprotectants. These are relatively small, nontoxic compounds that can stabilize proteins and cellular structures and increase the osmotic pressure of the cell (Yancey et al. 1982). This response is homeostatic for cell water status and protein integrity, which is perturbed in the face of soil solutions containing higher amounts of NaCl and the consequent loss of water from the cell. The accumulation of osmotically active compounds in the cytosol increases the osmotic potential to provide a balance between the apoplastic solution, which itself becomes more concentrated with Na^+ and Cl^- ions, and the vacuolar lumen, which in halophytes can accumulate up to 1 M Na^+ (and Cl^-). For short-term stress, this may provide the cells with the ability to prevent water loss; however, for continued growth under salinity stress, an osmotic gradient (towards the cytosol) must be kept to maintain turgor and water uptake and to facilitate cell expansion. Glycine betaine (GB) and proline are two major organic osmolytes that accumulate in a variety of plant species in response to environmental stresses such as drought, salinity, extreme temperatures, UV radiation, and heavy metals (Maris and Eduardo 2002). Many studies have indicated a positive relationship between accumulation of GB and proline and plant stress tolerance (Ozturk and Demir 2002; Yang et al. 2003).

2.3.4 Combating Antioxidant Stress

Salt stress induces production of reactive oxygen species (ROS) including superoxide radicals (O_2^-), hydrogen peroxide (H_2O_2), and hydroxyl radicals ($OH^•$). These species cause oxidative damage to different cellular components including membrane lipids, proteins, and nucleic acids. Low molecular mass antioxidants such as ascorbic acid and reduced glutathione and a diverse array of enzymes such as superoxide dismutases (SOD), catalases (CAT), ascorbate peroxidases (APX), glutathione *S*-transferases (GST), and glutathione peroxidases (GPX) are produced by

Table 2.2 Salt tolerance levels of various types of plants (McKenzie 1988)

Salt tolerance EC (ds/m)	Field crops	Forages	Vegetables	Trees, shrubs
Very high 20 (ds/m)		Beardless wild rye		
		Fulks altai grass		
		Levonns alkali grass		
		Alkali sucatan		
High 16 (ds/m)	Kochia	Altai wild rye		Siberian salt tree
	Sugar beets			Sea buckthorn
		Tall wheatgrass		Silver buffalo berry
		Russian wild rye		
		Sender wheatgrass		
8 (ds/m)	6-row barley	Bird's-foot trefoil	Garden beets	Hawthorn
	Safflower	Sweet clover	Asparagus	Russian olive
	Sunflower	Alfalfa	Spinach	American elm
				Siberian elm
	2-row barley	Bromegrass		Villosa lilac
	Fall rye			Laurel leaf willow
	Winter wheat			
	Spring wheat			
Moderate	Oats	Crested wheatgrass	Tomatoes	Spreading juniper
	Yellow mustard	Intermediate wheatgrass	Broccoli	Poplar
	Meadow fescue	Reed canary grass	Cabbage	Ponderosa pine
	Flax			Apple
	Canola			Mountain ash
4 (ds/m)	Corn		Sweet corn	Common lilac
			Potatoes	Siberian crab apple
				Manitoba maple
				Viburnum
Low	Timothy	White Dutch clover	Carrots	Colorado blue spruce
	Peas	Alsike clover	Onions	Rose
	Field beans	Red clover	Strawberries	Douglas fir
			Peas	Balsam fir
			Beans	Cottonwood
				Aspen, birch
				Raspberry
0 (ds/m)				Black walnut
				Dogwood
				Little-leaved linden
				Winged euonymus
				Spirea
				Larch

plants during stress (Haliwell and Gutteridge 1986). These play an essential role to maintain a balance between the overproduction of reactive oxygen species (ROS) and their scavenging to keep them at signaling level for reinstating metabolic homeostasis (Table 2.2).

2.4 Sodium Transporters at Cellular Level

Various transporters act at cellular level for either exclusion of sodium ions or for sequestering them in vacuoles (Fig. 2.2). Na^+/H^+ antiporter is present in tonoplast membrane of vacuoles, which sequester Na^+ in the vacuoles. The resulting vacuolar Na^+ sequestration protects essential enzymatic reactions in the cytoplasm from excess Na^+ levels while maintaining turgor (Glenn et al. 1999). Arabidopsis genome sequencing project led to the identification of the plant Na^+/H^+ antiporter gene, AtNHX1. At plasma level SOS1 acts as Na^+/H^+ antiporter. These mediate Na^+ extrusion that is energized by proton flux. SOS2 and SOS3 regulate Na^+ transport by SOS1 (Tomoaki and Julian 2004). High-affinity potassium transporters (HKT) control Na^+ transport in higher plants. HKT coding genes have been separated into two groups based on amino acid sequence. Group 1 HKT genes are involved in transport of Na^+ only and not in unloading of Na^+ from the xylem. The mechanism of action of transporters coded by group 2 gene is not clear whether the main function is to transport Na^+ or K^+ (Byrt et al. 2007).

2.5 Wheat Breeding for Salinity Tolerance

Breeding for salt tolerance is not specifically targeted because much of the focus is on other constraints, the major being drought. Comparatively little work has been done on breeding for salt tolerance. Targeted breeding has been largely confined to India and Pakistan. The most successful releases have been the Indian KRL1-4 and KRL-19, released by the Central Soil Salinity Research Institute (CSSRI) at Karnal; the Pakistani LU26S and SARC-1, released by the Saline Agriculture Research Cell (SARC)

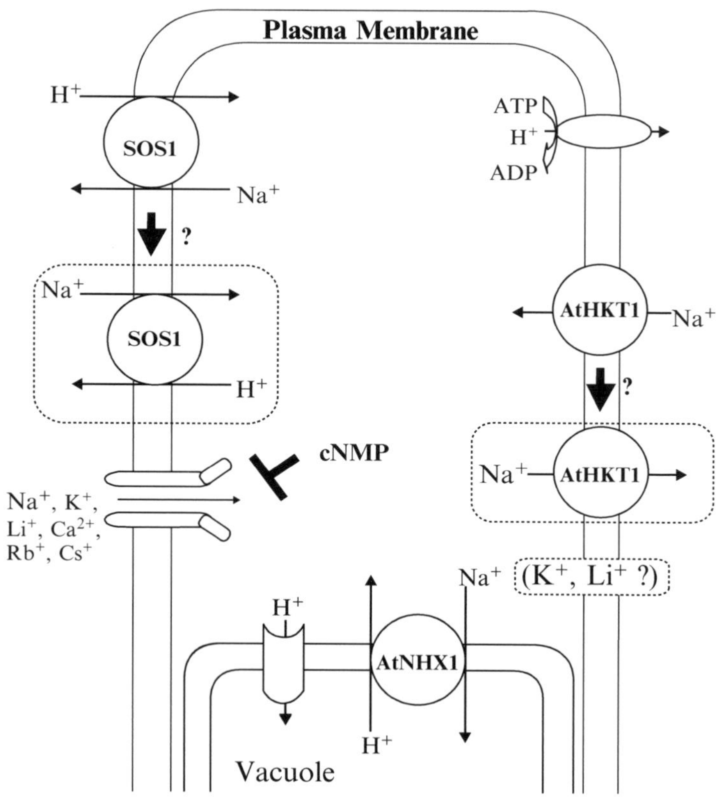

Fig. 2.2 Sodium transporters at cellular level (Tomoaki and Julian 2004)

at Faisalabad; and the Egyptian Sakha 8, released by the Agricultural Research Center at Giza (Munns et al. 2006). In India, almost all salt-tolerant wheat germplasm is derived from Kharchia 65, a line developed from selections from farmers' fields in the sodic-saline soils of the Kharchi-Pali area of Rajasthan (Rana 1986). KRL1-4, a cross of Kharchia 65 with WL711, has done well on the saline soils of northern India, but not in Pakistan, possibly because of the heavier soils and greater problems of water logging.

2.6 Genomic Approaches for Improving Salt Tolerance

Various genomic approaches are used nowadays for improving salt tolerance in plants. The basis of effective engineering strategies is provided by discovery of novel genes, determination of their expression pattern related to salt stress, and a better understanding of their role in stress adaptation.

2.6.1 Gene-Related Salt Tolerance Discovered in Wheat

Salinity tolerance in plants is a result of action of genes that limit the rate of salt uptake from the soil and the transport of salt throughout the plant and adjust the ionic and osmotic balance of cells in roots and shoots and genes that regulate leaf development and the onset of senescence (Munns 2005). Sodium exclusion trait present in bread wheat is due to the presence of K^+/Na^+ discrimination locus (Kna1) present on chromosome 5 (Dubcovsky et al. 1996). Kna1 is involved in Na^+ exclusion from leaves and discrimination of K^+ over Na^+ concentration in roots. Durum wheat (tetraploid, AB genomes) has higher rates of Na^+ accumulation and poor K^+/Na^+ discrimination (Gorham et al. 1987; Munns et al. 2000) and is less salt tolerant than bread wheat. Two genes for sodium exclusion, namely, Nax1 and Nax2, were mapped on chromosome 2A and 5A, respectively, in Line 149 (durum line) during international screening program of durum genotypes for improving salt tolerance in durum. These genes are involved in retrieval of sodium from xylem sap in the roots.

2.6.2 Specific Approaches Used for Gene Expression Studies of Salt Tolerance-Related Genes

Gene expression studies are important for giving an insight of gene function and its role in salt tolerance. Approaches such as serial analysis of gene expression (SAGE), nuclear expressed sequence tag (NEST), cDNA microarrays, subtractive cDNA hybridization, representational difference analysis, and subtractive suppression hybridization (SSH) are being used for studying expression (preferably differential expression). SAGE is used for rapid quantification of the occurrence of large number of transcripts in a particular population. It efficiently generates extremely large EST databases by sequencing multiple tags within each clone (Bertelsen and Valculescu 1998). NEST combines fluorescence-assisted nucleus sorting and cDNA generation (based on the expression of nucleus-targeted green fluorescent protein [GFP], which is controlled by a cell-specific promoter) from the RNA of isolated nuclei (Macas et al. 1998). Subtractive cDNA hybridization is used to identify and isolate cDNA of differentially expressed genes. It involves hybridization of cDNA from one population (tester) to excess of mRNA (cDNA) from other population and then selection of the unhybridized fraction (target) from the hybridized common sequence. But nowadays other techniques such as SSH are used. It is a PCR-based cDNA subtraction method, which is used to amplify target cDNA fragment (differentially expressed) and simultaneously to suppress nontarget DNA amplification. Representational difference is applied to enrich genomic fragments that differ in size or representation and to clone differentially expressed cDNA (Diatchenko et al. 1996). cDNA microarray is another high-throughput approach for obtaining comprehensive gene expression profiles. It is used to examine gene expression patterns in tissues including root, leaf, and flowers and for studying parallel gene expression (Lemieux et al. 1998).

References

Bertelsen AH, Valculescu VE (1998) High-throughput gene expression analysis using SAGE. Drug Discov Today 3:152–159

Byrt CS, Platten JD, Spielmeyer W, James RA, Lagudah ES, Dennis ES, Tester M, Munns R (2007) HKT 1,5- like cation transporters linked to Na^+ exclusion loci in wheat, Nax2 and KNa1. Plant Physiol 143:1918–1928

Diatchenko L, Lau YF, Campbell AP, Chenchik A, Moqadam F, Huang B, Lukyanov S, Lukyanov K, Gurskaya N, Sverdlov ED (1996) Suppression subtractive hybridization: a method for generating differentially regulated or tissue-specific cDNA probes and libraries. Proc Natl Acad Sci U S A 93:6025

Dubcovsky J, Santa Maria G, Epstein E, Luo MC, Dvořák J (1996) Mapping of the K^+/Na^+ discrimination locus *Kna1* in wheat. Theor Appl Genet 2:448–454

Glenn EP, Brown J, Blumwald E (1999) Salt tolerance and crop potential of halophytes. Crit Rev Plant Sci 18:227–256

Gorham J, Hardy C, Wyn Jones RG, Joppa LR, Law CN (1987) Chromosomal location of a K/Na discrimination character in the D genome of wheat. Theor Appl Genet 74:584–588

Haliwell B, Gutteridge JMC (1986) Oxygen free radicals and iron in relation to biology and medicine: some problems and concepts. Arch Biochem Biophys 246:501–514

Hollington PA (1998) Technological breakthroughs in screening/breeding wheat varieties for salt tolerance. In: Gupta SK, Sharma SK, Tyagi NK (eds) Proceedings of the National Conference 'Salinity management in agriculture'. Central Soil Salinity Research Institute, Karnal, pp 273–289

Husain S, Munns R, Condon AG (2003) Effect of sodium exclusion trait on chlorophyll retention and growth of durum wheat in saline soil. Aust J Agric Res 54:589–597

Lee KS, Choi WY, Ko JC, Kim TS, Gregoria GB (2003) Salinity tolerance of Japonica and Indica rice (Oryza sativa L.) at the seedling stage. Planta 216:1043–1046

Lemieux B, Aharoni A, Schena M (1998) Overview of DNA chip technology. Mol Breed 4:277–289

Macas J, Lambert GM, Dolezel DW (1998) Galbraith Nuclear expressed sequence tag (NEST) analysis: a novel means to study transcription through amplification of nuclear RNA. Cytometry 33:460–468

Maris PA, Eduardo B (2002) Engineering salt tolerance in plants. Curr Opin Biotechnol 13:146–150

McCauley A (2005) Salinity and sodicity management. Salt and water management module 2. Montana State University Extension Service, Bozeman, pp 1–15

McKenzie RC (1988) Tolerance of plants to soil salinity. In: Proceedings of the dryland salinity control workshop. Alberta Agriculture, Food and Rural Development, Conservation and Development Branch, Calgary, Alberta, pp 246–251

Munns R (2005) Genes and salt tolerance: bringing them together. New Phytol 167:645–663

Munns R, James RA (2003) Screening methods for salinity tolerance: a case study with tetraploid wheat. Plant Soil 253:201–218

Munns R, Hare RA, James RA, Rebetzke GJ (2000) Genetic variation for improving the salt tolerance of durum wheat. Aust J Agric Res 51:69–74

Munns R, James RA, Läuchli A (2006) Approaches to increasing the salt tolerance of wheat and other cereals. J Exp Bot 57:1025–1043

Ozturk L, Demir Y (2002) In vivo and in vitro protective role of proline. Plant Growth Regul 38:259–264

Poustini K, Siosemardeh A (2004) Ion distribution in wheat cultivars in response to salinity stress. Field Crop Res 85:125–133

Rana RS (1986) Evaluation and utilization of traditionally grown cereal cultivars on salt affected areas in India. Indian J Genet Plant Breed 46:121–135

Rengasamy P, Chittleborough D, Helyar K (2003) Root zone constraints and plant-based solutions for dryland salinity. Plant Soil 257:249–260

Seelig BD (2000) Salinity and sodicity in North Dakota soils. EB 57. North Dakota State University Extension Service, Fargo, ND

Shannon MC (1997) Adaptation of plants to salinity. Adv Agron 60:87–120

Tester M, Davenport R (2003) Na^+ tolerance and Na^+ transport in higher plants. Ann Bot 91:503–527

Tomoaki H, Julian I (2004) Sodium transporters in plants. Diverse genes and physiological functions. Plant Physiol 136:2457–2462

Whipker BE, Cavins J (2000) Electrical conductivity (EC): units and conversions. FLOREX. 002

Yancey PH, Clark ME, Hand SC, Bowlus RD, Somero GN (1982) Living with water stress: evolution of osmolyte systems. Science 217:1214–1222

Yang WJ, Rich PJ, Axtell JD, Wood KV, Bonham CC, Ejeta G, Mickelbart MV, Rhodes D (2003) Genotypic variation for glycine betaine in sorghum. Crop Sci 43:162–169

Zhang HX, Blumwald E (2001) Transgenic salt-tolerant tomato plants accumulate salt in foliage but not in fruit. Nat Biotechnol 19:765–768

Role of Biotechnology for Commercial Production of Fruit Crops

3

S. Kajla, A.K. Poonia, P. Kharb, and J.S. Duhan

Abstract

India is the second largest producer of fruit plants after China. Mango, banana, citrus, guava, grape, pineapple, and apple are the major fruit plants grown in India. Apart from these, other fruits like papaya, ber, phalsa, sapota, annona, jackfruit, and pomegranate in the tropical and subtropical group and peach, apricot, pear, almond, walnut, and strawberry in the temperate group are also grown in a remarkable area. Propagation in fruit plants through seed is primarily done to raise rootstocks, which are required for grafted plants. Some of the fruit plants like papaya are propagated through seeds, but dormancy, poor germination, low seed viability, and adverse environment conditions are major limiting factors. Advances in biotechnological techniques like plant tissue culture provided new methods for rapid production of high-quality, disease-free, and true-to-type planting material. The technique not only offers a valuable alternative in fruit trees propagation studies but is also useful for virus control and management of genetic resources. Nowadays the range of routine technologies of plant tissue culture has expanded to include somatic embryogenesis, somatic hybridization, virus elimination, in vitro mutagenesis, anther or microspore culture production of haploids, embryo rescue technique or embryo culture, protoplast culture, and somatic fusion. Out of the abovementioned techniques, the most exploited one for mass production of fruit plants is micropropagation. To produce virus-free plants, meristem culture and micrografting techniques have been

S. Kajla (✉) • A.K. Poonia • P. Kharb
Centre for Plant Biotechnology, HSCST,
Department of Science and Technology, Haryana,
CCSHAU Campus, Hisar 125004, India
e-mail: ksubhash73@hotmail.com

J.S. Duhan
Department of Biotechnology, Chaudhary Devi
Lal University, Sirsa 125055, India

R.K. Salar et al. (eds.), *Biotechnology: Prospects and Applications*,
DOI 10.1007/978-81-322-1683-4_3, © Springer India 2013

standardized in different fruit plants. The success varies with the plant species, variety, and the culture environment. Recently, attention has turned to the possible beneficial effects of microorganism's in vitro plant cultures. It has been observed that mycorrhiza enhances the survival percentage of in vitro-raised plant.

3.1 Introduction

India is the second largest producer of fruit plants after China. Mango, banana, citrus, guava, grape, pineapple, and apple are the major fruit plants grown in India. Apart from these, other fruits like papaya, ber, phalsa, sapota, annona, jackfruit, and pomegranate in the tropical and subtropical group and peach, apricot, pear, almond, walnut and strawberry in the temperate group are also grown in a remarkable area. Although fruit is grown throughout the country, the major fruit-growing states are Maharashtra, Tamil Nadu, Karnataka, Andhra Pradesh, Bihar, Uttar Pradesh, and Gujarat. The major fruit crops of India are given below.

Mango is the most important fruit accounting for 22 % production of total fruits in the country, which is highest in the world with India's share of about 54 %. India has the richest collection of mango cultivars. Major mango-growing states are Uttar Pradesh, Bihar, Andhra Pradesh, Orissa, West Bengal, Maharashtra, Gujarat, Karnataka, Kerala, and Tamil Nadu. The main varieties of mango grown in the country are Alphonso, Dashehari, Langra, Fazli, Chausa, Totapuri, Neelum, etc.

Banana comes next in rank accounting for about 38 % of the total production of fruits. India has the first position in the world in banana production. While Tamil Nadu leads other states with a share of 19 %, Maharashtra has highest productivity of 58.6 metric tonnes against India's average of 32.5 metric tonnes per ha. The other major banana-growing states are Karnataka, Gujarat, Andhra Pradesh, and Assam. The main varieties of banana are Dwarf Cavendish, Bhusaval Keli, Basrai, Poovan, Harichhal, Nendran, Safed Velchi, etc.

Citrus fruits rank 3rd accounting for 13 % of total production. Lime, lemons, sweet oranges, and mandarin cover bulk of the area under these fruits and are grown mainly in Maharashtra, Andhra Pradesh, Karnataka, northeastern states, Punjab, Orissa, and Madhya Pradesh.

Guava is the fourth most widely grown fruit crop in India accounting for 4 % of total production. The popular varieties of guava are Allahabad Safeda, Lucknow-49, Nagpur seedless, Dharwar, etc. Bihar is the leading state in guava production with 0.30 MT, followed by Andhra Pradesh and Uttar Pradesh. The other states where guava is grown widely are Gujarat, Karnataka, Punjab, and Tamil Nadu.

Grapes occupy the fifth position among fruit crops. The major varieties of grapes grown in India are: Thompson seedless, Sonaka, Anab-e-Shahi, Perlette, Bangalore blue, Pusa seedless, Beauty seedless, etc. Maharashtra occupies the first position with a production of 0.68 MT of grapes, followed by Karnataka. The other states growing grapes are Punjab, Andhra Pradesh, and Tamil Nadu.

3.2 Propagation Techniques

Plant propagation is the method of production of more than one plant from the mother plant or the tissue over a specific time period. True-to-type plant production from the mother plant is one of the major objectives of propagation. Plants can be propagated either by sexual or vegetative means. In sexual reproduction method, the progeny gets characteristics from both of its parents. It will not grow "true" to either parent and has an unpredictable combination of characteristics. Plant propagation has been a useful tool since centuries, for fruit plants since time immemorial. Plant propagation depends on the plant species, variety, method of propagation, and climatic and growth conditions. It is better to propagate fruit cultivars by vegetative propagation methods in order to ensure genetic reliability.

Plant propagation is primarily done by conventional methods, which include sexual and asexual means. However, in the recent past commercial plant propagation through biotechnological applications has made great contributions towards production of plants.

3.2.1 Sexual Method of Propagation

This is the method of raising the plants through seeds. The method is used for evolution of new varieties through breeding and production of hybrids.

3.2.1.1 Advantages

1. This is the most popular method of propagation in some fruit plants like papaya.
2. Seed propagated rootstocks are hardy and develop better root system.
3. Viruses do not transmit through seeds; thus mostly the seedlings are virus-free.
4. Occurrence of polyembryony in fruits leads to the development of uniform seedlings.

3.2.1.2 Disadvantages

1. Long juvenile period in seed-grown plants as compared to asexually raised plants.
2. The progeny is not true to type.
3. It is not economical to handle larger trees.

3.2.2 Asexual Method of Propagation

In this method, the plants are propagated from vegetative part of the mother plant instead of seeds.

3.2.2.1 Advantages

1. In some fruit plants, which do not bear seeds, this is the only propagation method.
2. The plants are generally true to type and uniform in growth, yield, and quality.
3. Short juvenile period as compared to seed-grown plants.
4. The advantages of rootstocks can be obtained by budding or grafting susceptible varieties on resistant/tolerant rootstocks.

3.2.2.2 Disadvantages

1. New variety or hybrid cannot be developed by this method.
2. Plants are not so vigorous and long lived as the seedling trees.
3. Conservation of these plants is expensive as compared to storage of seeds.

3.3 Propagation of Fruit Plant by Seed

Propagation in fruit plants through seed is primarily done to raise rootstocks, which are required for grafted plants. Some of the fruit plants like papaya are conventionally seed propagated. There may be following events may occur during seed propagation.

3.3.1 Dormancy

The dormancy in seeds may be due to hard seed coat, water and gas impermeability, immaturity of embryo, deficiency of some endogenous growth promoters, or excess of endogenous growth inhibitors. Different methods like stratification, scarification, and chemical treatment can be used for breaking of dormancy in seed to improve germination.

3.3.2 Germination

This refers to the emergence of a new plant from the mature seed. Germination requires seed viability and appropriate environmental conditions.

3.3.3 Apomixis

Apomixis occurs when an embryo is produced from a single cell of the saprophyte and does not develop from fertilization of two gametes. This is a natural mechanism during which vegetative embryo is produced instead of sexual or zygotic embryo.

3.3.4 Polyembryony

Polyembryony refers to the seeds having more than one embryo in the seed. One of the embryos arises from the union of male and female gametes and is called gametic or sexual embryo. The other embryos are produced by simple mitotic division of cells of nucellus without the help of male gamete in their formation. The phenomenon of nucellar embryo is of common occurrence in citrus and mango.

3.4 Vegetative Propagation

Vegetative or asexual propagation is propagation of plants by the method other than sexual propagation. It includes no change in genetic makeup or quality of the new plant. The plants produced by this method are true to type in growth, ripening, yield, and fruit quality.

Commercial multiplication of various fruit plants is done by using vegetative or asexual propagation. These methods include cutting, layering, budding, and grafting.

3.4.1 Cutting

It is the method of propagating fruit plants. In this method the stem (small part of plant) having at least few buds is detached from parent plant and placed under favorable conditions to develop into a complete plant. This method is commonly used in plants, which cause rooting easily and readily, thus multiplication of plants is very quick and cheap. The fruit plants like phalsa, baramasi lemon, and grapes are commercially propagated by cuttings. In case of deciduous fruit plants such as grape, pomegranate, phalsa, and fig, the cuttings are made after pruning. While in evergreen fruit plants like baramasi lemon, the cuttings can be prepared during the spring (February–March) and rainy season (August–September).

3.4.2 Layering

Layering is a vegetative propagation method, in which roots are induced on the shoots while they are still attached to the mother plants. This is an alternate propagation method in fruit plants which do not root easily when detached from the mother plants. Most commonly used methods of layering are air, ground, and mound layering. Air layering is done in air, and the rooting is done on the shoot itself when it is still attached to the mother plant. These layers can be planted in the fields during the following year in February or September–October. In ground layering method, a branch of plant, which is near the ground, is chosen, and a ring of bark is removed just below the bud. This branch is then bended and buried in soil when still attached to the mother plant. The soil is regularly watered to keep it moist. Within a few weeks, the roots are formed and new plant is separated from the mother plant. This method is commonly followed for propagation of baramasi lemon. In mound layering method, plant is headed back either in February or in July. The new shoots come out during April and September, from ground level. A ring of bark is removed from these shoots and they are covered with moist soil. The rooted stools of April stooling are separated during rainy season, and those of August are removed in the following spring. These stools, after separating from the parent plant are planted in the nursery fields. This method is also known as stool layering and is used for propagation of guava and apple rootstocks.

3.4.3 Budding

This is a method in which only one bud is inserted in the rootstock. The method is very easy and fast and saves budwood as compared to grafting. As soon as the bark starts slipping both on the stock and scion, this is considered to be the optimum time for budding. This shows that the cambium (xylem and phloem), which is the tissue responsible for union, is active. This method is generally employed during spring and rainy season. The budding may be T-budding, patch budding, and chip budding.

In most fruit trees T-budding is performed either in the spring (March–April) or in the rainy season (July–September). This is the most common method of propagation of citrus plants. Patch budding is done in guava during May and June. Chip budding method is usually employed just before the start of new growth, i.e., when the stock and scion are still dormant.

3.4.4 Grafting

Grafting is a vegetative propagation method of fruits, where two plant parts are joined together in such a manner that they unite and continue their growth as one plant. In this method, the scion twig has more than two buds on it. Grafting is commonly done in pear, peach, plum, almond, mango, etc. In temperate fruits like peach, plum, and almond, grafting is done when the plants are dormant, while in mango it is done when the trees are in active growth. The different methods of grafting are tongue grafting, cleft grafting, approach grafting, and side grafting. Tongue grafting is commonly used when the stock and scion are of equal diameter. Cleft grafting/wedge grafting is done during dormant period. Approach grafting/inarching is commonly followed in mango. Side grafting can be carried out successfully from March to October, but success during May and October is rather low. This method of propagation is commonly used in mango.

3.5 Propagation of Fruit Plants Through Specialized Organs

Propagation of fruit plants through specialized organs like runners, suckers, and slips. Runner is a specialized stem that develops from the axil of a leaf at the crown of a plant. It grows horizontally along the ground and forms a new plant at one of the nodes, e.g., strawberry. A shoot arising on an old stem or underground part of the stem is known as sucker. The capacity of a plant to form suckers varies from plant to plant, variety to variety, and is even climate dependent. The sucker formation is common in fruit plants like pear and banana. In banana, sword suckers are commonly used for propagation of plants. Slips are shoots just arising below the crown but above the ground. Pineapple is commercially propagated through this method of propagation.

3.6 Special Techniques of Propagation

The use of biotechnological technique like plant tissue culture in the regeneration and fast multiplication of economically important plants is a comparative recent and radical development. Advances in biotechnology provided new methods for rapid production of high-quality, disease-free, and true-to-type planting material. Biotechnological tools like in vitro culture and micropropagation not only offer a valuable alternative in fruit trees propagation studies but are also useful for virus control and management of genetic resources. The technique of in vitro micropropagation was employed for virus elimination from certain plant species by meristem culture, rapid propagation under aseptic conditions, and in vitro preservation of plant germplasm. Today plant tissue culture applications encompass much more than clonal propagation and micropropagation. The range of routine technologies has expanded to include somatic embryogenesis, somatic hybridization, virus elimination, as well as the application of bioreactors to mass propagation (Aitken-Christie et al. 1995; Paek et al. 2001). These include:

- Micropropagation
- Meristem culture
- Somaclonal variations
- In vitro mutagenesis
- Anther or microspore culture production of haploids
- Embryo rescue technique or embryo culture
- Protoplast culture – somatic fusion

Out of the abovementioned techniques, the most exploited one for mass production of horticultural crop plants is micropropagation. Thus this technology is dealt with details as follow.

3.6.1 Micropropagation

The vegetative propagation of fruit plants has been practiced for centuries, and many improvements in conventional methods have been made over the years. Recently, the tissue culture technique, i.e., micropropagation, has expanded their scope and potential on commercial scale. Micropropagation is suitable for the rapid and large-scale clonal multiplication of elite germplasm. The technique has been referred as micropropagation because the size of the tissue in culture is very small as compared to conventional method. The size of meristem tissue used for micropropagation is about 0.1–0.5 mm size having only one or two leaf primordia. First time in 1952 virus-free plants were obtained by culturing shoot meristems. Later on with the discovery of the hormonal control of organogenesis and finding of most commonly used tissue culture media by Murashige and Skoog (1962), the scope of micropropagation was further extended to huge scale of plant species, including fruit and plantation crops. With the advancement in science and technology, micropropagation technique has also been standardized for many plants, and it is now widely used for multiplication of many horticultural plants. Now micropropagation is perhaps the most popular and widely commercialized global application of plant biotechnology in horticulture. A large number of plants are being cloned and exploited commercially worldwide. Novel germplasm in horticultural crops, created using various biotechnological tools, also needs to be multiplied rapidly for quick dissemination. This is possible only by integrating in vitro culture and molecular biology techniques.

Micropropagation is well known as a means of producing millions of identical plants ("clones") under aseptic conditions, in a relatively short period of time, independent of seasonal constraints. Propagation of plants through tissue culture, including sophisticated techniques of meristem culture and molecular indexing of diseases, is of immense use in making available healthy propagules. Besides these, micropropagation is also applied advantageously for import and export of germplasm, obviating quarantine-related problems.

3.6.1.1 Advantages of Micropropagation

Micropropagation techniques have several advantages over conventional propagation techniques. These include:

1. Year-round availability of plants irrespective of seasonal constraints
2. Fast multiplication of true-to-type planting material
3. Conservation of plant diversity
4. Multiplication of hybrids
5. Disease-free plant production
6. Highly beneficial in dioecious fruit plant species (date palm and papaya), where large-scale production of female plants is possible
7. Export and import of germplasm become easy requiring minimum quarantine checks
8. Easy transport of propagation material

3.6.1.2 Ways of Micropropagation

The ways for the regeneration of whole plant from small excised plant parts include the following.

3.6.1.2.1 Regeneration from Existing Meristems

This is also known as axillary shoot proliferation. The existing meristems such as shoot tip or nodal bud are cultured on the different types of media fortified with various plant growth regulators, alone or in combination. The shoot proliferation depends on the type of hormones/regulator used. For axillary shoot proliferation, the commonly used growth regulators are benzylaminopurine (BAP), kinetin, and 2-isopentenyl adenine (2-ip). The regenerants are considered to be genetically stable. The process of shoot tip culture is successful in banana tissue culture.

3.6.1.2.2 Regeneration from Adventitious Meristems

Shoot multiplication either directly or by callus formation can be obtained by inducing adventitious shoot production on mature plant organs such as leaves, internodes/stems, and roots. For initiation of adventitious meristems, a proper combination of growth regulators is needed in culture medium. In general, shoots are formed when a high ratio of cytokinin to auxin is present, and reverse is true for root formation. The plants regenerated via this method are not always genetically stable, due to formation of mixoploids. The repeated subculture of callus also reduces its potential and regenerative capacity.

3.6.1.2.3 Regeneration by Somatic Embryogenesis

The somatic embryos are bipolar structures, possess both shoot and root meristem, and originate from somatic or vegetative cells. It requires a high level of auxin in culture medium for induction, followed by low auxin and cytokinin concentration in medium. Somatic embryos may arise directly on explants or via callus formation or liquid suspension cultures. The somatic embryos may act as synthetic seeds after encapsulation, which is an attractive alternative for propagation of plants. Among two methods, viz., hydrated or desiccated, for artificial seed production, the production of hydrated seeds is more popular. In this method, the individual somatic embryo is encapsulated in a water-based gel (hydragel such as calcium alginate). Embryos developed through tissue culture technique are mixed with sodium alginate and dropped with pipette into a calcium salt (calcium chloride) solution to form calcium alginate capsules. The capsules are washed in water and then placed on culture medium for germination. Artificial seeds have been produced in banana, citrus, mango, apple, olive, and kiwi fruit.

3.6.1.3 Stages/Steps Involved in Micropropagation

There are four main stages/steps involved in micropropagation of plants, such as explant establishment, shoot multiplication, rooting, and hardening and transfer to soil/field.

3.6.1.3.1 Explant Establishment

The establishment of explants depends on several factors such as the source of explants/genotype; size of explants; type of explants such as leaf, root, and stem from mature or immature plants/seedlings; explant sterilization process; types of sterilizing agents used; and the in vitro culture conditions such as culture media, composition, temperature, humidity, and light. The explants showing growth are considered established.

3.6.1.3.2 Shoot Multiplication

The established explants are subculture on shoot multiplication medium. The medium is fortified with such growth regulators/hormones which avoid callus formation and cause multiplication of established cultures. Thus the proper hormonal combinations result in multiplication. Hence careful use of auxins like IAA, NAA, and 2,4-D and cytokinins like BAP and kinetin is done in culture medium. It is a known fact that cytokinins enhance shoot multiplication.

3.6.1.3.3 Rooting of Shoots

The in vitro regenerated shoots are rooted in the medium containing auxins like IAA, NAA, and IBA. The rooting can also be induced on medium devoid of hormones due to stress conditions. The rooting should also be preferably without formation of callus, thus avoiding somaclonal variants.

3.6.1.3.4 Hardening and Transfer to Soil/Field

The in vitro rooted plantlets thus obtained are hardened/acclimatized before transferring to the field. The plants are hardened in greenhouse under high humidity. The various types of potting mixtures are used to enhance their survival rate in greenhouse.

3.6.1.4 Factors Affecting In Vitro Multiplication

3.6.1.4.1 Selection of Explants

Selection of healthy explant is important in micropropagation (William and Maheshwaran 1986). The suitability of which depend upon:

3.6.1.4.2 The Organ That Is to Serve as a Tissue Source

Different types of explants have been used by various scientists as the tissue source such as in vitro propagation of banana cultivars through the culture of excised shoot apices is well established as described by a number of researchers

(Doreswamy et al. 1983; Vuylsteke and De Langhe 1985; Banerjee et al. 1986; Pandey et al. 1993; Sudhavani and Reddy 1997; Kotecha and Kadam 1998; Kumar et al. 2005; Choudhary et al. 2013). Plant regeneration studies using male inflorescence (Krikorian et al. 1993; Escalant and Teisson 1994) have been reported in important banana cultivars. The shoot tip is considered to be most responsive and better explant as compared to other parts of the plant.

3.6.1.4.3 Size of Explant and Overall Quality of Parent Plant

Many workers have confirmed that the quality of explants primarily determines the establishment of in vitro cultures (John and Murray 1981; Kim et al. 1981; Keathley 1983). Madhulata et al. (2004) observed that larger explants regenerated earlier while too small shoot explants of banana failed to regenerate.

3.6.1.4.4 Physiological State

Physiological state of the plant from which explant is taken is an important factor in the process of micropropagation. Younger trees provide more suitable explants for regeneration than the aged trees (Bonga 1982; Sommer and Wetzstein 1984). The success in micropropagation from mature tree is very much dependent upon careful selection of explant (Murashige 1974; Sommer and Caldas 1981).

3.6.1.5 Methods of Disinfection of the Explants

Successful disinfection of explants is a prerequisite for in vitro culture and often involves a standard set of treatments, which vary with the type and species of explant in question (Thorpe and Patel 1984). Contamination in tissue culture can originate from two sources: either through carryover of microorganism on the surface of explant or in the tissue itself (endophytic microbes). For in vitro propagation of banana, bacterial contamination is a great problem. Huge numbers of explants are destroyed during micropropagation of banana due to endogenous bacteria (Hadiuzzaman et al. 2000).

The various types of disinfectants can be used to overcome the problem of contamination. The surface sterilizing agents like sodium hypochlorite, calcium hypochlorite, and mercury chloride can be used for sterilization of explants. The systemic sterilizing agents like Bavistin (fungicide) and streptocycline (bactericide) can also be useful for prevention of contamination. The above chemicals may be used alone or in combination to enhance the survival rate of explants during sterilization.

3.6.1.6 Problems Encountered During Micropropagation

The success of micropropagation is hampered by these problems. The main problems encountered during micropropagation are as follow.

3.6.1.6.1 Contamination

Bacterial/fungal contaminations in the cultures do not allow the explant/multiplied culture to grow. The problem can be overcome by growing donor plants in growth chambers, systemic fungicide spray prior to explant removal, effective explant sterilization with effective sterilizing agents, performing inoculations in sterilized laminar airflow cabinets, and using sterilized surgical instruments. Fumigation of inoculation room using dilute formaldehyde solution also helps to minimize the problem.

3.6.1.6.2 Release of Phenolic Compounds

The cultured explants of certain plant species secrete phenols into the medium, which cause browning due to oxidation of phenols and formation of quinones, the toxins which affect the growth of cultured explants. The use of antioxidants such as activated charcoal, citric acid/ascorbic acid, polyvinylpyrrolidone (PVP), and polyvinylpolypyrrolidone (PVPP) during sterilization and in the culture medium helps to prevent the browning. Date palm, banana, and guava are the fruit plants having the same problem.

3.6.1.6.3 Variations in Tissue Culture-Raised Plants

Variability occurs due to callusing and regeneration of plants from callus instead of direct shoot induction and proliferation. The variability is highly undesirable in the tissue culture-raised plants.

The plants regenerated via adventitious meristems as compared to axillary meristem are susceptible to mutations, as it is derived from either a single cell or a small group of cells. This may cause variation in regenerated plants. The variation due to callusing can be stopped by addition of growth regulators which inhibit callusing such as triiodobenzoic acid (TIBA), phloroglucinol, and phloridzin and also by reduction of inorganic salt concentration in the culture medium.

3.6.1.6.4 Mortality in Greenhouse

Tissue culture-raised plants have different leaf morphology, poor photosynthetic efficiency, malfunctioning of stomata (open), and reduced epicuticular waxes and thus are amenable to transplantation shock. Hardening of such plants is thus must before transplantation under field conditions. Conservation of moisture by creating high humidity around the plants, partial defoliation, and application of antitranspirants are useful for hardening of in vitro-raised plants.

3.6.1.7 Limitations

Limitations of micropropagation are as follows:

1. The facilities are costly.
2. Highly technical skill is required to carry out different procedures.
3. Pathogens, once appear in the system, multiply at a very fast rate and deteriorate the whole culture.
4. Plants having high levels of phenols (mango, date palm, coconut, etc.) usually do not respond to micropropagation techniques.
5. Establishment of laboratory-raised plants in the field is a very difficult task.

3.6.2 Meristem Tip Culture

This technique is widely used in fruit plants. In this method, the meristem tip consisting of one or two pairs of leaf primordia is cultured in a medium. After a few weeks, the plantlets are regenerated, and after hardening of the plantlets, these are transplanted in the soil under natural environmental conditions. Meristem tip-cultured plants give rise to polyploid plants instead of diploid plants. Moreover, meristem tip culture is very useful for the elimination of viruses from infected plant material. Rapid multiplication of the plants, which are otherwise not easily propagated by vegetative means, is also possible through meristem culture. Plants produced are free from pathogens and can be stored for longer period and in smaller space.

3.6.3 Micrografting

It is difficult to regenerate complete plants from meristem in woody species like most of fruit and forest plants; thus as an alternative micrografting is done to produce virus-free plants. The various steps in micrografting include scion preparation, rootstock preparation, in vitro grafting, and acclimatization/hardening of the plants. The in vitro-raised nucellar seedlings are used as rootstocks. The scion (meristem 0.1–0.4 mm) is obtained from either young growth of field-grown trees, defoliated glasshouse-grown plants, or in vitro proliferated nodal segments obtained from mature trees. The grafting is done with the help of stereomicroscope, under aseptic conditions. Several viruses have been eliminated via micrografting in fruit plants such as Citrus tristeza virus, peach latent mosaic viroid, and pear vein yellow virus.

3.6.4 In Vitro Mycorrhization

Recently, attention has turned to the possible beneficial effects of microorganisms in in vitro plant cultures. For example, the root endophyte *Piriformospora indica* promotes explant hardening (Sahay and Varma 1999); *Pseudomonas* spp. can reduce hyperhydricity (Bela et al. 1998); and *Bacillus pumilus, Alcaligenes faecalis*, and *Pseudomonas* spp. improve shoot multiplication (Monier et al. 1998). Mycorrhization in micropropagation, particularly the use of arbuscular mycorrhizal fungi (AMF), is now gaining momentum due to a demonstrated positive impact on posttransplant performance of in vitro-grown plants (Lovato et al. 1996; Rai 2001). Improved nutrient uptake, water relations, aeration, soil pH balance (Sylvia 1998), and their potential use as bioregulators (Lovato et al. 1996) have recently heightened research interest in AMF, contributing to the development of effective AMF production methods, mycorrhization of in vitro plants, and screening for efficient AMF strains. The potential of different AMFs for application in commercial micropropagation industries can now be tested using an array of tools. CPB has also completed the project on mycorrhiza. The mycorrhiza has been used as biohardening agent. It has been observed that mycorrhiza enhances the survival percentage of in vitro-raised plant. The study has been done in banana, sugarcane bamboo, etc.

3.7 Research Work on Fruit Propagation at the Centre for Plant Biotechnology (CPB), Hisar

CPB, Hisar, is engaged in the in vitro multiplication and production of elite germplasm of various medicinal, horticultural, and other crop plants. CPB has standardized the in vitro multiplication protocol for fruit crops like banana, guava, citrus, and strawberry. The center is also engaged in commercial production of two cultivars of banana, i.e., Robusta and Grand Naine (G-9), and supplying the planting material to the users.

References

Aitken-Christie J, Kozai T, Ann Lila Smith M (1995) Automation and environmental control. In: Plant tissue culture. Kluwer Academic Publishers, Dordrecht

Banerjee N, Vuylsteke D, De Langhe E (1986) Meristem tip culture of *Musa*: histological studies of shoot bud proliferation. In: Withers LA, Anderson PS (eds) Plant tissue culture and its agricultural applications. Butterworths, London, pp 139–148

Bela J, Ueno K, Shetty K (1998) Control of hyperhydricity in anise (*Pimpinella anisum*) tissue culture by *Pseudomonas* spp. J Herbs Spices Med Plants 6:57–67

Bonga JM (1982) Vegetative propagation of Jarrah (*Eucalyptus malginata*) by organ and tissue culture. Aust For Res 12:121–127

Choudhary D, Kajla S, Duhan JS, Poonia AK, Surekha, Kharb P (2013) Comparative study of various growth regulators on *in vitro* multiplication of commercial cultivar of banana cv. Grand naine (G-9). Ann Biol 29(3):288–293

Doreswamy R, Rao NKS, Chacko NK (1983) Tissue culture propagation of banana. Sci Hort 18:247–252

Escalant JV, Teisson C (1994) Amplified somatic embryogenesis from male flowers of triploid banana and plantain cultivars (*Musa spp.*). In Vitro Cell Dev Biol Plant 30:181–186

Hadiuzzaman S, Habiba, Reza S, Saha ML, Khan MR (2000) Development of a sustainable protocol for contamination free culture of table bananas and identification of associated endogenous bacteria. In: Proceedings of the 4th international plant tissue culture conference, Dhaka, Bangladesh, 1st–3rd Nov 2001. Abstracts, p 24

John A, Murray BW (1981) Micropropagation of sitka spruce (*Picea sitchensis*). In: AFOCEL (ed) Colloque international sur la culture *in vitro* des Essences Forestiers. AFOCEL, Fontainebleau, pp 65–70

Keathley DE (1983) Increased growth of hybrid spruce bud cultures on stationary liquid medium. North central tree improvement conference. Proceedings, vol 3, Wooster, Ohio, 17–19 August, pp 150–154

Kim JH, Lee SK, Chun YW (1981) Mass propagation of tree species through *in vitro* culture. Bud culture of *Populus alba x P. glandulosa*. For Res Rep Inst Genet Immok Jukchong Yonku-So Yongu Pogo 17:57–63

Kotecha PM, Kadam SS (1998) Minor vegetables. In: Salunkhe DK, Kadam SS (eds) Handbook of vegetable science and technology. Marcel Dekker Inc, New York, p 683

Krikorian AD, Irizarry H, Mitra SS, Rivera E (1993) Clonal fidelity and variation in plantain (*Musa* spp.) regenerated from vegetative stem and floral axis tips *in vitro*. Ann Bot 71:519–535

Kumar R, Sinha K, Kumar S (2005) Micropropagation of banana cv. Malbhog through meristem tip culture in consort with thermotherapy. Phytomorphology 55:17–22

Lovato PE, Gianinazzi-Pearson V, Trouvelot A, Gianinazzi S (1996) The state of mycorrhizas and micropropagation. Adv Hort Sci 10:46–52

Madhulata P, Anbalagan M, Jayachandran S, Sakthivel N (2004) Influence of liquid pulse treatment with growth regulators on *in vitro* propagation of banana (*Musa spp*. AAA). Plant Cell Tissue Organ Cult 76:189–191

Monier C, Bossis E, Chabanet C, Samson R (1998) Different bacteria can enhance the micropropagation response of *Cotoneaster lacteus* (Rosaceae). J Appl Microbiol 5:1047–1055

Murashige T (1974) Plant propagation through tissue cultures. Annu Rev Plant Physiol 25:135–166

Murashige T, Skoog F (1962) A revised medium for rapid growth and bioassays with tobacco tissue cultures. Physiol Plant 15:473–497

Paek KY, Hahn EJ, Son SH (2001) Application of bioreactors for large scale micropropagation systems of plants. In Vitro Cell Dev Biol Plant 37:149–157

Pandey RM, Doreswamy R, Sahijram L (1993) Tissue culture propagation of banana. Technical bulletin No. 11. Pub. Pandey RM, IIHR, Hesserghatta, Bangalore

Rai MK (2001) Current advances in mycorrhization in micropropagation. In Vitro Cell Dev Biol Plant 37:158–167

Sahay NS, Varma A (1999) *Piriformospora indica*: a new biological hardening tool for micropropagated plants. FEMS Microbiol Lett 181:297–302

Sommer HE, Caldas LS (1981) *In vitro* methods applied to forest trees. In: Thorpe TA (ed) Plant tissue culture: methods and applications in agriculture. Academic, New York, pp 349–358

Sommer HE, Wetzstein HY (1984) Hardwoods. In: Ammirato PV, Evans DA, Sharp WR, Yamada Y (eds) Handbook of plant cell culture, vol 3, Crop species. Macmillan, New York, pp 511–540

Sudhavani AK, Reddy GM (1997) Cost effective propagation of banana. In: Proceedings of the 20th of national symposium of plant tissue culture and molecular biology (PTCMB'97). Department of Genetics, Osmania University, Hyderabad, p 44

Sylvia D (1998) The promise (and obstacles) of AMF inoculation. In: Gianinnazzi S, Schuepp H (eds) *Arbuscular mycorrhizas* in sustainable soil-plant systems. Report of 1997 activities, Cost Action 821, Iceland, p 152

Thorpe TA, Patel KR (1984) Clonal propagation: adventitious buds. In: Vasil EK (ed) Cell culture and somatic cell genetics in plants, vol 1. Academic, New York, pp 49–60

Vuylsteke DR, De Langhe EA (1985) Feasibility of *in vitro* propagation of bananas and plantains. Trop Agric (Trin) 62:323–328

Williams EG, Maheswaran G (1986) Somatic embryogenesis: factors influencing coordinated behaviour of cells as an embryogenic group. Ann Bot 57:443–462

Starch: Its Functional, In Vitro Digestibility, Modification, and Applications

4

Maninder Kaur and Kawaljit Singh Sandhu

Abstract

Starch is a naturally occurring biopolymer widely available in nature. Amylose and amylopectin are two macromolecular components of starch granules. Starch can be characterized by using a variety of techniques including differential scanning calorimeter (DSC), rapid visco analyzer (RVA), rheometer, and X-ray diffraction. Native starches have limitations such as low shear resistance, thermal decomposition, and high tendency of retrogradation which limits their use in industrial food applications. These shortcomings can be easily overcome by starch modifications by a variety of physical, chemical, and enzymatic modifications. In recent years, glycemic index (GI) has become a potentially useful tool in planning diets for patients suffering from diabetes, dyslipidemia, cardiovascular disease, and even certain cancers. On the basis of digestibility, starches can be classified into readily digestible starch (RDS), slowly digestible starch (SDS), and resistant starch (RS). The starches from different botanical vary in their RDS, SDS, and RS contents. SDS and RS contents of starches have a variety of health benefits and these can be increased by different methods. Apart from variety of food applications, starch also has huge usage in non-food area.

4.1 Introduction

Starch is the most abundant reserve carbohydrate in plants, offering a range of desired technological properties and main carbohydrate in human nutrition. Starch is a valuable ingredient to the food industry, being widely used as thickener, gelling agent, bulking agent, and water retention agent (Singh et al. 2003). Identification of native starch

M. Kaur
Department of Food Science and Technology,
Guru Nanak Dev University, Amritsar 143005, India

K.S. Sandhu (✉)
Department of Food Science and Technology,
Chaudhary Devi Lal University, Sirsa, India
e-mail: kawsandhu@rediffmail.com

R.K. Salar et al. (eds.), *Biotechnology: Prospects and Applications*,
DOI 10.1007/978-81-322-1683-4_4, © Springer India 2013

sources is required for desired functionality and unique properties (Duxbury 1989). Amylose and amylopectin are two macromolecular units of starch granules. Normal starch consists of about 75 wt% branched amylopectin and about 25 wt% amylose that is linear or slightly branched. Amylose is composed of glucopyranose units linked through α-D-(1–4) glycosidic linkages, while the amylopectin is a branched polymer with one of the highest molecular weights known among naturally occurring polymers (Karim et al. 2000). Amylose has a degree of polymerization of 100–100,000 DP, whereas amylopectin has an average DP of two million. Starch is semicrystalline in nature with varying levels of crystallinity. The crystallinity is exclusively associated with the amylopectin component, while the amorphous regions mainly represent amylose (Zobel et al. 1988; Zobel 1988). The packing of amylose and amylopectin within the granules has been reported to vary among the starches from different species. X-ray diffraction diffractometry has been used to reveal the presence and characteristics of the crystalline structure of the starch granules (Hoover 2001). The cereal starches exhibit the typical A-type X-ray pattern, whereas the tuber starches show the B-form and legumes, the mixed state pattern "C." DSC has been used to study thermal properties associated with starch gelatinization and also in studying both the loss of crystalline order during gelatinization upon heating on presence of water and the reordering during aging. DSC can detect both first-order and second-order thermal transitions (Russel and Oliver 1989). DSC measures starch transition temperatures and gelatinization enthalpies which may be related to the degree of crystallinity (Krueger et al. 1987). Rheological and thermal techniques have been used to study gelatinization for variety of starch. Starch possesses unique viscosity behavior with change of temperature, shear rate, and concentration (Nurul et al. 1999) and can be measured in terms of curves obtained with RVA and rheometer. Rheometer is used for studying the viscoelastic or rheological properties of starches (Tsai et al. 1997; Lii et al. 1996). Scanning electron microscopy (SEM) has been used to relate granule morphology (Fannon et al. 1992a) and to relate paste structures to paste properties (Fannon et al. 1992b). Native starch is a good texture stabilizer and regulator in food systems (Cousidine 1982) but has limitations such as low shear resistance, thermal resistance, and thermal decomposition. So, they are often tailored by structural modification to overcome shortcomings of native starches and to develop desirable functional properties, such as solubility, texture, adhesion, dispersion, and heat tolerance, in order to be suitable for industrial applications (Rutenberg and Solarek 1984). Chemically starch modification is generally achieved through derivatization such as etherification, esterification, cross-linking, and grafting of starch, through decomposition (acid or enzymatic hydrolysis and oxidation of starch), by physical treatment of starch using heat or moisture, and also by using hydrolyzing enzymes. Food grade starches are chemically, physically, and enzymatically modified mainly to increase paste consistency, smoothness, and clarity and to impart freeze-thaw and cold storage stabilities (Shi and BeMiller 2000; Wu and Seib 1990; Xu and Seib 1997). The nutritional value of starch strongly depends on processing and state of starch. Glycemic index (GI) is most appropriately used to compare foods within a category of foods. Digestibility of human small intestine can be modified from a rapid digestion to indigestibility. Starch products vary in digestibility and the rate and extent of digestibility are reflected in the magnitude and duration of the glycemic response (Englyst and Hudson 1996). Depending on the digestibility timeline or glycemic index, starch is fractioned into three groups, i.e., rapidly digesting starch (RDS), slowly digesting starch (SDS), and resistant starch (RS).

4.2 Physicochemical Properties

The amylose content of the starch granules varies with the botanical source of the starch (Table 4.1) and is affected by the climatic conditions and soil type during growth (Inatsu et al. 1974; Morrison et al. 1984; Juliano et al. 1964). Amylose content of starches from different mango cultivars ranged from 9.1 to 16.3 % (Kaur et al. 2004a). The amylose content of chickpea starches is reported to

Table 4.1 Physicochemical properties of starches from different botanical sources

Starch source	Amylose content (%)	Swelling power (g/g)	Solubility (%)
Corn	16.9–21.3	13.7–20.7	9.7–15
Rice	4.1–16.4	17.2–38.8	–
Chickpea	28.6–34.3	11.4–13.6	13.2–14.9
Mango kernel	9.1–16.3	18–19.7	14.1–14.9

vary from 28.6 to 34.3 % (Singh et al. 2004a). Amylose content of various cornstarches ranged between 16.9 and 21.3 % (Sandhu and Singh 2007). The activity of the enzymes involved in starch biosynthesis may be responsible for the variation in amylose content among the various starches (Krossmann and Lloyd 2000). The amylose contents of different starches have been determined by colorimetric methods without prior defatting and/or by not taking into account the iodine complexing ability of the long external chains of tuber starches (Morrison and Karkalas 1990).

Starch molecules when heated in excess water, the crystalline structure is disrupted, due to which granule swelling and solubility is increased. Swelling power (SP) and solubility can be used to assess the extent of interaction between starch chains, within the amorphous and crystalline domains of the starch granule (Ratnayake et al. 2002). Sandhu and Singh (2007) reported swelling and solubility of different cornstarches in the range from 13.7 to 20.7 g/g and 9.7 to 15 %, respectively. Kaur et al. (2004a) reported swelling power of starches from different mango kernel cultivars in the range from 18.0 to 19.7 g/g, the highest for totapuri and the lowest for langra kernel starches was observed. The differences between swelling powers and solubilities of starches from different sources may also be due to differences in morphological structure of starch granules.

4.3 Thermal Properties

Starch gelatinization is the collapse (disruption) of molecular orders within the starch granule manifested in irreversible changes in properties such as granular swelling, native crystalline melting, loss of birefringence, and starch solubilization (Atwell et al. 1988). Thermal properties typically reported using DSC include gelatinization onset (T_o), peak (T_p), and conclusion (T_c) temperatures, peak height index (PHI), gentinization range (R), and enthalpy (ΔH_g). It measures first-order (melting) and second-order (glass transition) transition temperatures and heat flow changes in polymeric materials and gives information on order–disorder phenomena of starch granules (Biliaderis et al. 1986). It is an important starch functional property that varies with respect to the composition (amylose to amylopectin ratio, phosphorus, lipids, proteins and enzymes, etc.), the molecular structure of amylopectin (unit chain length), extent of branching, molecular weight and granule architecture (crystalline to amorphous ratio), granule morphology, and size distribution of starches (Krueger et al. 1987; Singh et al. 2003). The process causes disruption of the starch granule structure and swelling up to several times their original size. The process occurs in a nonequilibrium state, and thus, reaction kinetics pertaining to gelatinization provides a definite set of process parameters (temperature, time, viscosity, and mechanical strength) for a specific starch (Baik et al. 1997; Spigno and De Faveri 2004). DSC has been of great value in studying both the loss of crystallite order during gelatinization, which occurs when the starch paste materials are heated in the presence of water, and the reordering of such systems during aging. This technique can detect both first-order (melting) and second-order (glass) thermal transitions (Russel and Oliver 1989). Starch transition temperatures and gelatinization enthalpies, measured by DSC, may be related to characteristics of the starch granule, such as degree of crystallinity (Krueger et al. 1987). Retrogradation is a general term for the behavior of recrystallization of gelatinized starches on cooling and storage. Upon storage at a low temperature, gel firmness increases due to starch chain association in the amylopectin region (Karim et al. 2000). Broadening of gelatinization peaks in DSC curves of starches has been attributed to the change in the starch structure (Adebowale et al. 2009). Garcia et al. (1996) studied thermal transition of lintnerized tapioca starch in excess water and found that its gelatinization endotherm appears over a broader temperature range and is higher in onset and peak temperatures when compared with the native starch endotherms.

4.4 Rheological Properties of Starches

In industrial processes, slurry suspensions frequently require rheological characterization for flow equipment design and new formulations. Fully cooked and molecularly dispersed starches are used to provide viscosity and thickening and to impart texture (Raina et al. 2006). Many researchers have used rheological methods to study gelatinization for suspension of a variety of starches (Svegmark and Hermansson 1990; Tsai et al. 1997), as well as to determine viscoelastic properties of starch pastes (Evans and Lips 1992; Reddy et al. 1994). The viscosity parameters during pasting are cooperatively controlled by the properties of the swollen granules and the soluble materials leaching out from the granules (Doublier et al. 1987; Eliasson 1986). Starch exhibits unique viscosity behavior with change of temperature, concentration, and shear rate (Nurul et al. 1999). This can be measured in terms of rheological/pasting curves obtained with rapid visco analyzer (RVA) and rheometer. Information obtained from rheology/pasting curves is vital when considering a starch as a possible component of a food product (Adebowale and Lawal 2003). When starch is heated in water, the viscosity of the slurry increases. The increase in viscosity with temperature may be attributed to the removal of water from the exuded amylose by the granules as they swell (Ghiasi et al. 1982). The swelling behavior of starch is the property of its amylopectin content, and amylose acts as both a diluent and an inhibitor of swelling (Tester and Morrison 1990). Starch exhibits unique viscosity behavior with change of temperature, concentration, and shear rate (Nurul et al. 1999). Heat-induced viscoelasticity may affect the contribution of starch to texture in food systems and is of considerable importance to the acceptability of foods that contain starch (Paraskevopoulou and Kiosseoglou 1997). Temperature, chemical composition, solid matter contents, processing, interaction of food components, and others influence the rheological properties of food products (Hegedusic 1992). The pasting behavior is helpful for understanding the textural change or retrogradation potency of the applied products (Chen et al. 2003). Several workers have characterized the pasting properties of starches from corn (Seetharaman et al. 2001; Ji et al. 2003; Sandhu and Singh 2007) and legume (Huang et al. 2007b) using RVA. The viscosity parameters during pasting are cooperatively controlled by the properties of the swollen granules and the soluble materials leached out from the granules (Eliasson 1986; Doublier et al. 1987). Early in the pasting test, the temperature is below the gel temperature of the starch, and the viscosity is low. When the temperature rises above the gelatinization temperature, the starch granules begin to swell, and viscosity increases on shearing when these swollen granules have to squeeze past each other. The temperature at the onset of this rise in viscosity is known as the pasting temperature. Pasting properties of starches from different botanical source are shown in Table 4.2 and Fig. 4.1. Granule swelling is accompanied by leaching of granular constituents, predominantly amylose, into the external matrix resulting in a dispersion of swollen granules in a continuous matrix (Biliaderis 1992; Noel et al. 1993). When a sufficient number of granules become swollen, a rapid increase in viscosity occurs, known as peak viscosity. Peak viscosity occurs at the equilibrium point between swelling and polymer leaching. As the mixture is subsequently cooled, viscosity increases to a final viscosity. Miles et al. (1985) reported that increase in final viscosity might be due to the aggregation of the amylose molecules. The low setback value of a starch indicates its lower tendency to retrograde.

Starch pastes exhibit a viscoelastic character depending upon the type of starch, concentration, and temperature. Hence, it is an interesting system for rheological studies. The application of starches in foods also depends to a large extent on their rheological characteristics

Table 4.2 Pasting properties of starches from different botanical sources

Starch source	Pasting temperature (°C)[a]	Peak viscosity (cP)[b]	Breakdown viscosity (cP)[c]	Final viscosity (cP)[d]	Setback (cP)[e]
Corn	73	1,529	210	1,698	379
Rice	69.1	1,543	30	2,085	572
Mung bean	74.8	1,857	111	3,697	1,951
Mango kernel	83.6	2,091	444	2,385	738

[a] The temperature at the onset of rise in viscosity
[b] The equilibrium point between swelling and polymer leaching
[c] Peak viscosity–holding strength
[d] Increase in final viscosity upon cooling
[e] Setback

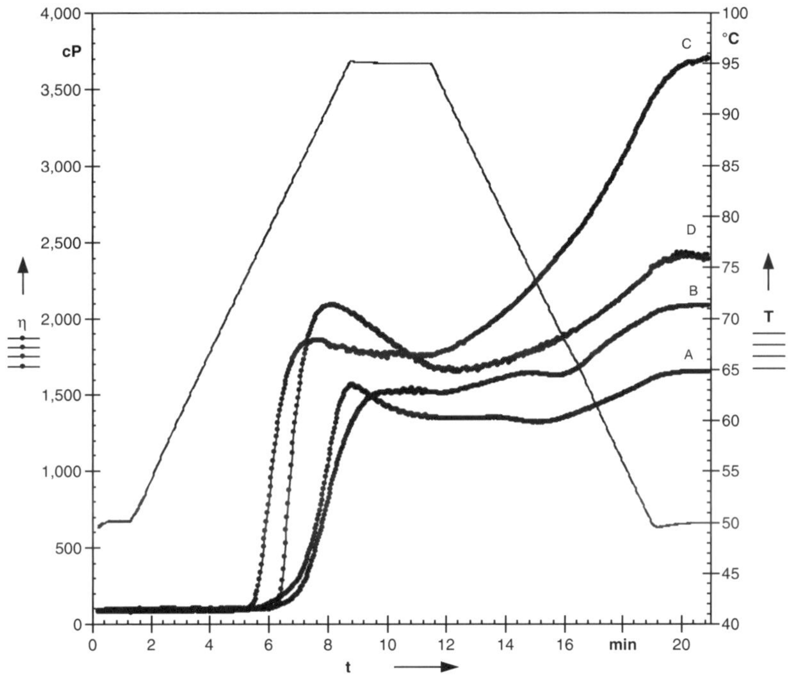

Fig. 4.1 Pasting properties of starches: *A* – Corn, *B* – Rice, *C* – Mung bean, and *D* – Mango kernel

(Moorthy 2002). The rheological properties of starch are very sensitive to several factors, including the type of starch, amylose/amylopectin ratio, temperature, starch concentration, pH, and the presence and concentration of other components, such as low-molecular-weight solutes (salts, acids and sugars) and macromolecules (Choi and Yoo 2008). The studies on the rheological properties of starch pastes proved that they are shear-thinning liquids, with a tendency to yield stress. The presence of yield stress at certain concentration of starch allows to classify them as viscoelastic dispersions (Doublier et al. 1987; Thebaudin et al. 1998). The yield stress exists because the hydrogen bonds in the helix structure form a stable configuration and resistant to flow (Harrison et al. 1999). During the gelatinization process, some changes can take place which can lead to the discrepancy in the viscosity values (Moorthy et al. 2008). The initial increase in G′ (storage modulus) may be attributed to swelling of the granules that filled

Table 4.3 Rheological properties of corn, rice, mung bean, and mango kernel starches

Starch source	*T* (°C) G′	Peak G′	Peak G″	Breakdown in G′	Peak tan δ
Corn	70	1,401	140	551	0.09
Rice	65	1,163	105	559	0.09
Mung bean	70	1,077	190	709	0.17
Mango kernel	77.6	684	69	86	0.10

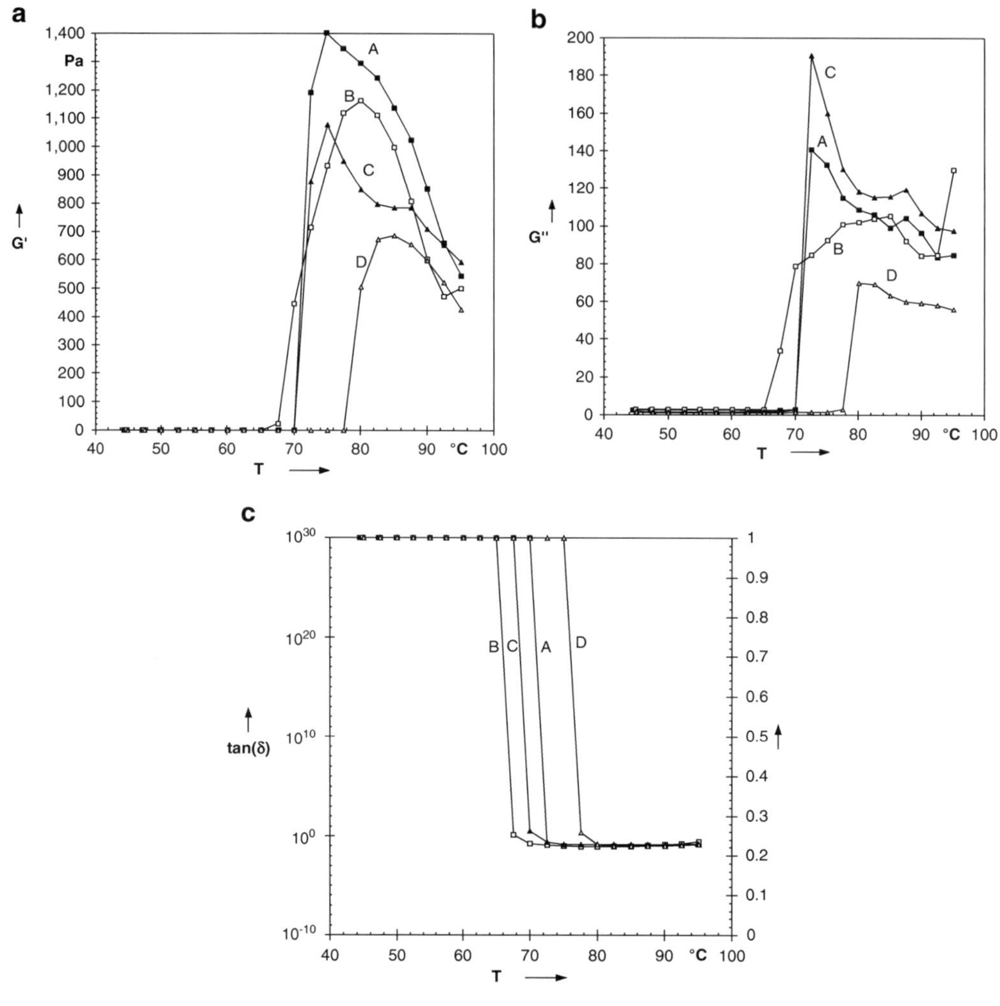

Fig. 4.2 (**a**) G′ of corn (*A*), rice (*B*), mung bean (*C*), and mango kernel (*D*) starches. (**b**) G″ of corn (*A*), rice (*B*), mung bean (*C*), and mango kernel (*D*) starches. (**c**) tan δ of corn (*A*), rice (*B*), mung bean (*C*), and mango kernel (*D*) starches

the entire available volume of the system and increased the peak G′ (Eliasson 1986). With further increases in temperature, G′ decreased, indicating that the gel structure is destroyed during heating (Tsai et al. 1997). This destruction is due to the melting of crystalline regions remaining in the swollen granules, allowing the granules to deform (Eliasson 1986). Several workers have characterized the rheological properties of starches from rice (Chun and Yoo 2007) and mango kernels (Kaur et al. 2004a) using rheometer. The rheological properties of starches from different botanical sources are shown in Table 4.3 and Fig. 4.2a–c. The dynamic rheological properties, such as storage modulus (G′), loss modulus (G″), and loss factor (tan δ), were determined using temperature sweep test.

4.5 Morphological Properties

Starch occurs naturally as discrete particles, called granules. Granule size and shape of starch are reported to be primarily affected by the germplasm from which the starch is isolated. Granule size can be determined by various techniques like microscopy (light microscopy, scanning electron microscopy [SEM]), sieving, electrical resistance, laser light scattering, and field flow fractionation (Lindeboom et al. 2004). Starch granules are packed in the form of semicrystalline areas and amorphous regions in an alternating fashion. SEM is often used to study starch morphology, both in native and modified starches (Fitt and Snyder 1984). Microscopy (light and SEM) has played an important role in increasing understanding of granular structure of modified starches. It has been used to detect structural changes caused by chemical modifications and the most substituted regions in starch granules (Kaur et al. 2004b; Kim et al. 1992). Morphological characteristics of starches from different plant sources vary with the genotype and cultural practices. The variation in the size and shape of starch granules is attributed to the biological origin (Svegmark and Hermansson 1993). The morphology of starch granules depends on the biochemistry of the chloroplast or amyloplast, as well as physiology of the plant (Badenhuizen 1969). Increasing knowledge of the granular structure and functional properties of starches has enabled chemists to process them in novel way to modify starches to meet special demands of the food and other industries (Mishra and Rai 2006).

4.6 In Vitro Starch Digestibility

The nutritional value of starch strongly depends on processing and state of starch. Glycemic index (GI) is most appropriately used to compare foods within a category of foods. Digestibility of starch in human small intestine can be modified from a rapid digestion to indigestibility. Starch products vary in digestibility and the rate and extent of digestibility are reflected in the magnitude and duration of the glycemic response (Englyst and Hudson 1996). Raw starch digestibility is greatly influenced by plant type and depends on physicochemical characteristics of the starch and plant microstructure and composition and is influenced by processing and storage condition (Ring et al. 1988). In human, starch and its derivatives are digested in several stages. In the mouth, in contact with saliva α-amylase, starch is cleaved into shorter oligosaccharides. Partially digested material is further hydrolyzed by human pancreatic α-amylase. The rate of reaction of hydrolysis decreases with increase in the degree of polysaccharide branching (Park and Rollings 1994).

Products of this digestion, maltose and branched dextrins, are converted into glucose by the brush border enzymes maltose-glucoamylase and isomaltase. On the basis of digestibility, timeline or glycemic index starch is fractioned into three groups (Table 4.4). Rapidly digesting starch (RDS) is defined as portion of starch that digests rapidly and amount of glucose released after 20 min. Slowly digesting starch (SDS) is a portion that digests slowly and releases glucose between 20 and 120 min of consumption, whereas resistant starch (RS) is defined as an amount of starch that remains un-hydrolyzed after 120 min. There are three distinct types of RS: physically inaccessible starch (RS_1), native starch granules of B-type (RS_2), and retrograded starch (RS_3) (Englyst et al. 1992). More recently, a fourth type (RS_4) has been identified which is associated to certain fraction of chemically modified starches (Tovar et al. 1999; Laurentin et al. 2003). Since RS is not digested and absorbed in upper gastrointestinal tract, it will reach the colon. The RDS, SDS, and RS contents of various starches vary depending upon their sources. The values for RDS determined in lentil were 5.2 %, 4.2 % in pigeon pea, and 9.7 % in mung bean starches (Sandhu and Lim 2008a) but were substantially lower for rice (32.4 %), wheat (40.1 %) (Zhang et al. 2006), and corn (27.4 %) starches (Sandhu and Lim 2008b). SDS values

Table 4.4 Starch classification

Fraction of starch	Digestion time and in vitro digestion place	Example	Main physiological properties	Structure
Rapidly digestible starch (RDS)	Within 20 min. mouth and small intestine	Freshly cooked food	Rapid source of energy	Mainly amorphous
Rapidly digestible starch (RDS)	20–120 min; small intestine	Native waxy maize starch, millet, legumes	Slow sand sustained source of energy and sustained blood glucose	Amorphous/crystalline
Resistant starch (RS)	>120 min; not in small intestine, main action in colon	Raw potato starch, staled bread	Effect on gut health (e.g., prebiotic, fermentation to butyrate with hypothesized anticarcinogenic effects)	Depending on the type, mainly crystalline

Adapted from Englyst et al. (1992)

have been reported for wheat (50.0 %), rice (43.8 %) (Zhang et al. 2006), corn (45.3 %) (Sandhu and Lim 2008b), bean (63.1–65.8 %) (Chung et al. 2008), pigeon pea (16.9 %), and mung bean (40.0 %) starches (Sandhu and Lim 2008a). RS contents in the range of 50.3–78.9 % for different legume starches have been observed (Sandhu and Lim 2008a).

4.7 Factor Affecting Starch Digestibility

One of the main factors affecting starch digestibility and its physiological response was attributed to its amylose/amylopectin ratio (Behall et al. 1988). Differences in the in vitro digestibility of native starches among and within species have been attributed to the interplay of many factors, such as starch source (Ring et al. 1988), granule size (Snow and O'Dea 1981), extent of molecular association between starch components (Dreher et al. 1984), degree of crystallinity (Hoover and Sosulski 1985), type of crystalline polymorphic (A, B, or C) form (Jane et al. 1997), and distribution of B-type crystalline in granules (Gerard et al. 2001). The digestion of starch granules is a complex process, which includes different phases: the diffusion of the enzyme towards the substrate with the impact porosity of the substrate, the adsorption of the enzyme to the starchy material, and the hydrolytic event. Interaction of starch with fiber, protein, and other food components can prevent effective diffusion and adsorption of enzyme (Colonna et al. 1992).

Gallant et al. (1992) proposed a favored digestion of amorphous region and stated that double helices cannot be digested unless they are untwisted. In millet, sorghum, and legume starches, one aspect of its slower digestibility can be attributed to its interaction with protein, which forms a protective network around the granule. In pulses, various factors which affect the rate and overall digestibility are cell-wall structure (Hoover and Zhou 2003); antinutrients such as amylase inhibitors, phytates, and polyphenolics (Bravo et al. 1998; Siddhuraju and Becker 2005; Yadav and Khetarpaul 1994); and high content of viscous soluble dietary fiber components (Tharanathan and Mahadevamma 2003).

4.8 Starch Modification

Modification of starch is generally carried out to overcome shortcomings of native starches and increase the usefulness of starch for industrial applications. Native starches when cooked can easily retrograde and there is a gelling tendency of pastes besides easily undergoing syneresis. Therefore, starch modification not only decreases retrogradation, gelling tendencies of pastes, and gel syneresis but also improves paste clarity, paste and gel texture, film formation, and adhesion (BeMiller 1997). Food grade starches are chemically modified mainly to increase paste consistency, smoothness, and clarity and to impart freeze-thaw and cold storage stabilities (Perera et al. 1997; Shi and BeMiller 2000). Starch modification, which involves the alteration of the physical and chemical characteristics of the native starch to improve its functional characteristics, can be used to tailor starch to specific food applications (Hermansson and Svegmark 1996). Starch modification is generally achieved through derivatization such as etherification, esterification, cross-linking, and grafting of starch; decomposition (acid or enzymatic hydrolysis and oxidization of starch); or physical treatment of starch using heat, moisture, etc. Acetylated starch with a low DS is commonly obtained by the esterification of native starch with acetic anhydride in the presence of an alkaline catalyst, whereas the food grade hydroxypropylated starches are generally prepared by etherification of native starch with propylene oxide in the presence of an alkaline catalyst. The hydroxypropyl groups introduced into the starch chains are capable of disrupting the inter- and intramolecular hydrogen bonds, thereby weakening the granular structure of starch, leading to an increase in motional freedom of starch chains in amorphous regions (Seow and Thevamalar 1993; Wootton and Manatsathit 1983). The characteristics of acetylated starch of diverse botanical sources have been reported (Elomaa et al. 2004; Xu and Hanna 2005; Chen et al. 2007; Huang et al. 2007a). The starch can also be modified by acid–alcohol treatment in which the starch granules are acid-treated in alcohols. The glycosidic linkages of starch, especially those in amorphous regions, are acid-hydrolyzed during treatment (Lin et al. 2003, 2005). The molecular weight and chain length of starch molecules decreased after treated (Lin and Chang 2006; Lin et al. 2005), but the recovery of starch granules after acid-treated in alcohols can be higher than 90 % (Chang et al. 2006; Lin et al. 2003, 2005). Starch oxidation has been practiced since the early 1800s, and various oxidizing agents have been introduced, for instance, hypochlorite, hydrogen peroxide, periodate, permanganate, dichromate, persulfate, and chlorite (Rutenberg and Solarek 1984). The main uses for oxidized starch are in paper and textile industries. However, the application of oxidized starches in the food industry is increasing because of their low viscosity, high stability, clarity, and binding properties. Oxidized starch has been widely used in many industries, particularly the paper, textile, laundry finishing, and building material industries to provide surface sizing and coating properties (Scallet and Sowell 1967). Oxidized starch also becomes increasingly important in the food industry for its unique functional properties such as low viscosity, high stability, clarity, film forming, and binding. The extent of oxidation can be affected by many factors, including pH, temperature, hypochlorite concentration, starch molecular structure, and starch source (Wurzburg 1986). Acid-modified, thin boiling starch is normally prepared by treatment of starch with hydrochloric acid and is used extensively in food, textile, and paper industries (Radley 1976; Wurzburg 1986). This modification allows the starch to be used at higher solid concentration for quick gelling and provides gum or jelly with shorter texture and flexible properties (Zallie 1988). Acid-thinned starches can be prepared by chemical, physical, and enzymatic methods (Agboola et al. 1991). Starch contains two types of hydroxyls, primary (6-OH) and secondary (2-OH and 3-OH). These hydroxyls are able to react with multifunctional reagents resulting in cross-linked starches. Cross-linking is done to restrict swelling of the starch granule under cooking conditions or to prevent gelatinization of starch.

The reagents permitted by FDA for making cross-linking food grade starch are phosphoryl chloride, sodium trimetaphosphate, adipic–acetic mixed anhydride, and mixtures of sodium trimetaphosphate and tripolyphosphates (Seib 1996). Cross-linked starches are used in salad dressings to provide thickening with stable viscosity at low pH and high shear during the homogenization process. Cross-linked starches with a slow gelatinization rate are used in canned foods where retort sterilization is applied; such starches provide low initial viscosity, high heat transfer, and rapid temperature increase, which are particularly suitable for quick sterilization (Rutenberg and Solarek 1984). Unlike ester linkage such as starch acetate, which tends to deacetylate under alkaline condition, ether linkages are more stable even at high pH. Etherification gives starch excellent viscosity stability. Hydroxyalkyl starches, including hydroxyethyl and hydroxypropyl, are mainly produced for industrial applications. Hydroxypropyl starches are being widely used in food products where they provide viscosity stability and freeze-thaw stability. These starches are normally combined with cross-linking to provide the desired viscosity, texture, and stability for processing and storage. Hydroxypropyl starches are used as thickeners in fruit pie fillings, puddings, gravies, sauces, and salad dressings.

Physical modification can be safely used as a modification process in food products as it does not involve the use of chemical. Heat-moisture treatment (HMT) is a physical modification technique applied to starches. It is considered to be natural and safe as compared to chemical modification. Heat-moisture treatment (HMT) commonly occurs at a low moisture content (<35 % w/w) and at elevated temperatures above the glass transition temperature (T_g) but below gelatinization temperature for a certain period of time (Gunaratne and Hoover 2002). Annealing is a hydrothermal treatment that can remove or minimize these defects from the starch granules (Qi et al. 2004). This treatment promotes a structural reorganization of starch granules in water at a temperature above the glass transition and below the gelatinization temperatures (Jacobs et al. 1998). Annealing is a physical treatment on starch granules in the presence of heat, temperature between the glass transition temperature and the gelatinization onset temperature (T_o), and excess water, generally higher than 60 % w/w (Tester et al. 2000). During annealing, the increase in molecular mobility of starch within the amorphous regions leads to a molecular reorganization (Atichokudomchai et al. 2002; Tester et al. 2000).

4.9 Applications

Starch is used in a wide range of foods for a variety of purposes including thickening, gelling, adding stability, and replacing or extending more costly ingredients. Starch products are used as crystal and texture controller. Starch can be hydrolyzed to make a variety of sweetness; high-maltose and high-conversion syrups control softness and freezing characteristics. High-fructose starch-based syrups (HFSS) are used for soft drinks as a sugar replacement with similar sweetness. High-maltose syrups find use as wort syrup in beer production. It is an excellent fermentation substrate and fermentation can be controlled by the sugar spectrum of the syrup. High-amylose cornstarch is also used in extruded and fried snack products to obtain crisp, evenly browned product and hampers penetration of cooking oils. Starch derivatives are used to encapsulate flavors, beverage clouds, creamers, and vitamins (Kenyon 1995). Starch-based fat replacers are commonly used in sauces, salad dressings, and dairy products to enhance mouthfeel and to give products the flow characters of the full fat product (Yackel and Cox 1992). Starch is also used in retorted baby foods to thicken and provide soft texture and in meats, including those reduced in fat, to increase moisture retention for succulence and purge control to reduce shrinkage and for firmness of bite (Demos et al. 1994). Modified food starches are used in a wide variety of dairy products to provide a variety of effects, including enhanced viscosity, cutability, mouthfeel, and stability. Very high dextrose equivalent (DE) glucose syrups are used as a fermentation booster in alcohol fermentation. Hydrophobic starches are used along with modified protein to sterilize dressing emulsions and canned fish products (Brueckner et al. 1987). Pregelatinized starch is used in cake mixes to soften the cake crumb and retain moisture in the baked product. They are also used to provide convenience (Takashima 2005). Starch also has a number of nonfood applications (Maurer and Kearney 1998). Worldwide, the majority of wheat starch is sold in unmodified form to manufacturers of paper and textile, where it has an influence on the production costs and quality of fine graphic and packaging paper, cardboard, corrugated board, acoustic fiber board, and single- or multiple-walled paper bags. It is also used in production of molds for the metal industry, as a binding agent in charcoal briquettes, and as an additive in various glues. In laundry sizing and cotton finishing, starch is considered to produce a superior stiff finish that is attributed to the bimodal granule size production.

4.10 Conclusions

Starch, the most abundant carbohydrate of nature, has a wide variety of food and non-food applications. It can be characterized by using physicochemical, thermal, morphological, and rheological methods. To improve the utilization of starch, it can be modified by a wide variety of physical, chemical, and enzymatic treatments. The digestibility of the starches can be tailored to increase or decrease their RDS, SDS, and RS contents. The tailored starch can be potentially beneficial for planning diets for patients suffering from diabetes, cardiovascular disease, and even certain cancers.

References

Adebowale KO, Lawal OS (2003) Functional properties and retrogradation behavior of native and chemically modified starch of mucuna bean (*Mucuna pruriens*). J Sci Food Agric 83:1541–1546

Adebowale KO, Henle T, Schwarzenbolz U, Doert T (2009) Modification and properties of African yam bean (Sphenostylis stenocarpa Hochst. Ex A. Rich.) Harms starch I: heat moisture treatments and annealing. Food Hydrocoll 23:1947–1957

Agboola SO, Akimbala JO, Oguntimein GB (1991) Physicochemical and functional properties of low DS cassava starch acetates and citrates. Starch-Starke 43:62–66

Atichokudomchai N, Varavinit S, Chinachoti P (2002) A study of annealing and freeze thaw stability of acid-modified tapioca starch by differential scanning calorimetry. Starch 54(8):343–349

Atwell WA, Hood LF, Lineback DR, Varriano-marston E, Zobel HF (1988) The terminology and methodology associated with basic starch phenomena. Cereal Foods World 33:306–311

Badenhuizen NP (1969) The biogenesis of starch granules in higher plants. Appleton Crofts, New York

Baik MY, Kim KJ, Cheon KC, Ha YC, Kim WS (1997) Recrystallization kinetics and glass transition of rice starch gel system. J Agric Food Chem 45:4242–4248

Behall KM, Scholfield DJ, Canary J (1988) Effect of starch structure on glucose and insulin responses in adults. Am J Clin Nutr 47:428–432

BeMiller JN (1997) Starch modification: challenges and prospects. Starch 49:127–131

Biliaderis CG (1992) Characterisation of starch networks by small strain dynamic rheometry. In: Alexander RJ, Zobel HF (eds) Developments in carbohydrate chemistry. AAOCC, St Paul, pp 87–135

Biliaderis CG, Page CM, Maurice TJ, Juliano BO (1986) Thermal characterization of rice starches: a polymeric approach to phase transitions of granular starch. J Agric Food Chem 34:6–14

Bravo L, Englyst HN, Hudson HJ (1998) Nutritional evaluation of carbohydrates in the Spanish diet: non-starch polysaccharides and in vitro starch digestibility of breads and breakfast products. Food Res Int 31:129–135

Brueckner J, Muschiolik G, Mieth G, Ackermann K (1987) DDR patent 250 048

Chang YH, Lin JH, Chang SY (2006) Physicochemical properties of waxy and normal corn starches treated in different anhydrous alcohols with hydrochloric acid. Food Hydrocoll 20:332–339

Chen JJ, Lai VMF, Lii C (2003) Effect of compositional and granular properties on the pasting viscosity of rice starch blends. Starch-Starke 55:203–212

Chen L, Li X, Li L, Guo S (2007) Acetylated starch-based biodegradable materials with potential biomedical applications as drug delivery systems. Curr Appl Phys 7S1:e90–e93

Choi HM, Yoo B (2008) Rheology of mixed system of sweet potato starch and galactomannans. Starch-Starke 60:263–269

Chun SY, Yoo B (2007) Effect of molar substitution on rheological properties of hydroxypropylated rice starch pastes. Starch-Starke 59:334–341

Chung HJ, Liu Q, Pauls KP, Fan MZ, Yada R (2008) In vitro starch digestibility, expected glycemic index and some physicochemical properties of starch and flour from common bean (Phaseolus vulgaris L.) varieties grown in Canada. Food Res Int 41:869–875

Colonna P, Leloup V, Buleon A (1992) Limiting factors of starch hydrolysis. Eur J Clin Nutr 46:S17–S32

Cousidine DM (1982) Foods and food production encyclopedia. Wiley, New York, p 142

Demos BP, Forrest JC, Grant AL, Judge MD, Chen LF (1994) Low-fat, no added salt in restructured beef steaks with various binders. J Muscles Foods 5:407–418

Doublier JL, Llamas G, Meur ML (1987) A rheological investigation of the cereal starch pastes and gels: effect of pasting procedures. Carbohydr Polym 7:251–275

Dreher ML, Berry JW, Dreher CJ (1984) Starch digestibility of foods: a nutritional perspective. Crit Rev Food Sci Nutr 20:47–71

Duxbury DD (1989) Modified starch functionalities-no chemicals or enzymes. Food Process 50:35–37

Eliasson AC (1986) Viscoelastic behavior during the gelatinization of starch I. Comparison of wheat, maize, potato and waxy-barley starches. J Texture Stud 17:253–265

Elomaa M, Asplund T, Soininen P, Laatikainen R, Peltonen S, Hyvarinen S (2004) Determination of the degree of substitution of acetylated starch by hydrolysis, 1H NMR and TGA/IR. Carbohydr Polym 57:261–267

Englyst HN, Hudson GJ (1996) The classification and measurement of dietary carbohydrates. Food Chem 57:15–21

Englyst HN, Kingman SM, Cummings JH (1992) Classification and measurement of nutritionally important starch fraction. Eur J Clin Nutr 46:S33–S50

Evans ID, Lips A (1992) Viscoelasticity of gelatinized starch dispersions. J Texture Stud 23:69–86

Fannon JE, Hauber RJ, BeMiller JN (1992a) Surface pores of starch granules. Cereal Chem 69:284–288

Fannon JE, Hauber RJ, BeMiller JN (1992b) In: Chanderasekaran R (ed) Use of low temperature scanning electron microscopy to examine starch granule structure and behaviour, vol 2, Frontiers in carbohydrate research. Elsevier Science, London, pp 1–23

Fitt LE, Snyder EM (1984) Photomicrographs of Starches. In: Whistler RL (ed) Starch chemistry and technology. Academic, New York, pp 675–689

Gallant DJ, Bouchet B, Buleon A, Perez S (1992) Physical characteristics of starch granules and susceptibility to enzymatic degradation. Eur J Clin Nutr 46:S3–S16

Garcia V, Colonna P, Lourdin D, Buleon A, Bizot H, Ollivon M (1996) Thermal transitions of cassava starch at intermediate water contents. J Therm Anal 47:1213–1228

Gerard C, Colonna P, Buleon A, Planchot V (2001) Amylolysis of maize mutant starches. J Sci Food Agric 81:1281–1287

Ghiasi K, Varriano-Marston K, Hoseney RC (1982) Gelatinization of wheat starch. II. Starch-surfactant interaction. Cereal Chem 59:86–88

Gunaratne A, Hoover R (2002) Effect of heat-moisture treatment on the structure and physicochemical properties of tuber and root starches. Carbohydr Polym 49:425–437

Harrison G, Franks GV, Tirtaatmadja V, Boger DV (1999) Suspension and polymers – common links in rheology. Korea-Aust Rheol J 3(11):197–218

Hegedusic V (1992) Progress in food rheology. In: Konja G, Lovric T, Strucelj D, Ttipalo B (eds) Advances in food process engineering. Faculty of Food Technology and Biotechnology, Zagreb, pp 13–29

Hermansson AM, Svegmark K (1996) Developments in the understanding of starch functionality. Trends Food Sci Technol 7:345–353

Hoover R (2001) Composition, molecular structure, and physicochemical properties of tuber and root starches: a review. Carbohydr Polym 45:253–267

Hoover R, Sosulski FW (1985) Studies on the functional characteristics and digestibility of starches from *Phaseolus vulgaris* biotypes. Starch 37:181–191

Hoover R, Zhou Y (2003) In vitro and in vivo hydrolysis of legumes starches by α-amylase and resistant starch formation in legumes- a review. Carbohydr Polym 54:401–417

Huang J, Schols HA, Klaver R, Jin Z, Voragen AGJ (2007a) Acetyl substitution patterns of amylose and amylopectin populations in cowpea starch modified with acetic anhydride and vinyl acetate. Carbohydr Polym 67:542–550

Huang J, Schols HA, Soest JJGV, Jin Z, Sulmann E, Voragen AGJ (2007b) Physicochemical properties and amylopectin chain profiles of cowpea, chickpea and yellow pea starches. Food Chem 101:1338–1345

Inatsu O, Watanabe K, Maida I, Ito K, Osani SJ (1974) Studies to improve the quality of rice grown in Hokkaido. I. Amylose contents of different rice starches. J Jpn Soc Starch Sci 21:115–117

Jacobs H, Eerlingen RC, Rouseu N, Colonna P, Delcour JA (1998) Acid hydrolysis of native and annealed wheat, potato and pea starches. DSC melting features and chain length distribution of lintnerized starches. Carbohydr Res 308(3/4):359–371

Jane JL, Wong KS, McPherson AE (1997) Branch structure differences in starches of A and B types X-ray patterns revealed by their Naegli dextrins. Carbohydr Res 300:219–227

Ji Y, Seetharaman K, Wong K, Pollak LM, Duvick S, Jane J, White PJ (2003) Thermal and structural properties of unusual starches from developmental corn lines. Carbohydr Polym 51:439–450

Juliano BO, Bautista GM, Lugay JC, Reyes ACJ (1964) Studies on the physico-chemical properties of rice. J Agric Food Chem 12:131–134

Karim AA, Norziah MH, Seow CC (2000) Methods for the study of starch retrogradation. Food Chem 71:9–36

Kaur M, Singh N, Sandhu KS, Guraya HS (2004a) Physicochemical, morphological, thermal and rheological properties of starches separated from kernels of some Indian mango cultivars (Mangifera indica L.). Food Chem 85:131–140

Kaur L, Singh N, Singh J (2004b) Factors influencing the properties of hydroxypropylated potato starches. Carbohydr Polym 55:211–223

Kenyon MM (1995) Encapsulation and controlled release of food ingredients. In: Risck and Reineccius (ed) ACS symposium series 42: 590

Kim HR, Hermansson AM, Eriksson CE (1992) Structural characteristics of hydroxypropyl potato starch granules depending on their molar substitution. Starch 44:111–116

Krossmann J, Lloyd J (2000) Understanding and influencing starch biochemistry. Crit Rev Biochem Mol Biol 35:141–196

Krueger BR, Knutson CA, Inglett GE, Walker CE (1987) A differential scanning calorimetry study on the effect of annealing on gelatinization behaviour of corn starch. J Food Sci 52:715–718

Laurentin A, Cardenas M, Ruales J, Perez E, Tovar J (2003) Preparation of indigestible pyrodextrins from different starch sources. J Agric Food Chem 51:5510–5515

Lii CY, Tsai ML, Tseng KH (1996) Effect of amylose content on the rheological property of rice starch. Cereal Chem 73:415–420

Lin JH, Chang YH (2006) Molecular degradation rate of rice and corn starches during acid–methanol treatment and its relation to the molecular structure of starch. J Agric Food Chem 54:5880–5886

Lin JH, Lee SY, Chang YH (2003) Effect of acid-alcohol treatment on the molecular structure and physicochemical properties of maize and potato starches. Carbohydr Polym 53:475–482

Lin JH, Lii CY, Chang YH (2005) Change of granular and molecular structures of waxy maize and potato starches after treated in alcohols with or without hydrochloric acid. Carbohydr Polym 59:507–515

Lindeboom N, Chang PR, Tyler RT (2004) Analytical biochemical and physicochemical aspects of starch granule size, with emphasis on small granule starches: a review. Starch-Starke 56:89–99

Maurer HW, Kearney RL (1998) Opportunities and challenges for starch in the paper industry. Starch-Starke 50:396–402

Miles MJ, Morris VJ, Orford PD, Ring SG (1985) The roles of amylose and amylopectin in the retrogradation of starch. Carbohydr Res 135:271–281

Mishra S, Rai T (2006) Morphology and functional properties of corn, potato and tapioca starches. Food Hydrocoll 20:557–566

Moorthy SN (2002) Physicochemical and functional properties of tropical tuber starches: a review. Starch-Starke 54:559–592

Moorthy SN, Larsson H, Eliasson AC (2008) Rheological characteristics of different tropical root starches. Starch-Starke 60:233–247

Morrison WR, Karkalas J (1990) Starch. In: Day PM, Harborne JB (eds) Methods in plant biochemistry, vol 2. Academic Press, London

Morrison WR, Milligan TP, Azudin MN (1984) A relationship between the amylose and lipids contents of starches from diploid cereals. J Cereal Sci 2:257–260

Noel TR, Ring SG, Whittam MA (1993) Physical properties of starch products: structure and function. In: Dickinson E, Walstra P (eds) Food colloids and polymers: stability and mechanical properties. Royal Society of Chemistry, Cambridge, pp 126–137

Nurul IM, Azemi BMNM, Manan DMA (1999) Rheological behaviour of sago (Metroxylon sagu) starch paste. Food Chem 64:501–505

Paraskevopoulou A, Kiosseoglou V (1997) Texture profile analysis of heat-formed gels and cakes prepared with low cholesterol egg yolk concentrates. J Food Sci 62:208–211

Park JT, Rollings JE (1994) Effects of substrate branching characteristics on kinetics of enzymatic depolymerization of mixed linear and branch polysaccharides: I. amylose/amylopectin a–amylolysis. Biotechnol Bioeng 44:792–800

Perera C, Hoover R, Martin AM (1997) The effect of hydroxypropylation on the structure and physicochemical properties of native, defatted and heat moisture treated potato starches. Food Res Int 30:235–247

Qi X, Tester RF, Snape CE, Yuryev VP, Wasserman LA, Ansell R (2004) Molecular basis of the gelatinization and swelling characteristics of waxy barley starches grown in the same location during the same season. Part II. Crystallinity and gelatinization characteristics. J Cereal Sci 39:57–66

Radley JA (1976) Industrial uses of starch and its derivatives. Applied Science Publishers, London

Raina CS, Singh S, Bawa AS, Saxena DC (2006) Rheological properties of chemically modified rice starch model solutions. J Food Process Eng 29:134–148

Ratnayake WS, Hoover R, Warkentin T (2002) Pea starch: composition, structure and properties – a review. Starch-Starke 54:217–234

Reddy KR, Subramanian R, Ali SZ, Bhattacharya KR (1994) Viscoelastic properties of rice-flour pastes and their relationship to amylose content and rice quality. Cereal Chem 71:548–552

Ring SG, Gee JM, Whittam M, Orford P, Johnson IT (1988) Resistant starch: its chemical form in foodstuffs and effect on digestibility in vitro. Food Chem 28:97–109

Russel PL, Oliver G (1989) The effect of pH and NaCl content on starch gel ageing. A study by differential scanning calorimetry and rheology. J Cereal Sci 10:123–138

Rutenberg MW, Solarek D (1984) Starch derivatives: production and uses. In: Whistler R, BeMiller JN, Paschall EF (eds) Starch: chemistry and technology. Academic Press, New York, pp 312–388

Sandhu KS, Lim ST (2008a) Digestibility of legume starches as influenced by its physical and structural properties. Carbohydr Polym 71:245–252

Sandhu KS, Lim ST (2008b) Structural characteristics and in vitro digestibility of mango kernel starches (*Mangifera indica L.*). Food Chem 107:92–97

Sandhu KS, Singh N (2007) Some properties of corn starches II: physicochemical, gelatinization, retrogradation, pasting and gel textural properties. Food Chem 101:1499–1507

Scallet BL, Sowell EA (1967) Production and use of hypochlorite-oxidized starches. In: Whistler RL, Paschall EF (eds) Starch chemistry and technology, vol 2. Academic, New York, pp 237–251

Seetharaman K, Tziotis A, Borras F, White PJ, Ferrer M, Robutti J (2001) Thermal and functional characterization of starch from Argentinean corn. Cereal Chem 78:379–386

Seib PA (1996) Starch chemistry and technology, Syllabus. Kansas State University, Manhattan

Seow CC, Thevamalar K (1993) Internal plasticization of granular rice starch by hydroxypropylation: effects on phase transitions associated with gelatinization. Starch 45:85–88

Shi X, BeMiller JN (2000) Effect of sulfate and citrate salts on derivatization of amylose and amylopectin during hydroxypropylation of corn starch. Carbohydr Polym 43:333–336

Siddhuraju P, Becker K (2005) Nutritional and antinutritional composition, in vitro amino acid availability, starch digestibility and predicated gylcemic index of differentially processed mucuna bean (*Mucuna pruriens var. utilis*): an under-utilised legumes. Food Chem 91:275–286

Singh N, Singh J, Kaur L, Sodhi NS, Gill BS (2003) Morphological, thermal and rheological properties of starches from different botanical sources: a review. Food Chem 81:219–231

Singh N, Sandhu KS, Kaur N (2004a) Characterization of starches separated from Indian chickpea (Cicer arietinum L.) cultivars. J Food Eng 63:441–449

Singh J, Kaur L, Singh N (2004b) Effect of acetylation on some properties of corn and potato starches. Starch 56:586–601

Snow P, O'Dea K (1981) Factors affecting the rate of hydrolysis of starch in foods. Am J Clin Nutr 34:2721–2727

Spigno G, De Faveri DM (2004) Gelatinization kinetics of rice starch studied by non-isothermal calorimetric technique: influence of extraction method, water concentration and heating rate. J Food Eng 62:337–344

Svegmark K, Hermansson AM (1990) Shear induced change in the viscoelastic behavior of heat-treated potato starch dispersion. Carbohydr Polym 13:29–45

Svegmark K, Hermansson AM (1993) Microstructure and rheological properties of composites of potato starch granules and amylose: a comparison of observed and predicted structure. Food Struct 12:181–193

Takashima H (2005) US patent 6:884 448

Tester RF, Morrison WR (1990) Swelling and gelatinization of cereal starches. Cereal Chem 67:558–563

Tester RF, Debon SJJ, Sommerville MD (2000) Annealing of maize starch. Carbohydr Polym 42(3):287–299

Tharanathan RN, Mahadevamma S (2003) Grain legumes – a boon to human nutrition. Trends Food Sci Technol 14:507–518

Thebaudin JY, Lefebvre AC, Doublier JT (1998) Rheology of starch pastes from starches of different origins: applications to starch-based sauces. Lebensm Wiss Technol 31:354–360

Tovar J, Herrera E, Laurentin A, Melito C, Perez E (1999) In vitro digestibility of modified starches. In: Pandalai SG (ed) Recent research developments in agricultural and food chemistry, vol 3. Research Signpost Co, Trivandrum, pp 1–10

Tsai ML, Li CF, Lii CY (1997) Effects of granular structure on the pasting behavior of starches. Cereal Chem 74:750–757

Wootton M, Manatsathit A (1983) The influence of molar substitution on the water binding capacity of hydroxylpropyl maize starches. Starch-Starke 35:92–94

Wu Y, Seib PA (1990) Acetylated and hydroxypropylated distarch phosphates from waxy barley: paste properties and freeze–thaw stability. Cereal Chem 67:202–208

Wurzburg OB (1986) Converted starches. In: Wurzburg OB (ed) Modified starches: properties and uses. CRC Press, Boca Raton, pp 17–41

Xu Y, Hanna MA (2005) Preparation and properties of biodegradable foams from starch acetate and poly (tetramethylene adipate-co-terephthalate). Carbohydr Polym 59:521–529

Xu A, Seib PA (1997) Determination of the level and position of substitution in hydroxypropylated starch by high resolution ^{1}H-NMR spectroscopy of alpha-limit dextrins. J Cereal Sci 25:17–26

Yackel WC, Cox C (1992) Applications of starch-based fat replacers. Food Technol 46:146–148

Yadav S, Khetarpaul N (1994) Indigenous legume fermentation: effect on some antinutrients and in vitro digestibility of starch and protein. Food Chem 50:403–406

Zallie J (1988) Benefits of quick setting starches. Manuf Confect 66:41–43

Zhang G, Ao Z, Hamaker BR (2006) Slow digestion property of native cereal starches. Biomacromolecules 7:3252–3258

Zobel HF (1988) Molecules to granules – a comprehensive starch review. Starch 40:44–50

Zobel HF, Young SN, Rocca LA (1988) Starch gelatinization. An X-ray diffraction study. Cereal Chem 65:443–446

Molecular Strategies for Improving Mineral Density and Bioavailability in Rice

5

Rajinder K. Jain, Jitender Kumar, Sunita Jain, and Vijay K. Chowdhury

Abstract

Globally, micronutrient malnutrition has become a major health problem affecting over three billion people. Of the various micronutrients, problems (anemia, mental retardation, stunted growth, decreased immune function, and increased mortality) resulting from iron and zinc deficiencies are most prevalent and devastating in the developing countries. Rice serves as a staple food for more than half of the world population, but it has insufficient levels of the key micronutrients (Fe and Zn) to meet daily dietary requirements. Biofortification, which refers to the breeding of plants/crops with high bioavailable micronutrient content using conventional breeding, genetic engineering, and molecular and genomic approaches, has the potential to provide coverage for remote rural populations, where supplementation and fortification programs may not reach, and it inherently targets the poor who consume high levels of staple food and little else. Biofortified rice can be an effective solution to combat micronutrient malnutrition in developing countries with limited resources. The facts that substantial genetic variation for Fe and Zn contents exists in rice and that traits for high nutrient content can be combined with superior agronomic characteristics and high yield have allowed many scientists to use conventional breeding approaches to develop micronutrient-rich rice genotypes. Alternatively, genomic, transformation, and molecular tools have been used to improve our understanding of factors regulating micronutrient contents/bioavailability and rapid discovery of genes involved in iron uptake and storage in

R.K. Jain (✉) • J. Kumar • V.K. Chowdhury
Department of Molecular Biology and Biotechnology, CCS Haryana Agricultural University, Hisar 125004, Haryana, India
e-mail: rajinder.jain@gmail.com

S. Jain
Bioinformatics Section, CCS Haryana Agricultural University, Hisar 125004, Haryana, India

R.K. Salar et al. (eds.), *Biotechnology: Prospects and Applications*, DOI 10.1007/978-81-322-1683-4_5, © Springer India 2013

target tissues and consequently to develop novel high-Fe and/or high-Zn transgenic plants in rice. At CCS Haryana Agricultural University, Hisar, we have assessed variability for iron and zinc in a collection of 220 rice genotypes and identified several iron- and zinc-rich genotypes which have been used subsequently to raise mapping population and used for identification of QTLs for minerals in brown rice. Material is being used to select mineral-rich high-yielding rice genotypes.

5.1 Introduction

Rice feeds more than half of the world population and is the number one staple food in Asia, where it provides 40–70 % of the total food calories consumed. Though late, both experts and authorities now agree that the micronutrient malnutrition is widespread, especially the poor people in the developing world thriving mainly on the staple food crops such as rice. Micronutrient malnutrition, particularly Fe, Zn, and vitamin A deficiency, affects over three billion people worldwide, mostly in developing countries (Welch and Graham 2004; Sperotto et al. 2010). Iron deficiency results in anemia, which is a worldwide public health problem, with global prevalence estimated at 24.8 % (Shaw and Friedman 2011). According to the WHO (2011) estimate, about two billion people in both developed and developing countries are affected by iron deficiency. Main consequences of iron deficiency include mental retardation, decreased immune function, and increased mortality of mother and child at birth (Puig et al. 2007). The severity of the problem for pregnant women is illustrated by the fact that around 200,000 deaths associated with childbirth each year can be attributed to this deficiency (Seymour 1996). Zinc is a component of more than 300 enzymes that are needed to repair wounds and maintain fertility. Symptoms of zinc deficiency are subtle and often include stunted growth, eczema, hair loss, delayed sexual maturation, and impaired mental development (Mares-Perlman et al. 1995). Vitamin A plays a vital role in visual system functioning and is important for normal growth and repair of different body tissues. More than 250 million people worldwide are deficient in vitamin A, putting them at risk of contracting various serious ailments including night blindness, xerophthalmia, and total blindness if it remains untreated. Vitamin A-deficient children are at a 23 % higher risk of death as a result of measles, diarrhea, or malaria (UNICEF 2009). Population affected by these nutrient deficiencies exhibit low productivity, decreased cognition, more morbidity, and higher rate of mortality. The costs of these deficiencies, in terms of diminished quality of life and lives lost, are staggering (www.harvestplus.org). Keeping in view staggering levels of micronutrient malnutrition, a HarvestPlus program has been initiated, which is coordinated by the International Center for Tropical Agriculture (CIAT) and the International Food Policy Research Institute (IFPRI). It focuses on three critical micronutrients in seven food crops including bean, cassava, maize, pearl millet, rice, sweet potato, and wheat. HarvestPlus envisions that in 50 years, millions of people suffering from micronutrient malnutrition will be eating new biofortified crop varieties.

During the "green revolution era," the main emphasis has been to improve the crop productivity with little concern on nutritional value or health-promoting qualities of food. Many crops such as spinach and leguminous plants are known to be iron rich; however, this useful trait usually exists simultaneously with oxalic acid and/or phytate-like substances that decrease the bioavailability of iron. Several kinds of agronomic approaches through fertilizer application in soil or by foliar spray have been attempted to improve the micronutrient profile of the crop produced but with little success. These are costly and may not target the iron accumulation in desired plant tissue/grain and could also be toxic to the plants reducing its growth, productivity, and commercial value. A number of other strategies, including supplementation with vitamin A capsules and zinc pills and the industrial fortification of flours and oils with iron and vitamin A or β-carotene, are currently employed in a number of countries to combat micronutrient deficiency-induced health problems. But these strategies are costly and have a marginal affect especially while targeting the poor masses in developing countries. Alternatively, "biofortification" of plants/crops with high Fe and high Zn contents using conventional breeding and molecular, genomic, and genetic engineering approaches has gained momentum in order to enhance the mineral nutritional qualities of crops at source (Bouis et al. 2003; Welch and Graham 2004; White and Broadley 2005; Pfeiffer and McClafferty 2007). Biofortification is likely to be more accessible than conventional interventions in the long term because it removes hurdles such as the reliance on infrastructure and compliance. Moreover, plants assimilate minerals into organic forms that are naturally bioavailable and contribute to the natural taste and texture of the food (Galera et al. 2010).

Several groups have examined and found that substantial useful genetic variation for micronutrient exists in key staple crops including rice (Graham et al. 1999; Gregorio et al. 1999; Garcia-Oliveira et al. 2009; Brar et al. 2011; Anuradha et al. 2012). Researchers at IRRI developed an improved line (IR68144-3B-2-2-3) with a high concentration of grain iron, ~21 ppm in brown rice from IR72/Zawa Bonday (traditional aromatic, tall plant stature) cross-combination (Gregorio et al. 2000). It need to be kept in mind that the acceptance of such biofortified crops by the farmers and consumers hinges on developing attractive trait packages without compromising agronomic (yield) and end-use (e.g., grain quality) characteristics. Alternatively, molecular tools have been employed to understand factors regulating iron and zinc contents/bioavailability and rapid discovery of genes involved in their uptake and storage in target tissues (Frossard et al. 2000). Genetic engineering has been carried out to tailor transgenic rice plants with high iron content and bioavailability by transferring "ferritin," "phytase," "ferric-chelate reductase," and "nicotianamine aminotransferase (*NAAT*)" genes driven by seed or endosperm-specific promoters (Lucca et al. 2001; Takahashi 2003; Guerinot 2007; Ishimaru et al. 2007). Overexpression of several genes such as *AtNAS1*, *Pvferritin*, *Afphytase*, *OsIRT1*, *OsNAS1*, *OsNAS2*, and *OsNAS3* significantly increased the Fe and Zn concentrations in unpolished rice grain (Wirth et al. 2009; Lee and An 2009; Johnson et al. 2011). Another landmark achievement relates to the engineering of β-carotene biosynthesis pathway in rice (popularly known as Golden rice and Golden rice 2) leading to β-carotene (provitamin A) formation in the endosperm of the rice which is retained even after the milling process (Ye et al. 2000; Paine et al. 2005). Research efforts are now being made by IRRI and various national institutions to transfer Golden rice 2 *psy/crtI* transgene combination into local rice cultivars via molecular breeding.

Several recent developments in genomics and molecular biology and access to rice genome sequence database in public domain shall have impeccable effect to advance our understanding of the molecular mechanisms but will also pave the way for efficient and targeted improvement in mineral content and bioavailability. In this chapter, we will discuss various molecular strategies mainly to improve iron and zinc in rice. Biofortification of vitamin A in rice has been discussed in detail in many recent reviews (Tang et al. 2012; Bhullar and Gruissem 2013).

5.2 Mineral Uptake and Transport in Rice

Grasses including rice primarily use a chelation-based strategy II for iron uptake (Palmer and Guerinot 2009). This strategy employs the release of chelators into the rhizosphere, known as phytosiderophores (PS), to bind Fe^{3+} for transport into the plant. Phytosiderophores are synthesized from methionine and are usually referred to collectively as belonging to the mugineic acid family (the MAs). Phytosiderophores form strong hexadentate chelates with Fe^{3+} to solubilize and transport it into the plant (Palmer and Guerinot 2009). The transporters belonging to the yellow stripe-like (YSL) family of proteins (Curie et al. 2009) play an important role by facilitating the transport of resulting Fe^{3+}–PS complexes into root cells. Members of the YSL family are also involved in metal distribution within the plant. Fe and Zn

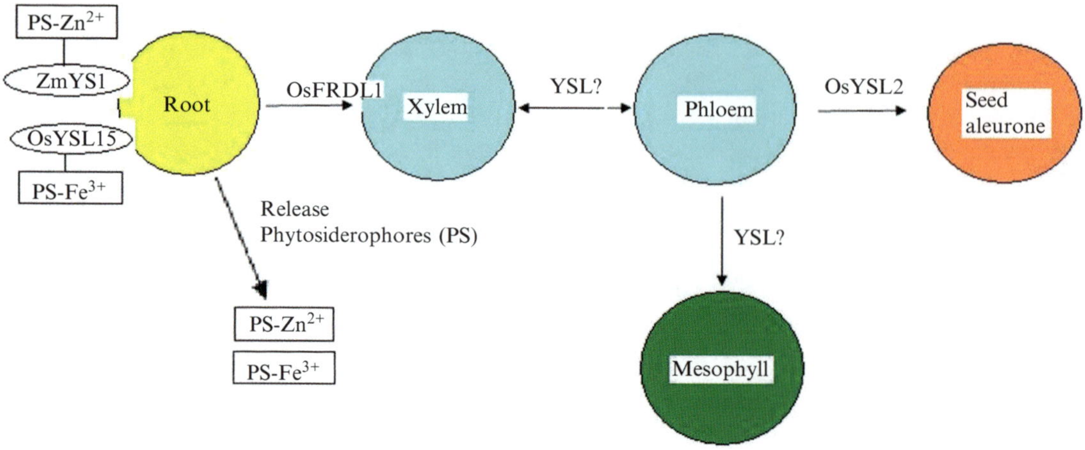

Fig. 5.1 Intercellular metal uptake and transport in grasses

are taken up as phytosiderophore chelates by YSL transporters. Metals move through the symplastic space to the vasculature, bypassing the waxy Casparian strip on the endodermis. The citrate effluxer FRDL1 is important for loading of citrate into the xylem and subsequent Fe transport to the shoot through the transpiration stream. YSL transporters also may play a role in unloading the xylem into the shoot and the phloem. Fe is unloaded from the phloem by OsYSL2 and OsIRT1 into shoot and seed tissue (Fig. 5.1). Rice secretes PS in relatively smaller amounts in response to iron deficiency and is, thus, susceptible to low iron availability. However, unlike other grasses, rice also has an efficient Fe^{2+} uptake mechanism and encodes homologs of IRT1 (OsIRT1 and OsIRT2), which are localized in the plasma membrane of root epidermal cells and specifically upregulated in iron-deficient plants (Cheng et al. 2007).

Grotz and Guerinot (2006) reported that genetic variation for mineral content in rice grains could be due to controlled homoeostatic mechanisms that regulate metal absorption from soil and then translocation and redistribution from roots to other parts of the plant. Several factors including grain characteristics such as the number of aleurone layers, size of the embryo and caryopsis, available levels of the micronutrient at the root–soil interface, genetic systems for mineral uptake by roots, and phloem sap loading and unloading rates within the reproductive organs determine the quantity of micronutrients in the grains (Welch and Graham 2004; Sellappan et al. 2009). More advance research is in progress to improve our understanding of metal uptake and homeostasis in plants, which can greatly help to design new strategies for development of mineral-dense crops.

5.3 Micronutrient (Fe and Zn) Levels and Bioavailability in Rice

Most of the commercially cultivated *indica* and *japonica* rice cultivars are deficient in iron and zinc compared to the other staple food crops such as wheat and maize (Table 5.1). In fact, most of

Table 5.1 Iron and zinc content in some major crops

Crop	Iron content (mg/kg)	Zinc content (mg/kg)
Rice	6–24	14–58
Wheat	25–73	25–92
Maize	10–63	13–58
Beans	34–92	21–59
Cassava	4–49	4–18
Pearl millet	30.1–75.7	24.5–64.8

Data taken from Gregorio et al. (2000), Monasterio and Graham (2000), Banziger and Long (2000), Islam et al. (2002), Maziya-Dixon et al. (2000), and Velu et al. (2007)

our present-day cultivated rice varieties with an exception to a few Basmati, local varieties, or landraces possess very low levels of iron and zinc contents in grains. This is not unexpected, given that in the past micronutrient density has never been a breeding objective in rice and latent variation may have been lost. Researchers at IRRI (Graham et al. 1999; Gregorio et al. 1999) reported wide range of Fe (6.3–24.4 μg/g) and Zn (13.5–58.4 μg/g) concentrations in brown rice within the eight sets of genotypes ($n=1{,}138$). Gowda et al. (2012) observed a large genetic variation for iron (100–170 μg/g) and zinc (>100 μg/g) concentration in 50 rice accessions. Anuradha et al. (2012) analyzed Fe and Zn concentration in brown rice samples of 126 accessions including cultivated *indica* and *japonica* rice cultivars, germplasm accessions, and wild rice genotypes; Fe concentration ranged from 6.2 to 71.6 ppm and Zn concentration ranged from 26.2 to 67.3 ppm. Both Fe and Zn were high in wild rice genotypes and least in *japonica*. Brar et al. (2011) evaluated iron and zinc contents in dehusked grains of 220 rice genotypes and reported large variation in grain Fe (5.1–441.5 μg/g) and Zn (2.12–39.4 μg/g) contents. Two of these genotypes, Taraori Basmati and Palman 579, had high iron (>180 μg/g) and zinc (>25 μg/g) content. Garcia-Oliveria et al. (2009) observed a positive correlation between grain iron and zinc content in 85 introgression lines (ILs) derived from a cross between an elite *indica* cultivar Teqing and the wild rice (*Oryza rufipogon*). Monasterio and Graham (2000) reported large variation for both Fe (28.8–56.5 μg/g) and Zn (25.2–53.3 μg/g) in grains of a wide range of wheat germplasm and revealed a high correlation between grain Fe and grain Zn concentrations in wheat. Thus, it should be plausible to improve Fe and Zn levels simultaneously in rice grain through plant breeding.

Milling and polishing are important operations during the production of white rice. The degree of milling and polishing has a significant effect on the nutritional aspects of white rice, especially on minerals (Chen and Siebenmorgen 1997; Chen et al. 1998). The removal of the outer layers of the rice seed by commercial milling dramatically reduces the level of Fe in the grains because most of the Fe is accumulated in the aleurone layer. The process of milling and polishing that converts brown rice into white rice destroys 67 % of the vitamin B3, 80 % of the vitamin B1, 90 % of the vitamin B6, half of the manganese, half of the phosphorus, 60 % of the Fe, and all of the dietary fiber, protein, and essential fatty acids (Rabbani and Ali 2009).

Bioavailability is that proportion of a mineral or trace element in a food, meal, or diet that is absorbed and utilized by the body for normal functions. It is not possible to measure the utilization of minerals and trace elements directly, except for iron which is incorporated in hemoglobin. Iron bioavailability in rice is closely related to Fe forms in the grains and the enhancers and inhibitors contained in diet (Brinch-Pedersen et al. 2007). Fe and Zn bioavailability in rice grain is greatly inhibited by phytate (Bohn et al. 2008). Phytate occurs widely in plant tissues but is concentrated in seeds or grains. Phytate, the naturally occurring form of phytic acid in seeds, is a salt of myo-inositol hexaphosphoric acid associated with a wide range of cations including K, Mg, Zn, and Fe and forms insoluble complexes. The absorption rate of Fe in plant source food is lower than 10 %, for example, 1 % in rice, 3 % in corn and black bean, 4 % in lettuce, and 5 % in wheat (Liu and Yu 1997).

Cereals store carbohydrates in the form of fructans (Brinch-Pedersen et al. 2007). Nondigestible oligosaccharides such as inulin and oligofructose are fermented by the local microflora and stimulate the growth of bifidobacteria and lactobacilli, which are considered to have health-promoting effects in the gastrointestinal tract (Gibson et al. 1995). Several studies in humans and animals have shown that inulin and oligofructose can increase the intestinal absorption of minerals (Scholz-Ahrens and Schrezenmeir 2002). A diverse range of other food components can counteract the adverse effects of inhibitors on mineral absorption, although in many cases their precise mechanism is not clear. They include vitamins (vitamin C, β-carotene, vitamin B2, and vitamin D), cysteine in proteins, a variety of amino and organic acids, and chemically similar ions competing for the same transport carriers

(Lopez et al. 2002; White and Broadley 2005). For example, ascorbic acid reduces conversion of ferric iron to ferrous iron which is then bound in an absorbable chelate that does not react with the inhibitors (Ballot et al. 1987; Gillooly et al. 1983). Similarly, β-carotene probably improves the bioavailability of iron by increasing the solubility (Garcia-Casal et al. 1998).

Several different strategies are being employed to increase the bioavailability of minerals in food crops, which will be discussed later in this chapter.

5.4 Biofortification

Biofortification focuses on enhancing the mineral nutritional qualities of crops at source, which encompasses processes that increase both mineral levels and their bioavailability in the edible parts of staple crops (Bouis et al. 2003; Welch and Graham 2004; Pfeiffer and McClafferty 2007; White and Broadley 2005). Biofortification is likely to be more accessible than conventional interventions in the long term because it removes hurdles such as the reliance on infrastructure and compliance. Moreover, plants assimilate minerals into organic forms that are naturally bioavailable and contribute to the natural taste and texture of the food (Galera et al. 2010). Economic studies have shown many potential health benefits of biofortification (Bouis 2002; Stein et al. 2008). To realize the potential benefits and applications of biofortification, a specific targeted "HarvestPlus program" was initiated by the Consultative Group on International Agricultural Research (CGIAR), which is now a global alliance of institutions such as the International Rice Research Institute (IRRI), the Centro Internacional de Mejoramiento de Maíz y Trigo (CIMMYT), the Centro Internacional de Agricultura Tropical (CIAT), and the International Institute of Tropical Agriculture (IITA). The major objective of this program is to improve human nutrition by breeding new varieties of staple food crops consumed by the poor. In the last two decades, a lot of research has been directed towards the discovery of genetic variation affecting heritable mineral traits and the feasibility of breeding for increasing mineral content and their bioavailability in edible tissues without affecting yields or other quality traits. Several strategies such as genetic transformation, molecular breeding, and genomics are being employed to increase the mineral content and bioavailability in cereal grains. A brief account of these strategies used to develop mineral-rich crops is described below.

5.5 Genetic Transformation

Mineral content in plants can be enhanced by using approaches that frequently focus on increasing mineral uptake, transport, and/or accumulation in edible portions of the plant. A number of genes related to the iron acquisition in plants such as *AHA1* (H^+-ATPase; Harper et al. 1989), *IRT1* (iron transporter; Eide et al. 1996; Bughio et al. 2002), and *FRO2* (ferric-chelate reductase; Robinson et al. 1999; Vasconcelos et al. 2006) have been isolated that may be useful in modulating Fe content via strategy 1 (reduction). The role of NRAMP (natural resistance-associated macrophage protein) family of cation transporters in iron homeostasis has been well established (Thomine et al. 2000). *Arabidopsis* NRAMP1 may have a role in iron subcellular transport and targeting to the storage compartments such as vacuoles or plastids (Curie et al. 2000). Genes/cDNAs related to biosynthesis of MAs and for a number of iron-holding molecules (e.g., ferritin, hemosiderin/phytosiderin, phytohemagglutinin) have been isolated (Goto and Yoshihara 2001; Guerinot 2007; Zhu et al. 2007). Of these, ferritin has received much attention for genetic transformation, because it can store up to 4,500 Fe atoms and easily release iron on demand with no other metabolic role than Fe storage and the prevention of active oxygen generation (antioxidant) (Bhullar and Gruissem 2013).

Attempts have been made to improve the Fe content and rice productivity by transferring the ferritin genes that greatly improves Fe^{3+} reduction process (strategy 1) (Guerinot 2007). Goto et al. (1999) reported the improvement in the iron content of rice by transferring the entire coding sequence of the soybean ferritin gene into the rice. The introduced ferritin gene was expressed under the control of a rice seed storage protein glutelin promoter, to mediate the accumulation of Fe specifically in the grain. The transgenic seeds stored up to three times more Fe than normal seeds. Iron levels in the whole (unmilled) seeds of the transformants varied from 13 to 38 mg/kg; pooled mean values were 23 mg/kg for transformants and 11 mg/kg for non-transformants. The average iron content in the endosperm was 3.4 mg/kg in the transformant and 1.6 mg/kg in the non-transformant. It was concluded that transgenic "ferritin rice" would be sufficient to provide 30–50 % of the daily adult requirement. Vasconcelos et al. (2003) reported enhanced iron and zinc accumulation not only in brown grains but also in polished grains of transgenic rice expressing the soybean ferritin gene driven by the endosperm-specific glutelin promoter. It must be noted that expression of Fe storage protein, ferritin, driven by seed-specific promoters in rice has yielded only modest increases (two- to three-fold) in seed.

Nicotianamine (NA) is a chelator of transition metals that plays important roles in long- and short-distance transport of metal cations, including Fe^{2+} and Fe^{3+}, in higher plants (Takahashi et al. 2003). NA is biosynthesized by trimerization of S-adenosylmethionine, a reaction catalyzed by the NA synthase (NAS) enzymes. Genes encoding NAS are known to be differentially regulated by iron status in a variety of plant species including maize, Arabidopsis, barley, and rice (Inoue et al. 2003) and show strong induction by Fe deficiency. Johnson et al. (2011) generated three populations of rice to constitutively overexpress rice NAS genes *OsNAS1*, *OsNAS2*, or *OsNAS3*, respectively. Nicotianamine, Fe, and Zn concentrations were significantly increased in unpolished grain of all three of the overexpression populations, relative to controls, with the highest concentrations in the *OsNAS2* and *OsNAS3* overexpression populations. Selected lines from each population had at least 10 μg/g Fe in polished grain and two *OsNAS2* overexpression lines had 14 and 19 μg/g Fe in polished grain, representing up to fourfold increases in Fe concentration. Twofold increases of Zn concentration were also observed in the *OsNAS2* population. Synchrotron X-ray fluorescence spectroscopy demonstrated that *OsNAS2* overexpression leads to significant enrichment of Fe and Zn in phosphorus-free regions of rice endosperm.

There are a few reports on engineering Fe-efficient plant genotypes by adding Fe^{3+} chelate-reductase gene to strategy I plants such as rice (Ishimaru et al. 2007). Ishimaru et al. (2007) reported enhanced tolerance to iron deficiency in calcareous soil in transgenic rice plants containing a yeast Fe^{3+} chelate-reductase gene (*refre1/372*) fused with the promoter of the Fe-regulated transporter, *OsIRT1*. Transgenic rice plants expressing the *refre1/372* gene showed higher Fe^{3+} chelate-reductase activity and a higher Fe uptake rate than vector controls under Fe-deficient conditions. Consequently, transgenic rice plants exhibited an enhanced tolerance to low-Fe availability and a 7.9-fold increase in the grain yield compared to the non-transformed plants in calcareous soils. However, there was no difference in the concentrations of Fe, Zn, Mn, or Cu in the seeds of transformed and vector control plants. To drive more Fe into the seed, it will probably be necessary to create an enhanced "sink" for Fe in the seed.

A recently reported strategy combining the genes facilitating Fe uptake and storage in rice showed more promising results. The rice plants constitutively expressing *A. thaliana* nicotianamine synthase gene (*AtNAS*) together with the endosperm-specific expression of *Phaseolus vulgaris* Ferritin and *Aspergillus fumigatus* Phytase showed a more than six-fold Fe increase in the polished rice grains (Wirth et al. 2009). The expression of *AtNAS1* increased the nicotianamine (NA) content in these plants, which ultimately facilitated Fe uptake and transport as Fe^{3+}-phytosiderophore chelates. Thus, the synergistic effect of NAS and ferritin was necessary to increase iron content in the rice endosperm.

Iron and zinc bioavailability can be greatly enhanced by reducing inhibitors. Phytic acid is a major inhibitor of Fe and Zn bioavailability (Bohn et al. 2008). In most plant seeds, phosphorus is stored primarily as *myo*-inositol 1,2,3,4,5,6-haxa*kis*phosphate (InsP6; phytic acid). Reducing the phytic acid content of seeds is a major target to increase bioavailability of mineral nutrients. The first step in phytic acid biosynthesis and inositol metabolism is catalyzed by 1D-*myo*-inositol 3-phosphate (*Ins(3)P₁*) synthase. Kuwano et al. (2006) aimed to reduce phytic acid levels in rice seeds by manipulating the expression of the rice $Ins(3)P_1$ synthase gene, *RINO1*, using transgenic methods. *RINO1* cDNA in the antisense orientation under the control of rice major storage protein glutelin *Glu B-1* promoter was used for rice transformation. Most of the transgenic T_5 rice seeds contained higher amounts of inorganic phosphates (Pi), without a reduction in total phosphorus levels, compared to non-transgenic seeds. Ion chromatography analysis suggested that the increase in available Pi is accompanied by a molar-equivalent decrease in phytic acid P.

Since most cereals are processed for human consumption by boiling or baking, the thermotolerance of the enzyme "phytase" is of great importance in reference to mineral bioavailability. Most phytases are not thermotolerant and start to lose activity around 60 °C (Ullah and Mullaney 1996). The *PhyA* phytase from *Aspergillus fumigatus* is known to possess a strong capacity for refolding into active confirmation after thermal denaturation. Lucca et al. (2001) reported the introduction of this *PhyA* phytase gene into rice and the transgenic showed the increased levels of phytase activity up to 130-fold. Unfortunately, after expressing the gene in grain, the phytase was no longer stable to heat and lost its activity on boiling.

Another approach is to improve iron absorption in the human intestine, which can be done by expression of peptides which form tridentate chelates with iron, a product well absorbed in the intestine (Gura 1999). Metallothionein was overexpressed in the rice endosperm, increasing the cysteine content sevenfold. Cysteine-containing peptides are responsible for the absorption of iron chelates in man.

Substantial efforts have been made to identify mutants with impaired phytate biosynthesis. Low-phytate mutants have been reported in several plant species, including wheat, barley, rice, and maize (Guttieri et al. 2003; Raboy 2001, 2002). For example, in barley, the low-phytate mutations lpa1-1, lpa2-1, lpa3-1, and M955 result in 50, 40, 65, and 95 % reductions of the phytate levels, respectively (Brinch-Pedersen et al. 2007). These reductions are accompanied by equivalent molar increases in inorganic phosphate or lower isomers of inositol phosphate, leaving the total P content almost unaffected. The reduction of phytate biosynthesis in rice has also been addressed by transformation with the Ins(3)P1 synthase (MIPS) gene (*RINO1*) in the antisense orientation (Kuwano et al. 2006). The MIPS-catalyzed conversion of glucose 6-phosphate to myo-inositol-3-phosphate is regarded as the first step in the biosynthesis of phytate, and there appears to be a close relationship between the biosynthesis of phytate and the formation of myo-inositol-3-phosphate by MIPS (Brinch-Pedersen et al. 2002). In the rice study, the phytate-P content was reduced considerably and this was associated with a similar increase in the level of Pi (up to 48 %). In comparison, the Pi fraction in lpa rice represented 32 % of the total seed phosphate (Larson et al. 2000).

5.6 Linkage Mapping and Marker-Assisted Selection

Molecular markers are of great value in applying genetic technologies to crop improvement such as determining genetic diversity, marker-assisted selection, gene pyramiding, QTL mapping, map-based cloning of important genes, monitoring introgression from exotic and wild species germplasm, and DNA fingerprinting of crop germplasm and pathogen populations (Khush and Brar 2002). One of the most important applications of DNA markers and molecular linkage maps is to dissect the genetic variation of quantitative traits into individual Mendelian factors through QTL mapping analyses (Li 2001). Molecular marker technologies can significantly increase the pace and prospects of success for breeding to improve the mineral content of staple food crops such as rice. Studies have shown the potential to exploit the genetic variation in seed concentration of iron and other minerals without the general negative effect on yield of adding new traits. The relationship between yield and mineral concentration may be positive, particularly in mineral-deficient soil (Gregorio 2002).

Gregorio et al. (2000) studied the genetics of Fe content using four traditional high-iron rice varieties, three advanced lines, and three released varieties (IR36, IR64, IR72) and reported highly significant differences between the crosses and parents. The genetic analysis of variance revealed the presence of additive gene action in addition to a significant nonadditive genetic variance. Environmental effects were also present, but their magnitude was smaller than the genetic effects. A high-Fe trait can be combined with high-yielding traits. Breeders at IRRI, Philippines, developed an improved line (IR68146-3B-2-2-3) with a high content of grain iron (about 21 ppm in brown rice) from a cross between a high-yielding variety (IR72) with a tall, traditional variety (Zawa Bonday) from India. The yield of the improved line was about 10 % below than IR72, but in compensation, maturity was earlier. This variety has good tolerance to soil deficient in minerals such as phosphorous, zinc, and iron. IR68146-3B-2-2-3 has no seed dormancy and has an excellent seedling vigor, suggesting that it would be a good direct-seeded rice (Gregorio et al. 2000).

Accumulation of minerals (iron and zinc) in seeds continues to be a complex phenomenon and controlled by many genes. Although, substantial genetic diversity for mineral content has been observed in rice, but linkage mapping of genes/QTLs for iron and zinc contents has gained momentum recently. Gregorio et al. (2000) reported three loci explaining a 19–30 % variation for iron content on chromosome 7, 8, and 9 in rice. Stangoulis et al. (2007) reported two major QTLs explaining 16.5 and 13.8 % phenotypic variation for iron content in rice grain on chromosome 2 and 12, respectively, using IR64/Azucena-derived double-haploid population. Lu et al. (2008) identified QTLs for the Fe, Ca, Zn, Mn, and Cu content of rice grain by using a set of 241 recombinant inbred lines (RILs) derived from a cross between Zhenshan 97 and Minghui 63. They used 221 marker loci (RFLP and SSR) covering a total of 1,796 cM for linkage mapping and QTL identification. Ten QTLs for five mineral elements were identified on 10 different intervals on six chromosomes.

In another study, Fe, Zn, Mn, Cu, Ca, Mg, P, and K contents of 85 introgression lines (ILs) derived from a cross between an elite *indica* cultivar Teqing and the wild rice (*Oryza rufipogon*) were measured by inductively coupled argon plasma (ICAP) spectrometry (Garcia-Oliveira et al. 2009). A total of 31 putative quantitative trait locus (QTLs) were detected for these eight mineral elements by single-point analysis. Wild rice (*O. rufipogon*) contributed favorable alleles for most of the QTLs (26 QTLs), and chromosomes 1, 9, and 12 exhibited 14 QTLs (45 %) for these traits. One major effect of QTL for Zn content accounted for the largest proportion of phenotypic variation (11–19 %) was detected near the simple sequence repeat marker, RM152, on chromosome 8. A QTL for grain iron content was mapped at a different location near the SSR marker 6641 on chromosome 2. Similar locations of QTLs for some mineral elements observed in this study suggested the possibility of simultaneous improvement of these traits in rice grain by marker-assisted selection. Norton et al. (2010) conducted multielement analysis on leaves and grain of the Bala×Azucena rice mapping population grown in the field. QTLs for the concentration of 17 elements were detected, revealing 36 QTLs for mineral content in leaves and 41 in grains. For selenium, lead, phosphorus, and magnesium, QTLs were detected in the same location for both tissues. Four QTLs for grain zinc content were identified on chromosome 6, 7, and 10. In general, there were no major QTL clusters, suggesting separate regulation of each element. QTLs for grain iron, zinc, molybdenum, and selenium are potential targets for marker-assisted selection to improve seed nutritional quality.

5.7 Genomics

A current approach to gene discovery that is most applicable to compounds of nutritional importance, synthesized or accumulated by plants and other organisms, has been termed as “nutritional genomics.” This discipline takes advantage of the homologies or similarities between the metabolic routes that lead to the generation of a given product in different organisms. In this context, previously characterized genes from various nutritional metabolic pathways can be found in public databases and may be used as a resource for further research for the improvement of the target mineral trait in a plant species.

A greater understanding of genomics and new biological technologies is enabling researchers worldwide to establish clear breeding objectives for nutritional quality improvement in cereal crops. Now there is a big focus on modifying grain nutritional quality traits and produce healthier food. Several QTLs for the grain micronutrient contents have been identified and mapped on different rice chromosomes using molecular markers, but their refinement and genetic dissection are yet underway to truly understand the quantitative variation and genes contributing to the trait, which limits their effective utilization in marker-assisted breeding programs. Candidate gene approach is now being widely used for characterizing QTLs as well as Mendelian traits. This approach is based on exploiting the information on the role and functions of a particular coding sequence and hypothesizing a plausible cause–effect relationship between the QTL and a feasible candidate gene mapped nearby (Pflieger et al. 2001; Tuberosa and Salvi 2007; Chandel et al. 2011). The candidate genes or DNA sequences with predicted function serve as an important source to generate novel molecular markers from within a given QTL region which are likely to show more stable association across the mapping populations or genetic stocks and thus will provide more useful tools for isolation and molecular characterization of QTLs (Thorup and Kearsey 2000). Expressed sequence tags (ESTs), complementary DNA sequences (cDNA), and exons circumvent other molecular marker-related limitations such as a large physical distance between genetically close markers and genes and also a recombination between the marker and gene under question.

By using genomic and bioinformatics tools, Chandel et al. (2011) characterized the genomic region encompassing five known QTLs identified in rice (Kaiyang et al. 2008), including two QTLs (qFE-1 and qFE-9) controlling Fe content and three QTLs (qZN-5, qZN-7, qZN11) for Zn content by searching for the presence of putative candidate genes responsible for the Fe and Zn traits and the spatiotemporal expression analysis of these genes in different developmental stages and plant tissues/organs using ESTs and MPSS signature tags.

5.8 Linkage Mapping and Molecular Breeding for Mineral Content: A Case Study

The first prerequisite for initiating a breeding program to develop mineral-rich genotypes is to screen available germplasm and identify the source of genetic variation for the target trait. At CCS Haryana Agricultural University, Hisar, Brar et al. (2011) used 220 rice genotypes for mineral content (Fe and Zn) analysis and identified several iron (HKR 95–130, HKR 95–157, and Palman 579)- and zinc (Taraori Basmati, TNG67, and Pusa 1460)-rich genotypes, which have been used in breeding programs for developing mineral-efficient genotypes. The Fe and Zn contents of dehusked grains differed significantly ($P=0.001$) among the various rice genotypes. It ranged between 5.1 and 441.5 μg/g (mean 67.8 μg/g) and 2.12 and 39.4 μg/g (mean 23.8 μg/g), respectively. In another study, Kumar et al. (2012) assessed the relationship between Fe and Zn contents in dehusked rice grains and leaf tissue in 20 rice genotypes. The grain/leaf ratio of iron and zinc content ranged between 0.03 and 1.56 and 0.16 and 0.87, respectively. Some of the rice genotypes (HKR 95–157, Palman 579, TNG 67, and Taraori Basmati) had higher grain/leaf ratio for iron and zinc. The results showed large variation for the uptake and transport of Fe and Zn, and genotypes showing greater iron and zinc accumulation in grains were further used in the breeding program.

Crosses were made between the iron-rich (Palman 579) and high-yielding (PAU201) rice genotypes, F_1 hybrids confirmed by microsatellite

Table 5.2 QTLs mapped for iron and zinc content in rice grains using PAU201 × Palman 579 F_2 population

Trait	QTL name	Chromosome number	Position (cM)	Flanking markers	Position of flanking markers (cM)	LR	LOD	Additive effect	PV (%)	DPE
Fe content	qFE1	2	54.4	RM53-RM521	32.7–58.4	53.8	11.3	135.26	21.4	PL
	qFE2	2	138.6	RM263-RM221	127.5–143.7	49.9	10.7	149.97	6.9	PL
	qFE3	2	159.2	RM221-RM208	143.7–186.4	49.9	10.8	147.47	26.8	PL
	qFE4	3	54.6	RM489-RM7	29.2–64.0	45.7	8.5	145.21	8.8	PL
	qFE5	7	31.3	RM481-RM418	3.2–42.1	17.7	3.8	110.58	2.4	PL
	qFE6	10	25.9	RM474-RM184	0.0–58.3	34.1	7.4	−143.23	9.2	P
	qFE7	10	108.0	RM228-RM496	96.3–113.0	28.4	6.1	149.91	18.1	PL
	qFE8	12	60.3	RM491-RM519	34.3–62.6	36.3	10.8	3.04	16.9	PL
Zn content	qZN1	2	65.5	RM521-RM29	58.4–68.9	125.6	12.5	−69.53	5.1	P
	qZN2	10	23.8	RM474-RM184	0.0–58.3	116.9	11.6	−70.72	19.1	P
	qZN3	10	115.9	RM496-RM591	113.0–118.3	108.2	10.8	−69.39	4.7	P

DPE (direction of phenotypic effect); PL and P indicate Palman 579 (iron rich) and PAU201 alleles, respectively. Additive effect is the effect of substituting a Palman 579 allele for a PAU201 allele; its positive value indicates that Palman 579 has a positive allele; *LR* likelihood ratio, *LOD* $\log_{10}$ of an odd ratio

marker analysis and F_2 population raised. In F_2 population consisting of 247 plants, large variations for Fe (18.6–475.4 μg/g) and Zn (4.4–157.4 μg/g) in brown rice grain samples were observed. Phenotypic correlation coefficient analysis showed a significant positive correlation (0.523, $p=0.01$) between Fe and Zn content in rice grain. Grain yield and other yield components showed no correlation with Fe content and Zn contents. Seventy-six microsatellite markers (SSRs) were used to map the QTLs associated with mineral content in grains using 50 PAU201 × Palman 579 F_2 plants displaying the entire range of variation in Fe and Zn contents. A total of 171 alleles were identified in 50 PAU201 × Palman 579 F_2 plants, and 14 new recombinant alleles (different than those in parent rice varieties) were identified. Composite interval mapping (CIM) analysis by WinQTL cartographer 2.5 revealed a total of 11 QTLs for mineral content in grains on chromosome 2, 3, 7, 10, and 12 (Table 5.2). Eight QTLs for Fe content were mapped on chromosome 2 (qFE1, qFE2, and qFE3 at map positions 54.4, 138.6, 159.2 cM, respectively), chromosome 3 (qFE4 at map position 54.6 cM), chromosome 7 (qFE5 at map position 31.3 cM), chromosome 10 (qFE6 and qFE7 at map positions 25.9, 108.0 cM, respectively), and chromosome 12 (qFE8 at map position 60.3 cM). In addition, three QTLs (qZN1, qZN2, and qZN3) for Zn content were identified on chromosome 2 (qZN1) and 10 (qZN2 and qZN3) with the map positions of 65.5, 23.8, and 115.9 cM, respectively.

5.9 Summary

Substantial genetic variation for Fe and Zn contents existing in key staple crops provides the basis for developing Fe- and Zn-dense rice by molecular breeding. A wide array of genes (ferritin, phytase, ferric-chelate reductase, nicotianamine aminotransferase) involved in iron acquisition, storage, and/or bioavailability have been identified and successfully transferred to rice with resultant increase in Fe content/bioavailability. Discovery of QTLs linked with grain Fe and Zn can be proved as a new dimension to the breeding approach. However, it needs to be kept in mind that the acceptance of biofortified crops by the farmers and consumers hinges on developing attractive trait packages without compromising agronomic (yield) and end-use (e.g., grain quality) characteristics. Development in new technologies like genomics, transcriptomics, and publically available databases and identification of genes controlling agronomically and nutritionally useful traits have potential utility in crop genetic improvement. More resources and joint efforts of the breeders, biotechnologists, soil scientists, food processing experts, and food nutritionists are required to lead these "biofortification" program and provide sustainable solutions to micronutrient malnutrition in the developing world.

References

Anuradha K, Agarwal S, Batchu AK, Babu AP, Swamy BPM, Longvah T, Sarla N (2012) Evaluating rice germplasm for iron and zinc concentration in brown rice and seed dimensions. J Phytol 4(1):19–25

Ballot D, Baynes RD, Bothwell TH, Gillooly M, Macfarlane BJ, Macphail AP, Lyons G, Derman DP, Bezwoda WR, Torrance JD, Bothwell JE, Mayet F (1987) The effects of fruit juices and fruits on the absorption of iron from a rice meal. Br J Nutr 57:331–343

Banziger M, Long J (2000) The potential of increasing the iron and zinc density of maize through plant breeding. Food Nutr Bull 21:397–400

Bhullar NK, Gruissem W (2013) Nutritional enhancement of rice for human health: the contribution of biotechnology. Biotechnol Adv 31:50–57

Bohn L, Meyers AS, Rasmussen K (2008) Phytate: impact on environment and human nutrition. A challenge for molecular breeding. J Zhejiang Univ Sci B 9:165–191

Bouis HE (2002) Plant breeding: a new tool for fighting micronutrient malnutrition. J Nutr 132:491S–494S

Bouis HE, Chassy B, Ochanda JO (2003) Genetically modified food crops and their contribution to human nutrition and food quality. Trends Food Sci Technol 14:191–209

Brar B, Jain S, Singh R, Jain RK (2011) Genetic diversity for iron and zinc contents in a collection of 220 rice (*Oryza sativa* L.) genotypes. Indian J Genet 71(1):67–73

Brinch-Pedersen H, Sorensen LD, Holm PB (2002) Engineering crop plants: getting a handle on phosphate. Trends Plant Sci 7:118–125

Brinch-Pedersen H, Borg S, Tauris B, Holm PB (2007) Molecular genetic approaches to increasing mineral availability and vitamin content of cereals. J Cereal Sci 46:308–326

Bughio N, Yamaguchi H, Nishizawa NK, Nakanishi H, Mori S (2002) Cloning an iron-regulated metal transporter from rice. J Exp Bot 53:1677–1682

Chandel G, Samuel P, Dubey M, Meena R (2011) *In silico* expression analysis of QTL specific candidate genes for grain micronutrient (Fe/Zn) content using ESTs and MPSS signature analysis in rice (*Oryza sativa* L.). J Plant Genet Transgenic 2(1):11–22

Chen H, Siebenmorgen TJ (1997) Effect of rice thickness in degree of milling and associated optical measurements. Cereal Chem 74:821–825

Chen H, Siebenmorgen TJ, Griffin K (1998) Quality characteristics of long-grain rice milled in two commercial systems. Cereal Chem 75:560–565

Cheng L, Wang F, Shou H, Huang F, Zheng L, He F, Li J, Zhao FJ, Ueno D, Ma JF, Wu P (2007) Mutation in nicotianamine aminotransferase stimulated the Fe(II) acquisition system and led to iron accumulation in rice. Plant Physiol 145:1647–1657

Curie C, Alonso JM, Le Jean M, Ecker JR, Briat JF (2000) Involvement of NRAMP1 from *Arabidopsis thaliana* in iron transport. Biochem J 347:749–755

Curie C, Cassin G, Couch D, Divol F, Higuchi K, Jean ML (2009) Metal movement within the plant: contribution of nicotianamine and yellow stripe 1-like transporters. Ann Bot 103:1–11

Eide D, Broderius M, Fett J, Guerinot ML (1996) A novel iron-regulated metal transporter from plants identified by functional expression in yeast. Proc Natl Acad Sci USA 93:5624–5628

Frossard E, Bucher M, Mächler F, Mozafar A, Hurrell R (2000) Potential for increasing the content and bioavailability of Fe, Zn and Ca in plants for human nutrition. J Sci Food Agric 80:861–879

Galera SG, Rojas E, Sudhakar D, Zhu C, Pelacho AM, Capell T, Christou P (2010) Critical evaluation of strategies for mineral fortification of staple food crops. Transgenic Res 19:165–180

Garcia-Casal MN, Layrisse M, Solano L, Baron MA, Arguello F, Llovera D, Ramirez J, Leets I, Tropper E (1998) Vitamin A and beta-carotene can improve nonheme iron absorption from rice, wheat and corn by humans. J Nutr 128:646–650

Garcia-Oliveira AL, Tan L, Fu Y, Sun C (2009) Genetic identification of quantitative trait loci for contents of mineral nutrients in rice grain. J Plant Biol 51(1):84–92

Gibson GR, Beatty ER, Wang X, Cummings JH (1995) Selective stimulation of bifidobacteria in the human colon by oligofructose and inulin. Gastroenterology 108:975–982

Gillooly M, Bothwell TH, Torrance JD, Macphail AP, Derman DP, Bezwoda WR, Mills W, Charlton RW, Mayet F (1983) The effects of organic acids, phytates and polyphenols on the absorption of iron from vegetables. Br J Nutr 49:331–342

Goto F, Yoshihara T (2001) Improvement of micronutrient contents by genetic engineering – development of high iron content crops. Plant Biotechnol 18:7–15

Goto FT, Yoshihara T, Shigemoto N, Toki S, Takaiwa F (1999) Iron fortification of rice seed by the soybean Ferritin gene. Nat Biotechnol 17:282–286

Gowda SJM, Randhawa GJ, Bisht IS, Firke PK, Singh AK, Abraham Z, Dhillon BS (2012) Morpho-agronomic and simple sequence repeat-based diversity in colored rice (*Oryza sativa* L.) germplasm from peninsular India. Genet Resour Crop Evol 59:179–189

Graham RD, Senadhira D, Beebe S, Iglesias C, Monasterio I (1999) Breeding for micronutrient density in edible portions of staple food crops: conventional approaches. Field Crops Res 60:57–80

Gregorio GB (2002) Progress in breeding for trace elements in staple crops. J Nutr 132:500S–502S

Gregorio GB, Senadhira D, Htut T, Graham RD (1999) Improving iron and zinc value of rice for human nutrients. Agric Dev 23(9):68–87

Gregorio GB, Senadhira D, Htut T, Graham RD (2000) Breeding for trace mineral density in rice. Food Nutr Bull 21:382–386

Grotz N, Guerinot ML (2006) Molecular aspects of Cu, Fe and Zn homeostasis in plants. Biochem Biophys Acta 1763:595–608

Guerinot ML (2007) It's elementary: enhancing Fe^{3+} reduction improves rice yields. Proc Natl Acad Sci USA 104:7311–7312

Gura T (1999) New genes boost rice nutrients. Science 285:994–995

Guttieri M, Bowen D, Dorsch JA, Raboy V, Souza E (2003) Identification and characterization of a low phytic acid wheat. Crop Sci 44:418–424

Harper JF, Surowy TK, Sussman MR (1989) Molecular cloning and sequence of cDNA encoding the plasma membrane proton pump (H^+-ATPase) of *Arabidopsis thaliana*. Proc Natl Acad Sci USA 86:1234–1238

Inoue H, Higuchi K, Takahashi M, Nakanishi H, Mori S, Nishizawa NK (2003) Three rice nicotianamine synthase genes, OsNAS1, OsNAS2, and OsNAS3 are expressed in cells involved in long-distance transport of iron and differentially regulated by iron. Plant J 36:366–381

Ishimaru Y, Kim S, Tsukamoto T, Oki H, Kobayashi T, Watanabe S, Matsuhashi S, Takahashi M, Nakanishi H, Mori S, Nishizawa NK (2007) Mutational reconstructed ferric chelate reductase confers enhanced tolerance in rice to iron deficiency in calcareous soil. Proc Natl Acad Sci USA 104:7373–7378

Islam FMA, Basford KE, Jara C, Redden RJ, Beebe SE (2002) Seed compositional and disease resistance differences among gene pools in cultivated common bean. Genet Resour Crop Evol 49:285–293. doi:10.1023/A:1015510428026

Johnson AAT, Kyriacou B, Callahan DL, Carruthers L, Stangoulis J, Lombi E, Tester M (2011) Constitutive overexpression of the *OsNAS* gene family reveals single-gene strategies for effective iron- and zinc-biofortification of rice endosperm. PLoS ONE 6(9):e24476. doi:10.1371/journal.pone.0024476

Kaiyang L, Lanzhi L, Zheng X, Zhihong T, Zhonglt H (2008) Quantitative trait loci controlling Cu, Ca, Zn, Mn and Fe content in rice grains. Indian Acad Sci 87:305–310

Khush GS, Brar DS (2002) Biotechnology for rice breeding: progress and impact. In: Sustainable rice production for food security. Proceedings of the 20th session of the international rice commission, Bangkok, Thailand, 23–26 July 2002

Kumar J, Chawla A, Kumar P, Jain RK (2012) Iron and zinc variability in twenty rice (*Oryza sativa* L.) genotypes. Ann Biol 28(2):90–92

Kuwano M, Ohyama A, Tanaka Y, Mimura T, Takaiwa F, Yoshida KT (2006) Molecular breeding for transgenic rice with low-phytic-acid phenotype through manipulating *myo*-inositol 3-phosphate synthase gene. Mol Breed 18:263–272

Larson SR, Rutger JN, Young KA, Raboy V (2000) Isolation and genetic mapping of a non-lethal rice (*Oryza sativa* L.) low phytic acid 1mutation. Crop Sci 40:1397–1405

Lee S, An G (2009) Over-expression of *OsIRT1* leads to increased iron and zinc accumulations in rice. Plant Cell Environ 32:408–416

Li ZK (2001) QTL mapping in rice: a few critical considerations. In: Brar DS, Hardy B, Khush GS (eds) Rice genetics IV. Science Publisher, Enfield

Liu ZC, Yu SX (1997) Nutrition and food sanitation, 2nd edn. Demotic Sanitation Publisher, Beijing

Lopez HW, Leenhardt F, Coudray C, Remesy C (2002) Minerals and phytic acid interactions: is it a real problem for human nutrition? Int J Food Sci Technol 37:727–739

Lu T, Huang X, Zhu C, Huang T, Zhao Q, Xie K (2008) RICD: a rice *indica* cDNA database resource for rice functional genomics. BMC Plant Biol 8:118

Lucca P, Hurell R, Potrykus I (2001) Genetic engineering approaches to improve the bioavailability and the level of iron in rice grains. Theor Appl Genet 102:392–397

Mares-Perlman JA, Subar AF, Block G, Greger JL, Luby MH (1995) Zinc intake and sources in the US adult population: 1976–1980. J Am Coll Nutr 14:349–357

Maziya-Dixon B, Kling JG, Menkir A, Dixon A (2000) Genetic variation in total carotene, iron and zinc contents of maize and cassava genotypes. Food Nutr Bull 21:419–422

Monasterio I, Graham RD (2000) Breeding for trace minerals in wheat. Food Nutr Bull 21:392–396

Norton GJ, Deacon CM, Xiong L, Huang S, Meharg AA, Price AH (2010) Genetic mapping of the rice ionome in leaves and grain: identification of QTLs for 17 elements including arsenic, cadmium, iron and selenium. Plant Soil 329:139–153

Paine JA, Shipton CA, Chaggar S, Howells RM, Kennedy MJ, Vernon G, Wright SY, Hinchliffe E, Adams JL, Silverstone AL, Rachel DR (2005) Improving the nutritional value of Golden Rice through increased pro-vitamin A content. Nat Biotechnol 23:482–487

Palmer CM, Guerinot ML (2009) Facing the challenges of Cu, Fe and Zn homeostasis in plants. Nat Chem Biol 5:333–340

Pfeiffer WH, McClafferty B (2007) Harvest Plus: breeding crops for better nutrition. Crop Sci 47(S3):S88–S105

Pflieger S, Lefebvre V, Causse M (2001) The candidate gene approach in plant genetics: a review. Mol Breed 7(4):275–291

Puig S, Andres-Colas N, Garcia-Molina A, Penarrubia L (2007) Copper and iron homeostasis in Arabidopsis: responses to metal deficiencies, interactions and biotechnological applications. Plant Cell Environ 30:271–290

Rabbani GH, Ali M (2009) New ideas and concepts, rice bran: a nutrient-dense mill-waste for human nutrition. The ORION Med J 32(3):694–701

Raboy V (2001) Seeds for a better future: 'low phytate', grains help to overcome malnutrition and reduce pollution. Trends Plant Sci 6:458–462

Raboy V (2002) Progress in breeding low phytate crops. J Nutr 132:503S–505S

Robinson NJ, Procter CM, Connolly EL, Guerinot ML (1999) A ferric-chelate reductase for iron uptake from soils. Nature 397:694–697

Scholz-Ahrens KE, Schrezenmeir J (2002) Inulin, oligofructose and mineral metabolism—experimental data and mechanism. Br J Nutr 87:S179–S186

Sellappan K, Datta K, Parkhi V, Datta SK (2009) Rice caryopsis structure in relation to distribution of micronutrients (iron, zinc, β-carotene) of rice cultivars including transgenic *indica* rice. Plant Sci 177:557–562

Seymour J (1996) Hungry for a new revolution. New Sci 149(2023):32–37

Shaw JG, Friedman JF (2011) Iron deficiency anemia: focus on infectious diseases in lesser developed countries. Anemia. Review Article 2011:10. doi:10.1155/2011/260380

Sperotto RA, Boffa T, Duartea GL, Santosb LS, Grusakc MA, Fett JP (2010) Identification of putative target genes to manipulate Fe and Zn concentrations in rice grains. J Plant Physiol 167:1500–1506

Stangoulis JCR, Huynh BL, Welch RM, Choi EY, Graham RD (2007) Quantitative trait loci for phytate in rice grain and their relationship with grain micronutrient content. Euphytica 154:289–294

Stein AJ, Sachdev HPS, Qaim M (2008) Genetic engineering for the poor: golden rice and public health in India. World Dev 3:144–158

Takahashi M (2003) Overcoming Fe deficiency by a transgenic approach in rice. Plant Cell Tissue Organ Cult 72:211–220

Takahashi M, Terada Y, Nakai I, Nakanishi H, Yoshimura E, Mori S, Nishizawa NK (2003) Role of nicotianamine in the intracellular delivery of metals and plant reproductive development. Plant Cell 15:1263–1280

Tang G, Hu Y, Yin S, Wang Y, Dallal GE, Grusak MA, Russell RM (2012) β-carotene in golden rice is as good as β-carotene in oil at providing vitamin A to children. Am J Clin Nutr 96:658–664

The World Health Organization (WHO) (2011) Micronutrient deficiencies: iron deficiency anemia. Available from: http://www.who.int/nutrition/topics/ida/en/index.html

Thomine S, Wang R, Ward JM, Crawford NM, Schroeder JI (2000) Cadmium and iron transport by members of a plant transporter gene family in *Arabidopsis* with homology to NRAMP genes. Proc Natl Acad Sci USA 97:4991–4996

Thorup GL, Kearsey FD (2000) The principles of QTL analysis (a minimal mathematics approach). J Exp Bot 49:1619–1623

Tuberosa R, Salvi S (2007) From QTLs to genes controlling root traits in maize: scale and complexity in plant systems research. Gene-Plant-Crop Relat 2:15–24

Ullah AHJ, Mullaney EJ (1996) Disulfide bonds are necessary for structure and activity in *Aspergillus ficuum* phytase. Biochem Biophys Res Commun 227:311–317

United Nations Children's Fund (UNICEF) (2009) Vitamin A deficiency: the challenge. Available from: http://www.childinfo.org/vitamina.html

Vasconcelos M, Datta K, Oliva N, Khalekuzzaman M, Torrizo L, Krishnan S, Margarida O, Goto F, Datta S (2003) Enhanced iron and zinc accumulation in transgenic rice with the ferritin gene. Plant Sci 164:371–378

Vasconcelos M, Eckert H, Arahana V, Graef G, Grusak MA, Clemente T (2006) Molecular and phenotypic characterization of transgenic soybean expressing the *Arabidopsis* ferric chelate reductase gene, *FRO2*. Planta 224:1116–1128

Velu G, Rai KN, Muralidharan V, Kulkarni VN, Longvah T, Raveendran TS (2007) Prospects of breeding biofortified pearl millet with high grain iron and zinc content. Plant Breed 126:182–185

Welch RM, Graham RD (2004) Breeding for micronutrients in staple food crops from a human nutrition perspective. J Exp Bot 55:353–364

White PJ, Broadley MR (2005) Biofortifying crops with essential mineral elements. Trends Plant Sci 10:586–593

Wirth J, Poletti S, Aeschlimann B, Yakandawala N, Drosse B, Osorio S (2009) Rice endosperm iron biofortification by targeted and synergistic action of nicotianamine synthase and ferritin. Plant Biotechnol J 7:631–644

Ye X, Babili AS, Kloeti A, Zhang J, Lucca P, Beyer P, Potrykus I (2000) Engineering the provitamin A (β-carotene) biosynthetic pathway into (carotenoid-free) rice endosperm. Science 287:303–305

Zhu C, Naqvi S, Gomez-Galera S, Pelacho AM, Capell T, Christou P (2007) Transgenic strategies for the nutritional enhancement of plants. Trends Plant Sci 12:548–555, www.harvestplus.org. Accessed 24 July 2013

Hairy Root Cultures of Medicinal Trees: A Viable Alternative for Commercial Production of High-Value Secondary Metabolites

6

Priyanka Siwach, Anita Rani Gill, and Khushboo Sethi

Abstract

Since the last two decades, a lot of attention has been given to the development of 'hairy root' system as potential source of desired secondary metabolites. To date, hairy root cultures have been developed for more than 100 medicinal plant species, including plants of various habits, habitat and belonging to different threat status also. Forest trees form a potential group among medicinal plants with some like *Taxus* spp. and *Cinchona* spp., yielding invaluable therapeutic molecules to the modern drug sector. The high-value secondary metabolites in trees are generally synthesized during the later stage of the life cycle and are produced in very less amount, making the development of stable in vitro source mandatory for commercial production of their metabolites. Hairy root systems have not been developed for many important tree species. Since the discovery of *Agrobacterium rhizogenes* as pathogenic bacteria causing hairy root disease, tremendous development towards establishment of hairy root system as biochemical factory has taken place. Diverse strategies can be developed to improve the yield so as to produce desired metabolites at large-scale and in eco-friendly conditions. Research inputs from metabolomics have facilitated the development of new strategies to manipulate the biosynthetic pathways while bioreactor design has allowed the scaling up of the hairy root systems. This chapter highlights the existing status of hairy root system for various medicinal trees and outlines different strategies in this direction.

6.1 Introduction

Plants were being used as the main therapeutic agents during ancient times in India, as described in *Susruta Samhita* and the *Charak Samhita* (Prajapati et al. 2003), as well as in many other known cultures of the past like the Egyptian, Babylonian, Jewish and Chinese. It was only

P. Siwach (✉) • A.R. Gill • K. Sethi
Department of Biotechnology, Chaudhary Devi Lal University, Sirsa, Haryana, India
e-mail: Psiwach29@gmail.com

R.K. Salar et al. (eds.), *Biotechnology: Prospects and Applications*,
DOI 10.1007/978-81-322-1683-4_6, © Springer India 2013

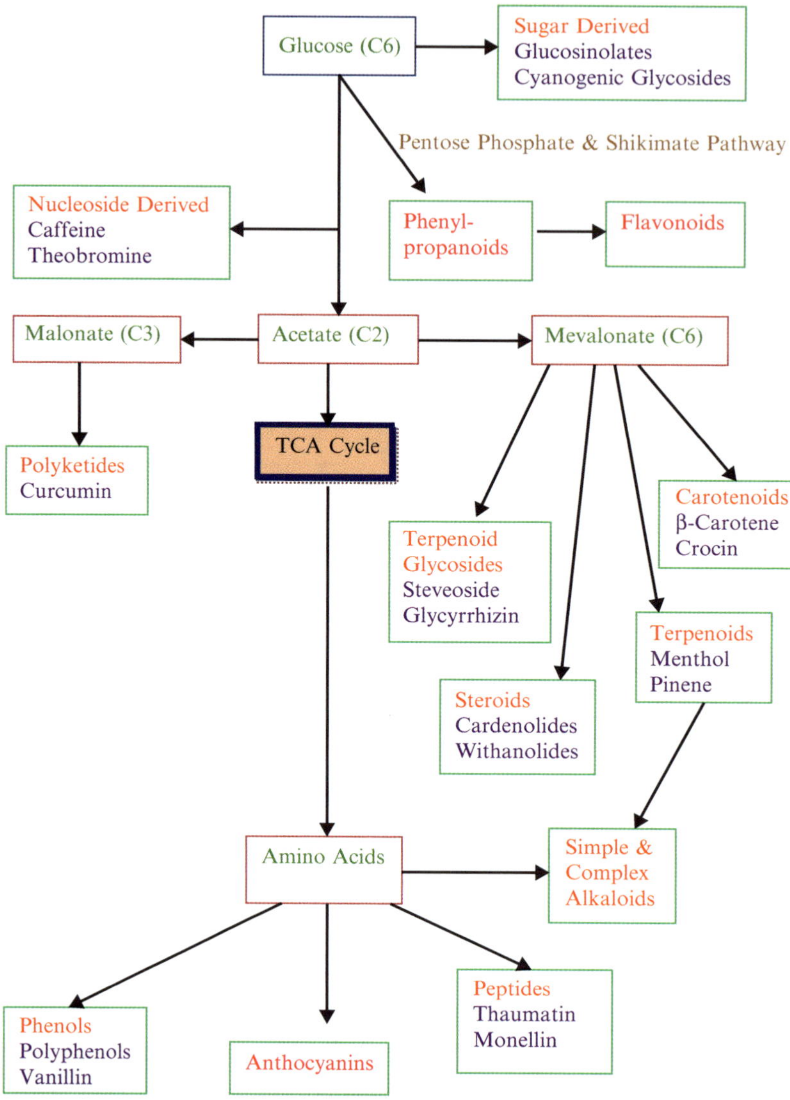

Fig. 6.1 A general scheme of different biosynthetic pathways for secondary metabolites in plant cell

during the last 100 years when the natural products have been partly replaced by synthetic drugs. Medicinal plants have come up as an irreplaceable entity in the modern drug discovery; nearly 20–30 % modern drugs being derived from plants while many are synthetic analogues of plant-derived natural compounds (Brower 2008). Up till now, 47 major drugs have been produced from the tropical forests, but it is believed that around 328 potential, major drugs and 125,000 flowering trees of potential pharmacological importance may be still hidden in the unexplored depths of these forests (Kulkarni 2000). Every year the market for herbs-based drugs is increasing at the rate of 12–15 % (http://www.phytochemistry.freeserve.co.uk). Plant-derived chemicals are not only valuable sources for pharmaceuticals but also for a variety of oils, flavours, dyes and resin (Parr 1988). These useful phytochemicals are predominantly secondary metabolites which are not essential to plant growth. These are produced in small amounts and generally accumulate in specialized tissues. The biosynthetic pathways of some of these secondary metabolites are depicted in Fig. 6.1. The chemical structure of these compounds is generally very complicated and exhibits chirality also. So in many cases, the organic synthesis is not cost-effective (Kim et al. 2002) and the compounds are generally extracted and purified from field-grown plants (DiCosmo and Misawa 1995).

There are nearly 250,000 higher plant species on earth. More than 80,000 plant species are medicinal in nature, of which perennial trees form a potential group. Many perennial trees have yielded therapeutic molecules for treatment of many dreaded diseases like Taxol from *Taxus* spp. (for cancer treatment), quinine from *Cinchona* spp. (for malarial treatment) and aspirin from *Willow* spp. (as analgesic) to quote a few. The high-value secondary metabolites in these trees are generally synthesized during the later stage of life and are produced in very less amount, not sufficient to meet the commercial demand, e.g. 14 kg of Yew bark yield only 1 g of 'Taxol', i.e. consumption of 12 Yew trees per cancer patient. The continuously increasing demand of these metabolites has posed a great threat to the existence of many of these trees and has pushed many of them under different threat status (Table 6.1).

In the search for alternatives to production of desirable medicinal compounds from plants, biotechnological approaches, particularly, plant tissue cultures, are found to have great potential as a supplement to traditional agriculture (Rao and Ravishankar 2002). The growing demand in market for natural products as well as a strong need for biodiversity conservation of natural resources has further accelerated the research on in vitro plant materials as potential factories for secondary metabolite products. Desired production of defined chemical products within carefully regulated, highly controlled microenvironment regimes are additional factors for pushing research in this direction. A number of efforts have been made to generate cell (callus and suspension) cultures for in vitro production of respective bioactive compound for different medicinal trees (Table 6.1). Some studies have elaborated different ways for generation of commercialized industrial scale plant cell cultures (Curtin 1983; Kieran et al. 1997), still these face certain problems. Some bioactive compounds are produced by differentiated cells (Balandrin et al. 1985) and so are not synthesized in the undifferentiated cells of suspension cultures (Charlwood and Charlwood 1991). Further suspension cultures are generally genetically unstable (Bais et al. 2001; Rhodes et al. 1994). Disappointing yields coupled with instability of undifferentiated cell cultures initiated the efforts for development of new stable methods for enhanced production of secondary products, which led to hairy root cultures. The latter have become popular in the last two decades as the most popular method of producing plant secondary metabolites.

6.2 Historical Background

In the early twentieth century, great attention was drawn towards two plant diseases, viz. crown gall and hairy root disease which were responsible for heavy losses in plum, peach, apple and other fruit nurseries. Scientific investigation for elucidation of molecular mechanism of these diseases led to the discovery of natural genetic engineers of genus *Agrobacterium* (Chilton et al. 1977, 1980).

Agrobacterium tumefaciens was found responsible for crown gall disease while *Agrobacterium rhizogenes* caused hairy root disease. The molecular mechanism was cracked by efforts of many groups (Chilton et al. 1982; Huffman et al. 1984; Nilsson and Olsson 1997; Sevon and Oksman-Caldentey 2002; Weising and Kahl 1996; White et al. 1985). Detailed information on mechanism of gene transfer by these natural genetic engineers is well discussed by some researchers (Tzfira et al. 2004). Since then, numerous efforts have been made to exploit these two species of genus *Agrobacterium* for the benefit of mankind. This chapter is an effort towards the applicability of hairy root cultures for harnessing the secondary metabolites of perennial trees.

6.3 Hairy Roots: An Excellent In Vitro System for Secondary Metabolites Production

Agrobacterium rhizogenes, the Gram-negative soil bacterium, infects higher plants at wound sites. After infection the bacteria transfers a DNA segment (T-DNA) from its plasmid (Ri plasmid)

Table 6.1 Some important medicinal trees with popular medicinal applications, bioactive ingredient, IUCN status and different in vitro sources generated so far

Scientific name	Medicinal uses	IUCN status	Culture type	Active ingredient	References
Ailanthus altissima	Antioxidant activity, antimalarial, intestinal haemorrhage, asthma, epilepsy, cancer	EN	Suspension	Alkaloids, canthinone alkaloids	Anderson et al. (1987) and Anderson et al. (1986)
Azadirachta indica	Anthelmintic, antifungal, antidiabetic, antibacterial, antiviral, contraceptive, skin diseases	NE	Suspension, hairy root	Azadirachtin	Sujanya et al. (2008) and Srivastava and Srivastava (2012)
Azadirachta indica	Anthelmintic, antifungal, antidiabetic, antibacterial, antiviral, contraceptive, skin diseases	NE	Suspension	Azadirachtin	Poornasri et al. (2008)
Betula utilis	Antispasmodic, antidysenteric, hemostatic	EN	Callus	Betulin, lupeol, oleanolic acid	Zaki et al. (2011)
Brucea javanica	Antitumoral, anti- inflammatory	NT	Suspension	Canthinone alkaloids, Cathin	Liu et al. (1990)
Cinchona L.	Haemorrhoids, influenza, swine flu, malaria, fever, cancer, enlarged spleen	VU	Suspension, hairy root	Alkaloids, quinine	Koblitz et al. (1983) and Hamill et al. (1989)
Cinchona robusta	Antimalarial, cold	VU	Suspension	Robustaquinone	Schripsema et al. (1999)
Cinchona spec.	Antimalarial, common cold	VU	Suspension	Anthraquinones	Wijnsma et al. (1985)
Cinchona succirubra	Antimalarial, common cold	VU	Suspension	Anthraquinones	Khouri et al. (1986)
Citrus sp.	Antioxidant properties, fever colic, as an astringent, diuretic	NE	Callus	Naringin, limonin	Barthe et al. (1987)
Eriobotrya japonica	Analgesic, antibacterial, antiviral, astringent, diuretic, expectorant	NE	Callus	Triterpenes	Taniguchi et al. (2002)
Eucalyptus tereticornis	Stimulant, anaesthetic, antiseptic, antibacterial, hepatoprotective	NE	Callus	Sterols and phenols	Venkateswara et al. (1986)
Ginkgo biloba	Alzheimer's disease, antioxidant, antitussive, astringent, expectorant	EN	Suspension, hairy root	Ginkgolide A	Carrier et al. (1991) and Ayadi and Tremouillaux-Guiller (2003)
Morinda citrifolia	Digestive, intestinal, respiratory, immune systems	NE	Suspension	Anthraquinones	Bassetti et al. (1995)
Picrasma quassioides	Antibacterial agent, Antileukemic	NE	Suspension	Quassin	Scragg and Allan (1986)
Taxus baccata	Cardiac remedy, breast and ovarian cancer	LC	Suspension	Taxol baccatin III	Cusido et al. (1999)
Taxus spp.	Cardiac remedy, breast and ovarian cancer	Least concern	Suspension	Taxol	Wu et al. (2001)

IUCN International union for conservation of nature, *EN* endangered, *NE* not evaluated, *NT* not threatened, *VU* vulnerable, *LC* least concern

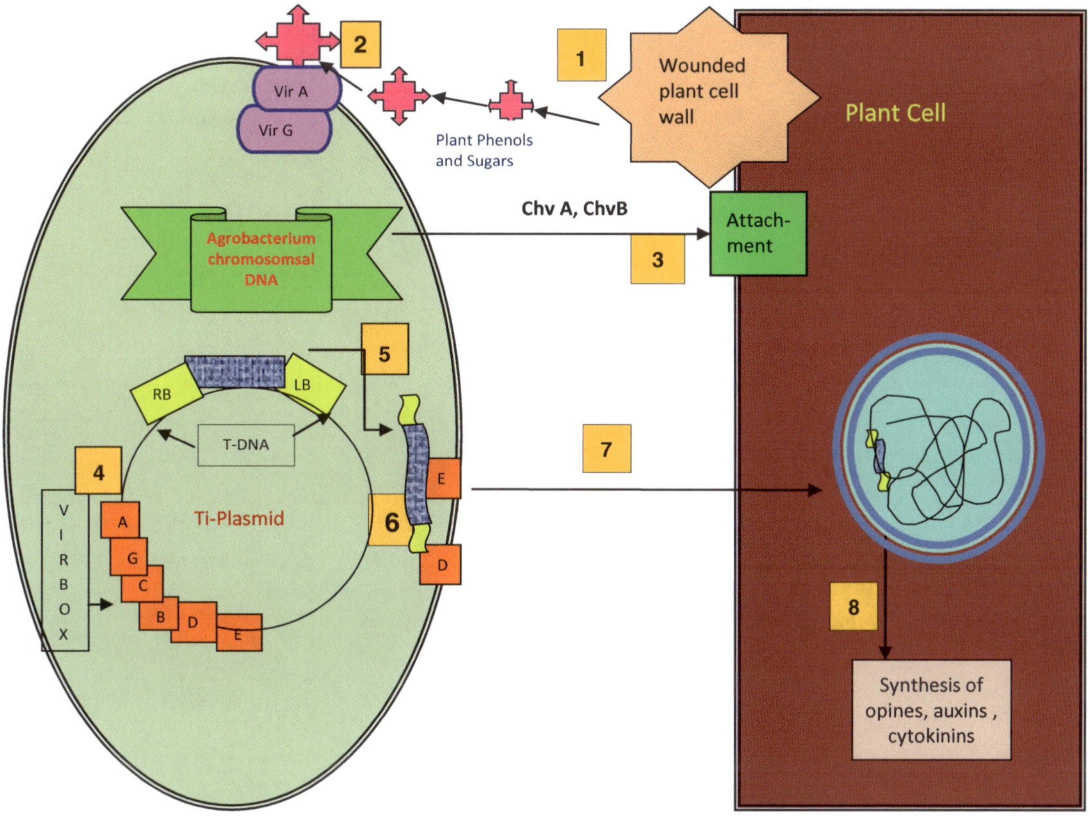

Fig. 6.2 *1* Secretion of acetosyringone, phenols and sugars from the wounded plant cell, *2* binding of acetosyringone to the receptor protein *Vir A* and *Vir G*, present on the bacterial wall, *3* attachment to the plant cell, important role being played by agrobacterium chromosomal gene *chv A* and *chv B*, *4* activation of vir box, *5* excision of T-DNA, *6* formation of complex of T-DNA and vir proteins (*Vir D*, *Vir E*), *7* transfer of T-DNA to the plant nucleus and integration in the plant genome, *8* synthesis of auxins, cytokinins and opines (auxins and cytokinins synthesis promote induction of hairy roots from the wound site while opines feed the agrobacteria)

into the plant cell genome. The T-DNA encodes enzymes which control auxin and cytokinin biosynthesis (Fig. 6.2). The new hormonal composition inside the plant cells induces the emergence of hairy roots at the site of wound. Once induced, these hairy roots can be isolated from the parent plant cell and can be propagated as axenic cultures in hormone-free medium. During the mid-1980s and early 1990s, there were several investigations on the capability of hairy root cultures for production of biologically active substances declaring hairy roots to have excellent biosynthetic capability (Kamada et al. 1986).

Hairy roots are characterized by fast plagiotropic growth with profuse branching and rapid root tip elongation (Tepfer 1984). Hairy roots are favoured over suspension cultures for many reasons like faster biomass doubling time, large size of individual clone, no hormones needed in the growth medium, organized structure, stable euploid to polyploid state, homogeneous, most often the products are released in the medium, growth independent of the inoculums size, minimum lag phase in growth initiation and easy continuous operation (Bhagyalakshmi and Ravishankar 1998; Charlwood and Charlwood 1991; Flores et al. 1999). Hairy root cultures often exhibit the same or greater biosynthetic capacity for secondary metabolites production compared to mother plants (Banarjee et al. 1998; Flores et al. 1987; Kittipongpatana et al. 1998). Certain bioactive compounds which accumulate only in

Table 6.2 Hairy root cultures producing novel secondary metabolites (not present in parent plants) (Kukreja et al. 1997)

Plant spp.	*Agrobacterium rhizogenes* strain	Novel secondary metabolite
Astragalus membranaceus	ATCC 15834	Agroastragaloside I, saponins and polysaccharides
Fagopyrum esculentum	ATCC 15834	Flavonoids/catechins procyanidins
Lobelia inflata	ATCC 15834	Polyacetylene/Robetyolinin
Rauvolfia serpentina	ATCC A4	12-hydroxyajmaline
Swertia japonica	ATCC 15834	8-o-primeverosyl bellidifolin

the aerial part of the plant have been found to be present in abundant amount in respective hairy root cultures, e.g. lawsone (accumulating in aerial part of *Lawsonia inermis*), was produced in the hairy root cultures of *Lawsonia inermis* under dark conditions (Bakkali et al. 1997). Similarly, artimisin was thought to accumulate only in the aerial parts of *Artemisia annua* plant (Wallaart et al. 1999); several groups exhibited its production in the hairy root cultures also (Weathers et al. 1994; Jaziri et al. 1995). Hairy root cultures have also been reported to produce foreign proteins, e.g. ribosome activation proteins (Thorup et al. 1994) and monoclonal antibodies (Sharp and Doran 2001), though many problems were faced in the process (Doran 2006). Many times, hairy roots also act as potential source of some novel natural compounds. For example, a new compound named licoagrodione was isolated from hairy root cultures of *Glycyrrhiza glabra*, and this compound was shown to possess strong antimicrobial activities (Li et al. 1998).

Similarly Berkov et al. (2003) observed the synthesis of a new tropane alkaloid ester in tetraploid hairy roots of *Datura stramonium*. Some other examples are given in Table 6.2.

Since the past two decades, studies have been conducted to raise commercially viable hairy root cultures of a number of plants including certain perennial trees also (Table 6.1). A number of efforts have been made to optimize the conditions for automated industrial production of secondary metabolites from hairy root cultures. Majority of the efforts are made for raising cultures of high-value medicinal herbs/shrubs, not much work is reported for enhancing the secondary metabolites production by the hairy root cultures of perennial trees. In the following section, a general discussion has been made about various strategies applied for enhancing the secondary metabolites production in the hairy root cultures of different medicinal plants including herbs, shrubs as well as trees.

6.3.1 Optimization of Culture Conditions

Variation in nutrient medium composition is a very simple way to affect the secondary metabolites production. The medium is generally modified with respect to concentration of carbon, nitrogen and phosphorous sources (Wilhelmson et al. 2006). Addition of fructose in the medium instead of sucrose was found to enhance the production of catharanthine in the cultures of *Catharanthus roseus* by two times (Jung et al. 1992). Ammonium ions were found to effectively alter the secretion pattern of shikonin from hairy root cultures of *Lithospermum erythrorhizon* (Fukui et al. 1998, 1999). Phosphate has also been shown to affect the secondary metabolite production (Pannuri et al. 1993; Dunlop and Curtis 1991).

The accumulation of secondary metabolites in plants is part of the defence response against pathogenic attack. Elicitors are chemicals or biofactors from various sources which are capable of inducing the physiological and morphological response in the plant cell (Zhao et al. 2005). Inclusion of elicitors in the medium is the most widely used strategy for enhancing the secondary

Table 6.3 Enhancement in secondary metabolite synthesis in the hairy root cultures through elicitation

Hairy root culture	Elicitor	Metabolite	References
Abiotic elicitors			
Brugmansia candida	$CdCl_2$	Hyoscyamine	Pita-Alvarez et al. (2000)
Panax ginseng	Jasmonic acid	Ginsenoside	Sivakumar et al. (2005)
Panax ginseng	Methyl jasmonate	Ginsenoside	Palazon et al. (2003a, b, c)
Panax ginseng	Vanadyl sulfate	Ginsenoside	Palazon et al. (2003a, b, c)
Ambrosia artemisiifolia	Vanadyl sulfate	Thiarubrine A	Bhagnath and Hjortso (2000)
Salvia miltiorrhiza	Silver	Tanshinone	Ge and Wu (2005)
Solanum tuberosum	Beta-cyclodextrin	Sesquiterpenes	Komaraiah et al. (2003)
Biotic elicitors			
Solanum tuberosum	*Rhizoctonia bataticola*	Sesquiterpenes	Komaraiah et al. (2003)
Panax ginseng	Chitosan	Ginsenoside	Palazon et al. (2003a, b, c)
Salvia miltiorrhiza	Yeast extract	Tanshinone	Yan et al. (2005)
Ammi majus	*Enterobacter sakazakii*	Umbelliferone	Staniszewska et al. (2003)
Tagetes patula	*Fusarium conglutinans*	Total thiophenes	Mukundan and Hjortso (1990)
Cichorium intybus	*Phytophthora parasitica* (medium filtrate)	Esculin and esculetin	Bais et al. (2000)
Scopolia parviflora	*S. aureus*	Scopolamine	Jung et al. (2003)
Brugmansia candida	Yeast extract	Scopolamine and hyoscyamine	Pita-Alvarez et al. (2000)

metabolites production in the in vitro cultures. The most frequently used elicitors are fungal carbohydrates, yeast extract, methyl jasmonate and chitosan. Presence of various abiotic elicitors like tannic acid, selenium and nickel in the hairy root culture medium of *Panax ginseng* enhanced ginseng saponin biosynthesis in the hairy roots (Jeong and Park 2007). Hairy roots of *Oxalis tuberosa* when cultured on medium having *Phytophthora cinnamon* elicitors, released large amount of harmine and harmaline into the medium (Bais et al. 2003). Abiotic factors like pH, sonication, temperature and oxygen stress stimulated the secretion of betaine from hairy roots of *Beta vulgaris* (Thimmaraju et al. 2003). Some more examples are presented in Table 6.3. Elaborate discussion about the increase in the yield of secondary metabolite by elicitation has been made by Georgiev et al. (2007). The secretion rate of metabolites can also be increased by use of some trapping system (Guillon et al. 2006). Inclusion of a mixture of two adsorbents (alumina and silica 1:1) in the nutrient medium enhanced the recovery of betaine efflux from red beet hairy root to 97.2 % (Thimmaraju et al. 2004).

6.3.2 Metabolic Engineering

Metabolic engineering allows desired production of desired metabolite in the in vitro cultures. It is based on inserting genes coding for enzymes of a given biosynthetic pathway between T-borders of Ri plasmid and transferring this construct into plant genome. Such efforts have been accelerated by the increasing research inputs in metabolome studies. It offers new opportunities to design metabolic engineering strategies to circumvent problems linked to precursor availability or negative regulatory loops (Memelink 2005). The alkaloid production in *Duboisia* hybrid, *Datura metel* and *Hyoscyamus muticus* hairy root could be enhanced by transforming the respective cultures with over-expressing genes encoding key enzymes of the tropane alkaloid biosynthesis pathway (Moyano et al. 2003; Palazon et al. 2003c).

Extensive efforts have been put to enhance the biosynthesis of terpenoid indole alkaloid (TIA) in *Catharanthus roseus*. The rate-limiting step of TIA biosynthesis was determined by precursor feeding method (Morgan and Shanks 2000).

Table 6.4 Large scale production of secondary metabolites from hairy root cultures in various bioreactors

Root culture	Secondary metabolite productivity	Reactor type	References
Artemisia annua	Artemisinin	Bubble column	Kim et al. (2001)
Artemisia annua	Artemisinin	Bubble column-nutrient mist	Kim et al. (2001)
Artemisia annua	Artemisinin	Erlenmeyer flask	Xie et al. (2000)
Atropa belladonna	Atropine	Isolated impeller- batch	Lee et al. (1999)
Atropa belladonna	Atropine	Segmented bubble column	Kwok and Doran (1995)
Datura stramonium	Hyoscyamine	Isolated impeller- continuous	Hilton and Rhodes (1990)
Datura stramonium	Hyoscyamine	Isolated impeller- batch	Hilton and Rhodes (1990)
Panax ginseng	Ginsenoside	2-1 wave system	Palazon et al. (2003b)
Beta vulgaris	Betalains	Temporary immersion system	Pavlov and Bley (2005, 2006)
Solanum chrysotrichum	Saponins	Airlift mesh – draught reactor with wire helix	Caspeta et al. (2005)

Metabolic engineering of indole pathway in *Catharanthus roseus* hairy roots resulted in increased accumulation of tryptamine and serpentine (Hughes et al. 2004).

6.3.3 Development of Bioreactors

The development of bioreactors for scaling up of hairy root cultures is very crucial for large-scale industrial production of the desired metabolite. However, difficulty in culturing of hairy roots in traditional bioreactors, because of their special morphological characteristics, was observed and a need was felt to develop special bioreactors for hairy root cultures (Wilson et al. 1987). The importance and usefulness of bioreactors for hairy root cultures has been well discussed earlier (Shanks and Morgan 1999; Giri and Narasu 2000). Different types of reactor systems suitable for hairy roots have been developed by different workers; each system having its own suitability and drawbacks (Kim et al. 2002). The bioreactors used to culture hairy roots are roughly divided into three types: liquid phase, gas phase and hybrid reactors (Kim et al. 2002). It would be difficult to select the best bioreactor design for cultivating hairy roots of a particular plant species. For a successful scale-up of hairy roots, several factors need to be considered, including the physiology, morphology, unusual rheology properties and high stress sensitivity of hairy roots (Wysokinska and Chmiel 1997). Table 6.4 gives a list of some of the metabolites produced from hairy root cultures in a bioreactor.

6.4 Conclusion

At present, the hairy root culture system is the most potential approach for large-scale synthesis of secondary metabolites produced by the perennial trees. The advances made in the recent past like engineering of secondary metabolism, the increased accumulation and excretion of metabolites after elicitation, designing and production of recombinant molecules and scaling up of the process have been largely applied to the cultures of medicinal herbs and shrubs; a lot needs to be done for the establishment of efficient hairy root culture systems for perennial medicinal trees. The genetic and biochemical stability of these differentiated cultures and the efficient scaling up of these to the production in bioreactors offers an eco-friendly way for harnessing the high-value secondary metabolites produced by long-lived trees.

References

Anderson LA, Hay CA, Roberts MF, Phillipson JD (1986) Studies on *Ailanthus altissima* cell suspension cultures. Plant Cell Rep 5:387–390

Anderson LA, Roberts MF, Phillipson JD (1987) Studies on *Ailanthus altissima* cell suspension cultures. The effect of basal media on growth and alkaloid production. Plant Cell Rep 6:239–241

Ayadi R, Tremouillaux-Guiller J (2003) Root formation from transgenic calli of *Ginkgo biloba*. Tree Physiol 23(10):713–718

Bais HP, Govindswamy S, Ravishanker GA (2000) Enhancement of growth and coumarin production in hairy root cultures of Witlof chicory (*Cichorium intybus* L.) under the influence of fungal elicitors. J Biosci Bioeng 90:648–653

Bais HP, Sudha G, George J, Ravishankah GA (2001) Influence of exogenous hormones on growth in secondary metabolite production in hairy root cultures of *Cichorium intybus* L. cv. lucknow local. In Vitro Cell Dev Biol Plant 37:293–299

Bais HP, Vepachedu R, Vivavco JM (2003) Root specific elicitation and exudation of fluorescent β-carbolines in transformed root cultures of *Oxalis tuberosa*. Plant Physiol Biochem 41:345–353

Bakkali AT, Jaziri M, Foriers A, Vander Heyden Y, Vanhaelen M, Momes J (1997) Lawsone accumulation in normal and transformed cultures of henna, *Lawsonia inermis*. Plant Cell Tissue Organ Cult 51:83–87

Balandrin MF, Klocke JA, Wurtele ES, Bollinger WH (1985) Natural plant chemicals: sources of industrial and medicinal materials. Science 228:1154–1160

Banarjee S, Rahman L, Uniyal GC, Ahuja PS (1998) Enhanced production of valepotriates by *Agrobacterium rhizogenes* induced hairy root cultures of *Valeriana wallichii* DC. Plant Sci 131:203–208

Barthe GA, Jourdan PS, McIntosh CA, Mansell RL (1987) Naringin and limonin production in callus cultures and regenerated shoots from *Citrus* sp. J Plant Physiol 127:55–65

Bassetti L, Hagendoorn M, Johannes T (1995) Surfactant-induced non-lethal release of anthraquinones from suspension cultures of *Morinda citrifolia*. J Biotechnol 39:149–155

Berkov S, Pavlov A, Kovacheva P, Stanimirova P, Philipov S (2003) Alkaloid spectrum in diploid and tetraploid hairy root cultures of *Datura stramonium*. Z Naturforsch C 58(1–2):42–46

Bhagnath SG, Hjortso MA (2000) Statistical analysis of elicitation strategies for thiarubrine A production in hairy root cultures of *Ambrosia artemisiifolia*. J Biotechnol 80:159–167

Bhagyalakshmi N, Ravishankar GA (1998) Natural compounds from cultured hairy roots. In: Khan IA, Khanum A (eds) Role of biotechnology in medicinal and aromatic plants, vol 1. Ukaaz Publications, Hyderabad, pp 166–182

Brower V (2008) Back to nature- extinction of medicinal plants threatens drug discovery. J Natl Cancer Inst 100(12):838–839

Carrier D, Chauret N, Mancini M, Coulombe P, Neufeld R, Weber M, Archambault J (1991) Detection of ginkgolide A in *Ginkgo biloba* cell cultures. Plant Cell Rep 10:256–259

Caspeta L, Quintero R, Villarreal ML (2005) Novel airlift reactor fitting for hairy root cultures: developmental and performance studies. Biotechnol Prog 21:735–740

Charlwood BV, Charlwood KA (1991) Terpenoid production in plant cell culture. In: Harborne JB, Tomas-Barberan FA (eds) Ecological chemistry and biochemistry of plant terpenoids. Clarendon, Oxford

Chilton MD, Drummond MH, Merlo DJ, Sciaky D, Montoya AL, Gordon MP, Nester EW (1977) Stable incorporation of plasmid DNA into plant cells: the molecular basis of crown gall tumorogenesis. Cell 11:263–271

Chilton MD, Saiki RK, Yadav N, Gordon MP, Quetier F (1980) T-DNA from *Agrobacterium* Ti plasmid is in the nuclear DNA of crown gall tumour cells. Proc Natl Acad Sci U S A 77:4060–4064

Chilton MD, Tepfer DA, Petit A, David C, Casse-Delbart F, Tempe J (1982) *Agrobacterium* inserts T-DNA into the genomes of the host plant root cells. Nature 295:432–434

Curtin ME (1983) Harvesting profitable products from plant tissue culture. Bio/Technology 1:649–657

Cusido RM, Palazon J, Navia-Osorio A, Mallol A, Bonfill M, Morales C, Pinol MT (1999) Production of taxol and baccatin III by a selected *Taxus baccata* callus line and its derived cell suspension culture. Plant Sci 146:101–107

DiCosmo F, Misawa M (1995) Plant cell and tissue culture: alternatives for metabolite production. Biotechnol Adv 13:425–453

Doran PM (2006) Foreign protein degradation and instability in plants and plant tissue cultures. Trends Biotechnol 24:426–432

Dunlop DS, Curtis WR (1991) Synergistic response of plant hairy root cultures to phosphate limitation and fungal elicitation. Biotechnol Prog 7:434–438

Flores HE, Hoy MW, Pickard JJ (1987) Secondary metabolites from root cultures. Trends Biotechnol 5:64–69

Flores HE, Vivanco JM, Loyola-Vargas VM (1999) Radicle biochemistry: the biology of root-specific metabolism. Trends Plant Sci 4:220–226

Fukui H, Hasan AFMF, Ueoka T, Kyo M (1998) Formation and secretion of a new brown benzoquinone by hairy root cultures of *Lithospermum erythrorhizon*. Phytochemistry 47:1037–1039

Fukui H, Hasan AFMF, Kyo M (1999) Formation and secretion of a new unique quinine by hairy root cultures of *Lithospermum erythrorhizon*. Phytochemistry 51:511–515

Ge X, Wu J (2005) Induction and potentiation of diterpenoid tanshinone accumulation in *Salvia miltiorrhiza* hairy roots by beta-aminobutyric acid. Appl Microbiol Biotechnol 68:183–188

Georgiev MI, Pavlov AI, Bley T (2007) Hairy root type plant in vitro systems as sources of bioactive substances. Appl Microbiol Biotechnol 74:1175–1185

Giri A, Narasu MI (2000) Transgenic hairy roots: recent trends and applications. Biotechnol Adv 18:1–22

Guillon S, Tremouillaux-Guiller J, Pati PK, Rideau M, Gantet P (2006) Hairy root research: recent scenario and exciting prospects. Curr Opin Plant Biol 9:341–346

Hamill JD, Robins RJ, Rhodes MJC (1989) Alkaloid production by transformed root cultures of *Cinchona ledgeriana*. Planta Med 55:354–357

Hilton MG, Rhodes MJC (1990) Growth and hyoscyamine production of hairy root cultures of *Datura stramonium* in a modified stirred tank reactor. Appl Microb Biotechnol 33:132–138

Huffman GA, White FF, Gordon MP, Nester EW (1984) Hairy root-inducing plasmid: physical map and homology to tumor-inducing plasmid. J Bacteriol 157:269–276

Hughes EH, Hong SB, Gibson SI, Shanka JV, San KY (2004) Metabolic engineering of the indole pathway in *Catharanthus roseus* hairy roots and increased accumulation of tryptamine and serpentine. Metab Eng 6(4):268–276

Jaziri M, Shimomura K, Yoshimatsu K, Fauconnier ML, Marlier M, Homes J (1995) Establishment of normal and transformed root cultures of *Artemisia annua* L. for artemisin production. J Plant Physiol 145:175–177

Jeong GA, Park DH (2007) Enhanced secondary metabolite biosynthesis by elicitation in transformed plant root system. Appl Biochem Biotechnol 130:436–446

Jung KH, Kwak SS, Kim SW, Lee H, Choi CY, Lin JR (1992) Improvement of the catharanthine productivity in hairy root cultures of *Catharanthus roseus* by using monosaccharides as a carbon source. Biotechnol Lett 14:695–700

Jung H-Y, Kang S-M, Kang Y-M, Kang M-J, Yun D-J, Bahk J-D, Yang J-K, Choi M-S (2003) Enhanced production of scopolamine by bacterial elicitors in adventitious hairy root cultures of *Scopolia parviflora*. Enzym Microb Technol 33:987–990

Kamada H, Okamura N, Satake M, Harada H, Shimomura K (1986) Alkaloid production by hairy root cultures in *Atropa belladonna*. Plant Cell Rep 5:239–242

Khouri HE, Ibrahim RK, Rideau M (1986) Effects of nutritional factors on growth and production of anthraquinone glucosides in cell suspension cultures of *Cinchona succirubra*. Plant Cell Rep 5:423–426

Kieran PM, MacLoughlin PF, Malone DM (1997) Plant cell suspension cultures: some engineering considerations. J Biotechnol 59:39–52

Kim Y, Wyslouzil BE, Weathers PJ (2001) A comparative study of mist and bubble column reactors in the in vitro production of artemisinin. Plant Cell Rep 20:451–455

Kim Y, Wyslouzil BE, Weathers PJ (2002) Secondary metabolism of hairy root cultures in bioreactors. In Vitro Cell Dev Biol Plant 38:1–10

Kittipongpatana N, Hock RS, Porter JR (1998) Production of solasodine by hairy root, callus and cell suspension cultures of Solanum aviculare Forst. Plant Cell Tissue Organ Cult 52:133–143

Koblitz H, Koblitz D, Schmauder HP, Groger D (1983) Studies on tissue cultures of the genus *Cinchona* L. alkaloid production in cell suspension cultures. Plant Cell Rep 2:122–125

Komaraiah P, Reddy GV, Reddy PS, Raghavendra AS, Ramakrishna SV, Reddanna P (2003) Enhanced production of antimicrobial sesquiterpenes and lipoxygenase metabolites in elicitor treated hairy root cultures of *Solanum tuberosum*. Biotechnol Lett 25:593–597

Kukreja AK, Zehra M, Sultana A, Banerjee S (1997) Hairy roots of medicinal plants: Potential and prospects. In: Ravishankar GA, Venkataraman LV (eds) Biotechnological applications of plant tissue and cell culture. Oxford/IBH Publishing Co. Pvt. Ltd., New Delhi/Calcutta, pp 74–90

Kulkarni AA (2000) Micropropagation and secondary metabolites studies in *Taxus* spp. And *Withania somnifera* (l.) Dunal. Ph.D thesis, Plant Tissue Culture Division, National Chemical Laboratory, Pune

Kwok KH, Doran PM (1995) Kinetic and stoichiometric analysis of hairy roots in a segmented bubble column reactor. Biotechnol Prog 11:429–435

Lee KT, Suzuki T, Yamakawa T, Kodama T, Igarshi Y, Shimonura K (1999) Production of tropane alkaloids by transformed root culture of *Atropa belladonna* in stirred tank bioreactors with stainless steel net. Plant Cell Rep 18:567–571

Li W, Asada Y, Yoshikawa T (1998) Antimicrobial flavonoids from *Glycyrrhiza glabra* hairy root cultures. Planta Med 64:746–747

Liu KCS, Yang SH, Roberts MF, Phillipson JD (1990) Production of canthin-6-one alkaloids by cell suspension cultures of *Brucea javanica* (L.) Merr. Plant Cell Rep 9:261–263

Memelink J (2005) The use of genetics to dissect plant secondary pathways. Curr Opin Plant Biol 8:230–235

Morgan JA, Shanks JV (2000) Determination of metabolic rate limitations by precursor feeding in *Catharanthus roseus* hairy root cultures. J Biotechnol 79:137–145

Moyano E, Jouhikainen K, Tammela P, Palazon J, Cusido RM, Pinol MT, Teeri TH, Oksman-Caldentey KM (2003) Effect of pmt gene overexpression on tropane alkaloid production in transformed root cultures of *Datura metel* and *Hyoscyamus muticus*. J Exp Bot 54:203–211

Mukundan U, Hjortso MA (1990) Effect of fungal elicitor on thiophene production in hairy root cultures of *Tagetes patula*. Appl Microbiol Biotechnol 33:145–147

Nilsson O, Olsson O (1997) Getting to the root: the role of the *Agrobacterium-rhizogenes* –genes in the formation of hairy roots. Physiol Plant 100:463–473

Palazon J, Cusido RM, Bonfill M, Mallol A, Moyano E, Morales C, Pinol MT (2003a) Elicitation of different *Panax ginseng* transformed root phenotype for an improved ginsenoside production. Plant Physiol Biochem 41:1019–1025

Palazon J, Mallol A, Eibi R, Lettenbaur C, Cusido RM, Pinol MT (2003b) Growth and ginsenoside production in hairy root cultures of *Panax ginseng* using a novel bioreactor. Planta Med 69:344–349

Palazon J, Moyano E, Cusido RM, Bonfill M, Oksman-Caldentey K-M, Pinol MT (2003c) Alkaloid production in *Duboisia* hairy roots and plants overexpressing the h6h gene. Plant Sci 165:1289–1295

Pannuri S, Reddy GR, McNeill D, Curtis WR (1993) Interpreting the role of phosphorous and growth rate in enhanced fungal induction of sesquiterpenes from *Hyoscyamus muticus* root cultures. Appl Microbiol Biotechnol 38:550–555

Parr AJ (1988) Secondary products from plant cell culture. In: Mizrahi A (ed) Advances in biotechnology progress, Vol 9: Biotechnology in agriculture. Alan R Liss, New York

Pavlov A, Bley T (2005) Betalains biosynthesis by *Beta vulgaris* L. hairy root culture in different bioreactor systems. Scientific Works, University of Food Technologies 52: 299–304

Pavlov A, Bley T (2006) Betalains biosynthesis in *Beta vulgaris* L. hairy root culture in temporary immersion cultivation system. Process Biochem 41:848–852

Pita-Alvarez SI, Spollansky TC, Guilietti AM (2000) The influence of different biotic and abiotic elicitors on the production and profile of tropane alkaloids in hairy root cultures of *Brugmansia candida*. Enzym Microb Technol 26:252–258

Poornasri DB, Vimala A, Sai I, Chandra S (2008) Effect of cyanobacterial elicitor on neem cell suspension cultures. Indian J Sci Technol 1:1–5

Prajapati ND, Purohit SS, Sharma AK, Kumar T (2003) A handbook of medicinal plants. Agrobios, Jodhpur

Rao SR, Ravishankar GA (2002) Plant cell cultures: chemical factories of secondary metabolites. Biotechnol Adv 20:101–153

Rhodes MJC, Parr AJ, Giuletti A, Aird ELH (1994) Influence of exogenous hormones on the growth and secondary metabolite formation in transformed root cultures. Plant Cell Tissue Organ Cult 38:143–151

Schripsema J, Ramos Valdivia A, Verpoorte R (1999) Robustaquinones, novel anthraquinones from an elicited *Cinchona robusta* suspension culture. Phytochemistry 51:55–60

Scragg AH, Allan EJ (1986) Production of triterpenoid quassin and cell suspension cultures of *Picrasana quassioides* Bennett. Plant Cell Rep 5:356–359

Sevon N, Oksman-Caldentey KM (2002) *Agrobacterium rhizogenes*- mediated transformation: root cultures as a source of alkaloids. Planta Med 68:859–868

Shanks JV, Morgan J (1999) Plant hairy root culture. Curr Opin Biotechnol 10:151–155

Sharp JM, Doran PM (2001) Strategies for enhancing monoclonal antibody accumulation in plant cell and organ cultures. Biotechnol Prog 17:979–992

Sivakumar G, Yu KW, Hahn EJ, Paek KY (2005) Optimization of organic nutrients for ginseng hairy roots production in large scale bioreactors. Curr Sci 89:641–649

Srivastava S, Srivastava AK (2012) Azadirachtin production by hairy root cultivation of *Azadirachta indica* in a modified stirred tank reactor. Bioprocess Biosyst Eng 35(9):1549–1553

Staniszewska I, Krolicka A, Malinski E, Lojkowska E, Szafranek J (2003) Elicitation of secondary metabolites in in vitro cultures of *Ammi majus* L. Enzym Microb Technol 33:565–568

Sujanya S, Poornasri DB, Sai I (2008) *In vitro* production of azadirachtin from cell suspension cultures of *Azadirachta indica*. J Biosci 33:113–120

Taniguchi S, Imayoshi Y, Kobayashi E, Takamatsu Y, Ito H, Hatano T, Sakagami H, Tokuda H, Nishino H, Sugita D, Shimura S, Yoshida T (2002) Production of bioactive triterpenes by *Eriobotrya japonica* calli. Phytochemistry 59:315–323

Tepfer D (1984) Genetic transformation of several species of higher plants by *Agrobacterium rhizogenes*: phenotypic consequences and sexual transmission of the transformed genotype and phenotype. Cell 37:959–967

Thimmaraju R, Bhagyalakshmi N, Narayan MS, Ravishankar GA (2003) Kinetics of pigment release from hairy root cultures of *Beta vulgaris* under the influence of pH, sonication, temperature and oxygen stress. Process Biochem 38(7):1069–1076

Thimmaraju R, Bhagyalakshmi N, Ravishanker GA (2004) In situ and ex situ adsorption and recovery of betalains from hairy root cultures of *Beta vulgaris*. Biotechnol Prog 20:777–785

Thorup JE, McDonald KA, Jackman AP, Bhatia N, Dandekar AM (1994) Ribosome-inactivating protein production from *Trichosanthes kirilowii* plant cell cultures. Biotechnol Prog 10:345–352

Tzfira T, Li J, Lacroix B, Citovsky V (2004) *Agrobacterium* T-DNA integration: molecules and models. Trends Genet 20:375–383

Venkateswara R, Sankara Rao S, Vaidyanathan CS (1986) Phytochemical constituents of cultured cells of *Eucalyptus tereticornis* SM. Plant Cell Rep 3:231–233

Wallaart TE, Pras N, Quax WJ (1999) Isolation and identification of dihydroartemisinic acid hydroperoxide from *Artemisia annua*: a novel biosynthetic precursor of artemisinin. J Nat Prod 62:1160–1162

Weathers PJ, Cheetham RD, Follansbee E, Teoh T (1994) Artemisinin production by transformed roots of *Artemisia annua*. Biotechnol Lett 16:1282–1286

Weising K, Kahl G (1996) Natural genetic engineering of plant cells: the molecular biology of crown gall and hairy root disease. World J Microbiol Biotechnol 2:327–351

White FF, Taylor BH, Huffman GA, Gordon MP, Nester EW (1985) Molecular genetic analysis of the transferred DNA regions of the root inducing plasmid of *Agrobacterium rhizogenes*. J Bacteriol 164:33–44

Wijnsma R, Go JTKA, Van Weerden IN, Harkes PAA, Verpoorte R, Svendsen AB (1985) Anthraquinones as phytolexins in cell and tissue cultures of *Cinchona* sp. Plant Cell Rep 4:241–244

Wilhelmson A, Hakkinen ST, Kallio PT, Oksman-Caldentey K-M, Nuutila AM (2006) Heterologous expression of Vitreoscilla haemoglobin (VHb) and cultivation conditions affect the alkaloid profile of *Hyoscyamus muticus* hairy roots. Biotechnol Prog 22:350–358

Wilson PDG, Hilton MG, Robins RJ, Rhodes MJC (1987) Fermentation studies of transformed root cultures. In: Moody GW, Baker PB (eds) International conference on bioreactors and bio-transformations. Elsevier, London

Wu J, Wang C, Mei X (2001) Stimulation of taxol production and excretion in *Taxus* spp cell cultures by rare earth chemical lanthanum. J Biotechnol 85:67–73

Wysokinska H, Chmiel A (1997) Transformed root cultures for biotechnology. Acta Biotechnol 17:131–159

Xie D, Wang L, Ye H, Li G (2000) Isolation and production of artemisinin and stigmasterol in hairy root cultures of *Artemisia annua*. Plant Cell Tissue Organ Cult 63:161–166

Yan Q, Hu Z, Tan EX, Wu J (2005) Efficient production and recovery of diterpenoid tanshinones in *Salvia miltiorrhiza* hairy root cultures with in situ adsorption, elicitation, and semi-continuous operation. J Biotechnol 119(4):416–424

Zaki M, Sofi MS, Kaloo ZA (2011) A reproducible protocol for raising clonal plants from leaf segments excised from mature trees of *Betula utilis* a threatened tree species of Kashmir Himalayas. Int Multidiscip Res J 1(5):7–13

Zhao J, Davis LC, Verpoorate R (2005) Elicitor signal transduction leading to production of plant secondary metabolites. Biotechnol Adv 23:283–333

Precision Phenotyping for Mapping of Traits for Abiotic Stress Tolerance in Crops

7

Ratan Tiwari and H.M. Mamrutha

Abstract

In any plant breeding programme in an area of climate change, plant phenotyping is required to produce high-yielding crops. The success of a breeding programme is dependent on the precise measurement of phenotyping traits and the ability to correlate these with gene or quantitative trait locus (QTL) expression. However, precision in phenotyping is a major obstacle because of the complexity associated with environment and genotype interaction. Plant phenotyping traits and their importance vary depending on the breeding objective. Phenotyping under abiotic stress includes a set of physiological traits that indirectly explain the stress tolerance and susceptibility of the genotype. Although a large amount of information is available on the methods of estimating these traits, there is minimum information on the significance of these traits under abiotic stress conditions. This chapter focuses on the different physiological traits available for phenotyping and their underlying principles and importance under stress conditions, and highlights the new temperature-controlled phenotyping facility.

7.1 Importance of Plant Phenotyping

Phenotyping, i.e. the collection of phenotypic data for different traits of individual plant genotypes, is a prerequisite for any crop improvement research. Multi-location and annual information is generated following specific experimental planting designs for the analysis of different components of trait variations. This is required not only for making decisions about how to breed better crops for the future, but also for the identification of the genomic regions that confer improved traits in

R. Tiwari (✉)
Principal scientist (Biotechnology), Directorate of Wheat Research, Karnal 132 001, Haryana, India
e-mail: tiwari1964@yahoo.com

H.M. Mamrutha
Scientist (Plant Physiology), Directorate of Wheat Research, Karnal 132 001, Haryana, India

R.K. Salar et al. (eds.), *Biotechnology: Prospects and Applications*, DOI 10.1007/978-81-322-1683-4_7, © Springer India 2013

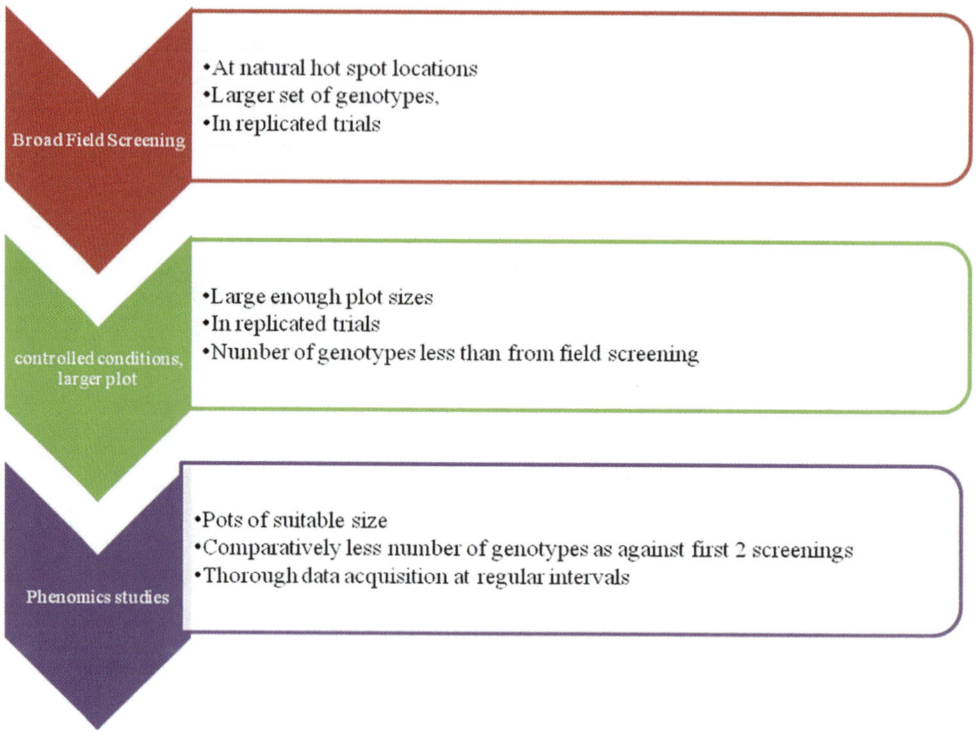

Fig. 7.1 Phenotyping approach for increased precision and application potential

the crop plants. Phenotypes may be due to single gene, or due to multiple genes, which can generate different phenotypic outcomes based on how they interact with each other and with the environment.

Crop improvement activities, via the creation of genetic variation by hybridization/mutation followed by selection of desirable genotypes to the green revolution, tremendously increased crop yields within decades. The next quantum jump in crop productivity, if it happens, will come from an integrated approach involving both genomics and phenomics. While newer generation genotyping techniques/tools are capable of capturing sequence variation at a required high throughput and resolution, phenotyping methodologies are limited. In view of the complex environmentally dependent plant responses, a comprehensive and interdisciplinary approach for phenotyping is urgently needed. This would require precision at all stages, including field preparation, laying out the experimental design and timely acquisition of and processing/analysis of data. This would hopefully culminate in a direct application to resolve the major problems currently limiting crop production (Zamir et al. 2013).

Global warming has enhanced the threat of terminal heat stress on winter crops such as wheat. In addition, increased variability and intensity of rainfall can significantly influence the spread and severity of diseases. Hence, strategies for climate change adaptation will be a major determinant for the sustainability of crop production in the future. One of the important components of such strategies is the genetically improved varieties that are resilient to thermal stress and resistant to diseases. Molecular approaches can complement the conventional approach for genetic enhancement of the stress tolerance in crops. Molecular mapping of the stress tolerance and disease resistance traits leading to identification of reliable molecular markers is an important step forward. Although the genotyping techniques have evolved quickly, our knowledge of the gene function is limited because of a lack of precision in large-scale plant phenotyping. As a result, the approach in plant phenotyping is to screen a large number of genotypes with a low-energy investment to generate more precise and intensive information capable of capturing even small and repeatable variations (Fig. 7.1).

7.2 Physiological Traits Associated with Precision Phenotyping

Phenotyping for abiotic stress tolerance can be broadly classified into two parts: shoot phenotyping and root phenotyping. Under shoot phenotyping, there are two levels: one at the plant level, which includes both destructive and non-destructive methods, and the other at the tissue/cellular level. Root phenotyping methods are classified into in situ (in the field) (Prior et al. 2004) and ex situ (in the pot). Different physiological traits and their importance and significance for drought stress tolerance are discussed below.

7.2.1 Non-destructive Traits at the Plant Level

7.2.1.1 Canopy Temperature

Canopy temperature has been extensively used in wheat breeding programmes for genotype screening under heat and drought stress, but this has its own advantages and disadvantages. A plant canopy emits long-wave infrared radiation as a function of its temperature. The infrared thermometer senses this radiation and converts it into an electrical signal that is displayed as temperature. This property of a plant canopy is used as a selection criterion under drought and heat stress because genotypes with a low canopy temperature are able to extract moisture from a deeper layer of soil or by minimising the stomatal water loss. There are several studies indicating this as a promising trait for screening a large number of genotypes under drought (Balota et al. 2008; Olivares et al. 2007). As it is non-destructive, it can be measured quickly under field conditions, covers larger genotypes in the field and indirectly allows for the selection of genotypes with better water use, deep roots and stomatal conductance under abiotic stress. There are different types of infra-red thermometers available. The preferred time for observation is at or just after noon to get maximum differences; atmospheric conditions during measurements should be stable and the sun should be at the back of the operator during observational recording. The angle of measurement should be constantly maintained so that it covers maximum leaf canopy and avoids a change in soil temperature. Frequent checking as to the reliability of the instrument is crucial.

7.2.1.2 Leaf Chlorophyll Content

The chlorophyll pigment absorbs all colours of light and as it reflects green colour it is green in colour. The canopy greenness is directly related to photosynthetic efficiency of the plants. One of the traits for genotype screening under drought and heat stress is the measurement of leaf chlorophyll content by optical sensing. The chlorophyll content can be estimated by the chemical dimethyl sulphoxide (DMSO):acetone extraction method, but measurement by optical sensing is more relevant under field studies (Dwyer et al. 1991). The instrument used is called the chlorophyll meter and the measurement taken is the chlorophyll content index (CCI). The CCI values vary depending on the type of chlorophyll meter used; however, the values are used in relative terms and observations can be taken at any time of the day.

7.2.1.3 Photosynthesis Using Portable Photosynthesis Meters

Although the difference in the rate of photosynthesis between the control and stress samples can be indirectly identified by chlorophyll content, the exact photosynthetic rate of plants can also be estimated. All heteroatomic gas molecules including CO_2 have a characteristic absorption spectrum at the infra-red region. CO_2 has its major absorption peak at 4.25 μm with secondary peaks at 2.66, 2.77 and 14.99 μm. In the photosynthesis meter, the difference in the amount of CO_2 pumped in and moving out of the closed leaf chamber is measured and the difference is used to find the amount of CO_2 fixed by photosynthesis using an infra-red gas analyser (Nataraja and James 1999). The higher the rate of photosynthesis, the higher the tolerance mechanism of the genotype under stress. The photosynthesis estimation is costly, time consuming and sensitive to atmospheric conditions. Hence, it is preferably used in high-precision phenotyping of a small number of genotypes rather than in a large population. There are different photosynthesis meters available; Licor is the most commonly used portable photosynthesis system. Photosynthesis should be measured at noon or prior or after noon on a bright sunny day to avoid errors, and it should be recorded at uniform positions in the leaf.

7.2.1.4 Chlorophyll Fluorescence (CFL)

CFL is one of the traits used extensively in field phenotyping after canopy temperature and chlorophyll content. It is used indirectly for measuring photosynthetic efficiency of the genotypes, mainly in terms of photo system II (PSII) function. CFL measures the Fv/Fm ratio, i.e. immediately after dark adaptation when the leaf is exposed to light. The maximum amount of photons used for photochemistry is estimated as the ratio of Fv/Fm in Fv-variable fluorescence and Fm-maximal fluorescence. When photons fall on the leaf surface, they have been dissipated mainly into two processes, i.e. photochemical quenching in the form of photosynthesis and non-photochemical quenching in the form of heat and fluorescence. Under non-stress conditions with maximum photon use for photochemistry, the ratio of Fv/Fm obtained is 0.79 to 0.84. Dark adaptation of the leaf, before CFL measurement is a prerequisite mainly to send all the electrons of PSII to photo system I for CO_2 fixation, so that PSII will be competent to capture maximum photons when exposed to light. A minimum of 15–20 min of dark adaptation is required for CFL measurement. The observation timing should be strictly followed as that of photosynthesis measurement. When the plant is stressed, the PSII efficiency will be reduced and hence the ratio will be of less value compared with tolerant genotypes (Kate and Giles 2000). Photosynthesis measurement using infrared gas analyzer (IRGA) under field conditions is a laborious process, hence, CFL measurement is used as a common process for assessing photosynthetic variability in wheat breeding programmes. A minimum of three to four readings should be taken in each treatment to assess variation.

7.2.1.5 Normalized Difference Vegetation Index (NDVI) and Digital Imaging

The green leaves absorb light in a visible region and emit it in the form of long-wave radiation at the near-infra-red region. NDVI measures light at the near-infrared region and in turn reveals the canopy coverage. The higher the canopy/vegetation, the higher the reflectance and NDVI value. This parameter is used in wheat breeding programmes to assess the variability in early vigour, ground cover, biotic/abiotic stress effect, senescence and nutrient use efficiency (nitrogen) (Verhulst and Govaerts 2010a). Alternatively, the digital picturing of the canopy can be done at particular height from the ground level for all the testing genotypes and pictures can be analysed with the available software (Adobe Photoshop CS3 extended or later version) to assess the early groundcover (Mullan and Reynolds 2010). The leaf area index and the green area index are two other parameters used for assessing green canopy cover in the field. Where portable NDVI machines are available, the machine should be held at the same distance from the ground level across the genotypes. Readings can be taken at any time of the day if the machine has its own light source, otherwise it should be taken preferably between 11.00 and 14.00 h. NDVI values *are typically in the* range of 0–1; zero represents no greenness and one represents maximum greenness.

7.2.1.6 Phenotyping for Crop Growth Stages

The different crop growth stages of wheat are defined by Zadoks (Zadoks et al. 1974) and are known as the Zadoks scale. The variation in time duration for attaining different crop growth stages under stress will also determine the variation for tolerance and susceptibility of the genotypes. In wheat crops under drought and heat stress, variabilities among the genotypes for heading days, anthesis days, grain filling duration and crop maturity were reported, which in turn reflects the variation in final crop yield among the genotypes.

7.2.1.7 Soil Moisture Content

Estimating soil moisture content is crucial to assess the potential of the genotypes under irrigated and rain-fed conditions. Random soil samples should be collected in both control and stress fields at different depths (0–30, 30–60 cm) and covered in aluminium foil to prevent moisture loss. The samples should be taken to the laboratory to measure the moisture levels either by the gravimetric method or with a potentiometer (Dew-point potentiometer-WP4) (Pearcy et al. 1989). The in situ soil moisture can also be measured with a tentiometer or with soil moisture probes. While taking the sample in the field, care should be taken to collect more soil samples and also to cover the entire field.

7.2.2 Destructive Traits

7.2.2.1 At Whole Plant Level

After crop harvesting, the whole plant weight, i.e. biological yield (excluding grain) and economic yield (grain yield), was taken under control and stress conditions. Later, the harvest index (HI) (economic yield/(biological + economic yield)) was estimated. There is a large variation among the genotypes for HI. The genotypes with high HI under stress are tolerant genotypes.

7.2.2.2 At the Tissue Level

Leaf Water Potential (LWP): The LWP measures the available free water in the leaf tissues for performing physiological functions. The scholander pressure chamber/pressure bomb is most commonly used for this estimation (Turner 1988). Under irrigated conditions, the value varies from −5 to −10 bars, whereas under drought conditions, the values are −20 to −40 bars. As LWP estimation is a costly and slow process, this trait is used less often in the breeding programme.

Leaf Relative Water Content (RWC): RWC is one of the easiest and most routinely used parameters to address the variation in water content among genotypes under drought and heat stress. RWC measures the capacity of a genotype to maintain a turgid state under stress for the smooth functioning of stomata and photosynthetic processes. Depending on the nature of the leaf sample, either leaflets/leaf disc/leaf pieces are collected before noon, fresh weight is recorded and the same is put in water overnight to record turgid weight (TW). The dry weight (DW) of the sample is recorded by oven drying the sample at 80 °C. The RWC is estimated using the following formula (Barr and Weatherley 1962)

$$\text{RWC}(\%) = (\text{FW-DW})/(\text{TW-DW})*100.$$

In a drought-tolerant turgid leaf, the RWC value is in the range of 95–98 % and under severe drought conditions it is 40 %.

Cell Membrane Integrity Test: All types of stress damage cell membrane integrity, resulting in the leakage of electrolytes (ions) from cells. These leaked ions usually have charges and these are measured by an electrical conductivity (EC) meter. Under stress conditions, there are genotypes that maintain membrane integrity; this genotypic variation is exploited for stress tolerance during breeding programmes (Blum and Ebercon 1981). This trait needs a simple instrument and is also commonly used in breeding programmes after RWC. The EC measured is directly related to membrane damage caused by stress. Weights of stressed and control samples are collected, washed in distilled water, blotted on Whatman filter paper and then incubated in 25 mL of water with continuous shaking for 2 h. Initial EC is taken using an EC meter. The leaf discs are then boiled for 30 min and then the final EC is taken. The ion leakage is computed using the formula.

$$\text{Leachate}(\%) = (\text{Initial EC/Final EC})*100$$

Osmotic Potential: Under stress conditions, the accumulation of solute molecules, mainly sugars, amino acids, organic acids, polyols and quaternary amines in the plant cell, increases the osmotic potential of the cell. This solute molecule protects the enzymes and membrane molecules from degradation due to dehydration and thus helps in the smooth functioning of cell physiological processes (Morgan 1983). Large variation has been observed across the genotypes and species for this trait. One gram of leaf tissue from both stress and control samples is taken and frozen in liquid nitrogen and thawed. The thawed tissue is placed in Eppendorf with a hole at the bottom. The tube with the tissue is placed in another tube and centrifuged at 4,300 g for 5 min to collect the cell sap, and the osmotic potential is measured using a vapour pressure osmometer. Sampling can be done at any time during the day for estimation and it is expressed in mega pascals.

Carbon Isotope Discrimination: The atmosphere has two forms of carbon: ^{12}C (98.9 %) and ^{13}C (1.1 %). The crop species during its CO_2 uptake from stomata discriminates for ^{12}C and ^{13}C. ^{12}C being a lighter molecule is more preferred than ^{13}C. This kind of discrimination is more prominent in C_3 crops than C_4 crops. The ratio of $^{13}C/^{12}C$ is compared with respect to the standard limestone Peedee Belemnite and the unit of expression is permil (‰), which is always negative. The ratio is in the range of −33 to −24‰ for C_3 plants and −17 to −9‰ for C_4 plants. Using this ratio, delta $(\Delta)^{13}C$ is calculated and is always positive in nature. It has been proven that $\Delta^{13}C$ is inversely related to water use efficiency (Farquar et al. 1989). This relationship is used in genotype screening and showed that high water use-efficient genotypes with low $\Delta^{13}C$ values perform better under drought. The dried leaf/grain sample at any stage is used for analysis. Analysis is normally outsourced from the central facilities available. As it is costly and time consuming, this trait is used less in breeding programmes.

7.3 Phenotyping Facility for Screening Heat Stress-Tolerant Genotypes

Several techniques and structures have been used throughout the world to evaluate crop genotypes for heat stress tolerance. However, there is often a lack of effective structures with controlled environments for in situ evaluation of genotypes for high-temperature tolerance. One option is a structure that can open and close the roof and side walls to expose the crop to high-temperature stress at desired crop growth stages. This approach, coupled with precise planting methods, ensures equal treatment for each plant and thereby reduces phenotyping errors and has the potential to identify very valuable germplasm. Automated image-based characterization of the germplasm under innovative facilities/techniques as well as an appropriate statistical layout is expected to improve the quality of phenotypic data essential for efficient molecular mapping programmes.

Novel facilities need to be created for screening genotypes in the larger plot size (as is in the fields) as well as under controlled conditions with a precise temperature-controlled facility. The structure should support natural growth conditions prevailing outside during initial growth stages, and hence no extra expenditure as compared with the field conditions should be required. However, it should have facilities to be instantly converted into a controlled facility, thereby ensuring stress treatment for the required duration at a specific crop stage, as and when required. The variation in specific growth conditions should be very precise and

Fig. 7.2 Precision phenotyping facility at Directorate of Wheat Research, Karnal, Haryana, India

better linked to the ambient conditions. Hence, efforts are needed to establish a phenotyping facility and efficient techniques to screen genotypes for heat tolerance.

A unique structure, on these lines, has been developed at the Directorate of Wheat Research, Karnal. This temperature-controlled phenotyping facility provides an opportunity for the national wheat programme to precisely identify genotypes under climate change, particularly for abrupt temperature variation at different growth stages (Fig. 7.2).

It is believed that high-throughput and precise phenotyping mechanisms will help crop scientists improve genomic sciences toward higher crop yields, thereby providing food security to millions.

References

Balota M, Payne WA, Evett SR, Peters TR (2008) Morphological and physiological traits associated with canopy temperature depression in three closely related wheat lines. Crop Sci 48(5):1897–1910

Barr HD, Weatherley PE (1962) A re-examination of the relative turgidity technique for estimating water deficit in leaves. Aust J Biol Sci 15:413–428

Blum A, Ebercon A (1981) Cell membrane stability as a measure of drought and heat tolerance in wheat. Crop Sci 21:43–47

Dwyer LM, Tollenaar M, Houwing L (1991) A nondestructive method to monitor leaf greenness in corn. Can J Plant Sci 71:505–509

Farquar D, Ehleringer JR, Hubick KT (1989) Carbon isotope discrimination and photosynthesis. Annu Rev Plant Physiol Plant Mol Biol 40:503–537

Kate M, Giles NJ (2000) Chlorophyll fluorescence: a practical guide. J Exp Bot 51:659–668

Morgan J (1983) Osmoregulation as a selection criterion for drought tolerance in wheat. Aust J Agric Res 34:607–614

Mullan DJ, Reynolds MP (2010) Quantifying genetic effects of ground cover on soil water evaporation using digital imaging. Funct Plant Biol 37:703–712

Nataraja KN, James J (1999) Clonal differences in photosynthesis in Hevea brasiliensis. Photosynthetica 36(1–2):89–98

Olivares-Villegas JJ, Reynolds MP, Mc Donald GK (2007) Drought adaptive attributes in the Seri-Babax hexaploid population. Funct Plant Biol 34:189–203

Pearcy RW, Ehleringer JR, Mooney HA, Rundel PW (1989) Plant physiological ecology: field methods and instrumentation. Chapman and Hall, London/New York, 457 pp

Prior SA, Runion GB, Tobert HA, Erbach DC (2004) A hydraulic coring system for soil-root studies. Agron J 96:1202–1205

Turner NC (1988) Measurement of plant water status by the pressure chamber technique. Irrig Sci 9:289–308

Verhulst N, Govaerts B (2010) The normalized difference vegetation index (NDVI) Green Seeker handheld sensor: toward the integrated evaluation of crop management. Part A: Concepts and case studies. CIMMYT, Mexico City

Zadoks JC, Chang TT, Konzak CF (1974) A decimal code for growth stages of cereals. Weed Res 14:415–421

Zamir D (2013) Where have all the crop phenotypes gone? PLoS Biol 11(6):e1001595. doi:10.1371/journal.pbio.1001595

Part II

Microbial Biotechnology

Physiology of Free and Immobilised Brewer's Yeast in Stress Conditions

8

D. Smogrovicova

Abstract

An increase in substrate concentration in ethanol fermentation results in higher metabolic activity and increased ethanol production; however, during fermentation of concentrated substrates, the yeasts are exposed to extreme conditions and show differences in metabolism accompanied with many negative consequences. A significant increase of fermentation rate of VHG wort for entrapped yeasts in contrast to free yeast cells was observed. The specific rate of ethanol production of yeast immobilised in calcium pectate in wort of concentration 24 % (w/w) was at the level of the specific rate of ethanol production of free yeast and yeast adsorbed on cellulose in wort of concentration 16 %. An increase in wort gravity resulted in significantly higher trehalose content in free cells, whereas trehalose content in entrapped cells did not increase with increasing initial wort gravity up to the concentration of 24 %. Higher degree of saturation of total fatty acids in the entrapped fermenting yeast compared to the free cells under VHG fermentation conditions correlated positively with ethanol tolerance and improved fermentation rate. Beer produced from VHG wort by entrapped yeasts had a suitable diacetyl concentration and higher alcohols-to-esters ratio, as opposed to beer produced by free yeasts.

8.1 Introduction

Very high gravity (VHG) fermentation technology has become popular in distillers and biofuel alcohol production, as well as in brewing industry due to a reduced capital cost and energy in the brew house, increased production capacities without making costly investments and improved quality of beer – flavour and haze stability. Very high gravity technology means brewing of beer with an increased original gravity from concentrated wort – having a dissolved solids concentration of more than 16 g per 100 g and subsequent dilution of fermented beer with carbonated water to a prescribed original gravity or alcohol concentration (Thomas et al. 1996; Smogrovicova et al. 2000; Pradeep et al. 2011).

D. Smogrovicova (✉)
Department of Biochemical Technology, Faculty of Chemical and Food Technology, Slovak University of Technology in Bratislava, 812 37 Bratislava, Slovakia
e-mail: daniela.smogrovicova@stuba.sk

R.K. Salar et al. (eds.), *Biotechnology: Prospects and Applications*,
DOI 10.1007/978-81-322-1683-4_8, © Springer India 2013

8.2 Yeast Stress Factors and Factors Influencing Alcoholic Fermentation

Brewer's yeast in general can tolerate only about 7–9 % of ethanol, quantities that are usually produced during the fermentation of wort ranging the gravity of 16–20 % and osmotic pressure of 1.3–3 MPa. During fermentation of concentrated substrates, the yeasts are exposed to extreme conditions and show differences in the metabolism accompanied with many negative consequences. An increase in the wort gravity results in a higher metabolic activity and increased ethanol production, but at a certain concentration of dissolved solids in the medium, the effect of osmotic pressure, toxicity of the produced ethanol, nutrient limitation and carbon dioxide concentration cause a growth inhibition and decrease in the yeast viability, rate and extent of fermentation as well as changes in the yeast cell composition.

There is some controversy as to which is the most detrimental factor to the yeast during very high gravity fermentations. Some authors have reported that ethanol toxicity is the main cause of poor performance of yeast under those conditions. Others have reported that nutrient limitation or a non-optimised supply of nutrients is the main cause (particularly dissolved oxygen and assimilable nitrogen). The main sites for ethanol effects in yeast are cellular membranes, hydrophobic and hydrophilic proteins and the endoplasmic reticulum (Casey and Ingledew 1986; Ciesarova et al. 1998; Birch and Walker 2000; Smogrovicova et al. 2000; Devantier et al. 2005).

8.3 Improvement of Alcoholic Fermentation

An improvement in the alcoholic fermentation and better fermentative activities are achievable by nutrient supplementation and by the addition of components, which microorganisms synthesise in response to stress.

Assimilable nitrogen is an important component of fermentation media and it plays a significant role in the course of fermentation; it is necessary for growing and multiplication and influences ethanol tolerance of yeasts and the rate of ethanol production. The addition of free amino nitrogen leads to higher final ethanol concentrations in the fermented media (Thomas et al. 1996; Bafrncova et al. 1999).

Trehalose and glycogen are the main reserve carbohydrates of yeast; however, trehalose is also synthesised in response to stress and may act as a general stress protectant for yeasts. High levels of trehalose in yeast cells have been associated with increased osmotic, thermic and ethanol stresses, limitation of assimilable nitrogen and cold shock (Bafrncova et al. 1999; Patkova et al. 2000).

Immobilised cells depending on the carrier used for immobilisation show various modifications in the physiology, morphology, biochemical composition and metabolic activity. A number of authors have published an enhanced ethanol tolerance and production of ethanol using immobilised cells compared to free cells; however, several authors have found a decreased ethanol production using immobilised cells (Norton et al. 1995; Ciesarova et al. 1998; Patkova et al. 2000; Smogrovicova et al. 2000; Kourkoutas et al. 2004; Pradeep et al. 2011).

In conventional brewing, the primary fermentation is considered to be complete, after 60 % of wort extract is fermented. This degree of wort attenuation in wort concentrations from 12 to 22 % (w/w) was reached (Table 8.1). However, in wort over 22 % (w/w), the fermentation ceased prematurely, reaching the attenuation only of 40 %. The maximum rate of ethanol production increased with increasing initial wort gravity up to 24 % (w/w), reaching the ethanol concentration of 6.22 % (v/v). Every following enhancement in wort gravity resulted in a decrease of the maximum rate of ethanol production. At wort concentration 30 % (w/w), the final ethanol concentration fell to 5.33 % v/v, the same value achieved at 12 % (w/w) wort fermentation (Patkova et al. 2000).

The specific rates of ethanol production (Table 8.2) of free yeast cells and cells immobilised on DEAE-cellulose were very similar at all concentrations of wort and were reduced, as compared to yeast cells immobilised in other carriers. The

Table 8.1 Parameters of wort fermentation of different gravity using free yeast cells, 13 °C

Wort concentration [% (w/w)]	12	20	22	24[a]	28[a]	30[a]
Final ethanol concentration (% v/v)	5.33	6.03	6.14	6.22	5.40	5.33
Final cell concentration (10^{-7} cells/ml)	2.8	3.5	3.8	3.9	3.2	3.0
Max. rate of ethanol production (g/l.h)	0.56	0.61	0.65	0.76	0.60	0.54
Specific rate of ethanol production (g/g.h)	0.08	0.09	0.08	0.10	0.10	0.09
Cell viability after fermentation (%)	74	70	67	61	50	54

[a]Fermentations were not complete

Table 8.2 Specific rate of ethanol production (*rp*) in wort fermentation of different gravity, 15 °C

Wort concentration (% w/w)	12	16	20	24
Yeast immobilisation carrier	Specific rate of ethanol production (g/l.h.g)			
Free yeast cells	0.051	0.049	0.047	0.044
DEAE-cellulose	0.051	0.047	0.046	0.043
κ-Carrageenan	0.054	0.051	0.048	0.046
Calcium pectate	0.054	0.052	0.052	0.048
Calcium alginate	0.053	0.051	0.051	0.047
Agar	0.054	0.051	0.047	0.045

rp from function d*P*/d*t* = *f*(*t*), *P* product, *t* time

specific rate of ethanol production of yeast immobilised in calcium pectate in 24 % wort (0.048 g/l.h.g) was at the level of the specific rate of ethanol production of free yeast (0.049 g/l·h·g) and yeast adsorbed on DEAE-cellulose (0.047 g/l.h.g) in wort of 16 % concentration (Smogrovicova et al. 2000).

There are differences between different types of cell binding. Yeast cells were immobilised either on the surface of cellulose, or entrapped in calcium pectate, alginate and κ-carrageenan. It is assumed that in the microenvironment of a gel

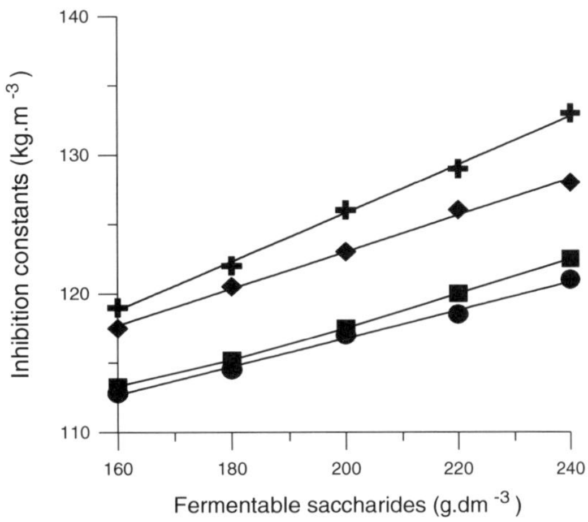

Fig. 8.1 Effect of increase in wort gravity on product inhibition constants (K_p) in wort fermentation (original extract of 22–30 % w/w, fermentable saccharides of 160–240 g/l) using free and immobilised yeast cells. ✚ free yeasts, ◆ κ-carrageenan, ■ calcium alginate, ● calcium pectate. K_p – see Eq. (8.1)

matrix, the ethanol stress is lower and the cells can produce ethanol more intensively. When the gel matrix contained calcium cations, the ethanol tolerance was improved even more.

The higher protective effect of calcium-contained carriers is evident from Fig. 8.1. We used a model sufficient for the study of ethanol fermentation kinetics suggested by Ghose and Tyagi (1979). This model follows the Monod-type limitation of biomass growth on sugar substrate and includes an inhibition term which predicts a linear decline in the specific growth rate with the ethanol concentration (Eq. 8.1):

$$\mu = \frac{\mu_m c_s}{K_s + c_s}\left(1 - \frac{c_p}{K_p}\right) \quad (8.1)$$

μ – specific growth rate
μ_m – maximum specific growth rate
c_s – substrate concentration
c_p – ethanol concentration
K_s – Monod constant
K_p – product inhibition constant

In Fig. 8.1 there is an effect of increase in wort gravity on product inhibition constants (K_p) in wort fermentation of concentrations from 22 % w/w (content of fermentable saccharides 160 g/l) to 30 %

Table 8.3 Degree of unsaturation (IU) of lipids in wort fermentation of different gravity

Wort concentration (% w/w)	12	16	20	24
Yeast immobilisation carrier	Index of unsaturation (IU)			
Free yeast cells	0.51	0.57	0.64	0.70
Cellulose	0.52	0.59	0.67	0.71
κ-Carrageenan	0.48	0.51	0.56	0.60
Calcium pectate	0.47	0.53	0.56	0.58
Calcium alginate	0.47	0.54	0.55	0.58
Agar	0.49	0.51	0.57	0.59

w/w (content of fermentable saccharides 240 g/l) using free or immobilised yeast cells. These ethanol inhibition constants were significantly lower when yeast cells were entrapped in calcium pectate or calcium alginate compared to the κ-carrageenan carrier. The value of the constant for wort fermentation of concentration 28 % w/w (content of fermentable saccharides 220 g/l) was 118 kg/m^3 using yeast immobilised in calcium pectate. On the other hand, the value of the constant for wort fermentation of concentration 22 % w/w (content of fermentable saccharides 160 g/l) was 119 kg/m^3 using free yeast (Smogrovicova et al. 2000).

As ethanol toxicity affects mainly the cell membrane and the yeast responds by altering the lipid bilayer, we also compared the influence of immobilisation on fatty acid composition of yeast lipids in fermenting yeast. The degree of unsaturation (IU) of a lipid can be simply quantified using the unsaturation index (Kates and Baxter 1962).

In Table 8.3 there are unsaturation indexes of lipids in free and immobilised yeast cells after batch fermentation of worts of increased gravity. Alginate and pectate immobilised cells contained higher percentages of saturated fatty acids compared to free cells and to cells immobilised on DEAE-cellulose.

The major fatty acids were found to be palmitic (16:0), palmitoleic (16:1), stearic (18:0) and oleic (18:1) acids. Increasing the concentration of produced ethanol decreased the relative percentage of saturated fatty acids, especially palmitic and stearic, and increased the content of monounsaturated fatty acids – palmitoleic and oleic acids, more in free and on DEAE-cellulose immobilised cells than in yeast entrapped in calcium

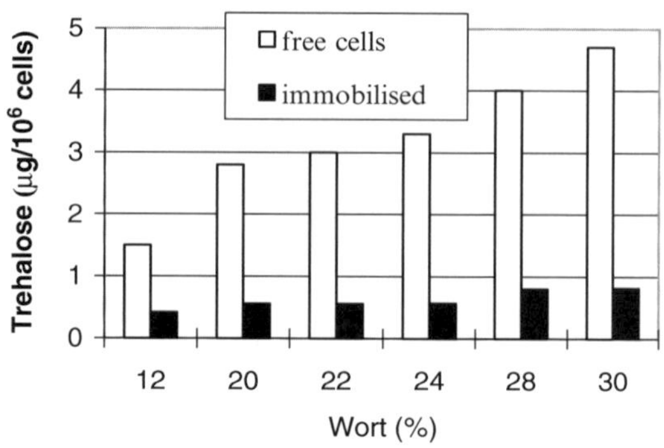

Fig. 8.2 Maximum trehalose content in free and calcium-alginate-entrapped yeast cells in wort fermentation (original extract of 12–30 % w/w)

pectate, calcium alginate or κ-carrageenan. Lower unsaturation index correlated with increased rate of fermentation and ethanol tolerance of yeasts immobilised in gels (Smogrovicova et al. 2000).

Trehalose and glycogen are the main reserve carbohydrates in yeast. However, trehalose is also synthesised in response to stress and may act as a general stress protectant for yeasts. High levels of trehalose in yeast cells have been associated with increased osmotolerance, thermotolerance and ethanol tolerance (Patkova et al. 2000).

In calcium-alginate-entrapped yeast, the trehalose content did not increase with increasing original wort gravity up to concentration of 24 % (w/w) and only slightly in 28 and 30 % (w/w) worts (Fig. 8.2). In free yeasts, an increase in wort gravity resulted in significantly higher cellular trehalose content, contrary to cells immobilised in calcium alginate. This confirmed that the gel matrix protects the cell also against osmotic and not only against ethanolic stress.

8.4 Beer Flavour Under High Gravity Fermentation

Yeast metabolism makes an important contribution to beer flavour. Immobilisation may lower the stress and increase the productivity of fermentation; however, it also influences the balance of sensory compounds to create an acceptable flavour profile of beer (Smogrovicova and Domeny 1999; Branyik et al. 2005; Verbelen et al. 2006; Willaert and Nedovic 2006).

Table 8.4 Primary wort fermentation in continuous "gas-lift" reactor

Beer volatiles (mg/l)	Ca-pectate-entrapped yeasts 24 % wort	Ca-pectate-entrapped yeasts 12 % wort	Free yeasts 24 % wort	Free yeasts 12 % wort
Diacetyl	0.43	0.48	0.61	0.68
Alcohols	128.0	104.3	124	101.5
Esters	31	22.5	36.3	21.8
Alcohols/esters ratio	4.12:1	4.63:1	3.41:1	4.65:1

Beers after dilution on original extract concentration of 12 %. Residence times: 24 % wort – immobilised yeasts 27 h, free 56 h; 12 % wort – immobilised yeasts 13 h, free 21 h
The optimum higher alcohols-to-esters ratio: *4.1 to 4.7:1*

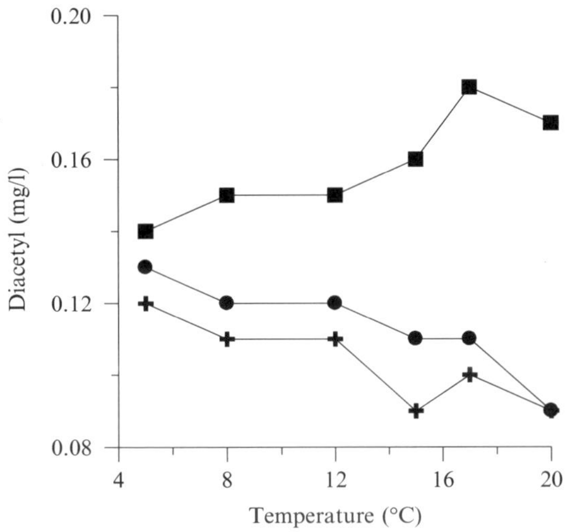

Fig. 8.3 Diacetyl concentration in beers produced by batch fermentation of wort as a function of temperature. ✚ PVAL, ● calcium alginate, ■ free cells. K_p – see Eq. (8.1)

Entrapped yeasts produced beers with lower diacetyl concentration than the control did. Figure 8.3 shows that yeast entrapped in both calcium alginate beads and PVAL produced beers with lower diacetyl concentration than the control (free yeast cells) and the concentration of diacetyl decreased as the temperature was increased. In contrast, the concentration of diacetyl increased with increasing temperature with free yeast. We reported similar results using calcium pectate or κ-carrageenan (Smogrovicova and Domeny 1999; Smogrovicova et al. 2001).

In beers produced in VHG fermentation (24 %), the content of esters and alcohols was higher, compared to normal gravity fermentation (12 %) in both cases – using free and immobilised yeasts. However, the flavour of a beer depends not only on the content of its compounds but also on their ratio (Polednikova et al. 1993). Beer produced from VHG wort by yeasts entrapped in calcium pectate had suitable higher alcohols-to-esters ratio, contrary to beer produced by free control yeasts (Table 8.4).

Primary beer fermentation with the immobilised cells was reported to produce excessive amounts of diacetyl, which has a "cheese-like" aroma and taste with very high off-flavour potential. This fact was not confirmed in our beers (Table 8.4, Fig. 8.3).

8.5 Conclusions

A significant increase of fermentation rate of VHG wort for entrapped yeasts in contrast to free yeast cells was observed. The results confirm that immobilisation protects the microbial cells against possible toxic effect of substrates or products, whereby entrapment in gels appears to be more favourable than adsorption (DEAE-cellulose). If the gel matrix contained calcium cations, the ethanol tolerance was even more improved.

According to our data, all carriers tested for immobilisation were suitable for beer fermentation. The behaviour of yeast immobilised by adsorption on DEAE-cellulose was very similar to that of free yeast at all temperatures, while entrapped yeast showed metabolic differences. The aroma and flavour of beers produced by yeast entrapped in calcium pectate or κ-carrageenan at temperatures around 15 °C were similar to beers produced at lower temperatures using free yeast.

Seven of ten tasters considered beer produced using calcium-pectate-entrapped or PVAL-entrapped yeast cells at 15 °C in our continuous system to be comparable to a beer produced by classical fermentation technology. This fact has high practical significance, because fermentation at higher temperatures is much cheaper and the process using immobilised cells in continuous fermentation can be carried out with significantly reduced residence times (Smogrovicova et al. 1997, 1998).

Acknowledgements This work was supported by the Scientific Grant Agency of the Ministry of Education of the Slovak Republic and the Academy of Sciences, registration number 1/0096/11.

References

Bafrncova P, Smogrovicova D, Slavikova I, Patkova J, Domeny Z (1999) Improvement of VHG ethanol fermentation by media supplementation using *Saccharomyces cerevisiae*. Biotechnol Lett 21:337–341

Birch RM, Walker GM (2000) Influence of magnesium ions on heat shock and ethanol stress responses of *Saccharomyces cerevisiae*. Enzyme Microb Technol 26:678–687

Branyik T, Vicente AA, Dostalek P, Teixeira JA (2005) Continuous beer fermentation using immobilized yeast cell bioreactor systems. Biotechnol Prog 21:653–663

Casey GP, Ingledew WM (1986) Ethanol tolerance in yeasts. CRC Crit Rev Biotechnol 13:219–280

Ciesarova Z, Domeny Z, Smogrovicova D, Patkova J, Sturdik E (1998) Comparison of ethanol tolerance of free and immobilized *Saccharomyces uvarum* yeasts. Folia Microbiol 43:55–58

Devantier R, Scheithauer B, Villas-Boas SG, Pedersen S, Olsson L (2005) Metabolite profiling for analysis of yeast stress response during very high gravity ethanol fermentations. Biotechnol Bioeng 90:703–714

Ghose TK, Tyagi RD (1979) Rapid ethanol fermentation of cellulose hydrolyzate. II. Product and substrate inhibition and optimization of fermentor design. Biotechnol Bioeng 21:1401–1420

Kates M, Baxter RM (1962) Lipid composition of mesophilic and psychrophilic yeasts (*Candida species*) as influenced by environmental temperature. Can J Biochem Physiol 40:1213–1227

Kourkoutas Z, Bekatorou A, Banat IM, Marchant R, Koutinas AA (2004) Immobilization technologies and support materials suitable in alcohol beverages production. Food Microbiol 21:377–397

Norton S, Watson K, D'Amore T (1995) Ethanol tolerance of immobilized brewers' yeast cells. Appl Microbiol Biotechnol 43:18–24

Patkova J, Smogrovicova D, Bafrncova P, Domeny Z (2000) Changes in the yeast metabolism at very high gravity wort fermentation. Folia Microbiol 45:335–338

Polednikova M, Voborsky J, Chladek L, Sruma T (1993) Zkusenosti svyrobou piva pomoci imobilizovanych kvasinek. Kvas Prumysl 39:2–7

Pradeep P, Smogrovicova D, Reddy OVS, Ko S (2011) Very high gravity (VHG) ethanolic brewing and fermentation: a research update. J Ind Microbiol Biotechnol 38:1133–1144

Smogrovicova D, Domeny Z (1999) Beer volatile by-product formation at different fermentation temperature using immobilised yeasts. Process Biochem 34:785–794

Smogrovicova D, Domeny Z, Gemeiner P, Malovikova A, Sturdik E (1997) Reactors for continuous primary beer fermentation using immobilised yeast. Biotechnol Tech 11:261–264

Smogrovicova D, Domeny Z, Svitel J (1998) Effect of immobilised cell concentration on beer quality in continuous beer fermentation. Food Biotechnol 12:123–137

Smogrovicova D, Patkova J, Domeny Z, Navratil M (2000) Improvement in beer fermentation under very high gravity conditions by entrapped yeast. Minerva Biotechnol 12:331–336

Smogrovicova D, Domeny Z, Navratil M, Dvorak P (2001) Continuous beer fermentation using polyvinyl alcohol entrapped yeast. In: Proceedings of European brewery convention, 28th congress, Budapest, pp 477–485

Thomas KC, Hynes SH, Ingledew WM (1996) Practical and theoretical considerations in the production of high concentrations of alcohol by fermentation. Process Biochem 31:321–333

Verbelen PJ, De Schutter DP, Delvaux F, Verstrepen KJ, Delvaux FR (2006) Immobilized yeast cell systems for continuous fermentation applications. Biotechnol Lett 28:1515–1525

Willaert R, Nedovic VA (2006) Primary beer fermentation by immobilised yeast – a review on flavour formation and control strategies. J Chem Technol Biotechnol 81:1353–1367

9 Biofuels from Green Microalgae

P. Sharma, M.B. Khetmalas, and G.D. Tandon

Abstract

Increasing demands of fossil fuels and pollution caused by them run parallel to hinder the ecological balance leading to green house gases effect. Biofuel production from renewable feedstocks has become a prerequisite for nations worldwide. Microalgal biomass serves as a renewable and eco-friendly source of clean biofuel production for complete replacement of fossil fuels in the near future. The current review reveals the credits of algal technology in the current era of upcoming white biotechnology with reference to the biofuels production and the process and methodology of their generation from green microalgae. With the landmark advancements in phytoplankton area of research, the capital investment cost for green fuel production from microalgal feedstock would come down to competing prices to make clean fuel at commercial scale a reality.

9.1 Introduction

Biofuels are the fuel derived from biomass. They are termed as biorenewable energy resources that have the potential to replace nonrenewable fossil fuels efficiently. Biomass, on the other hand, is the common name for organic materials used as renewable energy sources such as wood, crops, algae and other lignocellulosic, agricultural wastes. Biofuel feedstocks are the organic materials used in the production of biofuels.

Biofuels are not a new energy source. It was used in Germany and France way back in 1894 by the then developing industry of internal combustion engines. Brazil utilised ethanol as fuel since 1925. Biofuels include biomethane, B100 Biodiesel, E100 Ethanol, biogas and biohydrogen (Ghirardi et al. 2000; Frac et al. 2010) biochar. For example, biomethane is the natural gas that is produced from the anaerobic decomposition of organic material in a landfill. While biomethane normally contain about half the Btu (British thermal unit) content of typical natural gas sold by gas companies, biomethane is a substitute

P. Sharma • M.B. Khetmalas • G.D. Tandon (✉)
Dr. D. Y. Patil Biotechnology & Bioinformatics Institute, Dr. D. Y. Patil University, Pune, Maharashtra, India
e-mail: g.tandon@dpu.edu.in

R.K. Salar et al. (eds.), *Biotechnology: Prospects and Applications*,
DOI 10.1007/978-81-322-1683-4_9, © Springer India 2013

form and completely replaces natural gas. Likewise, B100 Biodiesel is a substitute and completely replaces petroleum diesel on a gallon for gallon basis as does E100 Ethanol, which replaces gasoline on a gallon for gallon basis. Biofuels like hydrogen, biomethane and biogas hold tremendous potential to lower the dependence of descendent generations on the conventional nonrenewable, fossil fuels like petroleum, coal, diesel and natural gas.

Instability of supply always created a series of petroleum crisis. Disruption in oil supply is always accompanied by a continuous increase in oil prices which threaten a country's economic and national security. Fuel demand is dependent on the nature of society, technical development, economic cycles and political conditions. Since 1998, global demand for oil rose by 2 % every year. The US Energy Information Administration forecasted that worldwide demand for oil would increase to 60 % by 2020.

9.1.1 Advantages/Benefits of Renewable Biofuels

Advantages of biofuels over non-alternative sources of energy could be summed up as follows: (1) Cost: with the peer advancements in technological sectors, biofuels will become more economical than fossil fuels. (2) Source material: biofuels can be manufactured from a wide range of materials including crop waste, manure and other by-products, making it an efficient step in recycling. (3) Renewability: it takes thousands of years for fossil fuels to be produced, but biofuels are much more easily renewable and collected. (4) Security: by reducing dependence on foreign fuel sources, countries can protect the integrity of their energy resources and make them safe from outside influences. (5) Economic stimulation: because biofuels are produced locally, biofuel-manufacturing plants can increase the employment opportunities especially in the rural areas. Biofuel production will also increase the demand for suitable biofuel crops, providing economic stimulation to the agriculture industry. (6) Biodegradability: biofuels are easily biodegradable and far safer to handle than traditional fuels, making spills less hazardous and much easier and less expensive to clean up. (7) Lower carbon emissions: when biofuels are burned, they produce significantly less carbon output and fewer toxins, making them a safer alternative to conserve the atmospheric quality and lower air pollution (Reda et al. 2010).

9.1.2 First-Generation Biofuels

Biofuels assigned to first generation are energy sources that are produced from biomass—the living matter of plants organic waste. Biofuel crops recycle carbon dioxide, one of the main GHGs, by absorbing it when they grow and then releasing it back into the atmosphere when they are burned. Theoretically, biofuels should not add to GHG emissions. In practice, the energy balance, i.e. the amount of fossil fuels required to make the crop and convert it into a biofuel versus the energy it produces, varies from crop to crop. First-generation biofuel includes two main types: (1) plant sugars or starches of biofuel crops, such as sugar cane and maize, being fermented to produce ethanol and (2) biodiesel produced using the oily seeds of rapeseed or soybean. New technologies have to be developed to replace the use of food crops as the biofuel because the turn over produced from them is comparatively low with the contrary of their pressure on agriculture feed.

9.1.3 Second-Generation Biofuels

Second-generation biofuels involve enzymatic hydrolysis of cellulose and its fermentation into alcohol. Nonedible plants and plant parts like grass, wood and agricultural residues can be used as substrates. This will therefore reduce the level of organic matter in crop lands, which may destabilise soil health. However, these nonfood crops could be tailored into ideal biofuel crops via genetic improvement, combining desirable characteristics including resistance to pests and diseases, and, most importantly, increase biomass yields by a factor of two or more.

9.1.4 Third-Generation Biofuels

These are the algae-based biofuels that hold great promise to dominate and suppress variations in oil prices with assuring relaxation in strong global concerns about climate change. Algae as feedstock for biodiesel production can provide significant environmental benefits by reducing the land, pollutant and water footprints of biofuel production, not competing with the food crops, benefitting farmers by creating employment, strengthening country's economy by decreasing dependency on import of fuels and mitigating climate change by playing a carbon neutral role.

9.2 Algae

Algae are the most primitive, robust and diverse forms of living organisms that capture light energy through photosynthesis and synthesise simple organic compounds using the captured solar energy with the help of inorganic salts. Algae range from single-celled organisms to multicellular organisms, some with a fairly complex differentiated form as in case of macroalgae. They can be freshwater living in freshwater systems or marine that thrives in salty water bodies.

The absence of vascular system in algae differentiates them from higher plants. Microalgae are classified as (1) Chlorophyceae (green algae), (2) Cryptophyceae, (3) Phaeophyceae (brown algae), (4) Rhodophyceae (red algae), (5) Xantho-phyceae (yellow-green algae), (6) Dinophyceae, (7) Bacillariophyceae (diatoms), (8) Chloromo-nadineae, (9) Euglenophyceae, (10) Chrysophyceae and (11) Myxophyceae, based on the number and mode of attachment of flagella in the motile cells, thallus structure, chemical nature of pigments, reserve food materials, method of reproduction and variation in the life cycles (Fritsch 1945). Over 40,000 algal species have been identified around the world (Hu et al. 2008). Around 8,000 species of green alga existing in fresh, marine and saline waters have been reported so far (Hannon et al. 2010).

9.2.1 Isolation and Characterisation of Microalgae

9.2.1.1 Natural Habitats

Algae can be isolated from a variety of natural aqueous habitats ranging from freshwater to brackish water, marine and hypersaline environments and soil (Round 1984). Furthermore, large-scale sampling efforts should be undertaken to ensure the maximum coverage of their natural habitats. Dynamic maps, geographic information systems (GIS) data and analysis tools are the few available tools that support to help in the selection and collection of algae from a specific site. Algal biodiversity occurs at aquatic ecosystems like oceans, lakes, rivers, streams and ponds and geothermal springs, which include fresh, brackish, hypersaline, acidic, alkaline and terrestrial environments at different latitudes and longitudes. Collection sites can also include public lands, pits, nearby waters of tunnels, drainage systems, as well as various sites within the sampling area. Sampling methodology should be aimed in a way that it accounts for spatial distribution along with temporal succession at all the respective seasons.

9.2.1.2 Isolation Techniques

For the isolation of new strains of algae from natural habitats, traditional cultivation techniques may be used such as enrichment cultures (Andersen 2005). However, some algal strains take weeks to months to be isolated by traditional methods. For large-scale sampling and isolation efforts, high-throughput automated isolation techniques involving fluorescence-activated cell sorting (FACS) have proved to be the right choice. Due to the resemblance in morphological characteristics, while comparing many algal species, actual strain identification should be based on molecular methods such as rRNA sequence comparison with the application of gene markers.

9.2.1.3 Algal Databases

With the developing interdisciplinary technologies involving role of computers in biosciences, databases of algae can be generated. Development of databases that involves featuring all the characteristic of algae; its collection, isolation and identification; and morphological and ecological characteristics can help a researcher to easily locate his/her algae of interest and the vast algal data procured can be viewed in a synchronised form. For example, WORMS, that is, World Register of Marine Species, provides data of numerous marine algal species. "Algal Database" (developed by Dr. D. Y. Patil Biotechnology & Bioinformatics Institute, Pune, India) available at the site http://algaedb.dpu.edu.in is also an example of one such database which enables the researchers to locate the diversity of freshwater green algae in a specific region (paper under publication). It provides information about algal species, their natural availability, flourishing season, optimum temperature, rainfall, soil type and physiology in relation to their ecology, therefore making it easier for a researcher to screen an alga and decide its probable classification and application.

9.2.1.4 Significance of Screening

For higher rates of metabolite production, primary screening involves determining the cellular composition of proteins, lipids and carbohydrates, with a second step of measuring the metabolites generation mainly for biofuel production in this context. The appropriate or desired strain obtained would decide its fate in terms of cultivation for biofuel production. For example, a strain isolated for oil production would allow for distinction between neutral and polar lipids and would provide fatty acid profiles. Nile red staining of the algal cells and observation under fluorescence gives a qualitative analysis of its lipid content to select the strain that reveals high lipid contents (Ramasamy et al. 2011). Furthermore, many strains may also secrete certain metabolites into the growth medium. Some of these could prove to be valuable coproducts, but simultaneous new approaches are needed to develop more precise screening methods.

9.2.2 Cultivation of Algae

For cultivation of microalgae at commercial scale, raceway ponds and closed photobioreactors are used. At laboratory scale, cultivation has been done in Erlenmeyer flasks/beakers/glass bottles containing strain-specific media. Feeding of CO_2 accelerates growth rate in most of the cultures. At commercial scale it is estimated that there are 50 thousand species of algae in the world, but only about 30,000 algal species have been identified and examined so far (Frac et al. 2010). Selection of the precise and appropriate strain of algae is very much important in context of biofuel generation. A strain should have higher biomass and lipid productivity with a robust growth and faster doubling time. It should be able to survive under stress conditions along with having a naturally admirable carbon sequestration capacity (Brennan and Owende 2009). Algae cultivation requires supplementary plant nutrients for its growth and multiplication. These nutrients are commonly represented as N, P and K as well as other macro- and micronutrients. Optimisation of growth and production conditions for enhancing biomass and biofuel productivity is a prerequisite for economising the fuel production process. The designed media should provide optimal growth nutrients to enhance the microalgal growth, and at the same time the media should serve at fairly lower costs (Chanakya et al. 2012). Some of the media used for the growth of Chlorophyceae in the past and to date are BBM, Bristol, BG-11, Fogg's, Kuhn's, etc. (Andersen 2005; Makarevičienė et al. 2011; Velichkova et al. 2012). Multi-tier cropping of algae with paddy is an equally effective way of algal cultivation in agricultural-based lands (Chanakya et al. 2012).

9.2.3 Growth and Metabolism in Algae with Reference to Lipid Production

In general, two carboxylation pathways are reported to be operating in alga: the first one being the C3 pathway and the second one being the C4 carbon fixation pathway. In the C3 pathway, the enzyme ribulose-bisphosphate carboxylase (Rubisco) catalyses the reaction of ribulose-bisphosphate, carbon dioxide and water to yield 3-phosphoglyceric acid, which enters the Calvin cycle resulting in the production of sugar and starch. It is assumed that most of the algae operate the C3 pathway to fix the inorganic carbon, but some follows an alternative C4 pathway, wherein CO_2 is first converted into a four-carbon organic compound and then CO_2 is released for fixation by Rubisco. Existence of occurrence of C4 pathway is available only in eukaryotic algae. Under the aquatic environments, algae use carbonic anhydrase located on the surface of the cell to promote the conversion of dissolved CO_2 to HCO_3^-, which is then transported into the cell and reversed to CO_2 by cytoplasmic carbonic anhydrase. Due to the higher dissolution factor of bicarbonate as compared to that of carbon dioxide in solution, the carbon metabolism enhances the photosynthetic efficiency in algal cell (Chen et al. 2009; Sharma et al. 2012).

Cell growth determines the maximal potential of the algal biomass production. Cell growth consists of two phases—cell proliferation and enlargement in cell volume—both of which contribute to accumulation of algal biomass. Thus, there lies the significance of cell growth in biomass production over a larger scale. In case of small and round microalgae where the size of the cell is in micrometres, the biomass production is calculated based on the proliferation rate. The biomass production from heterotrophic algae is usually two orders higher than that from autotrophic algae. In addition, heterotrophic cells are found to accumulate higher levels of oils than the autotrophic ones (Xu et al. 2004). Under light, some algal species can use either CO_2 or organic carbon through a photoheterotrophic or mixotrophic pathway. In such a case, the biomass yield lies between the autotrophic and heterotrophic biomass production.

The photosynthetic efficiency of the strain has to be improved so as to obtain the desired strain having higher production capability of the desired product may it be oil for biodiesel or carbohydrate for bioethanol or proteins for nutrient supplement. Major parts of the fatty acids are produced in chloroplasts for the construction of chloroplast membranes which may be attributed to the fact that chloroplast metabolism is predominant in green algae. Under nitrogen depletion, fatty acids in the form of triacylglycerols (TAGs) are accumulated in the cell. TGA synthesis occurs by de novo fatty acid synthesis in the chloroplasts. The rate-limiting step in fatty acid synthesis is the conversion of acetyl CoA to malonyl CoA, catalysed by acetyl-CoA carboxylase (ACCase). There is a great scope of maximising the TAG production yield via genetic engineering as not much data has been revealed in the light of lipid accumulation mechanism in algae. Stresses including nutrition starvation in N, P and Si promote TAG synthesis in algal cells. Also, osmotic stress has proved to enhance lipid biosynthesis in both with and without nitrogen starved microalgal cells (Wang et al. 2011). But increase in TAG leads to a decrease in algal cell growth and finally affects the overall TGA yield which is not so high. Thus, the higher rate of biomass production in hybrid systems is inversely proportional to the higher rate of oil production in algae. It is therefore advisable to screen and isolate naturally existing algae that have a capacity of higher TGA production under normal growth conditions (Table 9.1).

Table 9.1 Determination of lipid yield of some microalgae

Microalgae	Oil content (% dry wt.)
1. *Botryococcus braunii*	25–75
2. *Chlorella* sp.	28–32
3. *Crypthecodinium cohnii*	20
4. *Cylindrotheca* sp.	16–37
5. *Dunaliella primolecta*	23
6. *Isochrysis* sp.	25–33
7. *Monallanthus salina*	>20
8. *Nannochloris* sp.	20–35
9. *Nannochloropsis* sp.	31–68
10. *Neochloris oleoabundans*	35–54
11. *Nitzschia* sp.	45–47
12. *Phaeodactylum tricornutum*	20–30
13. *Schizochytrium* sp.	50–77
14. *Tetraselmis suecica*	15–23

Adapted from Chisti (2007)

9.2.4 Scaling Up of Algal Cultivation

With the developing technologies, presently three models of production systems are designed for growing the culture on a larger scale: the first

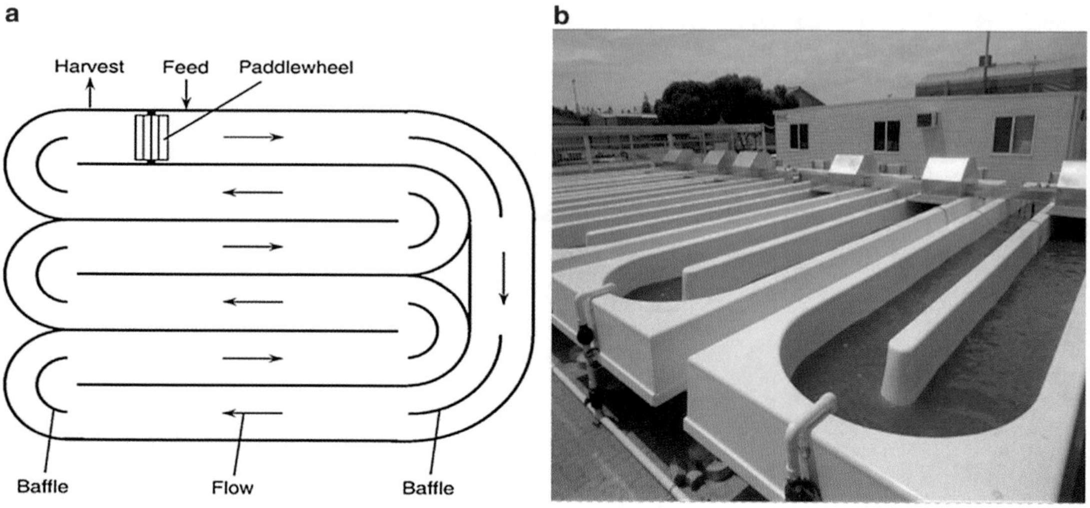

Fig. 9.1 (**a**) Arial view of a raceway pond as depicted by Chisti (2007). (**b**) Image of a raceway pond at SARDI, Australia

being raceway ponds, the second being closed photobioreactors and the third design being a combination of the first two called as hybrid systems.

9.2.4.1 Raceway Ponds

Since the 1950s, raceway ponds are being tried for mass culturing of microalgae. Raceway ponds are generally constructed either as a single unit or multiple joint units with agitation by means of a paddle wheel, propeller or air lift pumps. They can be circular, cascade, sloped and raceway (Chen et al. 2009). Mixing and circulation are done by a paddle wheel. Flow is guided around bends by baffles placed in the flow channel. Raceway channels are made of concrete or compacted earth that may be lined with white plastic. During daylight, the culture is fed continuously in front of the paddle wheel where the flow begins. Broth is harvested behind the paddle wheel, on completion of the circulation loop. The paddle wheel operates all the time to prevent sedimentation (Fig. 9.1) (Chisti 2007).

Raceway ponds are more expensive to construct due to the infrastructure required in constructing and providing stability of the unit (Table 9.2). Nevertheless, the choice of construction materials covers a wider range. Open ponds are easy to maintain since they have large open access to clean off the biofilm that builds up on surfaces. The main disadvantage encountered in this system is that since they are open to the atmosphere, there exists a greater probability of contamination by undesired algal species and fungal infections. Eradicating and making the pond free of the weedy species is a rather difficult, time-consuming and tedious job to deal with. Maintaining the water level is far more difficult due to constant loss of water from the unit in the form of evaporation (Schenk et al. 2008).

The large-scale cultivation of algae and cyanobacteria in outdoor open pond systems is well established (Becker 1994; Benemann and Oswald 1996; Borowitzka 1999). Open ponds are only suitable for a small number of algal species that can tolerate extreme environmental conditions. These species include fast growers, such as *Chlorella*, and species that require highly selective environments, such as *Spirulina* and *Dunaliella*, which thrive in highly alkaline or saline environments. There are four types of ponds:

1. *Unmixed open ponds*: Generally used for the culture of *Dunaliella salina*, these ponds have low productivities and are unsuitable for the culture of most algal species (Borowitzka 1999).
2. *Raceway ponds*: Widely used for the cultivation of *Spirulina*, *Haematococcus* and *Dunaliella* (Benemann and Oswald 1996), these ponds utilise paddle wheels for culture agitation and mixing.

Table 9.2 Brief note of the merits and demerits of microalgae cultivation in race ponds and photobioreactors

Production system	Advantages	Limitations
Raceway pond	Relatively cheap	Poor biomass productivity
	Easy to clean	Large area of land required
	Utilises nonagricultural land	Limited to a few strains of algae
	Low energy inputs	Poor mixing, light and CO_2 utilisation
	Easy maintenance	Cultures are easily contaminated
Tubular photobioreactor	Large illumination surface area	Some degree of wall growth
	Suitable for outdoor cultures	Fouling
	Relatively cheap	Requires large land space
	Good biomass productivities	Gradients of pH, dissolved O_2, CO_2 along tubes
Flat plate photobioreactor	High biomass productivities	Difficult scale-up
	Easy to sterilise	Difficult temperature control
	Low oxygen build-up	Small degree of hydrodynamic stress
	Readily tempered	Some degree of wall growth
	Good light path	
	Large illumination surface area	
	Suitable for outdoor cultures	
Column photobioreactor	Compact	Small illumination area
	High mass transfer	Expensive compared to open ponds
	Low energy consumption	Shear stress
	Good mixing with low shear stress	Sophisticated construction
	Easy to sterilise	
	Reduced photoinhibition and photo-oxidation	

Adapted by Brennan and Owende (2009)

3. *Circular ponds*: Used mainly in Asia for the production of *Chlorella*, these ponds are mixed by centrally located rotating arms.
4. *Thin layer, inclined ponds*: These consist of slightly inclined shallow trays, over which a very thin layer of algae flows to the bottom where the culture is collected and returned to the top.

9.2.4.2 Closed Photobioreactors (PBRs)

Due to the limitation of open pond systems, enclosed photobioreactors (PBRs) have evolved in the last 50 years. They can be located indoors and provided with artificial light or natural light. Two major types of enclosed PBR are tubular and plate types (Schenk et al. 2008).

Closed PBRs are gaining importance due to their advantages of saving water, energy and chemicals. They support up to higher productivity with respect to reactor volume and consequently have a smaller "footprint" on a yield basis (Table 9.2) (Barbosa et al. 2003). Based on the current energy costs and the given productivity, reactor costs should not exceed US$15 per m^2 (Fig. 9.2).

Tubular PBR constructed with transparent glass or plastic is one of the popular outdoor systems for mass algae cultivations. By shape, it can be horizontal, vertical, conical and inclined. By mixing, it can be airlift or pump system (Dayananda et al. 2007). Plate type of PBR can be vertical, horizontal and inclined. They may be airlift tubular, inclined tubular, horizontal tubular, undular row tubular, vertical column, outdoor helical tubular, parallel tubular, bubble column, flat plate and inclined plate (Al-Qasmi et al. 2012; Schenk et al. 2008).

The following are descriptions of types of PBR:

1. *Tubular PBRs*: Several serpentine, vertical, horizontal and inclined tubular PBRs have been designed and built in the last few decades (Lee and Low 1991; Lee et al. 1995). These systems include glass or plastic tubes with gas exchange vessels for the addition of CO_2

Fig. 9.2 Depiction of an outdoor tubular photobioreactor (Adapted from SARDI, Australia)

and the outgassing of O_2 and a recirculation pump for mixing.

2. *Vertical bubble columns and airlift reactors*: These cylindrical PBRs have gas bubbles introduced at the bottom of the columns and may be simple bubble columns, split cylinder airlifts or draft tube airlifts.
3. *Combined bubble column and inclined tubular PBR*: A patent application submitted by Berzin (2005) is for a right-angled triangular PBR which combines the principle of a bubble column with mixing by built-in static mixers in an inclined "down comer". The system also has a counter current of gas to increase mass transfer. Gas exchange occurs at a gas exchange vessel located at the apex of the triangle.
4. *Helical PBRs*: Helical PBRs are composed of parallel sets of flexible translucent tubes coiled helically around a cylindrical mesh frame. Gas exchange is accomplished via an incorporated gas exchange system at the top of the unit, and a heat exchange system may be included for temperature control.
5. *Flat plate PBRs*: Flat plate PBRs are made of thin rectangular translucent boxes, which are open at one end and may have ribs (alveolar) running vertically from bottom to top. Aeration and mixing are provided via perforated tube running along the entire bottom of FPR (Tredici et al. 1991; Hu et al. 1996).

9.2.4.3 Hybrid Systems

Open ponds are efficient but cost-effective and become contaminated with unwanted species very quickly. Photobioreactors are best designed for maintaining axenic cultures, but setup costs are around ten times higher than for open ponds. Therefore, a combination of both systems is probably the most optimum design for cost-effective cultivation of high product yielding strain of algae for biofuels. For large-scale microalgae biofuel production, there would need to be a series of photobioreactors of increasing size, from starter culture through to the final inoculum. As the bioreactors increase in size, the level of complexity should be reduced to minimise the cost per square metre.

9.2.4.3.1 Heterotrophic Culture

While most algae grow phototropically, some are capable of heterotrophic growth using organic substrates as sole carbon and energy sources. This mode of algal cultivation is well established (Barclay et al. 1994; Behrens and Kyle 1996) and has several advantages over phototrophic modes of growth including large, existing fermentation technology knowledge base, the high degree of process control for consistent, reproducible production, the elimination of light requirements, the independence from weather and climatic conditions and lower harvesting costs (Barclay et al. 1994). Sufficient oxygen is required for the catabolism of the organic substrates in heterotrophic cultivation of algae; hence, oxygen supply is often the most limiting factor preventing high cell concentration and high growth rate (Clark et al. 1995; Lee et al. 1995). Generally heterotrophic cultivation has been found to increase the total lipid content in algae compared to phototropically grown cells (Li et al. 2007; Miao and Wu 2006; Tan and Johns 1996; Wen and Chen 2000), and heterotrophic cultivation of algae usually results in higher yield (Ramasamy et al. 2011; Fang et al. 2004; Wen and Chen 2000; Yu et al. 2008).

9.2.4.3.2 Mixotrophic Culture

It is a nutritional mode in which photoassimilation of carbon dioxide and the oxidative catabolism of organic carbon sources proceed simultaneously, thereby offering the potential of greatly increased productivities. For the species that can utilise both light energy and chemical substrates, this mode of cultivation offers superior alternative to phototrophic and heterotrophic growth as both biomass and productivity increases have been reported (Chu et al. 1996; Day and Tsavalos 1996; Fang et al. 2004; Theriault 1965; Yu et al. 2009; Ngangkham et al. 2012).

9.3 Production of Biofuels

9.3.1 Biodiesel

Biodiesel is defined as the mono-alkyl esters of vegetable oils or animal fats. Biodiesel is produced by transesterifying the parent oil or fat to achieve a viscosity close to that of petro diesel (Karthikeyan 2012). Algal oils differ from higher plant oils in their high phospholipid and glycolipid concentrations. These lipid classes contain nitrogen, phosphorous and sulphur that may be problematic with regard to engine performance, if present in fuels. However, it is likely that the sulphur, phosphorus and nitrogen-containing compounds would end up in the water-soluble fraction following transesterification, so one would expect these elements to be low to nonexistent in algal biodiesel. Since about 30 % of the original lipid mass can be lost to the polar phase during esterification, the lipid class composition will also greatly affect the potential fuel yield by transesterification. Thus, triglycerides have a >99 % biodiesel yield compared to a <70 % yield for phospholipids (Nagle and Lemke 1990). The overall heating value of crude algal lipid is somewhat depressed to a value in the region of 36 kJ g^{-1}, due to the lower calorific value of the glyco- and phospholipids. However, as these fractions would be almost certainly separated off in the transesterification process, their low calorific value would have no effect on the properties of the final biodiesel. The chain length distribution of fatty acids occurring in microalgal species is more diverse compared with that of higher plants, making it possible that certain species could be cultured for selected fuel properties. For example, strains that primarily accumulate shorter chain fatty acids may be more amenable for use in the production of jet fuels as compared to the application of very long chain fatty acids in the lubricant market.

Fatty acids and fatty acid methyl esters (FAME) with four and more double bonds are susceptible to oxidation during storage, and this reduces their acceptability for use in biodiesel. Many microalgal oils do not comply with the European biodiesel standards, and so the extent of unsaturation of microalgal oil can be reduced by partial catalytic hydrogenation of the oil (Chisti 2007).

For algal biodiesel to be an accepted substitution fuel for fossil fuels, its properties must match or exceed the International Biodiesel Standard for Vehicles (EN14214). Algal biodiesel has several advantages over petroleum diesel such as it is derived from biomass and therefore is renewable, biodegradable and carbon neutral under sustainable production, is nontoxic and contains reduced levels of particulates, carbon monoxide, hydrocarbons and sulphur oxides. Algal biodiesel finds its significant role in the aviation industry where low freezing points and high energy densities are of prime importance (Brennan and Owende 2009).

Downstream Processing: It basically involves four steps: the first one being harvesting of the cells (collection of cells from continuous or batch cultures); at the second step, solid liquid separation takes place and the harvested cells are dewatered by means of drying. The third step involves the extraction of product (oil) from algal biomass, and the last step undertakes purification of product (transesterification in to biodiesel in case of biodiesel production). It involves harvesting algal cells and processing to end up with the resultant aimed by-product. Algae typically have a high water content and downstream harvesting and processing requires its removal. There is no single best method for harvesting microalgae and reducing their water content. Flocculation, microscreening and centrifugation are few of the harvesting processes. Cost-effective and energy-efficient harvesting methods are necessarily required to make the biofuels production process economical, and thus selection of a strain that can be easily harvested is essential (Schenk et al. 2008). Choice of harvesting technique is dependent on characteristics of microalgae, e.g. size, density and the value of the target products (Brennan and Owende 2009).

9.3.1.1 Harvesting Microalgal Cells

To conserve energy and reduce costs, algae are often harvested in a two-step process. In the 1st step the algae are concentrated often by flocculation which concentrates the dilute cultures to about 1–5 % solids. In the second step the cells are further concentrated by centrifugation, filtration or microstraining to get solid concentration of 15–25 %.

9.3.1.1.1 Flocculation

Algae carry negative cell surface charges (Golueke and Oswald 1965) which when neutralised lead to the agglomerisation of the algae into large clumps or “flocs”. These flocs can then be more readily separated from the culture.

Flocculation can be induced in the following various ways:

1. Algae may be induced to flocculate by the addition of inorganic chemicals, such as aluminium sulphate, ferric sulphate, ferric chloride or lime (Golueke and Oswald 1965; Heasman et al. 2000) which neutralise or reduce surface charge of the cells, causing the formation of flocs. The requirement for large doses of these chemicals led the Aquatic Species Program (ASP) to conclude that chemical flocculation was too expensive for the production of biofuels. The incorporation of the metal salts in the harvested biomass also limits its use for human and livestock feeds and creates disposal problems (Benemann and Oswald 1996; Molina et al. 2001).
2. Polymeric organic flocculants (polyelectrolyte) are highly charged organic macromolecule or aggregate formed in aqueous solution by dissociation of charged units of these macromolecules. In addition to neutralising the negative charges on algae cells, they also physically link the algae cells to each other (Tilton et al. 1972), thus producing more stable flocs. Lower levels of polyelectrolyte are required for flocculation compared to inorganic chemicals, and this together with their reported lack of toxicity has made them a more attractive flocculation option.
3. Some algae naturally flocculate after their transfer to settling ponds if left quiescent for some time. Environment stimuli or nitrogen limitation, pH and dissolved oxygen level attributes to biocoagulation/flocculation process. Electrocoagulation is based on the movement of electrically charged particles in an electric field in which active coagulant species are produced by oxidation of a metal anode. It involves the application of an electric current to a sacrificial anode (usually aluminium or iron), which then goes into solution generating metal ions which act as coagulating agents and releasing hydrogen gas at the cathode. The metal ions coagulate with the algae and bubbles produced at the cathode rise to the surface taking the flocs with them.

After flocculation, the flocs may be left to settle before recovery by pumping off the surface liquid layer (sedimentation) or the flocs may be removed by dissolved air floatation. This involves pressurising some of the liquid to dissolve additional air. The dissolved air comes

Table 9.3 Various methods of drying algae samples with their merits, limiting factors and distinguished feature

Method	Advantages	Limitations	Characteristics
Drum drying	Fast and efficient	Cost intensive	Ruptures cellulosic cell walls, sterilises the product, unsuitable for *Spirulina*
Spray drying	Fast and efficient	Cost intensive	Sterilises the product, breakage of cellulosic cell walls
Sun drying	Low fixed capital	Slow process Weather dependent	Biomass may ferment, sterilisation not possible, does not break cellulosic cell walls
Solar drying	Low capital costs	Weather dependent	Does not break cellulosic cell walls, sterilisation not possible
Crossflow drying	Faster than sun drying	Requires electricity	Does not break cellulosic cell walls, sterilisation is not possible
Vacuum-shelf drying	Gentle process	Cost intensive	Does not break cellulosic cell walls, product becomes hygroscopic, sterilisation not possible, preserves cell constituents
Freeze drying	Gentle process	Slow process Cost intensive	Does not break cellulosic cell walls, sterilisation not possible, preserves cell constituents

Adapted from Chen et al. (2009) and Becker (1994)

out in the form of bubbles that attach to the flocs, making them float.

9.3.1.1.2 Centrifugation

This is a well-established industrial process that uses gravitational force to achieve separation. The morphology and size of the cells to be harvested affects the product recovery and cost of the process as filamentous cells and large colonial cells will tend to settle more readily than single smaller cells. Centrifugation is energy intensive. The high capital and running costs associated with centrifuges limit their use to second-stage filtration in the processing of microalgae for bio fuels.

9.3.1.1.3 Filtration

The principle of filtration is introducing the particles onto a screen of given aperture sizes. The particles either pass through or are retained on the screen according to their size. Filtration can be performed under pressure or vacuum. Filtration can also be carried out by microstrainers which consist of a rotating drum covered by a straining fabric. A backwash spray collects the particles into an axial trough. Although the costs associated with filtration are low, screen clogging and membrane fouling limit its suitability to larger species of algae.

9.3.1.2 Drying

Harvested algae contain 97–99 % water which is not suitable for the product extraction, and thus removal of the water content of algal cells becomes mandatory for their use as a biofuel feedstock in downstream processes. Also, moisture contents in algae will favour microbial growth leading to contamination. Drying accounts for 30 % of the total production costs. Selection of the drying methods is based on the algal species, the scale of operation and final use of the dried products, costs and energy requirements (Table 9.3) (Chen et al. 2009; Becker 1994).

9.3.1.3 Lipid Extraction

9.3.1.3.1 Organic Solvent Extraction

Disruption of the microalgae cells releases the metabolites of interest. Several methods can be used depending on the microalgae wall and on the nature of the product to be obtained. They can be based on mechanical action like cell homogenisers, bead mills, ultrasounds, autoclave and spray drying or based on nonmechanical action like freezing, organic solvents and osmotic shock and acid, base and enzyme reactions (Mata et al. 2010).

Organic solvents, such as benzene, cyclohexane, hexane, acetone and chloroform, have shown to be effective on microalgal cell. They degrade microalgal cell walls and extract the oil that has a high solubility in organic solvents (Mercer and Armenta 2011). Extraction of algal lipids using a mixture of chloroform and methanol [1:2 (v/v)], Bligh and Dyer (1959) is generally followed at a larger scale, since among all the solvent systems examined for microalgal lipid extraction, chloroform-methanol system provided the highest extraction efficiency (Chen et al. 2009). This method gives 98 % extraction efficiency (Kanda and Li 2011; Mata et al. 2010).

In order to promote maximum penetration of the solvent into the cells and increase the lipid yield, some techniques, such as autoclave, bead beating, microwave, sonication, grinding and osmotic shock, are used to disrupt the cells before or during the lipid extraction from microalgae. Supercritical fluid extraction (SFE) and subcritical water extraction (SWE) are relatively new techniques featuring high selectivity and short operation time, and the solvent is free of toxicity (Chen et al. 2009).

9.3.1.3.2 Supercritical Fluid Extraction

Supercritical fluid extraction takes advantage of the fact that some chemicals behave as liquid and as a gas and exhibit increased solvating power when they are raised above their critical temperature and pressure points. Carbon dioxide is chosen because of its relatively low critical temperature (31.18 °C) and pressure (72.9 atm) (Mercer and Armenta 2011; Cooney et al. 2009). Supercritical CO_2 extraction efficiency is affected by factors like pressure, temperature and CO_2 flow rate and extraction time (Mercer and Armenta 2011).

9.3.1.3.3 Ultrasound/Sonication

Ultrasonic-assisted extractions can recover oils from microalgal cells by the means of cavitations. It occurs when vapour bubbles of a liquid form in an area (where pressure of the liquid is lower than its vapour pressure). These bubbles grow at negative pressure and compress under positive pressure causing a violent collapse of the bubbles. Altogether, ultrasound and microwave-assisted methods improve extractions of microalgae significantly, with higher efficiency, reduced extraction times and increased yields with very minute levels of toxicity also with an added advantage of being economical (Mercer and Armenta 2011).

9.3.1.4 Transesterification

Transesterification is a multiple step reaction, including three reversible steps in series, where triglycerides are converted to diglycerides, then diglycerides are converted to monoglycerides, and monoglycerides are then converted to esters (biodiesel) and glycerol (by-product) (Mata et al. 2010; Chen and Walker 2011). A homogeneous or heterogeneous, acid or basic catalyst can be used to enhance the transesterification reaction rate, although some processes use supercritical fluids like methanol or ethanol (Fig. 9.3) (Mata et al. 2010). Transesterification results into the conversion of crude oil extracted from algal cell into FAME, that is, fatty acid methyl esters, which is termed as the biodiesel from algae. Theoretically, 1 kg of oil yields about 1 kg of biodiesel.

9.3.2 Bioethanol

Based on cell wall structure, carbohydrates compose a major part of algae which can be exploited for its conversion to bioethanol. Cellulose and starch present in algae can be converted to their simple monomeric forms, that is, sugars, and these sugars could be utilised to produce bioethanol via the process of fermentation in the presence of yeast strain. The conversion of oligosaccharides into simple sugars is facilitated by the means of acidic/alkali pretreatments and enzymatic hydrolysis, consequently converting to alcohol by fermentation. The strains of algae rendering high stocks of carbohydrates can be selected as an optimal strain for bioethanol production (Agbor et al. 2011; John et al. 2011).

For those algae which produce substantial amounts of lipids and carbohydrates, the residual feedstock left after oil extraction can be turned to bioethanol (Sharma et al. 2013).Thus, from a single algae species, two biofuels can be generated

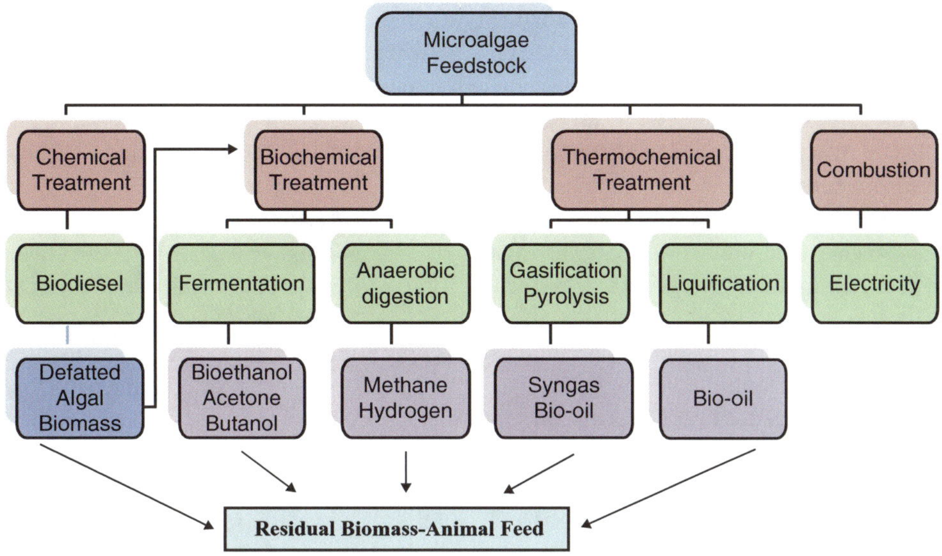

Fig. 9.3 Energy production by microalgal biomass conversion using biochemical, thermochemical, chemical and direct combustion processes (Adapted from Gouveia 2011; Wang et al. 2008)

(Fig. 9.3). The solid residue left after alcoholic fermentation of algal biomass could be used as a poultry feed/cattle feed or could be subjected to the gasification process to yield more pockets of energy (Miranda et al. 2012; Harun et al. 2011; Campbell 2008).

9.3.3 Hydrogen

Hydrogen is produced from microalgae under anaerobic conditions as an electron donor during the CO_2 fixation process. During photosynthesis, microalgae convert water molecules into hydrogen ions and oxygen. The hydrogen ions are then subsequently converted by hydrogenase enzymes into H_2. Due to reversibility of the reaction, hydrogen is either produced or consumed by the simple conversion of protons to hydrogen. Photosynthetic oxygen production causes rapid inhibition to the hydrogenase enzyme. Consequently, microalgae cultures for hydrogen production must be subjected to anaerobic conditions (Brennan and Owende 2009; Ghirardi et al. 2000).

9.3.4 Biomethane

The relatively high lipid, starch and protein contents make microalgae an ideal candidate for efficient biomethane production by fermentation in biogas plants. But the biomethane produced from algae costs higher than the biomethane produced from maize (Fig. 9.3) (Schenk et al. 2008; Lakaniemi et al. 2011).

9.3.5 Syngas, Bio-oil, Biochar, Biopolyols and Electricity Generation

9.3.5.1 Thermochemical Conversion

Algal biomass, in the form of whole algal cells or the residual algal biomass left after the production of biodiesel/bioethanol can serve as a suitable feedstock for thermochemical conversions like gasification, pyrolysis and hydrothermal liquefaction. Thus, resulting in the production of syngas, bio-oil, biopolyols and biochar (Fig. 9.3). Syngas can be combusted directly to produce heat or to generate electricity. Bio-oil can be used as heating oil or upgraded to liquid transportation fuels. Biopolyols are chemical stocks for material synthesis. Biochar can be used as active carbon, fertiliser and soil amendment agent. The biggest advantage of thermochemical conversion of algae over other conversion technologies is its high efficiency due to short retention time, ranging from seconds to minutes. There is a great scope of research in the thermochemical conversion of algal biomass. Over the years, less attention has being paid to it due to the increased focus on liquid biofuel production (Chen et al. 2009).

When biomass is processed under high temperature in the absence of oxygen, it produces three phases: the vapour phase, the liquid phase and the solid phase. The liquid phase being a complex mixture is called as bio-oil. The compositions of bio-oils vary significantly with the types of feedstock and processing conditions (Li et al. 2008).

9.3.5.2 Gasification

In gasification, biomass is converted to a combustible gas mixture called synthesis gas (syngas) through partial oxidation reactions at high temperature typically ranging from 700 to 1,100 °C. Syngas produced depends on the nature of feedstock: its moisture content, gasification conditions, etc. Syngas can be burned to produce heat or used in gas engines or gas turbines to produce electricity. Gasification units are commercially available in the market. Cleaning and conditioning of syngas has been a major drawback in its commercialisation (Li et al. 2008; Chen et al. 2009).

9.3.5.3 Pyrolysis

In pyrolysis, the biomass is degraded to bio-oil, syngas and biochar at medium high temperature (300–600 °C) in the absence of oxygen (Chen et al. 2009; Mohan et al. 2006). The bottleneck of pyrolysis of algae into bio-oil is the dewatering process prior to pyrolysis which is an energy-intensive process. Pyrolysis technology is expected to become a cost-effective conversion method only if dehydration/drying becomes inexpensive. Also, the components of bio-oil are very complex. Currently, much research has been focused on upgrading of bio-oil generated by reducing acidity and complexity as well as increasing stability (Chen et al. 2009).

9.4 Development of Biorefineries

Biorefining is a concept derived from petroleum refining. A biorefinery uses biomass as feedstock as opposed to fossil resources used in a petroleum biorefinery. The goal of biorefining is to produce a wide range of products such as fuels, materials and chemicals, from one or more biological resources. Development of new processes, design of the system and life cycle analysis are necessary for the development and implementation of algae-based biorefineries (Chen et al. 2009).

9.5 Genetic Engineering Aspects for Enhanced Biofuel Production from Algae

Intensive global research efforts for increasing and modifying the accumulation of lipids, alcohols, hydrocarbons, polysaccharides and other energy storage compounds in algae through genetic engineering are in progress. Although the application of genetic engineering to improve energy production in eukaryotic microalgae is in its infancy, significant advances in the development of genetic manipulation tools have been achieved with microalgal model systems. These are accomplished by manipulating central carbon metabolism of the organism.

Around 30 reported strains of microalgae have been transformed successfully. In many cases, transformation resulted in stable expression of transgenes, from either the nucleus or the plastid, but in some cases, only transient expression was observed (Snow and Smith 2012).

The efficiency of transformation is strongly species dependent, and the method of transformation is to be carefully selected and optimised for each microalgal strain. Transgene expression and protein localisation in the chloroplast are needed for the proper functioning of many metabolic genes of interest for biofuel production (Radakovits et al. 2010).

Nuclear transformation of microalgae generally results in the random integration of transgenes.

This may be suitable for transgene expression or for random mutagenesis screens, which makes it difficult to delete specific target genes. Some progress in homologous recombination has been made with the nuclear genome of *Chlamydomonas reinhardtii*, but the efficiency remains low. The quantity and quality of diesel precursors from a specific strain are closely linked to the control of lipid metabolism. Lipid biosynthesis and catabolism, as well as pathways that modify the length and saturation of fatty acids, have not been investigated for algae. But, many of the genes involved in lipid metabolism in terrestrial plants have homologs in the sequenced microalgal genomes. Therefore, there exists a probability that at least some of the transgenic strategies that have been adapted to modify the lipid content in higher plants might be effective with microalgae.

An early committing step in fatty acid synthesis is the conversion of acetyl coenzyme A (CoA) to malonyl CoA, catalysed by acetyl-CoA carboxylase (ACCase), which is considered the first committed step in fatty acid biosynthesis in many organisms. However, several attempts to utilise ACCase overexpression to increase lipid content in various systems have not yielded constructive results. It may be that ACCase levels are a limiting step in lipid biosynthesis mainly in cells that normally do not store large amounts of lipid. Another attempt to increase expression of a protein involved in fatty acid synthesis, 3-ketoacyl-acyl-carrier protein synthase III (KASIII), was also not successful in increasing lipid production.

9.5.1 Altering Lipid Catabolism

A complementary strategy to increase lipid accumulation is to decrease lipid catabolism, being provided in the following ways:

1. Cells operate β-oxidation of fatty acids for deriving energy under certain physiological conditions, thus knocking out lipid catabolism genes may result in increased lipid storage but also could have deleterious effects on cellular growth and proliferation.
2. Enzymes with overlapping functions also exist for many of the steps of β-oxidation, making it difficult to completely abolish lipid catabolism functions.
3. During dual light–dark cycles, many microalgae initiate TAG storage during the day and deplete those stores at night to support cell division. Consequently, inhibition of β-oxidation would prevent the loss of TAG during the night, but most likely at the cost of reduced growth. This strategy, therefore, may not be beneficial for microalgae grown in outdoor open ponds, but it may be a supportive strategy to increase lipid production in microalgae cultured in photobioreactors.

9.5.2 Modification of Lipid Characteristics

On the other hand, the quality of lipids, with regard to its suitability as a diesel fuel feedstock, can be modified to ensure increased production of lipids. The carbon chain length and degree of unsaturation of the fatty acids in microalgae can affect the cold flow and oxidative stability properties of a biodiesel fuel. Most microalgal fatty acids have a chain length between 14 and 20; major species are often 16:1, 16:0 and 18:1. Ideal fatty acids for diesel production should be 12:0 and 14:0. The chain lengths of fatty acids are determined by acyl-ACP thioesterases, which release the fatty acid chain from the fatty acid synthase. There are several acyl-ACP thioesterases from a variety of organisms that are specific for certain fatty acid chain lengths, and transgenic overexpression of thioesterases can be used to change fatty acid chain length. Fatty acids of even shorter chain lengths can also be used for production of gasoline and jet fuel. It is possible to use hydrocracking to break down longer hydrocarbons into shorter chain lengths that are more suitable as feedstocks for gasoline or jet fuel, but it may also be possible to reduce production costs through genetically engineering microalgae to directly produce these shorter chain lengths.

Increased lipid synthesis results in a reduction of cell division. In such a case, overexpression of lipid synthesis genes might be beneficial if they can be controlled by an inducible promoter that could be activated once the microalgal cells have grown to a high density and have entered stationary phase.

9.6 Industries Pioneering in Algal Technology

1. Aquaflow is a New Zealand-based company that aims at economical production of biofuel from wild algae harvested from open-air environments.
2. GreenFuel Technology Corp. Cambridge, Massachusetts, is a company working with power plants to build algae-producing photobioreactors. It captured about 80 % of the CO2 emitted during the day when sunlight is available.
3. HR BioPetroleum is a Hawaii-based company that intends to be a designer-builder of algae biofuel plants and to produce and market renewable fuel feedstock and animal nutritional supplemental protein.
4. LiveFuels Inc., Menlo Park, CA, is a research-based industry that describes itself as a mini-Manhattan project with a national alliance of labs and scientists dedicated to transforming algae into biocrude by the year 2010. Their strategy involves developing algae that will thrive in open ponds.
5. Imperium Renewables company in Seattle has gained popularity for producing traditional biodiesel by having a five million-gallon refinery for algae oil.
6. OTEC is a Francisco bay-area firm developing photobioreactors—enclosed systems that produce algae in layer upon layer of tubes or shallow ponds.
7. PetroSun is a diversified energy company specialising in the discovery and development of both traditional fossil fuels and renewable energy resources. Under the terms of a November 2007 agreement, PetroSun Biofuels will supply Bio-Alternatives 50 % of its raw algal oil production from planned algae farms and extraction plants in Louisiana, Alabama and Mississippi up to a maximum of 150 million gallons per year.
8. Solazyme is a San Francisco-based biotechnology company, considered one of the leading producers of biofuels, and has already harvested thousands of gallons of algal oil. They have engineered more than a dozen specialised strains and ramped up pre-commercial production.
9. OriginOil is located in Los Angeles. It focuses on its clean-tech process that harvests algae and cleans up oil and gas using a chemical-free, high-speed process.

9.7 Conclusion

Increasing demands for fuel, food and a clean atmosphere has made mankind to look towards the very ancient microalgal forms to exploit the energy power stored in them. Total biomass utilisation of microalgae ensures a substantial feedstock for generation of biofuels and feed for animals. With developing technologies the challenges faced in commercialisation of algal fuels have to be met, to devise an economically feasible process. Screening for the suitable algal strain, its cultivation in laboratory, harvesting of algal cells for product extraction, purification of product and scale up, each step involved has to be economised with further advancements in algal biotechnology. Algal research has to be encouraged in countries worldwide since many developing countries are still ignorant about its vast potential. Microalgae can serve as an ideal feedstock for the production of renewable fuels in the near future leading to decreased dependency on fossil fuels.

References

Agbor VB, Cicek N, Sparling R, Berlin A, Levin DB (2011) Biomass pretreatment: fundamentals toward application. Biotechnol Adv 29(6):675–685

Al-Qasmi M, Raut N, Talebi S, Al-Rajhi S, Al-Barwani T (2012) A review of effect of light on microalgae growth. In: Proceedings of the world congress on engineering, vol I, WCE 2012, London, UK, July 4–6

Andersen R (2005) Algal culturing techniques, 1st edn. Academic Press/Elsevier, San Diego, p 596. ISBN 0-12-088426-7

Barbosa B, Albrecht M, Wijffels R (2003) Hydrodynamic stress and lethal events in sparged microalgae cultures. Biotechnol Bioeng 83:112–120

Barclay WR, Meager KM, Abril JR (1994) Heterotrophic production of long chain omega-3-fatty acids utilising algae and algae-like microorganisms. J Appl Phycol 6(2):123–129

Becker EW (1994) Microalgae: biotechnology and microbiology. Cambridge University Press, Cambridge

Behrens PW, Kyle DJ (1996) Microalgae as a source of fatty acids. J Food Lipid 3:259–272

Benemann JR, Oswald WJ (1996) Systems and economic analysis of microalgae pond for conversion of carbon dioxide to biomass (Final report: Grant No. DE-FG22-93PC93204). Pittsburgh Energy Technology Center, Pittsburgh, PA, US Department of Energy

Berzin I (2005) Photobioreactors and process for biomass production and mitigation of pollutants in flue gases. United States Patent Application. Pub No: US2005/0260553 A1, USA, Publication

Bligh EG, Dyer WJ (1959) A rapid method of total lipid extraction and purification. Can J Biochem Physiol 37:911–917

Borowitzka MA (1999) Commercial production of microalgae: ponds, tanks, tubes and fermenters. J Biotechnol 70:313–321

Brennan L, Owende P (2009) Biofuels from microalgae—a review of technologies for production, processing, and extractions of biofuels and co-products. Renew Sustain Energy Rev. doi:10.1016/j.rser.2009.10.009

Campbell MN (2008) Algae as a renewable source for liquid fuel. Guelph Eng J 1:2–7

Chanakya HN, Mahapatra DM, Ravi S, Chauhan VS, Abitha R (2012) Sustainability of large-scale algal biofuel production in India. J Indian Inst Sci 92(1): 63–98

Chen YH, Walker TH (2011) Biomass and lipid production of heterotrophic microalgae Chlorella protothecoides by using biodiesel-derived crude Glycerol. Biotechnol Lett. doi:10.1007/s10529-011-0672-y

Chen P, Min M, Chen Y, Wang L, Li Y et al (2009) Review of the biological and engineering aspects of algae to fuels approach. Int J Agric Biol Eng 2(4):1–30

Chisti Y (2007) Biodiesel from microalgae. Biotechnol Adv 25(2007):294–306

Chu WL, Phang SM, Goh SH (1996) Environmental effects on growth and biochemical composition of Nitzschia inconspicua Grunow. J Appl Phycol 8:389–396

Clark GJ, Langley D, Bushell ME (1995) Oxygen limitation can induce microbial secondary metabolite formation: investigations with miniature electrodes in shaker and bioreactor culture. Microbiology 141:663–669

Cooney M, Young G, Nagle N (2009) Extraction of bio-oils from microalgae. Sep Purif Rev 38:291–325

Day JG, Tsavalos AJ (1996) An investigation of the heterotrophic culture of the green alga Tetraselmis. J Appl Phycol 8(1):73–77

Dayananda C, Sarada R, Usha Rani M (2007) Autotrophic cultivation of Botryococcus braunii for the production of hydrocarbons and exopolysaccharides in various media. Biomass Bioenergy 31:87–93

Fang X, Wei C, Zhao-Ling C, Fan O (2004) Effects of organic carbon sources on cell growth and eicosapentaenoic acid content of Nannochloropsis sp. J Appl Phycol 16:499–503

Frac M, Tys SJ, Tys J (2010) Microalgae for biofuels production and environmental applications: a review. Afr J Biotechnol 9(54):9227–9236

Fritsch FE (1945) The structure and reproduction of the algae, vol 2. Cambridge University Press, Cambridge

Ghirardi ML, Zhang L, Lee JW, Flynn T, Seibert M et al (2000) Microalgae: a green source of renewable H_2. Trends Biotechnol 18:506–511

Golueke CG, Oswald WJ (1965) Harvesting and processing sewage-grown planktonic algae. J Water Pollut Control Fed 37:471–498

Gouveia L (2011) Microalgae as a feedstock for biofuels. Springer briefs in microbiology. doi:10.1007/978-3-642-17997-6_1

Hannon M, Gimpel J, Tran M, Rasala B, Mayfield S (2010) Biofuels from algae: challenges and potential. Biofuels 1(5):763–784

Harun R, Jason WS, Cherrington T, Danquah MK (2011) Exploring alkaline pretreatment of microalgal biomass for bioethanol production. Appl Energy 88(10):3464–3467

Heasman M, Diemar J, Oconnor W, Sushames T, Foulkes L (2000) Development of extended shelf-life microalgae concentrate diets harvested by centrifugation for bivalve molluscs-a summary. Aquacult Res 31(8–9): 637–659

Hu Q, Gutterman H, Richmond A (1996) A flat incline, modular photobioreactor (FIMP) for outdoor mass cultivation of photoautotrophs. Biotechnol Bioeng 51:51–60

Hu Q, Sommerfeld M, Jarvis E, Ghirardi M, Posewitz M et al (2008) Microalgal triacylglycerols as feedstocks for biofuel production: perspectives and advances. Plant J 54:621–639

John RP, Anisha GS, Nampoothiri KM, Pandey A (2011) Micro and microalgal biomass: a renewable source for bioethanol. Bioresour Technol 102:186–193

Kanda H, Li P (2011) Simple extraction method of green crude from natural blue-green microalgae by dimethyl ether: extraction efficiency on several species compared to the Bligh-Dyer's method. World Renewable Energy Congress, Sweden, 8–11 May 2011

Karthikeyan S (2012) A critical review: microalgae as a renewable source for biofuel production. Int J Eng Res Technol (IJERT) 1:4

Lakaniemi AM, Hulatt CJ, Thomas DN, Tuovinen OH, Puhakka JA (2011) Biogenic hydrogen and methane production from Chlorella vulgaris and Dunaliella tertiolecta biomass. Biotechnology 4:34

Lee YK, Low CS (1991) Effect of photobioreactor inclination on the biomass production of an outdoor algal culture. Biotechnol Bioeng 38:995–1000

Lee YK, Ding SY, Low CS, Chang YC, Forday WL, Chew PC (1995) Design and performance of an α-type tubular photobioreactor for mass cultivation of microalgae. J Appl Phycol 7:47–51

Li X, Xu H, Wu Q (2007) Large-scale biodiesel production from microalga Chlorella protothecoides through heterotrophic cultivation in bioreactors. Biotechnol Bioeng 98(4):764–771

Li Y, Horsman M, Wu N, Lan CQ, Calero ND (2008) Articles: biocatalysts and bioreactor design. Biotechnol Prog 24:815–820

Makarevičienė V, Andrulevičiūtė V, Skorupskaitė V, Kasperovičienė J (2011) Environmental research. Eng Manage 3(57):21–27

Mata TM, Martins AA, Caetano NS (2010) Microalgae for biodiesel production and other applications: a review. Renew Sustain Energy Rev 14:217–232

Mercer P, Armenta RE (2011) Developments in oil extraction from microalgae. Eur J Lipid Sci Technol. doi:10.1002/ejlt.201000455

Miao X, Wu Q (2006) Biodiesel production from heterotrophic microalgal oil. Bioresour Technol 97(6):841–846

Miranda JR, Passarinho PC, Gouveia L (2012) Pretreatment optimization of Scenedesmus obliquus microalga for bioethanol production. Bioresour Technol 104:342–348

Mohan D, Pittman CU, Steele HP (2006) Pyrolysis of wood/biomass for bio-oil: a critical review. Energy Fuel 20:848–889

Molina E, Fernandez J, Fernandez AC, Chisti Y (2001) Tubular photobioreactor design for algal cultures. J Biotechnol 92(2):113–131

Nagle N, Lemke P (1990) Production of methyl ester fuel from microalgae. Appl Biochem Biotechnol 24–25(1):355–361

Ngangkham M, Ratha SK, Prasanna R, Saxena AK, Dhar DW et al (2012) Biochemical modulation of growth, lipid quality and productivity in mixotrophic cultures of Chlorella sorokiniana. Springer Plus 1:33

Radakovits R, Jinkerson R, Darzins A, Posewitz MC (2010) Genetic engineering of algae for enhanced biofuel production. Eukaryot Cell 9(4):486–501

Ramasamy Sakthivel R, Sanniyasi Elumalai S, Arif MM (2011) Microalgae lipid research, past, present: a critical review for biodiesel production, in the future. J Exp Sci 2(10):29–49

Reda AI, Shanab A, Jeon BH, Song H, Kim Y, Hwang JH (2010) Algae-biofuel: potential use as sustainable alternative green energy. Online J Power Energy Eng 1(1):4–6

Round FE (1984) The ecology of algae. Cambridge University Press, Cambridge. ISBN-10: 0521269067

Schenk PM, Thomas-Hall SR, Stephens E, Marx UC, Mussgnug JH, Posten C, Kruse O, Hankamer B (2008) Second generation biofuels: high-efficiency microalgae for biodiesel production. Bioenergy Res 1:20–43. doi:10.1007/s12155-008-9008-8

Sharma KK, Schuhmann H, Schenk PM (2012) High lipid induction in microalgae for biodiesel production. Energies 5:1532–1553

Sharma P, Tandon GD, Khetmalas MB (2013) Total biomass utilization of Spirogyra *singularis* for renewable biofuel production. IJBPAS 2(1):138–148

Snow AA, Smith VH (2012) Genetically engineered algae for biofuels: a key role for ecologists. BioScience 62:765–768

Tan CK, Johns MR (1996) Screening of diatoms for heterotrophic eicosapentaenoic acid production. J Appl Phycol 8:59–64

Theriault RI (1965) Heterotrophic growth and production of xanthophylls by Chlorella pyrenoidosa. Appl Microbiol 13:402–416

Tilton RC, Murphy J, Dixon JK (1972) The flocculation of algae with synthetic polymeric flocculants. Water Res 6:155–164

Tredici MR, Carlozzi P, Chini Zittelli G, Materassi R (1991) A vertical alveolar panel(VAP) for outdoor mass cultivation of microalgae and cyanobacteria. Bioresour Technol 38:153–159

Velichkova K, Sirakov I, Georgiev G (2012) Cultivation of *Botryococcus braunii* strain in relation of its use for biodiesel production. J BioSci Biotechnol 157–162. Special Edition,"Biotechnologies" National Youth Conference "Biological sciences for a better future", Plovdiv, October 19–20

Wang B, Li Y, Wu N, Lan CQ (2008) CO2 bio-mitigation using microalgae. Appl Microbiol Biotechnol 79(5): 707–718

Wang ST, Pan YY, Liu CC, Chuang LT, Chen CN (2011) Characterization of a green microalga UTEX 2219–4: effects of photosynthesis and osmotic stress on oil body formation. Bot Stud 52:305–312

Wen ZY, Chen F (2000) Production potential of eicosapentaenoic acid by the diatom *Nitzschia laevis*. Biotechnol Lett 22(9):727–733

Xu H, Miao X, Wu Q (2004) High quality biodiesel production from a microalga *Chlorella protothecoides* by heterotrophic growth in fermenters. J Biotechnol 126:499–507

Yu H, Jia S, Dai Y (2009) Growth characteristics of the cyanobacterium Nostoc flagelliforme in photoautotrophic, mixotrophic and heterotrophic cultivation. J Appl Phycol 21(1):127–133

Molecular Diversity of Rhizobial and Nonrhizobial Bacteria from Nodules of Cool Season Legumes

10

S.S. Dudeja and Nidhi

Abstract

A specific trait of legumes is the ability to form nodules with nitrogen-fixing rhizobia. At present plenty of reports are available regarding the presence of other microorganisms (associated or endophytic) in the nodules apart from nitrogen-fixing bacteria, as nodules are much richer in nutrients as compared to roots. There are 16 genera of bacteria which are able to form nodules in different legumes. These include *Rhizobium*, *Ensifer*, *Mesorhizobium*, *Phyllobacterium*, *Bradyrhizobium*, *Ochrobactrum*, *Methylobacterium*, *Azorhizobium*, *Allorhizobium*, *Aminobacter*, *Shinella* and *Devosia* belonging to α-proteobacteria and four genera, *Burkholderia, Microvirga, Cupriavidus* and *Herbaspirillum*, belonging to β-proteobacteria. About 120 species belonging to these genera form nodules in different legumes. In case of cool season legumes, *Mesorhizobium* form nodules in chickpea and 30 *Mesorhizobium* species are known, but *M. ciceri*, *M. mediterraneum*, *M. temperadae*, *M. tianshanens*, *M.* sp. (*Cicer*) and *Mesorhizobium muleiense* sp. nov. have been reported to form nodules in chickpea. Similarly *R. leguminosarum* bv. *viciae* nodulates crop legume pea (*Pisum sativum*), *Vicia* spp. and *Lens esculenta.* The legume root nodules, apart from fixing atmospheric nitrogen, mediate numerous underground interactions with beneficial microbes, such as rhizobia, nonrhizobial bacteria, mycorrhizae and parasitic interactions with other pathogenic microbes. Legume root nodules constitute vast and diverse niches for endophytic organisms, and there is not a single legume nodule devoid of endophyte. The population density of endophytes is highly variable, depending mainly on the bacterial species and host genotypes but also on

S.S. Dudeja (✉)
Department of Microbiology, CCS Haryana Agricultural University, Hisar 125 004, India
e-mail: ssdudeja@gmail.com

Nidhi
Department of Biotechnology, Maharishi Dayanand University, Rohtak 124 001, India
e-mail: nidhirajatdudeja@gmail.com

R.K. Salar et al. (eds.), *Biotechnology: Prospects and Applications*,
DOI 10.1007/978-81-322-1683-4_10, © Springer India 2013

the host developmental stage, inoculum's density and environmental conditions. Large number of culturable as well as nonculturable rhizobial and nonrhizobial endophytes in legume nodules has been reported. Commonly rhizobial and nonrhizobial genera existing as endophytic in legume tissues include *Agrobacterium, Bacillus, Curtobacterium, Enterobacter, Erwinia, Mycobacterium, Paenibacillus, Pseudomonas, Phyllobacterium, Ochrobactrum, Sphingomonas, Rhizobium, Ensifer, Mesorhizobium, Burkholderia, Phyllobacterium* and *Devosia*.

10.1 Introduction

Symbiotic nitrogen fixation is an important source of nitrogen, and the various legume crops and pasture species often fix as much as 200–300 kg nitrogen per hectare. However, consumption of fertiliser nitrogen not only depletes nonrenewable energy sources but also leads to environmental pollution. Concomitant with N_2 fixation, the use of legumes in rotations offers control of crop diseases and pests. Among other nitrogen fixers, rhizobia are agriculturally more important as these form root or stem nodules on leguminous plants, in which they can fix atmospheric nitrogen. A legume can drive 5–83 % of its N requirement by fixing atmospheric N_2 under field conditions and remaining from the soil N pool (Dudeja et al. 2011; Sheokand et al. 2012). The use of efficient and competitive rhizobial inoculants to replace the inefficient native rhizobia often fails. To combat the situation, different strategies are used.

In this context other bacterial or fungal inoculants are used. Mainly the use of other inoculants which enhances the activity of rhizobial inoculants and enhances N_2 fixation like PSB inoculants, PGPR inoculants, *Trichoderma*+vitavax, VA mycorrhiza or even another endophytic fungus *Piriformospora indica* is important (Dudeja et al. 2011). However, depending upon the environmental conditions, cultural practices and mainly the number, structure and effectivity of native rhizobia, the use of above-mentioned inoculants along with rhizobia varies. Newly introduced strains often fail to compete with well-adapted indigenous populations, even in the presence of these inoculants. It is important to investigate the genetic structure of natural populations and their dynamics in relation to the host plant. Actually there are different groups of rhizobia and nonrhizobia which may show host specificity and compatibility in the formation and exist as endophytes in nodules.

Chickpea (*Cicer arietinum* L.) and field pea (*Pisum sativum*) are the most important cool season legume crops of the arid zones of India. However, Lentil (*Lens esculenta*), faba bean (*Vicia faba*) and fenugreek or methi (*Trigonella foenugrecum*) are the other cool-season legumes. Legumes form nodules in symbiotic association with rhizobia, in which they can fix atmospheric nitrogen. At present, the common belief that rhizobia of genus *Rhizobium* are able to form symbiotic nodules with legume plants is no longer true; now there are 16 genera of bacteria which are able to form nodules in different legumes. These include *Rhizobium, Ensifer, Mesorhizobium, Phyllobacterium, Bradyrhizobium, Ochrobactrum, Methylobacterium, Azorhizobium, Allorhizobium, Aminobacter, Shinella* and *Devosia* belonging to α-proteobacteria, and four genera – *Burkholderia, Microvirga, Cupriavidus* and *Herbaspirillum* – belonging to β-proteobacteria have been identified (Dudeja et al. 2012a; Euzéby 2012; Weir 2012; ICPS 2013).

Large number of bacterial species even more than 120 belonging to these genera has been identified which forms nodules in different legumes. Further there are large numbers of species of these rhizobial genera that do not form nodules and do not fix nitrogen, but performs various other functions in the soil or may exist as endophytic in different plant tissues. Little is known about the single host legumes being grown in different parts of the world, and this may be nodulated and inhabited by a wide range of rhizobial and nonrhizobial genera and species. Rhizobia form symbiotic nodules and also exist as nodule endophytes. Nonrhizobial bacteria also exist as nodule endophytes and are helper to rhizobia. Worldwide the molecular diversity of about 150 species of legumes has been assessed. There are about 685 genera of legumes with about 19,000 species, so to understand the different rhizobia forming nodules and fixing N_2 in nodules and also existing as endophytes and not forming nodules and not fixing N_2 needs concerted efforts (Dudeja and Narula 2008; Dudeja et al. 2012b).

10.2 Molecular Diversity of Nodule-Forming Rhizobia of Cool Season Legumes

Interestingly, different species of mesorhizobia form nodules in chickpea, and *R. leguminosarum* bv. *viciae* nodulates crop legumes pea (*Pisum sativum*), *Vicia* spp. and *Lens esculenta*, while *Ensifer meliloti* is reported to nodulate fenugreek or methi. Therefore, in this review, rhizobia nodulating only these cool season legumes will be reviewed and discussed.

10.2.1 Molecular Diversity of Chickpea Nodule-Forming Rhizobia

The genus *Mesorhizobium* was described as in between *Rhizobium* and *Bradyrhizobium* by Jarvis et al. in 1997. Several species were moved from *Rhizobium* to this genus. *Mesorhizobium* species have been described in Africa, Asia, Australia, Europe, South and North America and even in the Arctic (Poinsot et al. 2001). Nearly 30 *Mesorhizobium* species, including *Mesorhizobium abyssinicae*, *M. albiziae*, *M. alhagi*, *M. amorphae*, *M. australicum*, *M. camelthorni*, *M. caraganae*, *M. chacoense*, *M. ciceri*, *M. gobiense*, *M. hawassense*, *M. huakuii*, *M. loti*, *M. mediterraneum*, *M. metallidurans*, *M. opportunistum*, *M. plurifarium*, *M. qingshengii*, *M. robiniae*, *M. sangaii*, *M. septentrionale*, *M. shangrilense*, *M. shonense*, *M. silamurunense*, *M. tamadayense*, *M. tarimense*, *M. temperatum*, *M. thiogangeticum* and *M. tianshanense*, have been reported to nodulate various legumes in the Mimosoideae and Papilionoideae subfamilies of the Leguminosae. *Mesorhizobium* has been reported to infect and form nodules in *Acacia* spp., *Albizia*, *Anagyris latifolia*, *Astragalus sinicus*, *Astragalus luteolus*, *Astragalus ernestii*, *Carmichaelia*, *Cicer arietinum*, *Clianthus*, *Cytisus scoparius* (broom), *Lotus berthelotii*, *Montigena*, *Robinia pseudoacacia*, *Sophora*, *Ulex europaeus and agroforestry legume trees* (Degefu et al. 2013; Euzéby 2012; Ramírez-Bahena et al. 2012; Weir 2012; Zhang et al. 2012; Zheng et al. 2012; Zhou et al. 2013; ICSP 2013).

Different species of Mesorhizobia like *M. ciceri*, *M. mediterraneum*, *M. temperadae*, *M. tianshanens*, *M.* sp. (*Cicer*) and *Mesorhizobium muleiense* sp. nov. have been reported to form nodules in chickpea (Dudeja et al. 2009; Zhang et al. 2012). The molecular diversity of nodule-forming chickpea mesorhizobia has been reported by several workers. Nour et al. (1994) studied the genetic and phenotypic diversity of *Rhizobium* isolated from chickpea. The diversity of 16 strains of chickpea infecting rhizobia was studied using multilocus enzyme electrophoresis, RFLP of amplified 16S and IGS (intergenic spacer) rRNA gene and assimilation of 147 carbon sources; antibiotic resistance and tolerance to NaCl and extreme pH values and temperatures were tested. These approaches had different discriminating powers. Esterase polymorphisms gave a unique pattern for each strain, allowing this method to be used for strain fingerprinting. Genetic distances between strains were estimated. The three approaches used in this study yielded consistent results. High heterogeneity was evidenced among the strains and made it possible to classify the strains into two clusters. Isozyme patterns for superoxide dismutase were particularly interesting, since they delineated the same two groups. The phenotypic tests clearly confirmed the existence of two genetic groups on the basis of 11 phenotypic characters.

Nour et al. (1995) studied the genetic diversity of 30 isolates obtained from nodules on chickpea growing in uninoculated fields over a wide geographical range. A number of taxonomic approaches were used such as DNA-DNA relatedness analysis, RFLP of 16S rDNA intergenic spacer (IGS) and total 16S rDNA sequence analysis. The division of chickpea-infective strains into two major phylogenetic groups (A and B) was confirmed by polymorphism of 16S IGS rDNA. A total of five genomic species including *R. ciceri* at present shifted to *M. ciceri* was identified. All of the group B strains except one were homogeneous and belonged to a single genomic species corresponding to *R. ciceri.* Group A was heterogeneous, containing three genomic species and five strains that remained unclassified, and its members had very different PCR-RFLP profiles.

Tan et al. (1997) studied the genetic and phylogenetic relationships for strains of *M. tianshanense* and its relatives. The phylogenetic analysis based on the 16SrRNA gene sequences showed that *M. tianshanense* was closely related to the *Mesorhizobium* phylogenetic branch.

Chaudhary et al. (2001) studied phenotypic characteristics and intrinsic antibiotic resistance (IAR) pattern of indigenous rhizobial population infecting high-nodulating (HN) and low-nodulating (LN) selections of two chickpea cultivars ICC4948 and ICC5003. Respective isolation efficiencies from nodules in HN selection were 77 and 66 %, while in LN selections were 56 and 23 %. Rhizobial growth from different nodules took 1–15 days. In different isolates, diversity in colony coloration, brightness or dullness, raised or flat, gum production and dye uptake was also observed. Based on IAR pattern, rhizobia were grouped into 12 groups. The isolates from cv. ICC4948 HN and LN selections belonged to 7–8 and 1–3 groups, respectively. Corresponding groups of nodulation variants of another cv. ICC5003 were 3–7 and 1–2 groups. IAR pattern of the isolates was not limited to 12 groups but was more heterogeneous, and HN selections were more promiscuous than LN selections. Further, to assess the heterogeneity of rhizobia, two criteria were used. One based on presence of proteins of different molecular weight and *Rm* value, while second was the variation in the presence of major protein bands in an isolate (Chaudhary et al. 2002). In majority of the isolates, both from HN and LN variants, proteins of more than 45kD were present, and slight variation in the presence of proteins of lower molecular weight in different isolates was observed. However, based on the presence of major and minor bands, large heterogeneity in rhizobial isolates was observed. In isolates from HN selections, all the 13 major bands were present, while in case of isolates from LN selections, two major bands (8th and 9th) were absent in all the isolates and two other bands (7th and 11th) were absent in majority of them. The ability to form nodules on HN and LN selection of same cultivars was also assessed to further ascertain the heterogeneity among the isolates. All the isolates from HN and LN selections were able to nodulate plants of their respective host HN and LN selections, except for isolate no. 855, isolated from LN selection of cv. ICC4948. Isolates selected from HN and LN selections, when cross infected to LN or HN plants, also formed nodules except the two HN isolates, which could be existing as endophytic in the chickpea nodules.

Laranjo et al. (2001, 2002, 2004) and Alexandre et al. (2006) evaluated differences between chickpea rhizobial populations from three geographical areas in southern Portugal. The genetic characterisation of the isolates was done by plasmid profiles and restriction analysis of the *nif*H gene. Symbiotic efficiency of the isolates was also determined. Soil from one of the location revealed a larger rhizobial population with higher symbiotic efficiency values. Genetic and phenotypic differences were detected between the natural rhizobial populations from the three locations.

Maatallah et al. (2002) determined the phenotypic and genotypic biodiversity of rhizobial strains nodulating chickpea from various areas of Morocco. Chickpea rhizobia isolated from Morocco soils were both phenotypically and genetically diverse. Most of these strains belonged to *Mesorhizobium* group. However, some strains appeared to have 16S rRNA genes similar to *Sinorhizobium* isolates.

Rachna (2005) reported the molecular diversity of chickpea rhizobia isolated from Haryana state and showed the presence of a considerable molecular diversity. Functional diversity was assessed as symbiotic effectiveness of the different rhizobial isolates with two chickpea cultivars HC1 and HC 5 and showed very wide symbiotic ratio (S/R) based on N uptake; it ranged from 2.0 to 4.2. Isolates like CP2381A, CP741, CP1423 and CP2437 were highly efficient as symbiotic ratio was >4.0 (Nandwani and Dudeja 2013). After sequencing of the amplified 16S rRNA gene of the two representatives, rhizobial strains showed homology with *Mesorhizobium* sp. (Jagiriti 2005).

Rivas et al. (2007) identified several strains of *Mesorhizobium amorphae* and *Mesorhizobium tianshanense* nodulating *Cicer arietinum* in Spain and Portugal and studied the symbiotic genes carried by these strains. The sequences of 16S–23S intergenic spacer (ITS), 16S rRNA gene and symbiotic genes *nod*C and *nif*H were analysed. According to their 16S rRNA gene and ITS sequences, the strains from this study were identified as *M. amorphae* and *M. tianshanense*. The type strains of these species were isolated in China from *Glycyrrhiza pallidiflora* and *Amorpha fruticosa* nodules, respectively, and are not capable of nodulating chickpea. These strains carry symbiotic genes, phylogenetically divergent from those of the chickpea isolates, whose *nod*C and *nif*H genes showed more than 99 % similarity with respect to those from *Mesorhizobium ciceri* and *Mesorhizobium mediterraneum*, the two common chickpea nodulating species in Spain and Portugal. The results from this study showed that different symbiotic genes have been acquired by strains from the same species during their coevolution with different legumes in distinct geographical locations. A new infrasubspecific division named biovar *ciceri* is proposed within *M. amorphae* and *M. tianshanense* to include the strains able to effectively nodulate *Cicer arietinum*.

Several phenotypic markers were used in this study to determine the biodiversity of rhizobial strains nodulating *Cicer arietinum* L. in various areas of Tunisia (L'taief et al. 2007). They include symbiotic traits, the use of 21 biochemical substrates and tolerance to salinity and pH. In addition, restriction fragment length polymorphisms (RFLPs) of PCR-amplified 16S rDNA were compared with those of reference strains. Numerical analysis of the phenotypic characteristics showed that the 48 strains studied fell into three distinct groups. This heterogeneity was highly supported by the RFLP analysis of 16S rRNA genes, and two ribotypes were identified. Chickpea rhizobia isolated from Tunisian soils are both phenotypically and genetically diverse. Results showed that 40 and 8 isolates were assigned, respectively, to *Mesorhizobium ciceri* and *Mesorhizobium mediterraneum*. Brigido et al. (2007) isolated chickpea rhizobia from 12 Portuguese soils which were moderately acidophiles and were identified as mesorhizobia by 16S rDNA partial sequence analysis.

Chickpea (*Cicer arietinum* L.) nodulation variants of two cultivars ICC 4948 and ICC 5003 were used as trap plants to isolate 385 native rhizobia from CCS Haryana Agricultural University, Hisar farm soil (Dudeja and Singh 2008). After authentication and considering growth characteristics, selected 110 rhizobia revealed immense molecular diversity using PCR-based amplification of ERIC sequences. Low-nodulating variants of cvs ICC 4948 and ICC 5003 were able to trap more numbers of rhizobial genotypes, viz. 7 as compared to 4–5 by high-nodulating variants of these cultivars. Overall eight rhizobial genotypes were trapped by the chickpea cultivars. Rhizobial isolates from the same nodule or same plants were present in the same or different clusters, and few isolates showed 100 % similarity also. Based on nodules from a plant, nodulation variant or cultivar rhizobia could not be differentiated, and no exclusive cluster was formed by either rhizobial isolates from low- or high-nodulating variants of both the cultivars. Two predominant rhizobial isolates LN 707b and LN 7007 isolated from low-nodulating cvs ICC 4948 and ICC 5003, respectively, were characterised by amplification and sequencing of 16S rRNA gene. Rhizobial isolate LN 707b and LN 7007 showed more than 98 and 99 % similarity, respectively, with *Mesorhizobium mediterraneum*, and other rhizobial spp. closely related to these chickpea rhizobial isolates showing less than 2 %

dissimilarity were *Mesorhizobium* sp. SH 2851 *Rhizobium mediterraneum*, *Mesorhizobium* sp. Rch 9, *Mesorhizobium* sp. STM 398 and *Mesorhizobium temperadae*. Relatedness with *Mesorhizobium ciceri* was about 97–98 %. So the chickpea rhizobia from Northern Indian subcontinent are proposed to be kept under *Mesorhizobium mediterraneum* strain LN707b and LN 7007.

Similarly, in total 137 mesorhizobial isolates were made from 32 different farmers' fields from eight districts of Haryana state in India (Nandwani and Dudeja 2009). Considerable molecular diversity in Haryana soils with high richness indices (0.61) was observed. Out of six mesorhizobial genotypes (MG), MG II was able to form 38 % of nodules. Mesorhizobial genotype III was invariably present in all the nodule samples tested. Zhang et al. (2012) reported that in China *Mesorhizobium muleiense* sp. nov. is also nodulating *Cicer arietinum* L.

10.2.2 Molecular Diversity of Field Pea Nodule-Forming Rhizobia

The genus *Rhizobium* for many years was a "catch all" genus for all rhizobia. Some species were later moved in to new genera based on phylogenetic analyses. The genus *Rhizobium* is found worldwide with about 58 species – *Rhizobium aggregatum*, *R. alamii*, *R. alkalisoli*, *R. borbori*, *R. cauense*, *R. cellulosilyticum*, *R. daejeonense*, *R. endophyticum*, *R. etli*, *R. fabae*, *R. galegae*, *R. gallicum*, *R. giardinii*, *R.grahamii*, *R. hainanense*, *R. halophytocola*, *R. helanshanense*, *R. herbae*, *R. huautlense*, *R. indigoferae*, *R. larrymoorei*, *R. leguminosarum*, *R. leucaenae*, *R. loessense*, *R. lupini*, *R. lusitanum*, *R. mesosinicum*, *R. mesoamericanum*, *R. miluonense*, *R. mongolense*, *R. multihospitium*, *R. nepotum*, *R. oryzae*, *R. petrolearium*, *R. phaseoli*, *R. pisi*, *R. pseudoryzae*, *R. pusense*, *R. qilianshanense*, *R. radiobacter*, *R. rhizogenes*, *R. rosettiformans*, *R. rubi*, *R. selenitireducens*, *R. skierniewicense*, *R. sphaerophysae*, *R. soli*, *R. sullae*, *R. tarimense*, *R. tibeticum*, *R. tropici*, *R.tubonense*, *R. undicola*, *R. vallis*, *R. vignae and R. vitis* (Kaur et al. 2011; López-Guerrero et al. 2012; López-López et al. 2011, 2012; Panday et al. 2011; Ren et al. 2011; Wang et al. 2011; Zhang et al. 2011; Althabegoiti et al. 2012; Bibi et al. 2012; Degefu et al. 2012; Euzéby 2012; Pulawska et al. 2012; Ribeiro et al. 2012; Ramírez-Bahena et al. 2012; Servín-Garcidueñas et al. 2012; Weir 2012; Xu et al. 2012; Yao et al. 2012; Zhang et al. 2012; Zheng et al. 2012; Zhou et al. 2013; ICSP 2013).

These rhizobial species are known to interact with large number of legumes to form nodules and fix nitrogen, and still large numbers of unidentified, unexplored legumes are there. These species of rhizobia nodulate *Astragalus* spp., *Calliandra calothyrsus*, *Campylotropis* spp., *Caragana intermedia*, *Cassia* spp., *Cytisus scoparius*, *Dalea purpurea*, *Galega* spp., *Gliricidia sepium*, *Gueldenstaedtia multiflora*, *Hedysarum coronarium*, *Kummerowia stipulacea*, *Lathyrus* spp., legume tree spp., *Leucaena leucocephala*, *Lotus edulis*, *Medicago* spp., *Mimosa affinis*, *Ornithopus compressus*, *Oxytropis ochrocephala*, *Phaseolus albescens*, *Phaseolus* spp., *Pisum sativum*, *Scorpiurus muricatus*, *Sesbania* spp., *Sphaerophysa salsula*, *Stylosanthes*, *Tepohrsia*, *Trifolium* spp., *Trigonella* spp., *Trisperma*, *Vicia faba* and *Vigna radiate* and other legumes.

It is important to point out that all the abovementioned species of rhizobia are not nodule forming; they may perform other functions. For example, in the *Rhizobium* genus, there are *Rhizobium radiobacter* and other species like *R. naphthalenivorans*, *R. subbaraonis*, *R. phenanthrenilyticum* and *R. tarimense* (Wen et al. 2011; Kaiya et al. 2012; Ramana et al. 2012; Turdahon et al. 2012). These species do not form nitrogen-fixing symbiotic root nodules, unless they contain a symbiotic (sym) plasmid (Velázquez et al. 2005).

R. leguminosarum bv. *viciae* nodulates crop legumes pea (*Pisum sativum*), *Vicia* spp., and *Lens esculenta*; bv. *trifolii* nodulates *Trifolium* spp.; and bv. *phaseoli* nodulates *Phaseolus*. However, according to the latest notification, *Rhizobium trifolii* is a later synonym of *R. leguminosarum*, *R. phaseoli* is a valid separate species, and some isolates formerly known as *R. leguminosarum* are now *R. pisi* (Ramirez-Bahena et al. 2008), indicating a fast change in the classification and number of species of *Rhizobium* with the exploration of unexplored legumes and using more advanced techniques.

Young et al. (1987) extracted 249 isolates from root nodules of pea. Electrophoretic type (ET) frequencies were significantly different among isolates from nodules on primary roots as opposed to lateral roots. The population on each individual plant was very diverse, but ET frequencies were similar from plant to plant. The ETs nodulating the primary roots were almost, although not perfectly, mixed, since the incidence of the same ETs in adjacent nodules was only about twice that expected by chance.

The genetic structure in soil populations of *Rhizobium leguminosarum* bv. *trifolii* and *viciae* was determined (Strain et al. (1994). The analysis of allelic variation at 13 enzyme-encoding loci by multilocus enzyme electrophoresis delineated 202 chromosomal types (ETs) among a total of 456 isolates representing two populations of *R. leguminosarum* bv. *trifolii* and two populations of *R. leguminosarum* bv. *viciae*. These results were further supported by repetitive extragenic palindromic and enterobacterial repetitive intergeneric consensus sequences and the PCR technique. Similarly 43 strains of *Rhizobium leguminosarum* biovars *viciae*, *trifolii* and *phaseoli* were characterised by PCR DNA fingerprinting of interrepeat sequences and RFLP analysis of PCR-amplified chromosomal and symbiotic gene regions and concluded that classifications of *R. leguminosarum* strains by the PCR-based methods were correlated with those previously obtained by conventional total DNA restriction profile comparisons and RFLP analysis using chromosomal and symbiotic gene probes (Laguerre et al. 1996).

Management regime affects pea-nodulating *Rhizobium leguminosarum* populations (Palmer and Young 2000). Usually population, and that *R. leguminosarum* biovar *trifolii* was approximately five times more abundant than biovar *viciae* in that pasture soil, whereas *Ensifer meliloti* was rare (Zézé et al. 2001). The stable persistence of the prevailing *R. leguminosarum* bv. *viciae* genotypes in the absence of its host plant was observed (Zhang et al. 2001; Duodu et al. 2005). Effective N_2-fixing strains of *Rhizobium leguminosarum* bv. *trifolii* nodulating red clover were common and genetically diverse in the soils (Duodu et al. 2007).

Mutch and Young (2004) characterised host specificity and genetic diversity for a soil population of *Rhizobium leguminosarum* biovar *viciae* sampled using six wild *Vicia* and *Lathyrus* species and the crop plants pea (*Pisum sativum*) and broad bean (*Vicia faba*). Host range tests of 80 isolates demonstrated that only 34 % of isolates were able to nodulate *V. faba*. By contrast, 89 % were able to nodulate all the local wild hosts tested. So, high genetic diversity of the rhizobial population cannot be ascribed directly to the diversity of host species at the site. Similarly 27 new *Rhizobium* isolates were obtained from root nodules of wild and crop legumes belonging to the genera *Vicia*, *Lathyrus* and *Pisum* (Moschetti et al. 2005). The analysis of symbiotic properties and stress tolerance tests revealed that wild isolates showed a wide spectrum of nodulation and a marked variation in stress tolerance compared with reference strains tested in the study. Blazinkov et al. (2007) also reported significant differences among *Rhizobium leguminosarum* bv. *viciae* isolates revealing considerable genetic diversity of rhizobial field populations.

Yang et al. (2008) studied the diversity of 42 isolates from effective nodules of *Pisum sativum*; the analysis divided the isolates into 18 genotypes and four groups. Whereas Wadhwa et al. (2011) using five field pea cultivars; HFP 4, HVP 3–5, HFP 9426, Jayanti and Hariyal, isolated 54 rhizobia. In total 13 genotypes of field pea rhizobia were identified, and only two pea rhizobial genotypes were detected in all the five pea cultivars. Furthermore, high strain richness index (0.43–0.5) of field pea rhizobia was observed by using molecular techniques.

10.3 Molecular Diversity of Rhizobial and Nonrhizobial Bacteria Existing as Nodule Endophyte

The legume root nodules, apart from fixing atmospheric nitrogen, mediate numerous underground interactions with beneficial microbes, such as rhizobia, nonrhizobial bacteria, mycorrhizae and plant growth-promoting rhizobacteria (PGPR) and parasitic interactions with other pathogenic microbes. Legume root nodules constitute vast and diverse niches for endophytic organisms, and there is not a single legume nodule devoid of endophyte. Endophytic populations of microorganisms may work jointly with nitrogen-fixing bacteroides. But benefits conferred by these endophytes are not well recognised. There is increasing interest in developing biofertilisers for enhancing crop productivity by these helper bacteria existing as nodule endophytes.

The need for an update of current work is especially valid for legume nodules which carry the specific trait to be able to fix N_2. The family Fabaceae is the third largest family of flowering plants. Nodules and legumes are benefited extensively by endophytic microbes; they promote plant growth and yield and confer enhanced resistance to various pathogens. Endophytes also produce unusual secondary metabolites of plant importance and activate the stress response system more rapidly and strongly in plants, leading to higher resistance against pathogens. Endophytes may also help to remove contaminants, solubilise phosphate or contribute assimilable nitrogen to plants. In addition, endophytic bacteria supply essential vitamins to plants. The production of auxin-like compounds increases seed production and germination along with increased shoot growth and tillering. Other effects of endophyte infection on the host plant include osmotic adjustment, stomatal regulation, modification of root morphology, enhanced uptake of minerals and alteration of nitrogen accumulation and metabolism (Dudeja et al. 2012b).

The population density of endophytes is highly variable, depending mainly on the bacterial species and host genotypes but also on the host developmental stage, inoculum density and environmental conditions. Endophytic N_2-fixing bacteria seem to constitute only a small proportion of the total endophytic bacteria, and increasing N_2-fixing populations in plants has been considered as a possibility to increase nitrogen fixation. It seems that the bacteria best adapted for living inside plants are naturally selected. Since soil represents a rich pool of inocula and roots are especially vulnerable to invasion, the soil- or rhizophere-derived endophytic community is of special importance to the plants.

A specific trait of legumes is the ability to form nodules. This tissue is a supreme habitat for bacteria, not only for nitrogen-fixing rhizobia. At present plenty of reports are available regarding the presence of other microorganisms (associated or endophytic) in the nodules as nodules are much richer in nutrients as compared to roots. During the present decade, large numbers of culturable as well as nonculturable rhizobial and nonrhizobial endophytes in legume nodules have been reported (Sun et al. 2008; Lei et al. 2008), and even in our laboratory, in roots and nodules of chickpea and pea, large numbers of bacterial endophytes were isolated (Narula et al. 2013; Saini et al. 2013). A total of 12 endophytic bacteria from roots and 76 from the nodules of chickpea were isolated. Out of 88 isolates, wide diversity among chickpea bacterial endophytes was observed, and in the university farm soils, 5 genotypes were present in chickpea roots, while 13 genotypes were present in the nodules. Similarly endophytic bacteria from root and nodules of field pea were isolated. Molecular studies of 75 isolates showed that 3 genotypes were present in field pea roots while 17 genotypes were present in the nodules.

Rhizobia though isolated from nodules, but upon reinfection they may not form the nodules, but are merely co-occurring as associated endophytes. They may potentially supply signals leading to nodulation by the plant but with insufficient invasion skills if not associated by helper bacteria.

In addition to the rhizobia, endophytic bacteria have been isolated from different legume plants such as alfalfa, clover, soybean, *Argyrolobium uniflorum, Vicia, Oxytropis, Medicago, Melilotus, Onobrychis*, peanut and *Acacia*. Large numbers of bacterial genera have been reported to be associated with different legumes. Over 24 nonrhizobial and about 10 rhizobial genera have been isolated from legume tissues including *Agrobacterium, Bacillus, Curtobacterium, Enterobacter, Erwinia, Mycobacterium, Paenibacillus, Pseudomonas, Phyllobacterium, Ochrobactrum, Sphingomonas,*

Rhizobium, *Ensifer*, *Mesorhizobium*, *Burkholderia*, *Phyllobacterium* and *Devosia* (Lei et al. 2008; Muresu et al. 2008).

Thirty-one bacterial species of 14 different genera were recovered from the nodules of red clover plants (*Trifolium pratense* L.) (Sturz and Christie 1995, 1996; Sturz et al. 1997; Sturz and Nowak 2000). *R. leguminosarum* bv *phaseoli* and *R. loti* were found in the nodules with 27.2 % of each. Clover root nodules were host to 12 bacterial species other than rhizobia, of which 8 were specific to this tissue. Similarly, Zakhia et al. (2006) described the association of 14 bacterial genera with wild legume nodules in Tunisia. *Agrobacterium tumefaciens* strains were reported to be endophytic bacteria in the roots, stems and root nodules of *Melilotus dentatus* and other legumes (Mahdhi et al. 2007; Stajković et al. 2009). Plant growth-promoting rhizobacteria (PGPR) were isolated from bean (*Phaseolus vulgaris* L.) nodules (Lee et al. 2005). The helper strain was a gram-positive, spore-forming and rod-shaped bacterium. Biolog test and phylogenetic analysis of 16S rRNA gene hypervariant region sequences demonstrated the strain to be *Bacillus subtilis*.

Forty-two bacterial isolates from root nodules of *Argyrolobium uniflorum* growing in the arid areas of Tunisia were characterised by phenotypic features, RFLP and sequencing of PCR-amplified 16S rRNA genes (Mahdhi et al. 2007). The isolates were found to be phenotypically diverse. The majority of the isolates tolerated 3 % NaCl and grew at temperatures up to 40 °C. Phylogenetically, the new isolates were grouped in the genera *Ensifer* (*Sinorhizobium*) (Nimnoi and Pongsilp 2009), *Rhizobium* (Hoque et al. 2011) and *Agrobacterium* (Rosenblueth and Martinez-Romero 2006). Except for the two *Agrobacterium* isolates, all strains induced nodulation on *Argyrolobium uniflorum*, but the number of nodules and nitrogen fixation efficiency varied among them, with *Ensifer* (*Sinorhizobium*) sp. strains being the most effective symbionts.

In a study, 61 root nodule isolates from *Vicia*, *Oxytropis*, *Medicago*, *Melilotus* and *Onobrychis* species grown in Qinghai-Tibet and a loess plateau were comparatively characterised (Kan et al. 2007). Based upon the results of numerical taxonomy, ARDRA, AFLP, DNA-DNA hybridisation and 16S rDNA sequencing, the isolates were classified as *Rhizobium leguminosarum*, *Ensifer* (*Sinorhizobium*) *meliloti*, *Ensifer* (*Sinorhizobium*) *fredii*, *Mesorhizobium* sp., *Phyllobacterium* sp., *Stenotrophomonas* sp. and two non-symbiotic groups related to *Agrobacterium* and *Enterobacteriaceae*.

Nonrhizobial endophytes from the surface-sterilised root nodules of alfalfa (*Medicago sativa* L.) were isolated and characterised (Stajković et al. 2009). Out of 15 endophytic nonrhizobial strains isolated, three isolates were gram-positive strains and identified as *Bacillus megaterium*, *Brevibacillus choshinensis* and *Microbacterium trichothecenolyticum*. None of these isolates was able to nodulate alfalfa when reinoculated under sterilised conditions. However, co-inoculation of all nonrhizobial strains with *Ensifer* (*Sinorhizobium*) *meliloti* positively influenced nodule number, while shoot and root parameters were comparable to those of uninoculated alfalfa plants.

Various rhizobial and nonrhizobial strains were isolated from root nodules of two widespread south eastern Australian tree legumes (*Acacia salicina*, *A. stenophylla*). This legume was nodulated primarily by *Bradyrhizobium*, while the results indicate significant associations with other root nodule-forming bacterial genera, including *Rhizobium*, *Ensifer*, *Mesorhizobium*, *Burkholderia*, *Phyllobacterium* and *Devosia*. Genetic analyses also revealed a diverse suite of non-nodulating bacterial endophytes, only a subset of which has been previously recorded (Hoque et al. 2011). Some of these may perform a helper function for nodule-inducing rhizobia, while others can be viewed as mainly opportunistic in the nitrogen-rich niche.

10.4 Conclusions

Earlier only one rhizobial genus, *Rhizobium*, with different species was known to form nodules on legumes and fix atmospheric nitrogen, but at present, with the development of advanced molecular techniques, identification of more and more nodule-forming bacteria has totally changed the scenario. To date about 150 symbiotic nodules forming bacterial species have been identified in 16 genera under α- and β-subclass of proteobacteria. Few species of rhizobia may not be nodules forming unless they contain a symbiotic (sym) plasmid. Moreover, large number rhizobial and nonrhizobial genera in nodules do not form nodules. The population density of rhizobia and nonrhizobial isolates is highly variable, depending mainly on the microbial species and host genotypes, plant tissue and environmental conditions. The quantum of benefit derived by plants from rhizobial and nonrhizobial bacteria and vice versa is still not clear. It seems that the rhizobial and nonrhizobial genus or species best adapted for the environment are naturally selected for association with the legume nodules. However, the processes of host-microbe signalling and colonisation and the mechanisms leading to mutual benefits is taking place is being studied, but still high amount of progress is required to fully understand the mechanism of establishment. Further enormous efforts are needed world over under different geographical regions to study the molecular diversity and identification of new genera and species of root nodule bacteria and existing as endophytic in nodules from the existing as well as from the unexplored legumes.

References

Alexandre A, Laranjo M, Oliveira S (2006) Natural populations of chickpea rhizobia evaluated by antibiotic resistance profiles and molecular methods. Microb Ecol 51:128–136

Althabegoiti MJ, Lozano LF, Torres-Tejerizo G, Ormeño-Orrillo E, Rogel MA, González V, Martínez-Romero E (2012) Genome sequence of *Rhizobium grahamii* CCGE502, a broad-host-range symbiont with low nodulation competitiveness in *Phaseolus vulgaris*. J Bacteriol 194:6651–6652

Bibi F, Chung EJ, Khan A, Jeon CO, Chung YR (2012) *Rhizobium halophytocola* sp. nov., isolated from the root of a coastal dune plant. Int J Syst Evol Microbiol 62:1997–2003

Blazinkov M, Sikora S, Uher D, Macesic D, Redzepovic S (2007) Genotypic characterisation of indigenous *Rhizobium leguminosarum* bv. *viciae* field population in Croatia. Agric Conspec Sci 72:153–158

Brigido C, Alexandre A, Laranjo M, Oliveira S (2007) Moderately acidophilic mesorhizobia isolated from chickpea. Lett Appl Microbiol 44:168–174

Chaudhary P, Dudeja SS, Khurana AL (2001) Chickpea nodulation variants as a tool to detect the population diversity of chickpea rhizobia in soil. Physiol Mol Biol Plant 7:47–54

Chaudhary P, Khurana AL, Dudeja SS (2002) Heterogeneity of rhizobia isolated from chickpea nodulation variants. Indian J Microbiol 42:195–199

Degefu T, Wolde-Meskel E, Liu B, Cleenwerck I, Willems A, Frostegård A (2013) *Mesorhizobium shonense* sp. nov., *Mesorhizobium hawassense* sp. nov. and *Mesorhizobium abyssinicae* sp. nov. isolated from root nodules of different agroforestry legume trees growing in southern Ethiopia. Int J Syst Evol Microbiol, 63:1746–1753

Dudeja SS, Narula N (2008) Molecular diversity of root nodule forming bacteria (Review). In: Khachatourians GG, Arora DK, Rajendran TP, Srivastava AK (eds) Agriculturally important microorganisms, vol II. Academic World International, Bhopal, pp 1–24

Dudeja SS, Singh PC (2008) High and low nodulation in relation to molecular diversity of chickpea mesorhizobia in Indian soils. Arch Agron Soil Sci 54:109–120

Dudeja SS, Nandwani R, Narula N (2009) Molecular diversity of mesorhizobia: particularly rhizobia infecting chickpea. In: Singh DP, Tomar VS, Behl RK, Upadhayaya SD, Bhale MS, Khare D (eds) Proceedings of international conference on sustainable agriculture for food bioenergy & livelihood security, 14–16 Feb 2007. Agrobios International Publishers, Jodhpur, pp 336–347

Dudeja SS, Singh NP et al (2011) Biofertilizer technology and pulses production. In: Singh A, Parmar N, Kuhad RC (eds) Bioaugmentation, biostimulation and biocontrol, Soil biology series. Springer, Berlin/Heidelberg/New York, pp 43–63

Dudeja SS, Sheokand S, Kumari S (2012a) Legume root nodule development and functioning under tropics and subtropics: perspectives and challenges. Legume Res 35:85–103

Dudeja SS, Giri R, Saini R, Suneja-Madan P, Kothe E (2012b) Interaction of endophytic microbes with legumes. J Basic Microbiol 52:248–260

Duodu S, Bhuvaneswari TV, Gudmundsson J, Svenning MM (2005) Symbiotic and saprophytic survival of three unmarked *Rhizobium leguminosarum* biovar *trifolii* strains introduced into the field. Environ Microbiol 7:1049–1058

Duodu S, Carlsson G, Huss-Danell K, Svenning MM (2007) Large genotypic variation but small variation in N_2 fixation among rhizobia nodulating red clover in soils of northern Scandinavia. J Appl Microbiol 102:1625–1635

Euzéby JP (2012) "Proteobacteria". List of Prokaryotic names with Standing in Nomenclature (LPSN). http://www.bacterio.cict.fr/classifphyla.html#Proteobacteria. Retrieved 2 May 2012

Hoque MS, Broadhurst LM, Thrall PH (2011) Genetic characterisation of root nodule bacteria associated with *Acacia salicina and Acacia stenophylla (Mimosaceae)* across south eastern Australia. Int J Syst Evol Microbiol 61:299–309

ICSP Subcommittee on the taxonomy of *Rhizobium* and *Agrobacterium* – diversity, phylogeny and systematics (2013) Rhizobial taxonomy up-to-date Submitted by vinuesa, 20 Jan 2013, 20:23

Jagiriti (2005) Characterization of chickpea, pigeonpea and mungbean rhizobia by RFLP of PCR amplified products. PhD thesis submitted to CCS Haryana Agricultural University, Hisar, India

Kaiya S, Rubaba O, Yoshida N, Yamada T, Hiraishi A (2012) Characterization of *Rhizobium naphthalenivorans* sp. nov. with special emphasis on aromatic compound degradation and multilocus sequence analysis of housekeeping genes. J Gen Appl Microbiol 58:211–224

Kan FL, Chen ZY, Wang ET, Tian CF et al (2007) Characterization of symbiotic and endophytic bacteria isolated from root nodules of herbaceous legumes grown in Qinghai-Tibet plateau and in other zones of China. Arch Microbiol 188:103–115

Kaur J, Verma M, Lal R (2011) *Rhizobium rosettiformans* sp. nov., isolated from a hexachlorocyclohexane dump site, and reclassification of *Blastobacter aggregatus* Hirsch and Müller 1986 as *Rhizobium aggregatum* comb. nov. Int J Genet Syst Evol Microbiol 61:1218–1225

L'taief B, Bouaziz S, Gi M, Mainassara ZA, Mokhtar L (2007) Phenotypic and molecular characterization of chickpea rhizobia isolated from different areas of Tunisia. Can J Microbiol 53:427–434

Laguerre G, Mavingui P, Allard MR, Charnay MP, Louvrier P, Mazurier SI, Rigottier-Gois L, Amarger N (1996) Typing of rhizobia by PCR DNA fingerprinting and PCR-restriction fragment length polymorphism analysis of chromosomal and symbiotic gene regions: application to *Rhizobium leguminosarum* and its different biovars. Appl Environ Microbiol 62:2029–2036

Laranjo M, Rodrigues R, Alho L, Oliveria S (2001) Rhizobia of chickpea from Southern Portugal: symbiotic efficiency and genetic diversity. J Appl Microbiol 90:662–667

Laranjo M, Branco C, Soares R, Alho L, Carvalho MD, Oliveria S (2002) Comparison of chickpea rhizobia isolated from diverse Portuguese natural populations based on symbiotic effectiveness and DNA fingerprinting. J Appl Microbiol 92:1043–1050

Laranjo M, Machado J, Young JPW, Oliveira S (2004) High diversity of chickpea *Mesorhizobium* species isolated in a Portuguese agricultural region. FEMS Microbiol Ecol 48:101–107

Lee KD, Bai Y, Smith D, Han HS, Supanjani (2005) Isolation of plant-growth-promoting endophytic bacteria from bean nodules. Res J Agric Biol Sci 1:232–236

Lei X, Wang ET, Chen WF, Sui XH, Chen WX (2008) Diverse bacteria isolated from root nodules of wild *Vicia* species grown in temperate region of China. Arch Microbiol 190:657–671

López-Guerrero M, Ormeño-Orrillo E, Velázquez E, Rogel MA, Acosta JL, González V, Martínez J, Martínez-Romero E (2012) *Rhizobium etli* taxonomy revised with novel genomic data and analyses. Syst Appl Microbiol 35:353–358

López-López A, Rogel MA, Ormeño-Orrillo E, Martínez J, Martínez-Romero E (2011) *Phaseolus vulgaris* seed-borne endophytic community with novel bacterial species such as *Rhizobium Endophyticum* sp. nov. Syst Appl Microbiol 33:322–327

López-López A, Rogel-Hernández MA, Barois I, Angel IB, Ceballos IO, Julio Martínez J, Ormeño-Orrillo E, Martínez-Romero E (2012) *Rhizobium grahamii* sp. nov., from nodules of *Dalea leporina*, *Leucaena leucocephala* and *Clitoria ternatea*, and *Rhizobium mesoamericanum* sp. nov., from nodules of *Phaseolus vulgaris*, siratro, cowpea and *Mimosa pudica*. Int J Syst Evol Microbiol 62:2264–2271

Maatallah J, Berraho EB, Munoz S, Sanjuan J, Lluch C (2002) Phenotypic and molecular characterization of chickpea rhizobia isolated from different areas of Morocco. J Appl Microbiol 93:531–540

Mahdhi M, Nzoué A, Gueye F, Merabet C et al (2007) Phenotypic and genotypic diversity of *Genista saharae* microsymbionts from the infra-arid region of Tunisia. Lett Appl Microbiol 45:604–609

Moschetti G, Peluso AL, Protopapa A, Anastasio M, Pepe O, Defez R (2005) Use of nodulation pattern, stress tolerance, *nod*C gene amplification, RAPD-PCR and RFLP-16S rDNA analysis to discriminate genotypes of *Rhizobium leguminosarum* biovar *viciae*. Syst Appl Microbiol 28:619–631

Muresu R, Polone E, Sulas L, Baldan B et al (2008) Coexistence of predominantly nonculturable rhizobia with diverse, endophytic bacterial taxa within nodules of wild legumes. FEMS Microbiol Ecol 63:383–400

Mutch LA, Young JP (2004) Diversity and specificity of *Rhizobium leguminosarum* biovar *viciae* on wild and cultivated legumes. Mol Ecol 8:2435–2444

Nandwani R, Dudeja SS (2009) Molecular diversity of mesorhizobia in Indian soils. J Basic Microbiol 49:463–470

Nandwani R, Dudeja SS (2013) Functional diversity of native mesorhizobial genotypes available in Indian soils of Haryana state which nodulates chickpea. Acta Agron Hung 61(3):207–217

Narula S, Anand RC, Dudeja SS, Kumar V, Pathak DV (2013). Molecular diversity of root and nodule endophytic bacteria from field pea (*Pisum sativum* L.). Leg Res 36(4):344–350

Nimnoi P, Pongsilp N (2009) Genetic diversity and plant-growth promoting ability of the indole-3-acetic acid (IAA) synthetic bacteria isolated from agricultural soil as well as rhizosphere, rhizoplane and root tissue of *Ficus Religiosa* L., *Leucaena leucocephala* and *Piper Sarmentosum* Roxb. Res J Agric Biol Sci 5:29–41

Nour SM, Cleyet-Marel JC, Back D, Effosse A, Fernandez MP (1994) Genotypic and phenotypic diversity of *Rhizobium* isolated from Chickpea (*Cicer arietinum* L.). Can J Microbiol 40:345–354

Nour SM, Cleyet-Marel JC, Normand P, Fernandez M (1995) Genomic heterogeneity of strains nodulating chickpeas (*Cicer arietinum* L.) and description of *Rhizobium mediterraneum* sp. nov. Int J Syst Bacteriol 45:640–648

Palmer KM, Young JPW (2000) Higher diversity of *Rhizobium leguminosarum* biovar *viciae* populations in arable soils than in grass soils. Appl Environ Microbiol 66:2445–2450

Panday D, Schumann P, Das SK (2011) *Rhizobium pusense* sp. nov., isolated from the rhizosphere of chickpea (*Cicer arietinum* L.). Int J Syst Evol Microbiol 61:2632–2639

Poinsot V, Bélanger E, Laberge S, Yang GP, Antoun H, Cloutier J, Treilhou M, Dénarié J, Promé JC, Debellé F (2001) Unusual methyl-branched α, ß-unsaturated acyl chain substitutions in the *Nod* factors of an Arctic *Rhizobium*, *Mesorhizobium* sp. Strain N33 (*Oxytropic arctobia*). J Bacteriol 183:3721–3728

Pulawska J, Willems A, Sobiczewski P (2012) *Rhizobium skierniewicense* sp. nov., isolated from tumours on chrysanthemum and cherry plum. Int J Syst Evol Microbiol 62:895–899

Rachna (2005) Determination of molecular diversity of chickpea rhizobia (*Mesorhizobium ciceri*). PhD thesis submitted to CCS Haryana Agricultural University, Hisar, India

Ramana CV, Parag B, Girija KR, Raghu Ram B, Ramana VV, Sasikala C (2012) *Rhizobium subbaraonis* sp. nov. an endolithic bacterium isolated from beach sand. Int J Syst Evol Microbiol 63(Pt 2):581–585

Ramirez-Bahena MH, Garcia-Fraile P, Peix A, Valverde A, Rivas R, Igual JM, Mateos PF, Martinez-Molina E, Velazquez E (2008) Revision of the taxonomic status of the species *Rhizobium leguminosarum* (Frank 1879) Frank 1889AL, *Rhizobium phaseoli* Dangeard 1926AL and *Rhizobium trifolii* Dangeard 1926AL. *R. trifolii* is a later synonym of *R. leguminosarum*. Reclassification of the strain *R. leguminosarum* DSM 30132 (=NCIMB 11478) as *Rhizobium pisi* sp. nov. Int J Syst Evol Microbiol 58:2484–2490

Ramírez-Bahena MH, Hernández M, Peix A, Velázquez E, León-Barrios M (2012) Mesorhizobial strains nodulating *Anagyris latifolia* and *Lotus berthelotii* in *Tamadaya ravine* (Tenerife, Canary Islands) are two symbiovars of the same species, *Mesorhizobium tamadayense* sp. nov. Syst Appl Microbiol 35:334–341

Ren DW, Chen WF, Sui XH, Wang ET, Chen WX (2011) *Rhizobium herbae* sp. nov. and *Rhizobium giardinii-related* bacteria, minor microsymbionts of various wild legumes in China. Int J Syst Evol Microbiol 61:1912–1920

Ribeiro RA, Rogel MA, López-López A, Ormeño-Orrillo E, Barcellos FG, Martínez J, Thompson FL, Martínez-Romero E, Hungria M (2012) Reclassification of *Rhizobium tropici* type A strains as *Rhizobium leucaenae* sp nov. Int J Syst Evol Microbiol 62:1179–1184

Rosenblueth M, Martinez-Romero E (2006) Bacterial endophytes and their interactions with hosts. Mol Plant Microbe Interact 19:827–837

Saini R, Dudeja SS, Giri R, Kumar V (2013) Isolation, characterization and evaluation of bacterial root and nodule endophytes from chickpea cultivated in Northern India. J Basic Microbiol 53:1–8

Servín-Garcidueñas LE, Rogel MA, Ormeño-Orrillo E, Delgado-Salinas A, Martínez J, Sánchez F, Martínez-Romero E (2012) Genome sequence of *Rhizobium* sp strain CCGE510, a symbiont isolated from nodules of the endangered wild bean *Phaseolus albescens*. J Bacteriol 194:6310–6311

Sheokand S, Dudeja SS, Kumari S (2012) Nitrogen fixation in tropical environments – adaptive response and benefits. Res Crops 13:743–753

Stajković O, Sofie DM, Bogić M, Anne W, Dušica D (2009) Isolation and characterization of endophytic non-rhizobial bacteria from root nodules of alfalfa (*Medicago sativa* L.). Bot Serbica 33:107–114

Strain SR, Leung K, Whittam TS, de Bruijn FJ, Bottomley PJ (1994) Genetic structure of *Rhizobium leguminosarum* biovar *trifolii* and *viciae* populations found in two Oregon soils under different plant communities. Appl Environ Microbiol 60:2772–2778

Sturz AV, Christie BR (1995) Endophytic bacterial systems governing red clover growth and development. Ann Appl Biol 126:285–290

Sturz AV, Christie BR (1996) Endophytic bacteria of red clover as agents of allelopathic clover-maize syndromes. Soil Biol Biochem 28:583–588

Sturz AV, Nowak J (2000) Endophytic communities of rhizobacteria and the strategies required to create yield enhancing associations with crops. Appl Soil Ecol 15:183–190

Sturz AV, Christie BR, Matheson BG, Nowak J (1997) Biodiversity of endophytic bacteria which colonized clover nodules, roots, stems and foliage and their influence on host growth. Biol Fertil Soils 25:13–19

Sun L, Qiu F, Zhang X, Dai X, Dong X, Song W (2008) Endophytic bacterial diversity in rice (*Oryza sativa* L.) roots estimated by 16S rDNA sequence analysis. Microb Ecol 55:415–424

Tan ZY, Xu XD, Wang ET, Gao JL, Martinez-Romero E, Chen WX (1997) Phylogenetic and genetic relationship of *Mesorhizobium tianshanense* and related rhizobia. Int J Syst Bacteriol 47:874–879

Turdahon M, Osman G, Hamdun M, Yusuf K, Abdurehim Z, Abaydulla G, Abdukerim M, Fang C, Rahman E (2012) *Rhizobium tarimense* sp. nov. isolated from soil in the ancient Khiyik river of Xinjiang, China. Int J Syst Evol Microbiol 63:2424–2429

Velázquez E, Peix A, Zurdo-Piñeiro JL, Palomo JL, Mateos PF, Rivas R, Muñoz-Adelantado E, Toro N, García-Benavides P, Martínez-Molina E (2005) The coexistence of symbiosis and pathogenicity-determining genes in *Rhizobium rhizogenes* strains enables them to induce nodules and tumors or hairy roots in plants. Mol Plant Microbe Interact 18:1325–1332

Wadhwa K, Dudeja SS, Yadav RK (2011) Molecular diversity of native field pea rhizobia trapped by five contrasting field pea genotypes in Indian soils. J Basic Microbiol 51:89–97

Wang F, Wang ET, Wu LJ, Sui XH, Li Y, Chen WX (2011) *Rhizobium vallis* sp. nov., isolated from nodules of three leguminous species. Int J Genet Syst Evol Microbiol 61:2582–2588

Weir BS (2012) The current taxonomy of rhizobia. New Zealand rhizobia website (http://www.rhizobia.co.nz/taxonomy/rhizobia.html). Updated 13 April 2012

Wen Y, Zhang J, Yan Q, Li S, Hong Q (2011) *Rhizobium phenanthrenilyticum* sp. nov., a novel phenanthrene-degrading bacterium isolated from a petroleum residue treatment system. J Gen Appl Microbiol 57:319–329

Xu L, Zhang Y, Deng ZS, Zhao L, Wei XL, Wei GH (2012) *Rhizobium qilianshanense* sp. nov., a novel species isolated from root nodule of *Oxytropis ochrocephala* Bunge in China. Antonie Van Leeuwenhoek 103(3):559–565

Yang C, Yang J, Li Y, Zhou J (2008) Genetic diversity of root-nodulating bacteria isolated from pea (*Pisum sativum*) in subtropical regions of China. Sci China Ser C-Life Sci 51:854–862

Yao Li Juan, Shen YY, Zhan JP, Xu W, Cui GL, Wei GH (2012) *Rhizobium taibaishanense* sp. nov., isolated from a root nodule of *Kummerowia striata*. Int J Syst Evol Microbiol 62:335–341

Young JPW, Demetriou L, Apte RG (1987) *Rhizobium* population genetics: enzyme polymorphism in *Rhizobium leguminosarum* from plants and soil in a pea crop. Appl Environ Microbiol 53:397–402

Zakhia F, Jeder H, Domergue O, Willems A et al (2006) Characterisation of wild legume nodulating bacteria (LNB) in the infraarid zone of Tunisia. Syst Appl Microbiol 27:380–395

Zézé A, Mutch LA, Young JPW (2001) Direct amplification of *nod*D from community DNA reveals the genetic diversity of *Rhizobium leguminosarum* in soil. Environ Microbiol 3:363–370

Zhang RJ, Hou BC, Wang ET, Li Y, Zhang XX, Chen WX (2011) *Rhizobium tubonense* sp. nov., a symbiotic bacterium isolated from root nodules of *Oxytropis glabra* grown in Tibet, China. Int J Syst Evol Microbiol 61:512–517

Zhang JJ, Liu TY, Chen WF, Wang ET, Sui XH, Zhang XX, Li Y, Li Y, Chen WX (2012) *Mesorhizobium muleiense* sp. nov., nodulating with *Cicer arietinum* L. in Xinjiang, China. Int J Syst Evol Microbiol 62 (Pt 11):2737–2742

Zheng WT, Li Y Jr, Wang R, Sui XH, Zhang XX, Zhang JJ, Wang ET, Chen WX (2012) *Mesorhizobium qingshengii* sp. nov., isolated from effective nodules of *Astragalus sinicus* grown in the Southeast of China. Int J Syst Evol Microbiol 63(Pt 6):2002–2007, PubMed PMID: 23041644

Zhou S, Li Q, Jiang H, Lindström K, Zhang X (2013) *Mesorhizobium sangaii* sp. nov., isolated from the root nodules of *Astragalus luteolus* and *Astragalus ernestii*. Int J Syst Evol Microbiol 63(Pt 8):2794–2799

Bacteriocins from Lactic Acid Bacteria

11

J.S. Duhan, K. Nehra, S.K. Gahlawat, P. Saharan, and Duhan Surekha

Abstract

Probiotics are the beneficial bacteria that are indigenous to a healthy digestive system. Basically, these are live microorganisms, which are administered in adequate amounts in order to provide health benefit to the host. They work by crowding out the bad bacteria within the digestive system and then attaching themselves to the intestinal wall where they increase the number of beneficial bacteria regulating and maintaining balance between the beneficial and harmful bacteria. The tradition of eating foods (such as yogurt, sauerkraut, fermented milk, miso and soy beverages) fermented with bacteria is solely based on health benefits provided by these bacteria. Thousands of different lactic acid bacteria (LAB) are used as probiotics, and the dominant members are *Bifidobacteria* (*Bifidobacterium bifidum*) and *Lactobacteria* (*Lactobacillus acidophilus*, *Lactobacillus plantarum*). The benefits associated with the use of probiotic bacteria are well documented, but their mechanism of action still remains very unclear. LAB protect food from spoilage and pathogenic microorganisms by producing organic acids, hydrogen peroxide, diacetyl, antifungal compounds such as fatty acids or phenyllactic acid and bacteriocins. Bacteriocins are a heterogeneous family of small, heat-stable peptides that are produced by

J.S. Duhan (✉) • S.K. Gahlawat • P. Saharan
Department of Biotechnology, Chaudhary Devi Lal University, Sirsa 125055, Haryana, India
e-mail: duhanjs68@gmail.com; duhanjs@rediffmail.com

K. Nehra
Department of Biotechnology, DCRUS&T, Murthal, Sonepat 131039, Haryana, India

D. Surekha
Department of Botany, Government National College, Sirsa 125055, Haryana, India

R.K. Salar et al. (eds.), *Biotechnology: Prospects and Applications*,
DOI 10.1007/978-81-322-1683-4_11, © Springer India 2013

many bacterial species, including many probiotic strains. Bacteriocins of LAB are considered as safe natural preservatives with potent antimicrobial activities and antagonistic affect. Bacteriocins mainly prevent food spoilage by inhibiting growth of food-borne pathogenic microorganisms. They also protect the body from cancer and effectively improve the immune system. These are classified into several groups in which classes I (Lantibiotics) and II (Bacteriocins) are studied most thoroughly. The present review is concerned with the diversity of bacteriocins, their expression system and their applications which lead to use of lactic acid bacteria as probiotics.

11.1 Introduction

Beneficial properties of fermented dairy products were proposed for the first time by Metchnikoff in 1908 (Metchnikoff 1908). Several nutritional and therapeutic benefits of lactic cultures and their fermented products to the consumers were suggested by many researchers (Taranto et al. 1998; Lee et al. 1999; Danone 2001; Lilly and Stillwell 1965; Gorbach et al. 1987; Mallett et al. 1989; De Vuyst and Vandamme 1994; Aso et al. 1995; Goldin et al. 1996; Campieri and Gionchetti 1999; Lee et al. 1999; Bengmark 2000; Caplan and Jilling 2000; Cunningham-Rundles et al. 2000; Gionchetti et al. 2000; Gorbach 2000; Guslandi et al. 2000; Kyne and Kelly 2001; Reid et al. 2001; Reid and Bruce 2001; MacFarlane and Cummings 2002). Lactic acid bacteria (LAB) have a long history of application in fermented foods because of their influence on nutritional, organoleptic and shelf-life characteristics (Wood and Holzapfel 1995; Leroy and De Vuyst 2004). They cause rapid acidification of the raw material through the production of organic acids mainly lactic acid. In addition, acetic acid, ethanol, aroma compounds, bacteriocins, expolysaccharides and several enzymes produced by LAB are of very importance (Leroy et al. 2006). Food fermented with lactic acid bacteria increases folic acid in yogurt, bifidus milk and kefir (Rajalakshmi and Vanaja 1967; Shahani and Chandan 1979; Deeth and Tamime 1981; Alm 1982). LAB release bacteriocins due to which it is capable of inhibiting various microorganisms in a food environment and display crucial antimicrobial properties with respect to food preservation and safety.

As LAB have a GRAS status, bacteriocins are generally recognised as safe LAB, which have received significant attention as a novel approach to the control of pathogens in foods (Klaenhammer 1993). Hence, bacteriocins of LAB are considered as safe natural preservatives, and it is assumed that they are degraded by the proteases in gastrointestinal tract (Cleveland et al. 2001). Nisin is the first antimicrobial polypeptide found in LAB (Rogers 1928). At the time of discovery, the producer strains of nisin were identified as *Streptococcus lactis*, but later on it was classified as *Lactococcus lactis* (Schleifer et al. 1985).

But some strains of LAB also combat gastrointestinal pathogenic bacteria such as *Helicobacter pylori*, *E. coli* and *Salmonella*. Bacteriocins actually are proteinaceous compounds with antibacterial activity, and they constitute a heterologous subgroup of ribosomally synthesised antimicrobial peptides (De Vuyst and Vandamme 1994).

In this chapter, the role of bacteriocins as fast acting, antimicrobial peptides in both food safety and gastrointestinal health has been discussed.

11.2 Histories of Bacteriocins

In 1925, Bacteriocins were discovered by Gratia when he was involved in the process of searching for ways to kill bacteria which also resulted in the development of antibiotics and the discovery of bacteriophage, all within a span of few years. His first discovery was named colicin because it was produced by killing *E. coli* (Gratia 1925; Gratia 2000). The first detailed characterisation of bacteriocinogenic activity of *Lactobacilli* was reported in 1961 (De Klrek and Coetzee 1961). Although not exactly defined, bacteriocins differ from classical antibiotics. They are a heterogeneous group of bacterial antagonists that vary considerably in molecular weight, biochemical properties and range of sensitive hosts and mode of action. Klaenhammer (1988) defined them as "Bacteriocins are proteins or proteins complexes with bactericidal activity directed against species that are usually closely related to the producer bacterium". In the past few years, many bacteriocin-like compounds have been reported and characterised (Stile and Hastings 1991). Many of these substances do not comply with the definition proposed above. Bacteriocins are generally active against closely related species, and thus the inhibitory spectrum is narrow. However, nisin from *Lactobacillus lactis* (Hurst 1981) and pediocin A from *Pediococcus pentosaceus* are active against broader range of food-borne pathogens (Daeschel and Klaenhammer 1985). An exception exists with regard to Klaehamer's definition of bacteriocins, including their bactericidal mode of action, the pure protein nature and the narrow spectrum of antibacterial activity. Most bacteriocins produced by LAB are bactericidal. However, lactocin 27 from *Lb. helveticus* (Upreti and Hindsdill 1975) and bacteriocins from *Lb. sake* 148 (Sobrino et al. 1991) possess bacteriostatic effect. The term bacteriocin-like substances was suggested for those antagonistic substances that do not fit the traditional definition of bacteriocins.

There is no clear cut boundary between antibiotics, bacteriocins and microcins. Like antibiotics, bacteriocins may be bacteriostatic or bactericidal with narrow or broad range of activity that could be included in a family of peptide antibiotics. Antibiotics are synthesised non-ribosomally by multistep enzyme pathways. There are many ribosomally synthesised bacteriocins. While many antibiotics promote the development of resistant strains, development of resistance to bacteriocins is rare. Many antibiotics are chemically synthesised. Till date, no reports are observed in which bacteriocins are chemically synthesised. However, it is appropriate to expect that, with genetic engineering, analogues of bacteriocins may be designed or constructed.

There is considerable overlap in the definition of antimicrobial substances. It is thus generally accepted that bacteriocins are heterogeneous group of proteinaceous that may vary in spectrum of activity, mode of action, molecular weight, genetic origin and biochemical properties. Their ability to inhibit other bacteria and the fact that all are proteinaceous may be the only common feature of this mixed group of substances (Stile and Hastings 1991). Lipid or carbohydrate moieties can be associated with these proteinaceous and be part of the bacteriocin complex. Bacteriocin can be either being cell bound or released extracellularly and may be produced early or late in the growth cycle. They are susceptible to proteases and have variable stability at different pH and temperature.

11.3 Bacteriocins from Lactic Acid Bacteria

Although bacteriocins may be found in many Gram-positive and Gram-negative bacteria, those produced by LAB have received particular attention in recent years due to their potential application in food industry as natural preservatives.

Bacteriocins produced by LAB are small ribosomally synthesised, antimicrobial peptides or proteins that possess activity towards closely related Gram-positive bacteria where producer cells are immune to their own bacteriocins (De Vuyst and Vandamme 1994; Cotter et al. 2005).

The antibacterial spectrum includes spoilage organisms and food-borne pathogens such as *Listeria monocytogenes* and *Staphylococcus aureus*. These bacteriocins act against Gram-negative bacteria (such as *E. coli* and *Salmonella*) only when the integrity of outer membrane has been compromised. For example, after osmotic shock or low pH treatment, in the presence of a detergent or chelating agent or after pulsed electric field or high-pressure treatment, bacteriocins are able to kill Gram-negative pathogenic bacteria (Stevens et al. 1991).

11.4 Classifications

Bacteriocins can be divided into four major classes:

Class I: Lantibiotics containing post-translationally modified amino acids such as lanthionine and *O*-methyl-lanthionine and nisin

Class II: Small unmodified thermostable peptides usually positively charged at a neutral pH

Class III: Large unmodified thermolabile peptides

Class IV: Complex bacteriocins which possess a carbohydrate or lipid component

11.5 Genetics of Bacteriocin Biosynthesis

11.5.1 Localisation of Genetic Determinants of Bacteriocins

Genetic determinants of bacteriocins are localised on:

1. Bacterial chromosome
2. Plasmids
3. Transposons

Chromosomal localisation of the bacteriocin operon has been described for both class-I lantibiotics and class-II small heat-stable peptides.

Most of the bacteriocin's operons are located on plasmids, and it has been suggested that this plasmid location helps intra- and interspecies phylogenetic dissemination of bacteriocins among LAB. For example, lacticin 481 and lecticin 3147 are produced by *Lactobacillus lactis* and are encoded by a gene located on a 70 and 63 kb plasmid, respectively (Rince et al. 1997). Structural genes of numerous non-lantibiotic bacteriocins are also located on plasmids; for example, operon of pediocin PA-1(Marugg et al. 1992) and pediocin *AcH* (Bukhtiyarova et al. 1994) bacteriocins produced by strains of *Pediococcus acidilactici* is located on 3.4 kb plasmid. Pediocin *AcH* genes are plasmid located, and they are arranged in a gene cluster of 3,500 bp sharing a common promoter and rho-dependent stem loop terminator. The four genes, each with independent ribose-binding sites (rbs), initiation and termination codon and spacer sequences in between, were designed as *papA*, *papB*, *papC* and *papD* (Motlagh et al. 1994). *PapA* is the structural gene for the pre-pediocin, while *papB* encodes the immunity protein. *PapC* and *papD* form the export machinery and are required for the membrane translocation and the removal of the leader peptide. *PapD* shares a double-glycine protease domain with other ABC export proteins active in transport of bacteriocins.

Sakicin A is another pediocin-like bacteriocin; its genes are organised in two divergent operons (Fig. 11.1) divided by an insertion sequence *ISI163*. There are six genes participating in sakicin A synthesis: *sapA*, the gene for the pre-sakicin A; *salA*, the gene encoding the sakicin A immunity

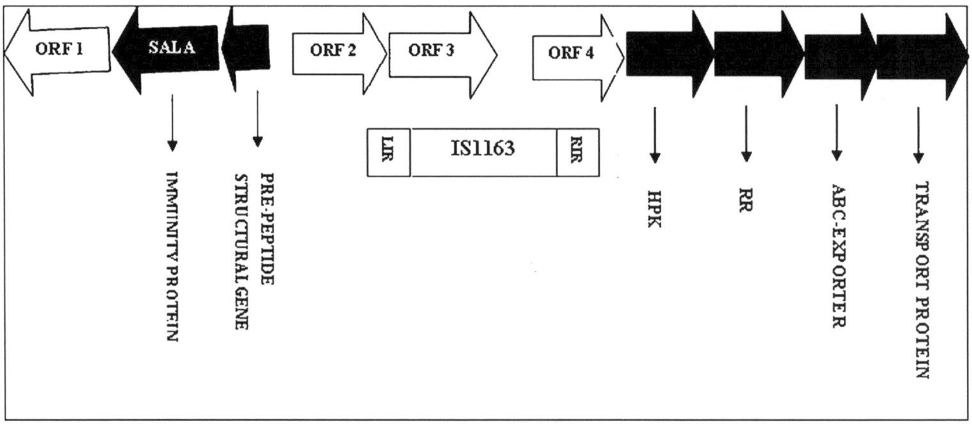

Fig. 11.1 Schematic structure of sakicin A operon. The genes related to bacteriocin production are marked with *black* colour (Adapted from Dimov et al. 2005)

protein; *sapK*, the histidine protein kinase; *sapR*, the response regulator; *sapT*, the ATP-binding cassette exporter; and *sapE*, the transport protein. *SapA* and *sapE* are dedicated to the export of the bacteriocin, while *sapK* and *sapR* play a role in the regulation of the expression via signal transduction mechanism.

In class-I bacteriocins, nisin is another lantibiotic produced by *Lactococcus lactis*, and its genetic determinants are located on the conjugative transposon *Tn5276* within bacterial chromosomes (Vander et al. 1993). In this case, the bacteriocin genes are flanked by intact insertion sequences (IS) or inverted repeats (IR) (Quadri et al. 1994).

11.5.2 Production of More than One Bacteriocin

Usually LAB produces only one bacteriocin, but there are many reports for such strains of LAB which can produce more than one bacteriocin. For example, *Lactococcus lactis* produces two bacteriocins, i.e. *LsbA* and *LsbB*, and both of these act synergistically. *Enterococcus faecium* also produces enterocin A and B. Total three bacteriocins are produced by *Lactobacillus plantarum*, out of which two bacteriocins (i.e. plantaricin E/F and J/K) consist of two peptides, and the third one is a single-peptide bacteriocin (plantarum N). (Diep et al. 1995). It is possible that when more than one bacteriocin is produced, the peptides can belong to different classes. For example, in case of *Staphylococcus aureus* C55, two peptides, i.e. staphylococcins C55α and C55β, belong to different classes but still show synergetic lantibiotic effect (Navaratna et al. 1998).

11.5.3 Organisation of Bacteriocin Genes

Till now, the scientific data show that almost all genetic determinants of bacteriocins are clustered in operons or regulons. This is not unexpected because in the simplest case, the bacteriocin expression needs at least two genes. One is the structural gene and another one that encodes for an immunity protein specific to the produced bacteriocin. Bacteriocin production also needs a specific export machinery and is subjected to some regulation factors.

In general, the lantibiotic operons are more complex than those encoding for non-lantibiotics because they need additional genes encoding enzymes for post-translational modifications.

One of the best characterised bacteriocin operons is that encoding the biosynthesis of nisin by *Lactococcus lactis*. The nisin gene cluster contains 11 genes and is usually designated as *nisABTCIPRKFEG* (Fig. 11.2). Of these genes:

- *NisA* encodes nisin A precursor peptide.
- *NisB* and *NisD* encode putative enzymes involved in the post-translational modifications.
- *NisT* encodes a putative transporter protein of the ABC translocator family that is probably involved in the extrusion of the modified nisin precursor.
- *NisP* encodes extracellular protease involved in precursor processing.
- *NisI* encodes a lipoprotein involved in the producer's self-protection against nisin.
- *NisFEG* encodes putative transporter protein that has also been implied in immunity.
- The proteins encoded by *NisR* and *NisK* have shown to be involved in the regulation of nisin biosynthesis. *NisR* is a response regulator and *NisK* is a sensor histidine protein kinase which belongs to the class of the signal transduction regulatory system proteins.
- The gene cluster contains two promoters. The first one is regulated upstream to the *nisA* gene and second between the *nisP* and *nisR* genes. It is interesting that an inverted repeat located between the genes *nisA* and *nisB* could act as a rho-independent terminator suggesting sophisticated regulation of the expression of the genes within the nisin cluster (Kuipers et al. 1995).

11.5.4 Biosynthetic Pathway

The proposed pathway for the nisin biosynthesis has been shown in Fig. 11.2:

- *NisA* is translated to pre-nisin A.
- Pre-nisin A is transformed to precursor nisin A by the products of the genes *nisB* and *nisC*. At this stage several disulfide bridges are

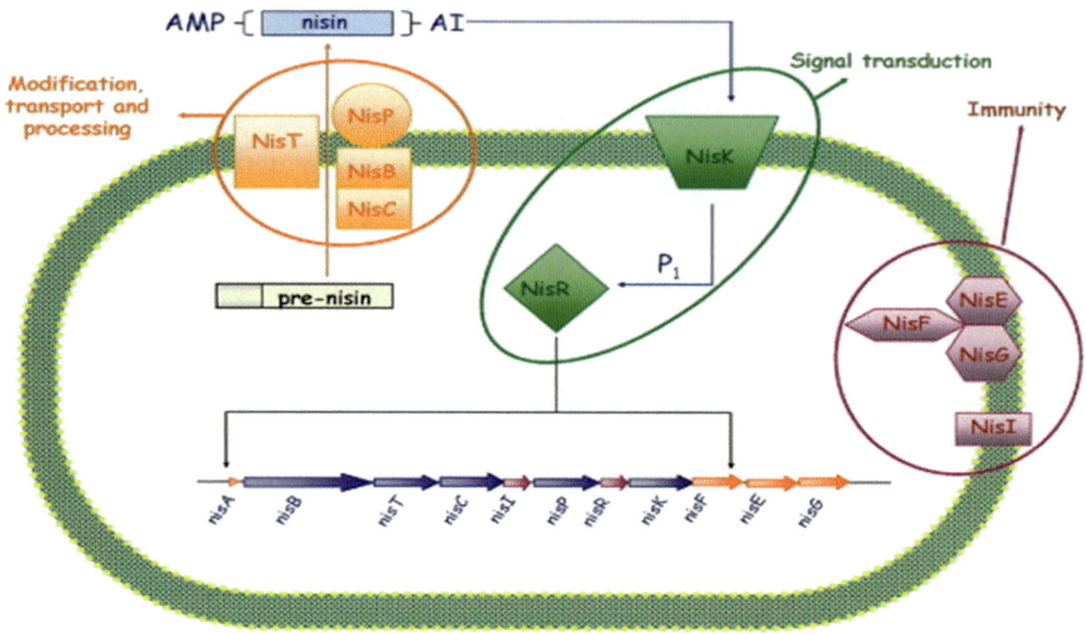

Fig. 11.2 Quorum-sensing regulation of class-I antimicrobial peptides (AMP) in lactic acid bacteria. *NisABTCIPRKFEG*, gene cluster encoding nisin; NisB and NisC, proteins involved in the intracellular post-translational modification reactions; NisT, putative transport protein of the ABC translocator family; NisP, extracellular protease for removing the leader peptide; AI, autoinducer; NisK, transmembrane associated signal transducer; NisR, response regulator; NisF, NisE and NisG, ABC exporter system that generates immunity through active cell extrusion from the cell; and NisI, lipoprotein that contributes to producer immunity. For the quorum-sensing mechanism, see the text adapted from Quadri (2002)

made and some amino acids are transformed to unusual ones as well.

- Finally, the precursor nisin A is exported out of the cell by the products of the genes *nisT* and *nisP* at the same time while the leader peptide is cleaved and the final product nisin A is obtained (Kuipers et al. 1995).

11.5.5 Regulation of Bacteriocin Expression

Expressions of bacteriocin genes are usually subjected to regulation by external induction factors (IF). In most of the cases, small peptides are secreted by the producer strain itself. However, in some cases, bacteriocin production depends on the environmental conditions (temperature, pH, etc.) or even it can be constitutive.

In many LAB, bacteriocin production is controlled by a “quorum-sensing” mechanism. The quorum-sensing system controls a plethora of different important biological processes in bacteria such as natural genetic transformation, virulence and sporulation. In Gram-positive bacteria, quorum sensing is predominantly mediated by peptide pheromones or IF. IF are the first of three components of signal transduction pathway (Fig. 11.2).

The induction factor is believed to bind specifically to the correspondent histidine protein kinase and to activate it to phosphorylate the response regulator, which then stimulates transcription of the target genes, most probably by binding to specific imperfect direct repeats found in many bacteriocin genes cluster. IF for a given bacteriocin operon can be produced either by the bacteriocin producer strain or by

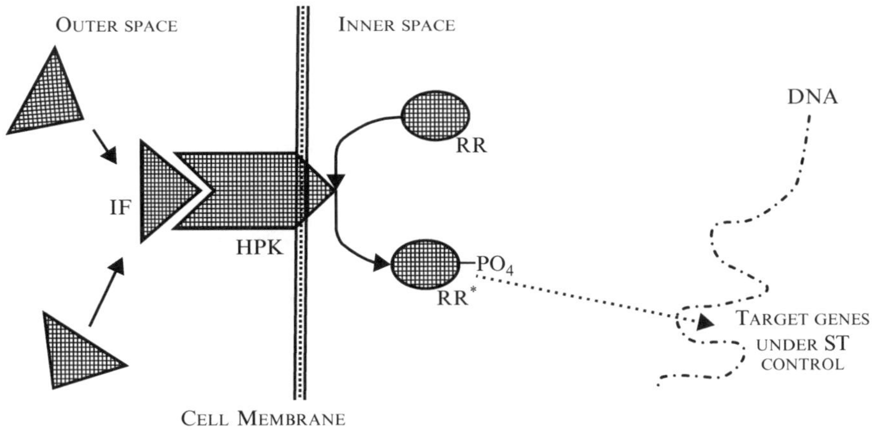

Fig. 11.3 Three component signal transduction pathway used by many bacteriocin operons in quorum-sensing regulation. Abbreviations: *IF* induction factor; *HPK* histidine kinase; *RR* and *RR** response regulators (Adapted from Dimov et al. 2005)

other strain belonging to the same or other species or genera.

When the inducer peptide is produced by the bacteriocin producer itself, there is an autoregulation of bacteriocin biosynthesis. The IF can be a dedicated peptide encoded by a respective gene or the bacteriocin molecule itself (Dimov et al. 2005) (Fig. 11.3). Typical case in which the bacteriocin molecule itself plays a role of an IF is nisin. In this case, the mature nisin molecule interacts with a sensor histidine protein kinase (HPK) produced by the gene *nisK* and disposed on the cell membrane. The HPK phosphorylates the response regulator (RR) product of the gene *nisR*, followed by binding to the imperfect direct repeats found in the promoter regions. This event leads to transcription activation of the nisin gene cluster and finally to the secretion of the more nisin molecules, thus closing the cycle of auto activation (Kuipers et al. 1995).

11.6 Bacteriocin Productions in Laboratory

There are many ways to demonstrate bacteriocin production depending upon the sensitivity and labour intensiveness desired. To demonstrate their production, technicians stab inoculate multiple strains on separate multiple nutrient agar petri dishes, incubate at 30 °C for 24 h, overlay each plate with one of the strains and incubate again at 30 °C for 24 h. After this process, the presence of bacteriocins can be inferred if there are zones of growth inhibition (Fig. 11.4). This is the simplest and least sensitive way.

Bacteriocins by LAB can be produced in much higher amounts during in vitro fermentations under optimal physical and chemical conditions (Leroy and Vuyst 2005). The higher in vitro production is due to the absence of limiting factors such as strong diffusion limitations, inactivation by proteases and adsorption to food particles (Leroy and Vuyst 2000). However, the yield was influenced by decrease in pH which results in a decreased adsorption of the bacteriocin molecules to the precursor cells and hence increased bioavailability (Yang et al. 1992; De Vuyst et al. 1996). In addition, temperature and nutrient availability seem to play a crucial role in bacteriocin production (Callewaert and De Vuyst 2000; Leroy and De Vuyst 2001; Verluyten et al. 2004a). Presence of sodium chloride usually decreases production levels (Leroy and De Vuyst 1999a; Verluyten et al. 2004b). Hence, we can say that bacteriocin production is a growth-dependent physiological trait and follows primary metabolite kinetics (Leroy and De Vuyst 1999b).

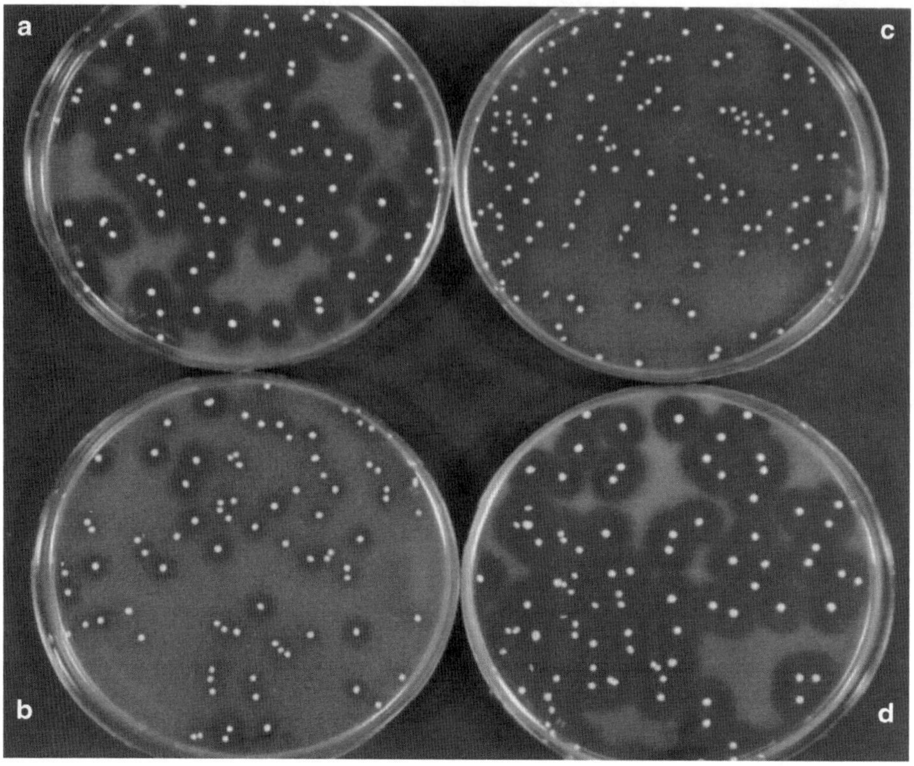

Fig. 11.4 Colonies of lactic acid bacteria obtained from salt-fermented cucumber. *Arrow* indicates a bacteriocin-producing colony with zone of inhibition produced on a lawn of the target strain *Leuconostoc mesenteroides* NRRL B640 (Adapted from Singh and Ramesh 2008)

11.7 Purification

There are three methods for the purification of bacteriocins by LAB:

- First, purification can be done by a conventional method that is based on a rather laborious series of subsequent steps of ammonium sulphate precipitation, ion exchange, hydrophobic interaction, gel filtration and reversed phase high-pressure liquid chromatography (Mortvedt et al. 1991; Tichaczek et al. 1992; Parente and Ricciardi 1999).
- Second, a three-step protocol has been developed, which includes:
  1. Ammonium sulphate precipitation
  2. Chloroform/methanol extraction/precipitation
  3. Reversed phase high-pressure liquid chromatography (Callewaert et al. 1999)
- Third, bacteriocins can be isolated through a unique unit operation, i.e. expanded bed adsorption using a hydrophobic interaction gel, after maximising the bioavailable bacteriocin titer through pH adjustment of the crude fermentation medium (Callewaert and Vuyst 1999; Foulquie et al. 2001).

Later two methods are fast than the first conventional one and successfully several bacteriocins with interesting potential have been purified by these methods. According to Juodeikiene et al. (2012), bacteriocins produced by LAB possess the following attractive features that make them suitable for use in food preservatives:

- Protein nature, inactivation by proteolytic enzymes of gastrointestinal tract
- Nontoxic to laboratory animals tested and generally nonimmunogenic
- Inactive against eukaryotic cells
- Generally thermo-resistant due to which they can maintain their antimicrobial activity even after pasteurisation and sterilisation

- Broad bactericidal activity affecting most of the Gram-positive bacteria and some damaged, Gram-negative bacteria including various pathogens such as *L. monocytogenes*, *Bacillus cereus*, *S. aureus* and *Salmonella*
- Genetic determinants generally located in plasmid, which facilitates genetic manipulation to increase the variety of natural peptide analogues with desirable characteristics

11.8 Bacteriocin Productions in Foods: Application Possibilities

- Bacteriocins can be used as food additives, presently; nisin is a permitted preservative in at least 48 countries, in which it is used in a variety of products including cheese, canned food and cured meat (Delves-Broughton 1990). Nisin is commercially prepared in partially purified form (Twomey et al. 2002), and a marketed preparation with the pediocin PA-1(*AcH*) producer is also available (Rodriguez et al. 2002).
- Nisin is being exploited as a food preservative mainly in the dairy foods. It is nontoxic and digested by intestinal enzymes. It is heat stable and does not contribute to off flavours (Hurst 1981).
- As an alternative to the addition of bacteriocins to foods, bacteriocins may be produced directly in the food as a result of starter culture or coculture activity. The application of bacteriocin-producing starter culture (in sourdough, fermented sausage and cheese production) had been studied by Leroy et al. 2005; Foulquie et al. 2003; Anastasiou et al. 2006 in both *in vitro* lab fermentations as well as on pilot scale level. Results of these studies were promising and underline the important role that functional bacteriocinogenic strains of LAB may play in the food industry as starter cultures, cocultures or bioprotective cultures to improve food quality and safety (Leroy and De Vuyst 2004; Ennahar et al. 1999).
- Many bacteriocins of LAB are safe and effective natural inhibitors of pathogenic and food spoilage bacteria in various foods, e.g. nisin prevents *Clostridial* spoilage of processed and natural cheeses, extends the shelf life of milk and prevents the growth of spoilage *Lactobacilli* in beer and against *Bacillus* and *Clostridial* spores in canned food (Vandenberg 1993; Delves et al. 1996).
- *Listeria monocytogenes* is common contaminant of raw foods such as milk, meat and vegetables and results in the serious food-borne illness in consumers (Ryser and Marth 1991). Pediocin PA-1 has the ability to control *L. monocytogenes* in cheese, vegetables and meat (Vandenberg 1993; Stiles 1996).
- Bioactive packaging is a further potential application in which bacteriocins or the producing strains can be incorporated into packaging destined to be in contact with food (Fig. 11.5) (Castellano and Vignolo 2006; Schobitz et al. 2006; De Martins et al. 2003).

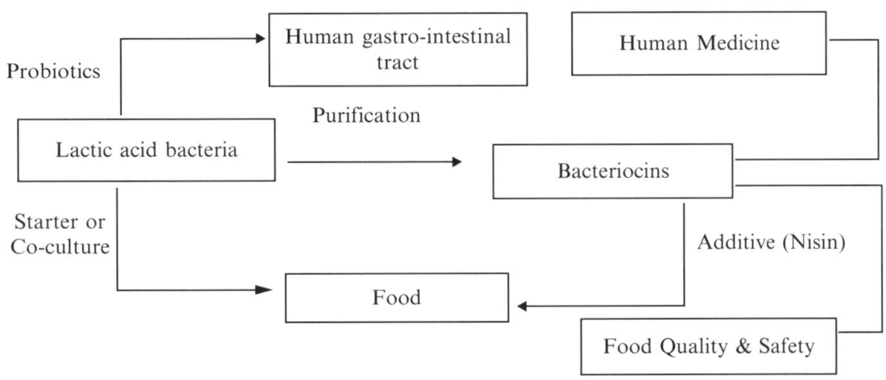

Fig. 11.5 Overview of application potential of bacteriocin production by LAB in food quality, safety and in medicine emphasises their role as food ingredients and in human gastrointestinal tract (De Vuyst and Leroy 2007)

11.9 Bacteriocins from Probiotics

As we enter into new millennium, people are aware that for spending a healthy lifestyle, diet plays a major role in preventing diseases and promoting health. Therefore, there is an increasing trend for foods containing probiotic culture.

Probiotic is a preparation of live microorganisms which, when applied to man or animals, beneficially affects the host by improving the properties of the indigenous microbiota (Havenaar and Huis 1992).

An experimental focus on bacteriocin production by probiotic lab strains has indicated that this potential might play a considerable role during in vivo interactions occurring in the human gastrointestinal tract, e.g. towards *Helicobacter pylori* (De Vuyst and Vandamme 1994; Avonts and De Vuyst 2001; Kim et al. 2003). Whereas bacteriocins in food are degraded by the proteolytic enzymes of the stomach, probiotic bacteria may lead to in situ production of bacteriocins in the gastrointestinal tract. Up to now bacteriocins have been isolated from the commercial probiotic strains *Lactobacillus casei* and *L. johnsonii* (Avonts et al. 2004).

11.10 Medical Significance

Bacteriocins are of interest in medicine because they are made by nonpathogenic bacteria that normally colonise the human body. Loss of these harmless bacteria following antibiotic use may allow opportunistic pathogenic bacteria to invade the human body.

They have been suggested as a cancer treatment (Farkas and Yu 1985; Baumal et al. 1982). They have shown distinct promise as a diagnostic agent for some cancers (Saito et al. 1979; Cruz et al. 2006; Sand et al. 2007; Farkas et al. 1995; Musclow et al. 1987). Bacteriocins have also been tested as AIDS drugs (Farkas et al. 1991). Potential use and veterinary applications of some bacteriocins are given in Table 11.1.

Table 11.1 Potential use and veterinary applications of some bacteriocins

Bacteriocin	Producer	Potential use	References
Gram-positive bacteria			
Nisin	*L. lactis* subsp. *Lactis*	Treat peptic ulcer disease Antimicrobial activity in medical devices like in catheters Treat *S. pneumonia* infections Treat mastitis in cattle Vaginal contraceptive agent	Bower et al. (2002), Goldstein et al. (1998), Hancock (1997), Reddy et al. (2004), and Sears et al. (1992)
Lacticin 3147	*L. lactis* subsp. *Lactis*	Treat mastitis in cattle	Ryan et al. (1998)
Galliderm	*Staphylococcus gallinarum*	Treat skin infections such as acne	Kellner et al. (1988)
Epidermin	*S. epidermidis*	Treat skin infections such as acne	Allgaier et al. (1986)
Mutacin B-Ny266	*Staphylococcus mutans*	Bacterial infection caused by methicillin-resistant staphylococci	Mota-Meira et al. (2005)
Tomicid	*Staphylococcus* sp. TOM-1606	Staphylococcal respiratory infections like Scarlet fever in children	Briko and Zhuravlev (2004) and Golshmid and Landsman (1989)
Gram-negative bacteria			
Microcins J25 and J24	*E. coli*	Treat *E. coli* and salmonella infections in chickens	Sable et al. (2000) and Wooley et al. (1999)
Colicins E1, E4, E7, E8, k and S4	*E. coli*	Treat haemorragic colitis and haemolytic uremic syndrome caused by *E. coli* 0157:H7	Jordi et al. (2001)

11.11 Limitations of Bacteriocins

- Unable to withstand thermal processing: Thermal processing is used extensively within the food manufacturing process and can have adverse effects on the bioactive capability of a bacteriocin, potentially rendering it less effective.
- The chemical and physical properties of food, e.g. pH and fat content, can also have a significant role in the suitability of a particular bacteriocin.
- Some food ingredients like sodium chloride, sodium nitrite, ascorbic acid, alginate and sodium lactate may interfere with bacteriocin activity (Vignolo et al. 1998).
- Development of resistant strains: The induction of bacteriocin-resistant strains and mutants may pose further problems in the use of bacteriocins in biopreservation (Rekhif et al. 1994; Rasch and Knochel 1998; Guinane et al. 2006).
- However, these limitations can be overcome by the use of hurdle technology which refers to the manipulation of multiple factors (intrinsic and extrinsic) designed to prevent bacterial contamination or control growth and survival in food (Deegan et al. 2006).

11.12 Conclusions

Bacteriocins produced by LAB have the potential to cover a very broad field of application including both the food industry and medical sector. Regarding their use in food, bacteriocin-producing starter cultures or cocultures have been successfully applied in pilot scale experiments, yielding food quality and safety advantages.

With respect to medical applications, antimicrobials produced by probiotic LAB might play a role during in vivo interactions occurring in the human gastrointestinal tract, hence contributing to gut health. Further research is needed to unravel the precise role of LAB bacteriocins in this process.

References

Allgaier H, Jung G, Werner RG, Schneider U, Zahner H (1986) Epidermin: sequencing of a heterodetic tetracyclic 21-peptide amide antibiotic. Eur J Biochem 160:9–22

Alm L (1982) Effect of fermentation on lactose, glucose and galactose content in milk and suitability of fermented milk products for lactose-deficient individuals. J Dairy Sci 65:346–352

Anastasiou R, Georgalaki M, Manolopoulou E, Kandarakis I, De Vuyst L (2006) The performance of *Streptococcus macedonicus* ACA-DC 198 as starter culture in Kasseri Cheese production. Int Dairy J 17:208–217

Aso Y, Akazan H, Kotake T, Tsukamoto T, Imai K (1995) Preventive effect of a *Lactobacillus casei* preparation on the recurrence of superficial bladder cancer in a double-blind trial. Eur Urol 27:104–109

Avonts L, De Vuyst L (2001) Antimicrobial potential of probiotic lactic acid bacteria. In: Proceedings of the 15th forum for applied biotechnology. Meded Rijksuniv Gent Fak Landbouwkd Toegep Biol Wet 66(3b):543–550

Avonts L, Van Uytven F, Vuyst D (2004) Cell growth and bacteriocin production of probiotic *Lactobacillus* strains in different media. Int Dairy J 14:947–955

Baumal R, Musclow E, Farkas HH, Marks A (1982) Variants of an interspecies hybridoma with altered tumorigenicity and protective ability against mouse myeloma tumors. Cancer Res 42(5):1904–1908

Bengmark S (2000) Colonic food: pre- and probiotics. Am J Gastroenterol 95:S5–S7

Bower CK, Parker JE, Higgins AZ, Oes ME, Wilson JT, Valentine BA, Bothwell MK, Mc Guire J (2002) Protein antimicrobial barriers to bacterial adhesion: *in vitro* and *in vivo* evaluation of nisin-treated implantable materials. Coll Surf Biointerf 25:81–90

Briko NI, Zhuravlev MV (2004) Use of tomicid in prophylaxis of respiratory streptococcal infection in the organised groups of children of pre-school age. Zh Mikrobiol Epidemiol Immunobiol 4:17–20

Bukhtiyarova M, Yang R, Ray B (1994) Analysis of the pediocin AcH gene cluster from plasmid pSMB74 and its expression in a pediocin-negative *Pediococcus acidilactici* strain. Appl Environ Microbiol 60(9): 3405–3408

Callewaert R, Vuyst D (1999) Expanded bed adsorption as a unique unit operation for the isolation of bacteriocins from fermentation media. Bioseparation 8:159–168

Callewaert R, Vuyst D (2000) Bacteriocin production with *Lactobacillus amylovorus* DCE 471 is improved and stabilised by fed batch fermentation. Appl Environ Microbiol 66:606–613

Callewaert R, Holo H, Devreese B, Van Beeumen J, Nes I, Vuyst D (1999) Characterization and production of amylovorin L471, a bacteriocin purified from

Lactobacillus amylovorus DCE471 by a novel three step method. Microbiology 145:2559–2568

Campieri M, Gionchetti P (1999) Probiotics in inflammatory bowel disease: new insight to pathogenesis or a possible therapeutic alternative. Gastroenterology 116:1246–1249

Caplan MS, Jilling T (2000) Neonatal necrotizing enterocolitis: possible role of probiotic supplementation. J Pediatr Gastroenterol Nutr 30:S18–S22

Castellano P, Vignolo G (2006) Inhibition of *Listeria innocua* and *Brochothrix thermosphacta* in vacuum packaged meat by addition of bacteriocinogenic *Lactobacillus curvatus* CRL 705 and its bacteriocins. Lett Appl Microbiol 143:193–199

Cleveland J, Montville TJ, Nes IF, Chikindas ML (2001) Bacteriocins: safe, natural antimicrobials for food preservation. Int J Food Microbiol 71:1–20

Cotter PD, Hill C, Ross RP (2005) Bacteriocins: developing innate immunity for food. Nat Rev Microbiol 3:177–788

Cruz CL, Puertollano MA, Peurtollano E, De Cienfuegos GA, De Pablo MA (2006) *In vitro* biological activities of magainin along or in combination with nisin. Peptides 27(6):1201–1209

Cunningham-Rundles S, Ahrne S, Bengmark S, Johann Liang R, Marshall F, Metakis L, Califano C, Dunn AM (2000) Probiotics and immune response. Am J Gastroenterol 95:S22–S25

Daeschel MA, Klaenhammer TR (1985) Association of a 13, 6 megadalton plasmid in *Pediococcus pentosaceous* with bacteriocin activity. Appl Environ Microbiol 50:1538–1541

Danone (2001) Fermented foods and healthy digestive functions. Danone Publications, John Libbey Eurotext, Montrouge

De Klrek HC, Coetzee JN (1961) Antibiosis among *Lactococci*. Nature 192:340–341

De Martins ECP, Satarosa PR, Freitzs FZ (2003) Preliminary characterization of bacteriocins produced by six lactic acid bacteria strains isolated from vacuum packaged meat products. Cienc Tecnol Ailm 23:195–199

De Vuyst L, Leroy F (2007) Bacteriocins from lactic acid bacteria: production, purification and food applications. J Mol Microbiol Biotechnol 13:194–199

De Vuyst L, Vandamme EJ (1994) Bacteriocins of lactic acid bacteria: microbiology, genetics and application. London, Blackie Academic and Professional 56:117–137

De Vuyst L, Callewaert R, Crabbe K (1996) Primary metabolite kinetics of bacteriocin biosynthesis by *Lactobacillus amylovorus* and evidence for stimulation of bacteriocin productions under unfavourable growth conditions. Microbiology 142:817–827

Deegan LH, Cotter PD, Hill C, Ross P (2006) Bacteriocins: biological tools for bio-preservation and shelf life extension. Int Dairy J 16:1058–1071

Deeth HC, Tamime AY (1981) Yogurt, nutritive and therapeutic aspects. J Food Prot 44:78–86

Delves BJ, Blackburn P, Evans RJ, Hugenholtz J (1996) Application of bacteriocin nisin. Antonie Van Leewenhoek 70:193–202

Delves-Broughton J (1990) Nisin and its uses as a food preservative. Food Technol 44:100–117

Diep DB, Havarstein LS, Nes IF (1995) A bacteriocin-like peptide induces bacteriocin synthesis in *Lactobacillus plantarum* C11. Mol Microbiol 18:631–639

Dimov S, Ivanova P, Harizanova N (2005) Genetics of bacteriocin biosynthesis by lactic acid bacteria. Biotechnol Biotechnol Eq 19:3–10

Ennahar S, Sonomoto K, Ishijaki A (1999) Class II a bacteriocins from lactic acid bacteria: antibacterial activity and food preservation. J Biosci Bioeng 87:705–716

Farkas HH, Yu H (1985) Purified colicin as cytotoxic agent of neoplasia: comparative study with crude colicin. Cytobios 42:193–207

Farkas H, Freedman J, Read SE, Asad S, Kardish M (1991) Bacterial proteins cytotoxic to HIV-I-infected cells. AIDS 5(7):905–907

Farkas HH, Hill R, Rosen B, Arab S, Lingwood CA (1995) The bacterial colicen active against tumor cells *in vitro* and *in vivo* is verotoxin-1. Proc Natl Acad Sci U S A 92(15):6996–7000

Foulquie MMR, Callewaert R, Vuyst D (2001) Isolation of bacteriocins through expanded bed adsorption using a hydrophobic interaction medium. Bioseparation 10:45–50

Foulquie MMR, Rea MC, Cogan TM, Vuyst D (2003) Applicability of a bacteriocin producing *Enterococcus faecium* as a co-culture in cheddar cheese manufacture. Int J Food Microbiol 81:73–84

Gionchetti P, Rizzello F, Venturi A, Campieri M (2000) Probiotics in infective diarrhoea and inflammatory bowel diseases. J Gastroenterol Hepatol 15:489–493

Goldin BR, Gualtieri L, Moore RP (1996) The effect of *Lactobacillus* GG on the initiation and promotion of dimethylhydrazine-induced intestinal tumors in the rat. Nutr Cancer 25:197–204

Goldstein BP, Wei J, Greenberg K, Novick R (1998) Activity of nisin against *Streptococcus pneumoniae*, *in vitro*, and in mouse infection model. J Antimic Chemother 42:277–278

Golshmid VK, Landsman NM (1989) The properties of the antibacterial preparation tomicid and the prospects of its use for treating staphylodermas. Vestn Dermatol Venerol 6:62–64

Gorbach SL (2000) Probiotics and gastrointestinal health. Am J Gastroenterol 95:S2–S4

Gorbach SL, Chang TW, Goldin B (1987) Successful treatment of relapsing *Clostridium difficile* colitis with *Lactobacillus* GG. Lancet 2:1519 (letter)

Gratia A (1925) Sur unremarquable example d' antagonisme enter deux souches de colibacille. Compt Rend Soc Biol 93:1040–1042

Gratia JP (2000) Andre Gratia: a forerunner in microbial viral genetics. Genetics 156(2):471–476

Guinane CM, Cotter PD, Hill C, Ross RP (2006) Spontaneous resistance in *Lactococcus lactis* IL 1403 to the lantibiotic lacticin 3147. FEMS Microbiol Lett 260:77–83

Guslandi M, Mezzi G, Sorghi M, Testoni PAL (2000) *Saccharomyces boulardii* in maintenance treatment of Crohn's disease. Dig Dis Sci 45:1462–1464

Hancock REW (1997) Peptide antibiotics. Lancet 349:418–420

Havenaar R, Huisin't Veld J (1992) Probiotics: a general view in wood (ed). The lactic acid bacteria in health and disease. Elsevier, London. Appl Sci 209–224

Hurst A (1981) Nisin. Adv Appl Microbiol 27:85–123

Jordi BJ, Boutaga K, van Heeswijk CM, van Knapen F, Lipman LJ (2001) Sensitivity of Shiga toxin-producing *Escherichia coli* (STEC) strains for colicins under different experimental conditions. FEMS Microbiol Lett 204:329–334

Juodeikiene G, Bartkiene E, Viskelis P, Urbonaviciene D, Eidukonyte D, Bobinas C (2012) Fermentation processes using lactic acid bacteria producing bacteriocins for preservation and improving functional properties of food products. In: Marian P (ed) Advances in applied biotechnology. InTech, Croatia. ISBN 978-953-307-820-5

Kellner R, Jung G, Horner T, Zahner H, Schnel N, Entian KD, Gotz F (1988) Galldermin: a new lanthionine-containing polypeptide antibiotic. Eur J Biochem 177:53–59

Kim TS, Hur JW, Yu MA, Cheigh CI, Kim KN, Hwang JK, Pyun YR (2003) Antagonism of *Helicobacter pylori* by bacteriocins of lactic acid bacteria. J Food Prot 66:3–12

Klaenhammer TR (1988) Bacteriocins of lactic acid bacteria. Biochemie 70:337–349

Klaenhammer TR (1993) Genetics of bacteriocins produced by lactic acid bacteria. FEMS Microbiol Rev 12:39–86

Kuipers OP, Beerthuyzen MM, Pascalle G, de Ruyter GA, Luesink EJ, de Vos WM (1995) Autoregulation of nisin biosynthesis in *Lactococcus lactis* by signal transduction. J Biol Chem 270:1–6

Kyne L, Kelly CP (2001) Recurrent *Clostridium* difficile diarrhoea. Gut 49:152–153

Lee YK, Nomoto K, Salminen S, Gorbach SL (1999) Handbook of probiotics. Wiley, New York

Leroy F, De Vuyst L (2004) Lactic acid bacteria as functional starter cultures for food fermentation industry. Trends Food Sci Technol 15:67–78

Leroy F, Vuyst D (1999a) The presence of salt and a curing agent reduces bacteriocin production by *Lactobacillus sakei* CTC 494, potential starter culture for sausage fermentation. Appl Environ Microbiol 65:5350–5356

Leroy F, Vuyst D (1999b) Temperature and pH conditions that prevail during the fermentation of sausage are optimal for the production of the antilisterial bacteriocin sakasin K. Appl Environ Microbiol 65:974–981

Leroy F, Vuyst D (2000) Sakacin. In: Naidu AS (ed) Natural food antimicrobial systems. CRC Press LLC, Boca Raton, pp 589–610

Leroy F, Vuyst D (2001) Growth of the bacteriocin producing *Lactobacillus sakei* strain CTC 494 in MRS broth is strongly reduced due to nutrient exhaustion a nutrient depletion model for the growth of lactic acid bacteria. Appl Environ Microbiol 67:4407–4413

Leroy F, Vuyst D (2005) Simulation of the effect of sausage ingredients and technology on the functionality of the bacteriocin producing *Lactobacillus sakei* CTC 494 strain. Int J Food Microb 100:141–152

Leroy F, Lievens K, De Vuyst L (2005) Modelling bacteriocin resistance and inactivation of *Listeria innocua* LMG 13568 by *Lactobacillus sakei* CTC 494 under sausage fermentation conditions. Appl Environ Microbiol 71:7567–7570

Leroy F, Verluyten J, De Vuyst L (2006) Functional meat starter culture for improved sausage fermentation. Int J Food Microbiol 106:270–285

Lilly DM, Stillwell RH (1965) Probiotics: growth promoting factors produced by microorganisms. Science 147:747–748

MacFarlane GT, Cummings JH (2002) Probiotics, infection and immunity. Curr Opin Infect Dis 15:501–506

Mallett AK, Bearne CA, Rowland IR (1989) The influence of incubation pH on the activity of rat and human gut flora enzymes. J Appl Bacteriol 66:433–437

Marugg JD, Gonzalez CF, Kunka BS, Ledeboer AM, Pucci MJ, Toonen MY, Walker SA, Zoetmulder LCM, Vandenbergh PA (1992) Cloning, expression, and nucleotide sequence of genes involved in production of pediocin PA-1, a bacteriocin from *Pediococcus acidilactici* PAC1.0. Appl Environ Microbiol 58:2360–2367

Metchnikoff E (1908) The prolongation of life. Putnam's Sons, New York/London

Mortvedt CI, Nissen MJ, Sletten K, Nes IF (1991) Purification and amino acid sequence of lactocin S, a bacteriocin produced by *Lactobacillus sakei* 45. Appl Environ Microbiol 57:1829–1834

Mota-Meira M, Morency H, Lavoie MC (2005) *In vivo* activity of mutacin B-Ny266. J Antimicrob Chemother 56:869–871

Motlagh A, Bukhtiyarova M, Ray B (1994) Complete nucleotide sequence of pSMB 74, a plasmid encoding the production of pediocin AcH in *Pediococcus acidilactici*. Lett Appl Microbiol 18(6):305–312

Musclow CE, Farkas HH, Weigtzman SS, Herridge M (1987) Acute lymphoblastic leukemia of childhood monitored by bacteriocin and flowcytometry. Eur J Cancer Clin Oncol 23(4):411–418

Navaratna M, Sahl HG, Tagg JR (1998) Two-component Anti *Staphylococcus aureus* lantibiotic activity produced by *Staphylococcus aureus* C55. Appl Environ Microbiol 64(12):4803–4808

Parente E, Ricciardi A (1999) Production, recovery and purification of bacteriocins from lactic acid bacteria. Appl Microbiol Biotechnol 52:628–638

Quadri LEN (2002) Regulation of antimicrobial peptide production by autoinducer-mediated quorum sensing in lactic acid bacteria. Antonie van Leeuwenhoek 82:133–145

Quadri LEN, Sailer M, Roy KL, Vederas JC, Stiles ME (1994) Chemical and genetic characterization of bacteriocins produced by *Carnobacterium piscicola* LV17B. J Biol Chem 269(16):12204–12211

Rajalakshmi R, Vanaja K (1967) Chemical and biological evaluation of the effects of fermentation on the nutritive value of foods prepared from rice and grams. Br J Nutr 21:467–473

Rasch M, Knochel S (1998) Variations in tolerance of *Listeria Monocytogenes* to nisin, Pediocin PA-1 and Bavaricin A. Lett Appl Microbiol 27:275–278

Reddy KVR, Aranha C, Gupta SM, Yedery RD (2004) Evaluation of antimicrobial peptide nisin as a safe vaginal contraceptive agent in rabbits: *in vitro* and *in vivo* studies. Reproduction 128:117–126

Reid G, Bruce AW (2001) Could probiotics be an option for treating and preventing urogenital infections? Medscape Womens Health 6:9

Reid G, Beuerman D, Heinemann C, Bruce AW (2001) Probiotic *Lactobacillus* dose required to restore and maintain a normal vaginal flora. FEMS Immunol Med Microbiol 32:37–41

Rekhif N, Atrih A, Lefebvre G (1994) Selection and properties of spontaneous mutants of *Listeria monocytogenes* ATCC 15313 resistant to different bacteriocins produced by lactic acid bacteria strains. Curr Microbiol 28:237–241

Rince A, Dufour A, Uguen P, Le Penec JP, Haras D (1997) Characterization of the lacticin 481 operon: the *Lactococcus lactis* genes lctF, lctE, and lctG encode a putative ABC transporter involved in bacteriocin immunity. Appl Environ Microbiol 63(11):4252–4260

Rodriguez JM, Martinez MI, Kok J (2002) Pediocin PA-1, a wide spectrum bacteriocin from lactic acid bacteria. Crit Rev Food Sci Nutr 42:91–121

Rogers LA (1928) The inhibitory effect of *Streptococcus lactis* on *Lactobacillus bulgaricus*. J Bacteriol 16:321–325

Ryan MP, Meaney WJ, Ross RP, Hill C (1998) Evaluation of Lacticin 3147 and a teat seal containing this bacteriocin for inhibition of mastitis pathogens. Appl Environ Microbiol 64:2287–2290

Ryser E, Marth EH (1991) Food born Listeriosis. In: Ryser E, Marth EH (eds) *Listeria*, Listeriosis and food safety. Marcel Dekker Inc, New York, pp 240–287

Sable S, Pons AM, Gendron-Gaillard S, Cottenceau G (2000) Antibacterial activity evaluation of microcin J25 against diarrheagenic *Escherichia coli*. Appl Environ Microbiol 66:4595–4597

Saito H, Watanabe T, Osasa S, Tada O (1979) Susceptibility of normal and tumor cells to mycobacteriocin and mitomycin C. Hiroshima J Med Sci 28(3):141–146

Sand SL, Haug TM, Nissen MJ, Sand O (2007) The bacterial peptide pheromone plantaricin A permeabilizes cancerous, but not normal, rat pituitary cells and differentiates between the outer and inner membrane leaflet. J Memb Biol 216(2–3):61–71

Schleifer KH, Kraus J, Dvorak C, Kilpper-Balz R, Collins MD, Fischer W (1985) Transfer of *Streptococcus lactis* and related streptococcus to the genus of *Lactococcus* gen nov. Syst Appl Microbiol 6:183–195

Schobitz RP, Borquez PA, Costa ME, Ciampi LR, Brito CS (2006) Bacteriocins like substances produced by *Carnobacterium piscicola* in a continuous system with three culture broth. Study of antagonism against *Listeria monocytogenes* in vacuum packaged salmon. Braz J Microbiol 37:52–57

Sears PM, Smith BS, Stewart WK, Gonazalez RN (1992) Evaluation of a nisin-based germicidal formulation on teat skin of live cows. J Dairy Sci 75:3185–3190

Shahani KM, Chandan RC (1979) Nutritional and healthful aspects of cultured and culture-containing dairy foods. J Dairy Sci 62:1685–1694

Singh AK, Ramesh A (2008) Succession of dominant and antagonistic lactic acid bacteria in fermented cucumber: Insights from a PCR-based approach. Food Microbiol 25(2):278–287

Sobrino OJ, Rodriguez JM, Moreira WL, Fernandez MF, Bernabesanz MF, Hernandez PE (1991) Antibacterial activity of *Lactobacillus sake* isolated from fermented sausages. Int J Food Microbiol 13:1–10

Stevens KA, Sheldon BW, Klapes NA, Klaenhammer TR (1991) Nisin treatment for inactivation of *Salmonella* spp. and other Gram negative bacteria. Appl Environ Microbiol 57:3613–3615

Stile ME, Hastings JW (1991) Bacteriocin production by lactic acid bacteria: potential for use in meat preservation. Trends Food Sci Technol 2:235–263

Stiles ME (1996) Bio preservation by lactic acid bacteria. Antonie Van Leewenhoek 70:235–249

Taranto MP, Medici M, Perdigon G, Ruiz Holgado AP, Valdez GF (1998) Evidence for hypocholesterolemic effect of *Lactobacillus reuteri* in hypercholesterolemic mice. J Dairy Sci 81:2336–2340

Tichaczek PS, Mayer JN, Nes IF, Vogel RF, Hammes WP (1992) Characterisation of the bacteriocins curvacin A from *Lactobacillus curvatus* LTH 1174 and sakacin P from *L. sake* LTH 673. Syst Appl Microbiol 15:460–468

Twomey D, Ross RP, Ryan M, Meaney B, Hill C (2002) Lantibiotics produced by lactic acid bacteria: structure, function and applications. Antonie van Leewenhoek 82:165–185

Upreti GC, Hindsdill RD (1975) Production and mode of action of lactocin27: bacteriocin from homofermentative *Lactobacillus*. Antimicrob Agents Chemother 7:139–145

Vandenberg PA (1993) Lactic acid bacteria, their metabolic products and interference with microbial growth. FEMS Microbiol Rev 12:221–238

Vander Meer JR, Rolman J, Beer Thuyzen MM, Siezen RJ, Kuipers OP, de Vos WM (1993) Characterization of the *Lactococcus lactis* nisin A operon genes nisP, encoding a subtilisin – like serine protease involved in precursor processing, and nisR, encoding a regulatory protein involved in nisin biosynthesis. J Bacteriol 175(9):2578–2588

Verluyten J, Leroy F, Vuyst D (2004a) Influence of complex nutrient source on growth of and curvacin A production by sausage isolate *Lactobacillus curvatus* LTH 1174. Appl Environ Microbiol 70:5081–5088

Verluyten J, Messens W, Vuyst D (2004b) Sodium chloride reduces production of curvacin A, a bacteriocin produced by *Lactobacillus curvatus* LTH 1174, originating from fermented sausage. Appl Environ Microbiol 70:2271–2278

Vignolo G, Fadd S, Dekairuz MN, Holgado APD, Oliver G (1998) Effects of curing additives on the control of *Listeria monocytogenes* by lactocin 705 in meat slurry. Food Microb 15:259–264

Wood B, Holzapfel WH (1995) The genera of lactic acid bacteria. Blackie Academic and Professional, London

Wooley RE, Gibbs PS, Schott EB (1999) Inhibition of *Salmonella typhimurium* in chicken intestinal tract by transformed avirulent avian *Escherichia coli*. Avian Dis 43:245–250

Yang R, Johnson MC, Ray B (1992) Novel method to extract large amounts of bacteriocins from lactic acid bacteria. Appl Environ Microbiol 58:3355–3359

12 Third Generation Green Energy: Cyanobacteria, Key to Production of Sustainable Energy Through Metabolic Engineering

Namita Singh and Ritika Chanan

Abstract

Biofuels are important as dependency on fossil fuels has resulted in economic instability in the world and heavy environmental damage. Burning of fossil fuel releases heavy amounts of carbon dioxide in the atmosphere, raising the concern of global warming. Development of alternative energy forms, sustainable and renewable in nature, is thus the need of the hour. In this context, agricultural production of biofuels has gained utmost importance, and more recently industrial biofuel production through cyanobacteria at large scale has almost stabilized the current scenario of global warming and current fuel demands. Modulation in the cyanobacterial biochemical and metabolic pathways at the genetic level for attractive biofuel yields is a challenge for upcoming scientists to offset petroleum and mineral oil usage and dependency.

12.1 Introduction

Fossil fuel has lost its significance as an energy resource. Global warming, depletion of oil and petroleum reserves in the near future (Nashawi et al. 2010), economic and energy crisis due to heavy oil demand at commercial scale, etc. are some of the serious issues that call for a change and replacement in the technology. Bioenergy production, in this regard, has emerged as an innovative substitute for fossil fuels in the past few years to generate heat, power and chemicals (Hall et al. 1993; Goldemberg 2000). Sustainable development of energy using biomass constitutes 75 % of renewable energy sources (Hall and Moss 1983), and this has stabilized environmental, social and economic issues. First- and second-generation biofuel productions using agricultural crops and lignocellulosic wastes, respectively, present a promising solution for heavy yields of clean energy. Agricultural crops such as sugarcane, cereal grains, oilseeds etc. can be processed to produce biofuels through various biochemical or thermochemical routes such as fermentation (sugar to alcohol), gasification, chemical synthesis etc. (Elam 1996). Use of lignocellulosic biomass to produce high amounts of bioenergy can play an important role in addressing both greenhouse gas emissions and dependency on mineral oil. Lignin content interferes with the biofuel

N. Singh (✉) • R. Chanan
Department of Bio & Nano Technology, Guru Jambheshwar University of Science and Technology, Hisar 125001, Haryana, India
e-mail: namitasingh71@gmail.com

R.K. Salar et al. (eds.), *Biotechnology: Prospects and Applications*,
DOI 10.1007/978-81-322-1683-4_12,

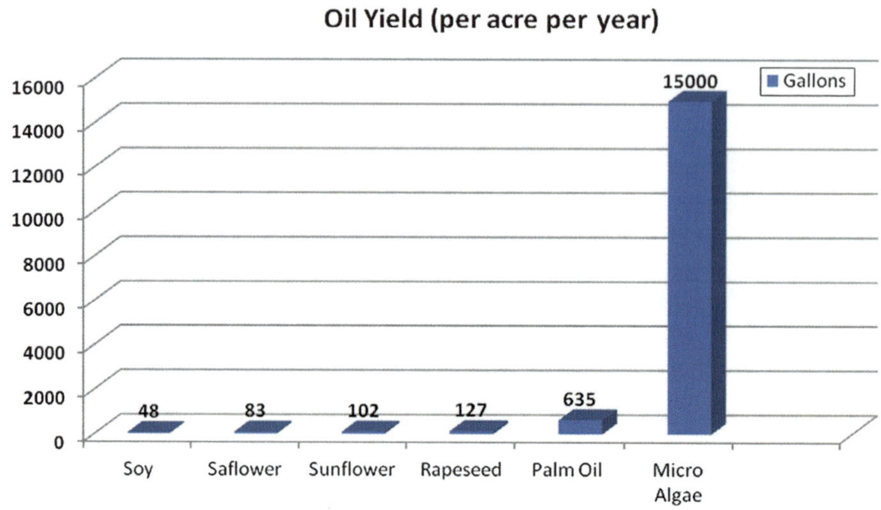

Fig. 12.1 Biodiesel yield of some feedstock (Source: The Green Chip Stocks 2008. http://www.greenchipstocks.com/articles/investing-algae-biofuel/253)

production. It has been observed that plant form with high cellulose content and low or modified lignin content is easily digestible in biofuel reactors by the respective enzymes which lead to enhanced production of biofuels (Gressel 2008; Barriere et al. 2004; Gressel and Zilberstein 2003). Lignin-degrading peroxidases are known, but they cannot be used in bioreactors as they have high energy requirements and these enzymes work in living microbes only in direct association with lignocelluloses. Transgenic modification of the lignin content and up-regulation of cellulose biosynthesis can help in clean production of biofuel in large amounts, and this can be achieved through partial silencing of the phenyl-propanoid pathway enzymes leading to reduced lignin synthesis. Many new molecular techniques such as antisensing and RNAi strategies can be helpful. Biofuel production using annual crops such as rapeseed, cereals, beet, etc. in comparison to perennial crops presents certain limitations—high production costs, low net energy yield (100–200 GJ/ha year), requirement of valuable agricultural land, high fertilizer requirement, etc. (Berndes et al. 2003). Other controversial issues like (a) competition between edible and energy crops for available land and water, (b) decline in the ecological value of the area with the plantation of energy crops and (c) influence on ecosystem diversity also limit the use of annual crops for biofuel production. In this respect, lignocellulosic biomass (e.g. wood and grasses) offers a better promising future with the use of biofuels than agricultural crops. Second-generation biofuel production using lignocellulosic wastes (e.g. ethanol production from lignocellulosic biomass) leads to higher overall energy conversion efficiencies and offers lower overall costs for longer term (Arthur 1999). Moreover, better-projected economics and less energy requirement by their feedstocks for growth and harvest are additional benefits of second-generation biofuels.

Green Chip Stocks 2008 special report states that the biofuel production from microalgae accounts for 20 % higher production efficiency than from palm oil without any competition with other feedstocks (Fig. 12.1). Microalgae can be grown in enclosed systems, and all the input requirements can be monitored and regulated accordingly for higher production of biofuel stock.

Cyanobacteria as a microalgae can be a good source of third-generation biofuel resource in contrast to energy crops and lignocellulosic wastes. Cyanobacteria have emerged as an innovative and promising platform for sustainable biofuel production. This viable alternative energy resource is devoid of the major setbacks associated with the first- and second-generation biofuels

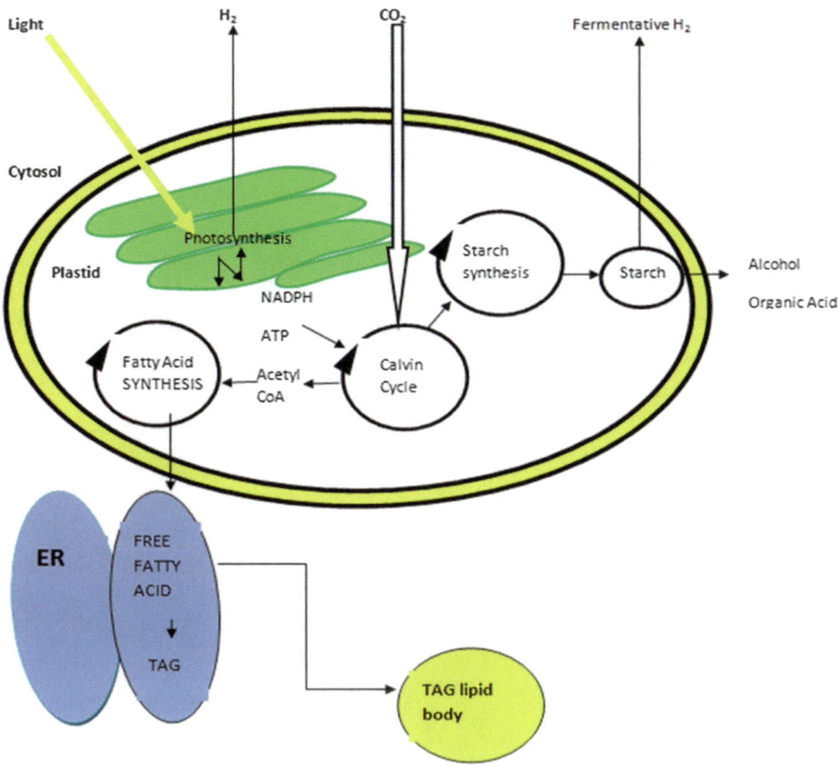

Fig. 12.2 Microalgal metabolic pathways that can be leveraged for biofuel production. *ER* endoplasmic reticulum, *TAG* triacylglycerol

(Mata et al. 2010). High carbon components in the atmosphere released from the automobiles and industries are sunk down by these plant forms through oxygenic photosynthesis in the aquatic environments to usable energy products. Cyanobacteria trap down the sun's energy through phycobilisomes present in the thylakoid membranes of their chloroplast systems converting light energy into chemical (renewable) form of energy (Griffiths and Harrison 2009). There are enormous advantageous features of cyanobacteria that make these species best suited for biofuel development. Non-requirement of bio-productive lands and freshwater for their growth, zero competition with agriculture, nonfood-based feedstock for the generation of biofuel, very short harvesting cycle, significant high growth rates and photosynthetic levels, basic nutritional requirements, inexpensive cultivation, high yields, considerable amounts of lipid content in their membrane systems, potential to thrive in elevated CO_2 conditions, ability to mitigate the progression of climatic change and production of other eco-friendly coproducts are some of the unique attributes of cyanobacterial forms that attract scientists to use this living biomass for the development of bioenergy (Parmar et al. 2011; Rittmann 2008) (Fig. 12.2). CO_2 emissions from automobiles and industrial exhaust gases get fixed up by these living forms photosynthetically in a carbon-neutral way to produce metabolites which act as potential feedstocks from which biofuel can be extracted. The burning of the biofuel in turn generates equal amounts of CO_2 maintaining the overall environmental integrity and nature's balance.

Eukaryotic algae- and cyanobacteria-based fuels are leading candidates for the replacement of fossil fuels. However the cost of harvesting the algae and extracting intracellular lipid from them offers certain limitations at the large scale which demand fuel production from engineered cyanobacteria (Pienkos and Darzins 2009; Yang et al. 2011). Engineering of cyanobacteria at the gene level is important as it results in successful bulk amount of biofuel yields at large scale. This introduces a new dimension in the current scenario of biofuel production—*metabolic engineering*. Genetic engineering of metabolic pathways at some important steps in cyanobacteria for improved and efficient fuel productivities is possible. With the advent of many new technologies and most importantly with the available full sequence data of most of the cyanobacterial species, it is easy to modify the enzymatic pathways at the gene level for higher production of free fatty acid and other lipid/sugar compounds for improved yields of biofuel. *Genetic engineering* in cyanobacteria can be focused on the following metabolic pathways for substantial yield of biofuel at commercial scale:

1. Carbon metabolism, genetic strategies to increase:
   (a) Glucan storage
   (b) Starch biosynthesis
   (c) Starch biodegradation
2. Lipid metabolism:
   (a) Lipid biosynthesis
   (b) Lipid catabolism
3. Modification of lipid characteristics

12.2 Carbon Metabolism

Carbon dioxide fixation to usable chemical products (sugar/lipid) and their extraction to several distinct energy forms is the key for sustainable biofuel development in cyanobacteria. Full-length genome sequences of most of the cyanobacterial species are known, knowledge of which helps to manipulate the rate-limiting steps in the carbohydrate metabolic pathways in cyanobacteria for maximum bioenergy yield. Also relatively small genome of cyanobacteria makes it possible to readily genetically engineer the genome to enhance biofuel production.

Rubisco plays an important role in central carbon metabolism in higher plants. Rubisco fixes carbon dioxide during photosynthesis. Cyanobacteria are also photosynthetic forms that fix atmospheric carbon dioxide with cyanobacterial Rubisco, which has a lower affinity for carbon dioxide. In view of the above problem, a carbon dioxide concentrating mechanism (CCM) is developed (Badger et al. 2006) in which two carbon-fixing enzymes Rubisco and carbonic anhydrase are stored in organized micro-compartments—carboxysomes—through the cell away from competing oxygen (Savage et al. 2010). This leads to enhanced carbon dioxide fixation and hence improved biofuel yield. Also, certain manipulations at the gene level in the carbon dioxide concentrating mechanism are possible. Genetic modification can be introduced by:

1. By over-expressing carbonic anhydrase enzyme
2. By inserting more efficient Rubisco
3. By inserting multiple copies of inorganic carbon transporters (responsible for intracellular delivery of carbon dioxide and bicarbonates)

12.2.1 Glucan Storage

Sugars represent the main source of energy. Genetic modulations to increase glucan storage in cyanobacteria can help lead to heavy accumulation of storage compounds. Biosynthetic pathways of starch and other storage compounds can be genetically modulated, and also certain desirable traits can be added to the metabolic pathways for the same. Resulting bulk amount of storage compounds can be extracted readily to distinct forms of biofuels.

12.2.2 Starch Biosynthesis

The rate-limiting step of starch synthesis (storage product) is the ADP-glucose pyrophosphorylase (AGPase) catalyzed reaction of glucose-1-P with ATP (Stark et al. 1992).

$$\text{Glc-1-P} + \text{ATP} \xrightarrow{\text{AGPase}} \text{ADP-glc} + \text{PPi}$$

Manipulations at the genetic level controlling the rate-limiting step of starch synthesis can improve starch accumulation. AGPases, the

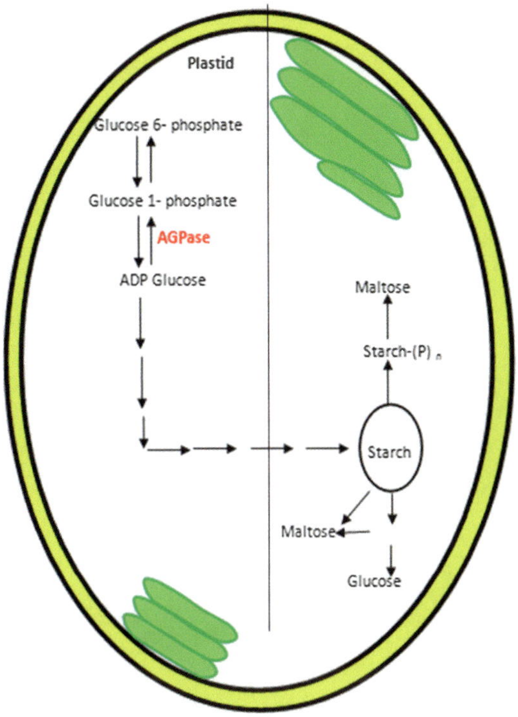

Fig. 12.3 Rate limiting step of starch synthesis catalyzed by ADP-glucose phosphorylase. *AGPase* ADP-glucose pyrophosphorylase

important enzymes, have been studied for their biocatalytic properties (Fig. 12.3). Allosteric structure and interactions of this enzyme with the respective substrate have also been studied (Smith 2008). Designer AGPases have been studied to increase starch content in higher plants (Giroux et al. 1996). These enzymes can be introduced and expressed in cyanobacteria also with no native AGPase activity. Another approach to increase glucan storage is to introduce genes responsible for starch synthesis into the protoplast through genetic engineering.

12.2.3 Starch Degradation

Starch content in the protoplast can also be increased by checking the starch degradation processes. Hydrolytic and phosphorolytic mechanisms are known that control starch degradation processes in cyanobacteria (Smith et al. 2005). By controlling the starch catabolic pathways at the gene level, heavy level of starch content can be achieved in the protoplast which later can be recovered and extracted to different biofuel forms. Gene knockout technologies are found useful for controlling starch degradation processes. Phosphorylation and hydrolytic steps demonstrate important gene knockout targets which help to display a starch accumulation phenotype in cyanobacteria. It is important here to refer that oxidative pentose phosphate (OPP) pathway is the major route of sugar catabolism in cyanobacteria. One of the key enzymes of OPP pathway is glucose-6-phosphate dehydrogenase. Its regulation and control at the level of gene expression can increase carbon dioxide fixation and hence can improve biofuel production (Kaplan et al. 2008; Stanier and Cohenbazire 1977; Vanderoost et al. 1989).

Most of the cyanobacteria also produce certain polymeric substances in their extracellular environment, for example, cellulose (Pereira et al. 2009). These species can be genetically modified for significant yields of extra-polymeric substances which can be extracted to produce biofuels.

12.3 Lipid Metabolism

Facile genetic manipulation in the lipid metabolic pathway for potential fuel excretion in cyanobacteria is possible. Tailor designing of biocatalyst for large-scale fuel production through genetic engineering in cyanobacteria also plays an important role. Understanding the lipid synthetic and catabolic pathways in cyanobacteria is of great interest for the ultimate production of fuel surrogates. There are really important pathways that control the saturation and length of fatty acids that also have been studied in cyanobacteria. Cyanobacteria have this unique ability to excrete fuel precursors such as free fatty acids extracellularly. This makes the recovery process and the further downstream processing steps relatively easy and simple (Kaczmarzyk and Fulda 2010).

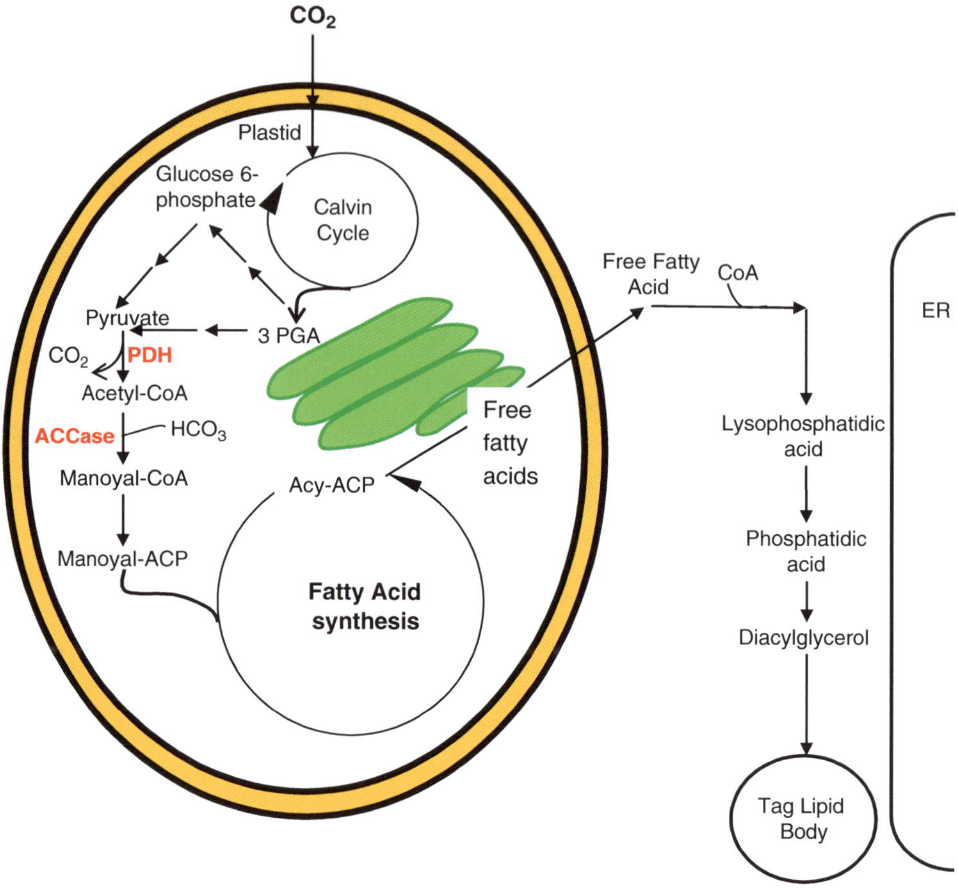

Fig. 12.4 Schematic overview of the metabolites and representative pathways in microalgal lipid biosynthesis shown in *black* and enzymes shown in *red*. *ACCase* acetyl CoA carboxylase, *ACP* acyl carrier protein, *CoA* coenzyme A, *G3PDH* gycerol-3-phosphate dehydrogenase

12.3.1 Lipid Biosynthesis

In recent years, much of the work has been done on cyanobacteria to understand the biochemical pathways and regulation of the enzymatic pathways at the gene level. Strategies have been established to increase the lipid content in the cell. Transgenic over-expression strategies have successfully resulted in heavy lipid production and accumulation in cyanobacteria. One early rate-limiting step in fatty acid synthesis is the conversion of acetyl-coenzyme A (CoA) to malonyl-CoA, catalyzed by acetyl CoA carboxylase (ACCase). Genetic modification of this committed enzyme can increase the fatty acid production and can result in quantitative yield of biofuels (Dunahay et al. 1995) (Fig. 12.4). Over-expression of the lipid biosynthesis genes encoding acetyl CoA carboxylase controlled by an inducible promoter can be activated once the microalgal cells have grown to a high density and have entered stationary phase. Similarly, over-expression of genes involved in triacylglycerol (TAG) assembly has shown some meaningful results. Glycerol-3-phosphate dehydrogenase (G3PDH) catalyzes the formation of glycerol-3-phosphate, which is needed for TAG formation. Over-expression of G3PDH can result in improved fuel production (Vigeolas et al. 2007). Another possible approach to increase cellular lipid content is to block starch synthetic pathways. For example, starchless mutant of *Chlorella pyrenoidosa* has been shown to have elevated polyunsaturated fatty acid content (Ramazanov and Ramazanov 2006).

12.3.2 Lipid Catabolism

Lipid catabolic pathways can be silenced to increase cellular lipid accumulation. This is one strategy wherein the genes involved directly in β-oxidation of fatty acids can be inactivated through gene knockout technology to achieve quantitative yields of biofuel (De Riso et al. 2009; Molnar et al. 2009; Zhao et al. 2009). Quoting one example—in *S. elongatus* 7942–any free fatty acid (FFA) released by membrane degradation is recycled for membrane synthesis via acyl-ACP synthetase (aas) (Fulda 2010). Cellular FFA accumulation can reach high peaks if this acyl-ACP synthetase enzyme is targeted for gene knockout. Silencing of this respective enzyme can lead to heavy accumulation of acyl-ACP in the cytosol followed by effective production of biofuel.

12.3.3 Modification of Lipid Characteristics

Not only the quantity but the quality of biofuel also matters. Quality of the lipids is checked by few properties such as carbon chain length and degree of unsaturation of the fatty acids. These affect the oxidative, cold flow and stability characteristics of the biofuel/biodiesel. The chain length of fatty acids is determined by acyl-ACP thioesterases, which release the fatty acid chain from the fatty acid synthase enzyme. Suitable expression of thioesterase can change the lipid profiles in cyanobacteria. Ideal fatty acid chain length for biofuel production should be 12:0 and 14:0. Appropriate thioesterase expression can improve the suitability of microalga-derived diesel feedstock (Voelker and Davies 1994).

12.3.4 Biosynthesis of Ethanol

Bioethanol is one of the renewable energy resources which is produced from fermentation of agricultural crops (Goldemberg 2007). In contrast to edible crops, cyanobacteria can be used for bioethanol production without the addition of yeast starter culture (for fermentation) (Heyer and Krumbein 1991). To maximize bioethanol generation in cyanobacteria, genetic manipulations are also possible, for example, genes encoding pyruvate decarboxylase and alcohol dehydrogenase (enzymes controlling ethanol biosynthesis) from bacterium *Z. mobilis* have been introduced into the cyanobacterial genome for high bioethanol production during photoautotrophic growth (Deng and Coleman 1999). Cyanobacteria have been studied to produce cellulose extracellularly at a yield of up to 25 % of the cell dry volume (Dewinder et al. 1990). Modification of *Synechococcus* (cyanobacteria) with cellulose synthesis gene from *Gluconobacter* can yield maximum amount of extracellular non-crystalline cellulose, an ideal feedstock for ethanol production (Nobles and Brown 2008).

12.3.5 Biosynthesis of Hydrogen

Renewable energy is produced from natural sources and cyanobacterial biomass stands out as one such natural resource in today's generation for bioenergy development. Biohydrogen has emerged as one of the promising renewable green energy sources, and nitrogenase-based photobiological hydrogen production from cyanobacteria offers a major platform for its high rate production under aerobic conditions (Tamagnini et al. 2002, 2007). Optimization of cyanobacterial strains is central to improved biofuel/biohydrogen yields. For commercial viability and economic feasibility, efficient biohydrogen production directly using the sun's energy has been studied in a number of photosynthetic species possessing nitrogenase and hydrogenase enzymes. *Cyanothece* strains maintain marine nitrogen cycle and grow in basic nutritional environment limited in dissolved inorganic nitrogen compounds (Zehr et al. 2001). In one of the works on *Cyanothece* sp. ATCC 51142 strain, it has been described that its nitrogenase enzyme can be modulated to convert solar energy into bioenergy at unexpected high rates under aerobic conditions. *Cyanothece* strains derive their nutritional requirements by performing photosynthesis during daytime and storing carbon in the form of glycogen. During dark period suboxic intracellular environment created as a result of respiration process drives oxygen-sensitive nitrogenase and hydrogenase enzymes to carry out nitrogen fixation and hydrogen production energy-rich processes at the expense of glycogen accumulated during light period (Schneegurt et al. 1997). Diurnal cycling patterns and other unique metabolic traits of this strain have made this strain an ideal organism for photobiological biohydrogen production (Stockel et al. 2008). *Cyanothece* 51142, a unicellular diazotrophic cyanobacterium, possesses a two-phase intracellular biohydrogen production pathway (Toepel et al. 2008): (a) growth phase, cells are grown aerobically under 12 h light/12 h dark cycles, followed by (b) incubation phase, cells are collected at the end of 12 h light period and are incubated at a specific temperature under continuous white light source for 12 h. The physiological activities of the cells get in sync with the dark condition during this latter phase, enhancing nitrogen fixation and hydrogen production processes. This particular experiment has led to a significant conclusion that it is possible to achieve a high-fold increase in biohydrogen production in *Cyanothece* 51142 strain during sub-aerobic/anoxic dark period by oxygen-sensitive hydrogenase and nitrogenase enzymes under photoactive conditions. Determination of hydrogen levels at the end of the experiment provided strikingly new results as most unicellular microbial strains till now utilize anaerobic environment for hydrogen production, but this case was different as more than 150 um of hydrogen per mg of chlorophyll per hour was produced.

Continued work on this particular strain has helped scientists to assess the effect of varying environmental conditions on cyanobacterial photobiological activity for improved biohydrogen yields. Cells grown under elevated CO_2 concentrations and under the presence of additional carbon sources have shown higher growth rates.

12.4 Biological Limitations

No doubt, lots of research to increase cellular lipid content and fat accumulation for maximum bioenergy yield has been done, but there are certain limitations also, to use such species at commercial scale for heavy biofuel production.

1. Risk of Contamination: Laboratory organisms when grown under field conditions for large-scale biofuel production can come under the risk of contamination by other indigenous local organisms.
2. Production and Harvesting Costs: High production cost is required for algal growth and for separate lipid producing ponds as most algae either grow or produce fat bodies but not both simultaneously. Harvesting of algae is a cost-prohibitive factor.
3. Carbon Dioxide Enrichment: Most of the higher plants and animals grow well when aerated with 5 % CO_2, but algae and cyanobacteria grow happily when enriched with 100 %CO_2. Still they pose a problem and their response to added CO_2 is not as good as it can be. Conversely, at such high CO_2 concentrations high algal growth can surely occur, but at such CO_2 levels it has been studied that overabundance of RUBISCO enzyme (scavenger of rare CO_2 present in air or water) at the expense of more needed rate-limiting enzymes occurs.
4. Light Penetration: At high light intensities in photo-bioreactors, algal growth can be inhibited due to photoinhibition.
5. Seasonality: Algal growth and high yields in ponds is usually seasonal and temperature dependent. Very low temperatures and extremely high temperatures both pose a problem in algal growth.

12.5 Conclusion and Future Prospects

Biofuel production is important as (a) it contributes in reducing carbon dioxide levels in the atmosphere; (b) it cuts off petroleum dependency; (c) its production is cost effective, sustainable and renewable in nature; and (d) it stabilizes economical and environmental issues. Fuels from lignocellulosic biomass (second-generation biofuels) and beyond have made a real dent in the petroleum usage and have created a strong impact in this field as a sustainable clean energy resource. First-generation biofuels consist of other food crops which not only have implications for the world's food supply but also impede the amount of land available to grow it on. Second-generation biofuels solve the food problem but not the land problem. As the generation feedstock microalgae (cyanobacteria) solve many of the previous problems, it solves the food vs. fuel issue, the land issue, the footprint issue and the scalability issue. Moreover, with the introduction of new molecular techniques, genetic tools, high-throughput analytical techniques, systematic biological approaches, etc., modification of the genomes of small microorganisms such as cyanobacteria at the genetic level has relatively become easy which has led to cost-effective optimum production of renewable and sustainable form of bioenergy. Lipid and carbohydrate metabolic pathways of these microorganisms have been regulated at the rate-limiting biochemical/enzymatic steps for successful biofuel generation. The problems which have been faced in this field till date can be solved in the near future. The strength of biofuel production can be substantially increased by expansion of cultivation areas for growing feedstock crops. This expansion can be achieved by converting waste uncultivated lands such as forests lands to growing new sources of biofuel feedstocks. Transgenics and the use of other modern technologies are big practical solutions to overcome such limitations. Risk of contamination, algal growth, light penetration, temperature, carbon dioxide enrichment, etc., can be sorted out by genetic modification technology. For example, cyanobacterial growth under high light intensities can cause the problem of photoinhibition, but the use of optical fibres to diffuse the light throughout the depth of the culture can offer a practical solution for the same (Sheehan et al. 1998). Also, metabolic pathways controlling lipid/carbohydrate metabolism can be manipulated by gene technology using novel molecular technologies for enriched bioenergy generation.

References

Arthur DL (1999) Analysis and integral evaluation of potential carbon dioxide neutral fuel chains. GAVE reports (Management, summary, sheet presentation and appendices). Netherlands Agency for Energy and the Environment, Utrecht

Badger MR, Price GD, Long BM, Woodger FJ (2006) The environmental plasticity and ecological genomics of the cyanobacterial CO_2 concentrating mechanism. J Exp Bot 57:249–265

Barriere Y et al (2004) Genetic and molecular basis of grass cell wall biosynthesis and degradability. II. Lessons from brown-midrib mutants. Comptes Rendus Biol 327:847–860

Berndes G, Hoogwijk M et al (2003) The contribution of biomass in the future global energy supply. Biomass Bioenergy 25:1–28

Deng MD, Coleman JR (1999) Ethanol synthesis by genetic engineering in cyanobacteria. Appl Environ Microbiol 65:523–528

Dewinder B, Stal LJ, Mur LR (1990) *Crinalium epipsammum* sp. nov.: a filamentous cyanobacterium with trichomes composed of elliptical cells and containing poly-β-(1,4) glucar (cellulose). J Gen Microbiol 136:1645–1653

Dunahay TG, Jarvis EE et al (1995) Genetic transformation of the diatoms *Cyclotella cryptica* and *Navicula saprophila*. J Phycol 31:1004–1011

Elam N (1996) Automotive fuels survey, Part 2: Raw materials and conversion. International Energy Agency, Breda

Fulda M (2010) Fatty acid activation in cyanobacteria mediated by acyl-acyl carrier protein synthetase enables fatty acid recycling. Plant Physiol 152(3):1598–1610

Giroux MJ, Shaw J, Barry G et al (1996) A single mutation that increases maize seed weight. Proc Natl Acad Sci 93:5824–5829

Goldemberg J (2000) World energy assessment, preface. United Nations Development Programme, New York

Goldemberg J (2007) Ethanol for a sustainable energy future. Science 315:808–810

Gressel J (2008) Genetic glass ceilings: transgenics for crop biodiversity. Johns Hopkins University Press, Baltimore

Gressel J, Zilberstein A (2003) Let them eat (GM) straw. Trends Biotechnol 21:525–530

Griffiths MJ, Harrison STL (2009) Lipid productivity as a key characteristic for choosing algal species for biodiesel production. J Appl Physiol 21:493–507

Hall DO, Moss PA (1983) Biomass for energy in developing countries. Geojournal 7(1):5–14

Hall DO, Rosillo-Calle F, Williams RH et al (1993) Biomass for energy: supply prospects. In: Johansson TB, Kelly H, Amulya KNR, Williams RH (eds) Renewable energy, sources for fuels and electricity. Island Press, Washington, DC

Heyer H, Krumbein WE (1991) Excretion of fermentation products in dark and anaerobically incubated cyanobacteria. Arch Microbiol 155:284–287

Kaczmarzyk D, Fulda M (2010) Fatty acid activation in cyanobacteria mediated by acyl-acyl carrier protein synthetase enables fatty acid recycling. Plant Physiol 152(3):1598–1610

Kaplan A, Hagemann M, Bauwe H, Kahlon S, Ogawa T (2008) Carbon acquisition by cyanobacteria: mechanisms, comparative genomics and evolution. In: Herrero A, Flores E (eds) The cyanobacteria: molecular biology, genomics and evolution. Caister Academic Press, Sevilla

Mata TM, Martins AA, Caetano NS (2010) Microalgae for biodiesel production and other applications: a review. Renew Sustain Energy Rev 14:217–232

Molnar AA, Bassett E et al (2009) Highly specific gene silencing by artificial microRNAs in the unicellular alga *Chlamydomonas reinhardtii*. Plant J 58:165–174

Nashawi IS, Malallah A, Al-Bisharah M (2010) Forecasting world crude oil production using multicyclic Hubbert model. Energy Fuels 24:1788–1800

Nobles DR, Brown RM (2008) Transgenic expression of *Gluconacetobacter xylinus* strain ATCC 53582 cellulose synthase genes in the cyanobacterium *Synechococcus leopoliensis* strain UTCC 100. Cellulose 15:691–701

Parmar A, Singh NK, Pandey A, Gnansounou E, Madamwar D (2011) Cyanobacteria and microalgae: a positive prospect for biofuels. Bioresour Technol 102:10163–10172

Pereira S, Zille A, Micheletti E, Moradas-Ferreira P, Philippis R, Tamagnini P (2009) Complexity of cyanobacterial exopolysaccharides: composition, structures, inducing factors and putative genes involved in their biosynthesis and assembly. FEMS Microbiol Rev 33:917–941

Pienkos PT, Darzins A (2009) The promise and challenges of microalgal derived biofuels. Biofuels Bioprod Biorefin 3(4):431–440

Ramazanov A, Ramazanov Z (2006) Isolation and characterization of a starchless mutant of *Chlorella pyrenoidosa* STL-PI with a high growth rate and high protein and polyunsaturated fatty acid content. Phycol Res 54:255–259

Riso D, Raniell VR et al (2009) Gene silencing in the marine diatom *Phaeodactylum tricornutum*. Nucleic Acids Res 37:e96

Rittmann BE (2008) Opportunities for renewable bioenergy using microorganisms. Biotechnol Bioeng 100(2):203–212

Savage DF, Afonso B, Chen AH, Silver PA (2010) Spatially ordered dynamics of the bacterial carbon fixation machinery. Science 327:1258–1261

Schneegurt MA, Sherman DM, Sherman LA (1997) Growth, physiology and ultrastructure of the diazotrophic cyanobacterium, *Cyanothece* sp. strain ATCC 51142 in mixotrophic and chemotrophic cultures. J Phycol 33:632–642

Sheehan J, Dunahay T et al (1998) A look back at the U.S. Department of Energy's Aquatic Species Program – biodiesel from algae. National Renewable Energy Laboratory, Golden, pp 1–328

Smith AM (2008) Prospects for increasing starch and sucrose yields for bioethanol production. Plant J 54:546–558

Smith AM, Zeeman SC, Smith SM (2005) Starch degradation. Annu Rev Plant Biol 56:73–98

Stanier RY, Cohenbazire G (1977) Phototrophic prokaryotes—Cyanobacteria. Annu Rev Microbiol 31:225–274

Stark DM, Timmerman KP, Barry GF, Preiss J, Kishore GM (1992) Regulation of the amount of starch in plant tissues by ADP glucose pyrophosphorylase. Science 258:287–292

Stockel J et al (2008) Global transcriptomic analysis of Cyanothece 51142 reveals robust diurnal oscillation of central metabolic processes. Proc Natl Acad Sci U S A 105:6156–6161

Tamagnini P et al (2002) Hydrogenases and hydrogen metabolism of cyanobacteria. Microbiol Mol Biol Rev 66:1–20

Tamagnini P et al (2007) Cyanobacterial hydrogenases: diversity, regulation and applications. FEMS Microbiol Rev 31:692–720

The Green Chip Stocks (2008) http://www.greenchipstocks.com/articles/investing-algae-biofuel/253

Toepel WJ et al (2008) Differential transcriptional analysis of the cyanobacterium *Cyanothece* sp. strain ATCC 51142 during light–dark and continuous-light growth. J Bacteriol 190:3904–3913

Vanderoost J, Bulthuis BA, Feitz S, Krab K, Kraayenhof R (1989) Fermentation metabolism of the unicellular cyanobacterium *Cyanothece* PCC 7822. Arch Microbiol 152:415–419

Vigeolas HP, Waldeck T et al (2007) Increasing seed oil content in oil-seed rape (*Brassica napus* L.) by overexpression of a yeast glycerol-3-phosphate dehydrogenase under the control of a seed-specific promoter. Plant Biotechnol J 5:431–441

Voelker TA, Davies HM (1994) Alteration of the specificity and regulation of fatty acid synthesis of *Escherichia coli* by expression of a plant medium-chain acyl-acyl carrier protein thioesterase. J Bacteriol 176:7320–7327

Yang J, Xu M et al (2011) Life cycle analysis on biodiesel production from microalgae: water footprint and nutrients balance. Bioresour Technol 102(1):159–165

Zehr JP et al (2001) Unicellular cyanobacteria fix nitrogen in the subtropical North Pacific Ocean. Nature 412:635–638

Zhao T, Wang W et al (2009) Gene silencing by artificial microRNAs in *Chlamydomonas*. Plant J 58:157–164

Part III

Environmental Biotechnology

Bioremediation of PCB-Contaminated Sediments and Adaptive Mechanisms of Bacterial Degraders Exposed to Polychlorinated Biphenyls (PCBs)

13

Katarína Dercová, Hana Dudášová, Lucia Lukáčová, Slavomíra Murínová, Pavel Hucko, Lívia Tóthová, and Juraj Škarba

Abstract

In the recent decades, several hundred tons of polychlorinated biphenyls (PCBs) have been released into the environment. Due to their hydrophobic properties, PCBs tend to be adsorbed by natural organic matter in the aqueous bottom sediments. Sediment is an essential, integral part of the hydrological system. However, because sediments are the ultimate reservoir for the numerous chemical contaminants, they have the potential to pose ecological and human health hazard. Environmental and economic reasons have urged the development of bioremediation technologies for PCB removal from the contaminated areas. This contribution is focused on biodegradation of PCBs in contaminated sediments using biostimulation and bioaugmentation approaches of bioremediation, as well as on the effects of PCBs as the potential stress factors on the cell membrane of the bacterial degraders and on the membrane adaptation mechanisms.

K. Dercová (✉) • H. Dudášová • L. Lukáčová
J. Škarba
Faculty of Chemical and Food Technology, Institute of Biotechnology, Slovak University of Technology, Radlinského 9, 812 37 Bratislava, Slovakia
e-mail: katarina.dercova@stuba.sk

S. Murínová
Faculty of Chemical and Food Technology, Institute of Biotechnology, Slovak University of Technology, Radlinského 9, 812 37 Bratislava, Slovakia

P. Hucko • L. Tóthová
Water Research Institute, Nábrežie arm. gen. L. Svobodu 5, 812 49 Bratislava, Slovakia

R.K. Salar et al. (eds.), *Biotechnology: Prospects and Applications*,
DOI 10.1007/978-81-322-1683-4_13, © Springer India 2013

Determination of ecotoxicity and genotoxicity of PCB-contaminated sediments represents an important part of our research together with isolation and identification of PCB-degrading bacterial strains from the sediments. The obtained results indicate beneficial effect of both biostimulation and bioaugmentation strategies during biodegradation of PCBs in the contaminated sediments. PCBs affected saturation of membrane fatty acids, cis–trans isomerization, caused pronounced adaptation changes, and altered membrane fluidity of the cells of the bacterial degraders. This phenomenon is thought to be the major adaptive mechanism in microorganisms exposed to toxic aromatic pollutants. Study of ecotoxicity demonstrated that sediments sampled from industrial canal and water reservoir, both located in the vicinity of the former producer of PCBs in Slovakia, were toxic for the tested bioindicators. It has been established that PCB-contaminated sediments represent a source of adverse effects on life functions of the biota. Eleven bacterial strains were isolated and identified in the contaminated sediments using 16sRNA method. Detection of *bphA* gene encoding biphenyldioxygenase, the important starting enzyme of PCB degradation, was performed. Data obtained from microcosm studies might be useful in the preliminary design of a site-specific biostimulation/bioaugmentation strategy.

13.1 Introduction

The widespread use of dangerous, toxic anthropogenic compounds and a long history of poor disposal practices led to pollution of the natural environment and created an urgent need for remediation technologies. In many cases, biological approach to decontamination of the polluted environment provides significant advantages over physicochemical technologies. Bioremediation can facilitate the complete destruction of hazardous compounds mostly without generating toxic emission or toxic by-products. Physicochemical remediation technologies are mainly used at strongly contaminated locations. However, predominant part of agricultural soil, industrial soil, or river and lake sediments is not extremely contaminated. Bioremediation technologies therefore represent a prospective and perspective ecological and economical alternative of physicochemical methods of pollutants elimination at the abovementioned contaminated areas. Biological techniques are based on biodegradation of pollutants using degradative ability of the microorganisms, predominantly bacteria and fungi (Mouhamadou et al. 2013).

A hierarchy of in situ treatment approaches has been developed to address different contaminant types, site-specific conditions, and regulatory factors. Natural attenuation, biostimulation, and bioaugmentation are the three accepted practices for enhancing bioremediation. Bioaugmentation is used when the competent degradative natural microbial population is not present among the indigenous population or when population may be sufficient to achieve a biostimulation remediation to increase the rate of degradation and shorten the time frame for full-scale remediation (Unterman et al. 2000). As with biostimulation, bioaugmentation can be combined with other technologies to improve effectiveness of pollutant degradation. Improved knowledge on the specific effects of enhancement methodologies and on the toxicological profiles of metabolites produced by indigenous or exogenous microorganisms will further increase the extent of application and acceptance of biological approaches of remediation.

Chlorinated organic compounds are among the most significant pollutants in the world. There is a great concern over chlorinated organic compounds because of their toxicity, persistence, and bioaccumulation. Among these compounds, polychlorinated biphenyls (PCBs) and chlorinated organic solvents are the major target for bioremediation. Sequential use of anaerobic halorespiring bacteria, which play a crucial role in biological dehalogenation processes, and aerobic bacteria whose oxygenases are modified by directed evolution could lead to efficient and total degradation of highly chlorinated organic pollutants (Furukawa 2003).

13.1.1 Bacteria with PCB-Degradative Ability

Polychlorinated biphenyls (PCBs) are highly hydrophobic, toxic, bioaccumulable, ubiquitous, and low degradable pollutants found in the environment and also present in food chain (Vrana et al. 1996a; Vasilyeva and Strijakova 2007). Serious pollution of the environment with PCBs stimulated intensive research on PCB-degrading microorganisms able to degrade PCB under a variety of conditions (Mrozik and Piotrowska-Seget 2010). Presence of PCBs markedly influences biodegradation ability, survival, and colonization of the microorganisms at their application in bioremediation technologies during decontamination of the environment. Reasons for slow or negligible biodegradation of PCBs in the environment may include insufficient number and poor survival of PCB degraders and environmental stress caused by PCBs to their cell membrane. Therefore, successful soil augmentation requires not only application of a strain or a consortium with degradative ability but also the microorganisms able to survive in the adverse environment (Pritchard 1992; Unterman et al. 2000; Ohtsubo et al. 2000; Dercová et al. 2008; Mrozik and Piotrowska-Seget 2010; Megharaj et al. 2011). Poor survival of the inoculated microorganisms and low bioavailability of the pollutant are usually the main obstacles to the successful inoculum amendment.

The microbial degradation of PCBs has been extensively studied, and the following general observations have been made: PCBs are degraded oxidatively by aerobic bacteria and other microorganisms such as white rot fungi. Many types of bacteria can transform PCBs under aerobic conditions, including Gram-negative bacteria in the genera *Alcaligenes*, *Burkholderia*, and *Pseudomonas* and Gram-positive bacteria in the genera *Bacillus*, *Corynebacterium*, and *Rhodococcus*. These populations typically metabolize PCBs by a combination of oxidative and reductive enzymatic steps that yield chlorodihydroxybiphenyl and chloroquinone compounds, chlorobenzoates, and aliphatic acids (Reineke 1984; McCullar et al. 2002). Under aerobic conditions, PCBs are transformed by a cometabolic process using biphenyl-catabolizing enzymes. Four enzymes are involved in this reaction, including biphenyldioxygenase, which introduces molecular oxygen to one of the biphenyl rings, usually at the 2 and 3 positions; a dehydrogenase; dihydroxybiphenyldioxygenase, which cleaves the biphenyl ring; and a hydrolase (Furukawa 2003). The dehalogenation of aromatic substrates may occur before, during, or after cleavage of an aromatic ring. Recently, genetic and biochemical studies have yielded new insight into these mechanisms (Pieters et al. 2001). PCBs are also reductively dehalogenated by anaerobic bacteria (during halorespiration process) belonging to genera *Desulfitobacterium* or *Dehalococcoides*. Anaerobic bacteria are generally hardly isolated because they are working mainly in consortium with some other indigenous microorganisms.

However, bioremediation strategies based solely on the use of naturally occurring PCB-degrading soil bacteria are very slow and not viable for three major reasons: (1) there are several bottlenecks along the biphenyl catabolic pathway that need to be overcome to expand their PCB substrate range; (2) full expression of the biphenyl pathway enzymes requires the presence of biphenyl or of some other fortuitous inducers; (3) PCBs are strongly hydrophobic and poorly bioavailable, and both PCBs and bacteria are unevenly distributed in soil (Sylvestre 2013).

To eliminate PCBs, the top priority is to lower their lipophilicity–hydrophobicity using enzyme dioxygenase. PCB concentration decreased mainly due to degradation of PCB congeners with one to three chlorine atoms. The content of higher chlorinated PCBs can be noticeably decreased only under sequential anaerobic/aerobic treatment. However, in particular cases, mainly at the laboratory experiments with freshly spiked PCBs, it was possible to reduce their content to a level permissible for technogenic soils (Megharaj et al. 2011).

In our previous papers, a simple apparatus for simultaneous effective monitoring of the PCB evaporation kinetics and biodegradation in batch experiments was described (Vrana et al. 1995, 1996a) together with a simple mathematical model taking evaporation and biodegradation of PCBs into account (Dercová et al. 1999a, b). The biphenyl/PCB metabolic pathway and chlorobenzoic acids (mainly 3-CBA) as toxic metabolites were extensively studied and reviewed (Vrana et al. 1996b; Tříska et al. 2004; Field and Alvarez 2007; Macková et al. 2010; Tandlich et al. 2013). Although PCB-degrading activity is in some cases higher when using pure cultures, the ability of mixed cultures to survive in a non-sterile environment is a key issue in the field applications of biodegradation. Bioaugmentation using reinoculation by massive inoculum of indigenous degraders proliferated at the laboratory conditions may potentially improve biodegradation of the pollutant (Gilbert and Crowley 1998). Biostimulation of autochthonous microflora using nutrients or inducers, e.g., biphenyl or terpenes (Tandlich et al. 2001; Dercová et al. 2003a, b; Tandlich et al. 2011; Dudášová et al. 2013), may enhance biodegradation of PCBs as well.

With recent advances in biomolecular engineering, the bioremediation of persistent organic pollutants (POPs) using for bioaugmentation purposes genetically modified microorganisms has become a rapidly growing area of research for environmental protection. Ang et al. (2005) described the most recent developments and applications of these biomolecular tools for enhancing the capability of microorganisms to bioremediate three major classes of POPs – polycyclic aromatic hydrocarbons (PAHs), polychlorinated biphenyls (PCBs), and obsolete pesticides. Most of the examples focused on the redesign of various features of the enzymes involved in the bioremediation of POPs, including the enzyme expression level, enzymatic activity, and substrate specificity. Overall, the rapidly expanding potential of biomolecular engineering techniques has created the exciting potential of remediating some of the most recalcitrant and hazardous compounds in the environment.

13.1.2 Background of PCB-Contaminated Sediments

In the recent decades, several hundred tons of PCBs have been released into the environment. Due to their hydrophobic properties, PCBs tend to be adsorbed by natural organic matter in the aqueous bottom sediments. Sediment is an essential integral part of the hydrological system. However, because sediments are ultimate reservoirs for the numerous chemical and biological contaminants that may be contained in effluents originating from agricultural and industrial lands, contaminated sediments in rivers and lakes have the potential to pose ecological and human health risks (Apitz et al. 2006).

Slovakia belonged to the eight largest world producers of commercial mixtures of PCBs produced mainly under the brand names Delor, Delotherm, and Hydelor. In the locality of the former PCB producer Chemko Strážske, a big amount of these substances is still persisting in sediments and soils. PCBs were manufactured mainly in the 1960s and 1970s. From 1959 to 1984, production of PCBs in Slovakia was approximately 21,500 t. About 46 % of the PCB production was exported mainly to the former East Germany. The rest (11,613 t) was used in the territory of former Czechoslovakia (later divided to Czech and Slovak Republic) in dielectric fluids for transformers and power capacitors, as heat exchanger and hydraulic fluids, as paint additives, and as lubricants. Nowadays, the use of PCB formulations in open systems is not allowed

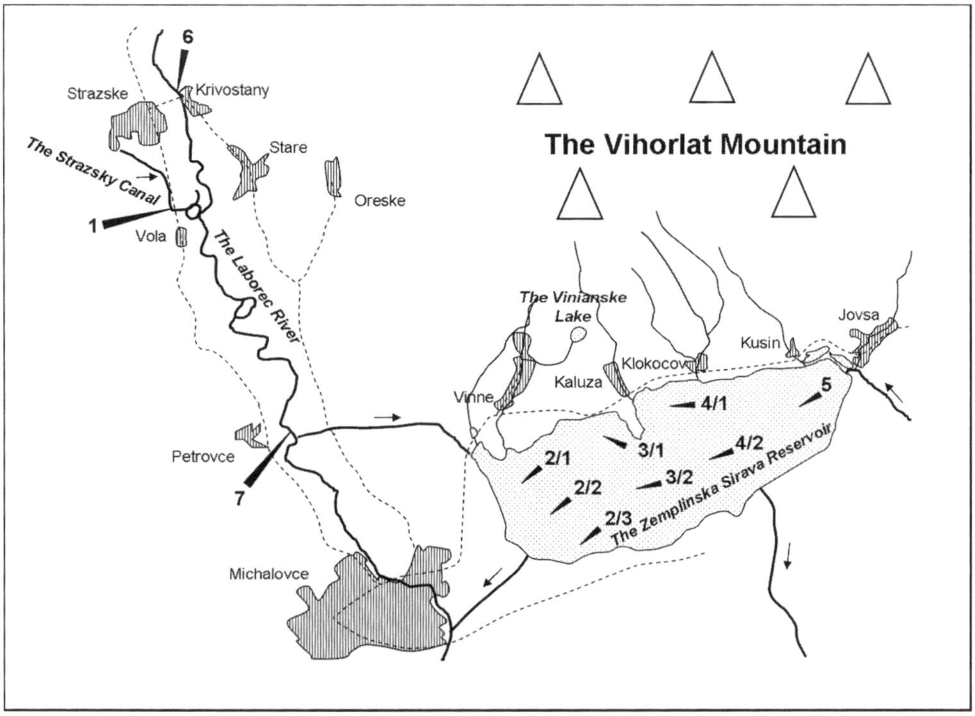

Fig. 13.1 PCB-contaminated area in Eastern Slovakia. Location of the sampling sites in the Zemplínska Šírava water reservoir and vicinity of Strážsky canal

in Slovakia; however, in closed systems such as transformers and capacitors, they still can be used. It is estimated that about 1,600 t of PCB wastes was generated during the production, 1,000 t is in closed systems (transformers, capacitors, and other equipment), 1,500 t is in various wastes, and 900 t is in waste dump Pláne (SHMÚ 2004).

Contamination at the factory Chemko Strážske surroundings belongs to the so-called old environmental burdens. It directly relates to the former PCB production. Contaminated areas are found inside the factory area as well as in wider surroundings. The contamination is primarily spread through surface water by gradual release from contaminated sediments of an open sewer that leads from the factory to Laborec River and subsequently through the inlet canal that leads to the water reservoir Zemplínska Šírava Lake (Fig. 13.1).

A thin layer of mud in the 5.3-km-long effluent canal of the former PCB producer still contains about 3 kg PCBs in 1 t of dry effluent canal mud 15 years after the termination of the production and thus represents a serious environmental threat by PCBs. This sediment is an abundant PCB source causing the long-term contamination of the waters of Eastern Slovakia (Kočan et al. 1999). Figure 13.2 documents PCB concentrations determined in sediment samples taken in the polluted area (Michalovce District) and in a background one (Stropkov District). As expected, the highest value was found in a muddy part of the effluent canal flowing from the Chemko factory (one of the samples contained up to 5 g of PCBs·kg^{-1}). It is doubtless that the polluted effluent canal emptying into the Laborec River has caused the contamination. The Zemplínska Šírava water reservoir (33.5 km^2 surface area) that is partly filled from the Laborec River contained several hundred times higher PCB levels (100–2,000 times) in comparison to a similar water reservoir (Domaša) in the background area (1.7–3.1 mg·kg^{-1} compared to 0.007–0.01 mg·kg^{-1}).

As regards PCBs, the established highly polluted sites, such as the former producer's effluent

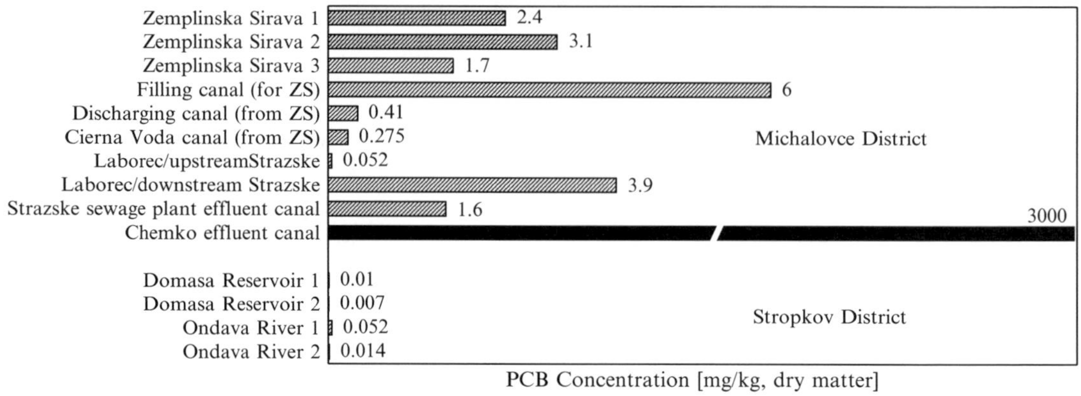

Fig. 13.2 PCB levels (the sum of 6 indicator PCB congeners) in sampled sediment taken from some watercourses in the districts of Michalovce (contaminated location) and Stropkov (control location) (Kočan et al. 2001; Dercová et al. 2010)

canal and some watercourses contaminated by this canal (Laborec River and Zemplínska Šírava water reservoir), can be considered as hot spots and present an urgent problem in Slovakia. Area of contamination by PCBs is partly documented by the previous studies performed within the project *Burden of the Environment and Human Population in the Area Contaminated with PCBs* (Kočan et al. 1998). Complex monitoring of PCB contamination in the area surrounding a former production site was not carried out until now. Only partial data are available (Kočan et al. 2001). For this reason, it is not possible to exactly estimate the amount of contaminated sediment and soils in precise concentration layers. Initial approximation indicates that the canal, river, and lake may contain up to 40,000 t of highly contaminated material (Dercová et al. 2010). For a final resolution of the problem with contamination of soils and sediment, it is needed to develop a complex strategy considering various concentrations of PCBs, technical and economical aspects of decontamination, as well as the timetable.

PCBs are emitted into atmosphere via evaporation from freely stored materials and refuse heaps contaminated with PCBs. In water, they are generally contained in sediment due to their strong adsorption properties. In soils, they have no tendency to a noticeable spread due to adsorption and low solubility. Degradation of PCBs under the environmental conditions is very slow – the more chlorinated substance, the slower is the degradation. Their half-life is estimated to be 3–21 days in atmosphere, more than 5 days in water, and more than 40 days in soil (which implies that complete degradation may last up to several years).

In humans, PCBs are cumulated mostly in adipose tissue and can be gradually long-term released from adipose cells into blood. The important issue is that they are contained in adipose cells of mother's milk, so that these toxic chemicals can be received already by newborns. Moreover, the adverse health effects are also ascribed to a concomitant content of dioxins in mixtures with PCBs. Toxic effects result in respiratory, digestive, and liver functional disorders and neurological and developmental processes (growth inhibition of children, low birth weight, psychomotor process arrest, as well as potential intelligence quotient decrease). They may also lead to changes in skin pigmentation and to rashes of acne type. There was no carcinogenicity for humans found; however, PCBs are considered to be potential carcinogens. Finally, PCBs are referred to as the substances that impair human immune system, so-called endocrine disruptors (Langer et al. 2006; Zhiwei et al. 2007; Svobodová et al. 2009).

Figure 13.3 shows the mean PCB profiles (mean percentage of each group within the sum of 12 PCBs) in sediments from Zemplínska Šírava

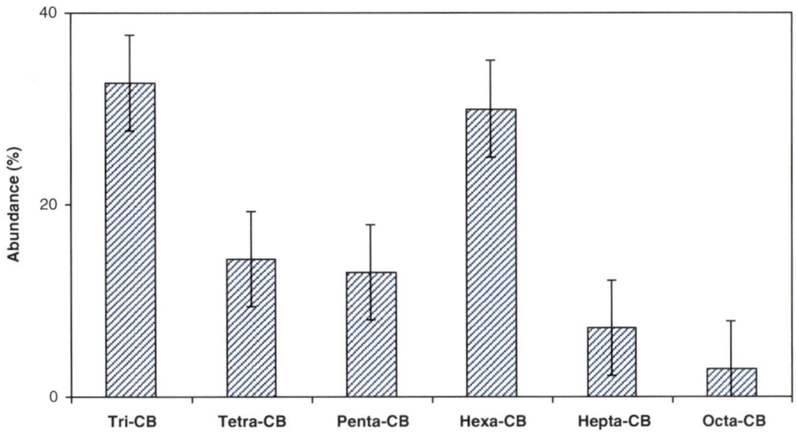

Fig. 13.3 PCB profiles (mean percentage of each group within the sum of 12 PCBs) in contaminated water reservoir bed sediments sampled from the contaminated locality (Adapted from Hiller et al. 2010; Dercová et al. 2011)

water reservoir. In general, tri-CBs (CBs 18, 28, and 31) and hexa-CBs (CBs 138, 149, and 153) are the most abundant PCB congeners in the sediments regardless of the reservoir locations. In sediment samples from Zemplínska Šírava, tri-CBs and hexachlorobiphenyls are equally the most dominant congeners, together accounting for more than 60 % of the total PCB content (Hiller et al. 2010).

13.1.3 Bioavailability of Pollutants

The process of bioremediation depends on the metabolic potential of bacteria to detoxify or transform the pollutant molecule, which is dependent on its accessibility and bioavailability. There is a considerable debate in the literature on bioavailable fraction and the methods of its measurements (Vasseur et al. 2008; Megharaj et al. 2011). Supercritical fluid extraction (SFE) has been applied to extract PCB bioavailable fraction from sediment (Björklund et al. 2000; Nilsson et al. 2003; Hallgren et al. 2006). The latter study showed that SFE at 40 °C and 120 bars for 60 min removed substantial part of the bioavailable fraction of 11 PCB congeners. Nilsson and Björklund (2005) tested new conditions, at which the bioavailable fraction was more completely removed, thereby demonstrating the applicability of selective SFE for the determination of bioavailable PCB fractions in contaminated sediments. The sequestration of pollutants in long-term contaminated soil and sediments may occur due to the contact and interactions of soil or sediment with pollutant molecules. Sequestration can reduce bioavailability. Factors such as organic mater, cation exchange capacity, micropore volume, soil and sediment structure, and surface area affect the pollutant sequestration (Alexander 1994). The pollutant sequestration due to the prolonged contact between soil particles and chemical molecules, however, poses less risks and threat to the environmental health. In general, difficulties with analytical measurements for determining low levels of organic pollutants, especially the new ones (nine chemicals from the appendix of the Stockholm Convention list, pharmaceuticals as new environmental pollutants in waste waters, etc.), in soil and sediment geographical and distribution patterns and the complexity of their toxicological interactions make the bioavailability measurements of organic pollutants exigent (Megharaj et al. 2011). Bioavailability of PCBs can be enhanced in several ways: maximal dispersion and agitation, as well as application of surfactants or biosurfactants. However, the presence of aged PCBs in soil and sediments may retard bioremediation, because the capacity of the metabolic residues in soil or sediments to bind PCBs is very high. A critical view on the knowledge gaps and limitations in field application strategies and the use of assays for monitoring and testing the efficacy of bioremediation of polluted sites including use of surfactants is provided by Megharaj et al. (2011).

13.1.4 Adaptive Responses of Bacterial Cell Membrane on the Environmental Stress Caused by the Presence of Toxic Organic Pollutants

Aromatic compounds such as chlorinated aromatics, including PCBs, affect the saturation of membrane fatty acids of various bacterial strains in a uniform way. There is a correlation between an increase in the degree of saturation of membrane fatty acids and increased tolerance towards the toxic compounds in phenol-degrading strains (Mrozik et al. 2010). This phenomenon is thought to be the major adaptive mechanism in microorganisms exposed to toxic aromatic compounds. Due to it, the bacterial membranes become more resistant to the fluidizing action of aromatic compounds, which allows the cells to survive in hydrocarbon-contaminated sites.

Microbial cells have evolutionary developed rapid molecular responses and efficient adaptation mechanisms for elimination of undesirable factors in order to survive in adverse environment (Šajbidor 1997). Since the membrane is the site of the primary contact of the cell with its environment, many vital cell activities find their origin in membrane function. Environmentally induced perturbations in membrane structure may result in significant disruption of cell physiological function, thus membrane flexibility and adaptation capability largely determine the survival of the cell. Since fatty acids are the major constituents of membrane glycerolipids, modulation of number and position of double bonds of acyl chains by individual fatty acid desaturases plays a crucial role in preserving suitable dynamic state of the bilayer during environmental impact (Šajbidor 1997; Čertík et al. 2003; Dercová et al. 2004).

Some bacteria are able to grow in extreme environments, such as high concentrations of chemicals, and very often display remarkable biodegradation properties. Our previous work dealt with the study of toxicity and microbial degradation of mono-, di-, and tri-chlorinated phenols and pentachlorophenol and investigated the factors that predominantly influence their biodegradation as well (Dercová et al. 2004). The major problem is to maintain bacterial ability to survive in the presence of higher concentration of pollutants. Thus, the aim of the part of our work was to determine the effect of chlorinated phenols (2, 4-dichlorophenol and pentachlorophenol, PCP) and PAHs on the cell membrane, especially on the content and composition of membrane lipids and fatty acids, and study of biodegradation of PCP (Čertík et al. 2003; Dercová et al. 2004; Sejáková et al. 2009; Vítková et al. 2011).

Different types of environmental and physiological stress conditions constantly challenge all microorganisms including bacterial degraders. Some of them present in long-term contaminated area, developed efficient adaptation mechanisms to survive in polluted environment. Adverse environmental conditions initiate in the microbial cells several processes that minimize the negative effects. All adaptation mechanisms are synchronized to provide necessary physiological function with low energy consumption. Being at the interface between the cell and the environment, the membrane modulates many cell functions. Since membranes constitute the main target for solvent effects, most adaptive mechanisms are concerned with maintenance of the membrane fluidity and lipid phase stability (Weber et al. 1994). Fluidity of cytoplasmic membrane is a very important aspect of membrane structure and is defined as the reciprocal of its viscosity. It can be modulated mainly by the alteration of fatty acids creating membrane phospholipids.

It seems likely that PCBs, known as very hydrophobic compounds (log *P* move in the range of 5–8), are localized and accumulated within the cytoplasmic membrane of bacterial degraders, predominantly in the membrane lipids (phosphatidylcholine and phosphatidylethanolamine). It is presumable that the entrapment of PCBs within the membrane reduces both the accessibility of PCBs to the degradation enzyme and the efficiency of PCB degradation. As the accumulation of hydrophobic compounds in the cytoplasmic membrane fluidizes and disrupts its function, it also affects cell viability.

13.1.4.1 Mechanism of Disturbance of Cytoplasmic Membrane by Organic Pollutants

Many organic pollutants increase the membrane fluidity. This increase leads to the loss of membrane function and to damage of bacterial cell. Organic compounds are able to partition into the inner layer of biological membrane resulting in swelling of the membrane and loss of its physiological functions and activity. Heipieper et al. (1994) and Heipieper and de Bont (1994) established the existence of a systematic relationship between the values of log *P* in the range of 1–5 and values for the partitioning of solvents in membrane–buffer systems. When the solvent has dissolved in a membrane, it disturbs its integrity and hence impairs its function, leading to, e.g., the uncontrolled proton efflux. The leakage of protons (and potassium ions), stimulated by the accumulation of solvent molecules in membrane interior, causes decrease of the proton motive force and leads to the perturbation in the energy conservation (Heipieper et al. 1992). The experiments involving nine organic compounds (e.g., benzene and toluene), each with log *P* values between 2 and 5, proved that their concentration in the membrane of up to 0.5 $\mu mol \cdot mg^{-1}$ phospholipids resulted in an increase of the surface area of the membrane (Sikkema et al. 1994).

13.1.4.1.1 Increase in Membrane Saturation

Increase in membrane saturation was the first described mechanism of cell adaptation to adverse conditions. Several papers described an increase in saturation of membrane phospholipids in the presence of toxic organic compounds (Heipieper et al. 2003; Dercová et al. 2004; Mrozik et al. 2004; Zorádová et al. 2011). Previous reports prove that aromatic compounds such as benzene, biphenyl, phenol, PCBs, and toluene cause that fatty acid alkyl chains of the phospholipids in the cell membrane become stacked more compactly, thus increasing the rigidity of the membrane and contributing to a tolerance against aromatic compounds (Sikkema et al. 1995; Mrozik et al. 2004; Duldhardt et al. 2010; Zorádová-Murínová et al. 2012; Murínová et al. 2012). The increasing degree of membrane lipid saturation is one of the major adaptive mechanisms of bacterial cells to the presence of many aromatic compounds (Heipieper et al. 1994; Gutierrez et al. 1999).

The reason for the ability of saturated fatty acids tightly packing is their transition temperatures. For long chain saturated fatty acids, these temperatures are very high (e.g., 63 °C for palmitic acid). Corresponding monounsaturated fatty acids with the *cis* configuration of double bond have lower transition temperatures (0 °C for C 16:1cis). Increase of saturated fatty acids content in cytoplasmic membrane leads to higher membrane rigidity that can counteract the fluidizing effect of most organic compounds (Hazel and Williams 1990; Cronan 2002; Kabelitz et al. 2003). This mechanism has limitation due to the condition of synthesis of saturated fatty acids. In bacteria, only the energy-dependent de novo biosynthesis of saturated fatty acids allows the increase in the degree of saturation. This is the major reason why alteration in the degree of saturation was observed in growing cells only. Therefore, under growth-inhibiting conditions, lipid biosynthesis is stopped due to stringent-response regulation, and that is why only growing cells can perform such kind of membrane adaptation (Heipieper and de Bont 1994; Heipieper et al. 2007).

13.1.4.1.2 Cis–Trans Isomerization

Cis–trans isomerization was described in many papers as adaptation mechanism of the bacterial cells under growth inhibition conditions (Heipieper et al. 1994; Duldhardt et al. 2010; Zorádová et al. 2012). Various bacterial strains mainly *Pseudomonas* and *Vibrio* can adapt to toxic compounds and their fluidizing properties by isomerization of *cis*-unsaturated fatty acids to their appropriate *trans* configuration. These two forms of unsaturated fatty acids have different steric structures. Configuration of *trans* isomers allows them to be tightly packed together and protect membrane against the fluidizing molecules.

This is the reason why the transformation of *cis* to *trans* fatty acid leads to the decrease of membrane fluidity. Heipieper et al. (2004) attested the activation of the *cis–trans* isomerase in resting cells by the addition of 3-nitrotoluene. This activation resulted in the conversion of the *cis*-unsaturated fatty acids into the corresponding *trans* isomers. The intensity of the conversion depended on the rate of *cis–trans* isomerization and the type and concentration of the added toxic compound. A relationship was found between the activation of this system and the induction/activation of other stress-response mechanisms. *Cis–trans* isomerization correlates with the toxicity and the amount of hydrophobic compound accumulated in cytoplasmic membrane (Heipieper et al. 1996).

The enzyme that is responsible for this adaptation mechanism is isomerase and belongs to cytochrome c type proteins and carries Cti polypeptide with a heme-binding site. This polypeptide was found in all tested *Pseudomonas* strains. Moreover, comparison of the amino acid sequences of the seven known Cti proteins identified them as heme-containing proteins of the cytochrome c type (von Wallbrunn et al. 2003). Cti polypeptide is responsible for the localization of *cis–trans* isomerase in periplasmic space. This is the reason why only fatty acids with *cis* double bond in specific depth of the membrane can reach the active site of isomerase (Heipieper et al. 2001). This enzyme has been purified from the periplasmic fraction of *Pseudomonas oleovorans* for the first time by Pedrotta and Witholt (1999). The *cis–trans* isomerase gene cloned and sequenced from *P. putida* DOT-T1E (Junker and Ramos 1999) made evident that the isomerase had an N-terminal hydrophobic signal sequence. This sequence is cleaved off after targeting the enzyme to the periplasmic space. The observations proved that *cis–trans* isomerase is constitutively present, does not require ATP or other cofactors including NAD(P)H and glutathione, and works in the absence of de novo synthesis of lipids. The occurrence of heme-binding site of the cytochrome c type strongly supports a mechanism of *cis–trans* isomerization by forming an enzyme–substrate complex. These findings favor a mechanism for the enzyme in which electrophilic iron (Fe^{3+}) provided by a heme domain directly attacks *cis* double bond of fatty acid and removes electron of the *cis* double bond thereby transferring the sp^2 bond into sp^3. Double bond is then transformed to *trans* configuration (Heipieper et al. 2004). *Cis–trans* isomerization is exergonic reaction because the energetic difference between *cis* and *trans* configuration is 3.1 kJ·mol^{-1} (Heipieper et al. 1992; Diefenbach and Keweloh 1994; Heipieper and de Bont 1994).

13.1.4.1.3 Increase in Cardiolipin Synthesis

Increase in cardiolipin synthesis strongly augments the adaptation ability of membrane lipids to the presence of organic solvents. This mechanism was observed in Pseudomonas species (Ramos et al. 1997).

Cardiolipin is a lipid that can be found abundantly in many bacterial strains. Cardiolipin synthase catalyzes the transfer of phosphatidyl group between two phosphatidylglycerol molecules. This enzyme is known as phospholipase D. Accumulation of cardiolipin increases at the beginning of stationary phase of growth. Cardiolipin is the most stable membrane phospholipid and is essential for the survival upon longtime starving. Only de novo synthesis of cardiolipin was described in bacteria species (Schlame 2008). Prokaryotes can change the amount of this lipid depending up their physiological status and growing conditions. Trace amount of this lipid can be found in bacterial cell during the exponential growth phase. Increase in the amount of this phospholipid is one of the known adaptation mechanisms in stress environment (Ramos et al. 1997). Cardiolipin stimulates changes in physical properties of cytoplasmic membrane. Even small amounts of cardiolipin decrease the lateral interaction within the monolayer leaflet, which reduces energy required to stretch the membrane and could favor the creation of membrane folds (Nichols-Smith et al. 2004). Also, cardiolipin has the ability to form clusters. The ability of cardiolipin to trap protons may have implications for the distribution of the proton motive force in energy-converting membranes. Wallbrunn et al. (2003) used a mutant that was not able to synthesize cardiolipin to investigate whether the *cis–trans* isomerization is able to compensate cardiolipin in adaptation mechanisms. Their results show that the mutant was not able to grow which proves that *cis–trans* isomerization is not fully able to replace cardiolipin adaptation influence.

13.1.4.1.4 Changes in the Content of Cyclopropyl and Branched Fatty Acids

Higher concentrations of organic pollutants stimulated production of cyclopropane fatty acids (C17-CP and C19-CP) in bacterial membrane. This effect was observed under the exposure to the polyaromatic hydrocarbons, phenols, and polychlorinated biphenyls (Čertík et al. 2003; Mrozik et al. 2010; Zorádová et al. 2011).

The role of these fatty acids in membrane adaptation mechanisms is still not clear. Some authors indicated that cells containing these fatty acids survived better in a strongly stressed environment. Moreover, it is known that cyclopropyl fatty acid formation is one of the most important mechanism that protects bacterial cells against many chemicals (aromatic compounds, organic solvents, alcohols) and environmental factors (salinity, pressure, and temperature) (Denich et al. 2003; Kabelitz et al. 2003; Chihib et al. 2005; Mrozik et al. 2004).

Several papers described that the content of branched fatty acids in fatty acid methyl esters (FAMEs) profiles obtained from the bacterial strains present in the contaminated environment was significantly higher than in control samples. Lipids of Gram-positive bacteria often contain high proportion of *iso*- and *anteiso*-branched fatty acids. Branched fatty acids can only be created by de novo synthesis from valine, leucine, and isoleucine precursors. Such synthesis results in *iso*-branched-even-chain, *iso*-branched-odd-chain, and *anteiso*-branched-odd-chain fatty acid species formation that demonstrate different physicochemical properties because of the structural differences. The transition temperatures of the branched fatty acids are lower for the *anteiso* fatty acids (51.7 °C for C15:0 *iso* and 23 °C for C15:0 *anteiso*) (Kaneda 1991). This difference causes a remarkable change in the fluidity of the membrane when the species of branched fatty acids are changed from one to the other. The effect on transition temperature caused by a change from *anteiso* to *iso* branching is comparable to the isomerization of *cis* to *trans* unsaturated fatty acids in many Gram-negative bacteria. G^+ and G^- bacteria that contain branched fatty acids adapt to differences in temperature and toxic organic substances by altering the *anteiso*/*iso* ratio in the cell membrane. According to the different physicochemical properties of the two species of branched fatty acids, the bacteria revealed a decreased content of *anteiso* fatty acids when grown under adverse conditions to decrease the viscosity of their cytoplasmic membranes. At higher growth temperatures as well as in the presence of toxic concentrations of phenol, 4-chlorophenol, 4-nitrophenol, and PCBs, the cells adapted their membrane by a dose-dependent decrease in the *anteiso*/*iso* ratio, leading to a more rigid membrane and counteracting the fluidity increase.

At present, very few studies have been carried out with a historically contaminated soil or sediments under field conditions or in laboratory conditions simulating the field ones. Therefore, our experiments were carried out with long-term PCB-contaminated sediments with the commercial mixture Delor 103 (42 % w/w of chlorine) and Delor 106 (48 % w/w of chlorine) sampled in the surroundings of the former PCB producer Chemko Strážske.

The objectives of this work were to provide characterization of PCB-contaminated bottom sediments sampled from the Zemplínska Šírava water reservoir and industrial effluent Strážsky canal from the point of view of bioavailability of the individual PCB congeners, their ecotoxicity, and biodegradation of PCBs in these sediments by indigenous (autochthonous) and inoculated (allochthonous) bacterial strains and to investigate the effect of PCBs on membrane fluidity of PCB degraders. Our research is focused mainly at bioremediation of PCB-contaminated sediments under the laboratory conditions using biostimulation and/or bioaugmentation strategy as well. Some results and the particular methods are described in more detail in cited papers yet published.

13.2 Materials and Methods

13.2.1 Cultivation Media

Luria–Bertani (LB) medium was prepared according to the instruction for use. To congelate the medium, 15 g l^{-1} agar was added (Noble Agar, Difco, UK). Minimal mineral medium (MM): 1 g $(NH_4)_2SO_4$; 2.7 g KH_2PO_4; 10.95 g $Na_2HPO4{\cdot}12\ H_2O$ filled with distilled water to 1 l. After sterilization of the solution, salts containing particular trace elements sterilized by filtration were added: 250 μl $FeSO_4{\cdot}7\ H_2O$ (2 g l^{-1}); 250 μl, Ca $(NO_3)_2{\cdot}4\ H_2O$ (6 g l^{-1}); and 250 μl $MgSO_4{\cdot}7\ H_2O$ (40 g l^{-1}), each in 50 ml of the prepared medium. Minimal solid medium: 5.37 g $Na_2HPO_4{\cdot}12\ H_2O$; 1.30 g KH_2PO_4; 0.50 g NH_4Cl; and 0.20 g $MgSO_4{\cdot}7\ H_2O$ filled with distilled water to 1 l. To congelate the medium, 15 g l^{-1} agar was added (Nobel Agar, Difco, UK). Plate count agar (PCA): 23.5 g PCA was dissolved in 1 l distilled water in accordance to the guide of Oxoid Company, UK. Phosphate buffer (pH 7.4): Solution A: 13.8 g $NaH_2PO_4\ H_2O$ filled with distilled water to 500 ml; solution B: 14.2 g Na_2HPO_4 filled with distilled water to 500 ml. A total of 60 ml of solution A and 440 ml of solution B were mixed.

13.2.2 Microorganisms

The isolate from long-term PCB-contaminated soil *Pseudomonas stutzeri* (Dercová et al. 1996), purified, identified, and maintained as safe keeping in the Czech Collection of Microorganisms (Masaryk University, Brno, Czech Republic), was used for the purpose of bioaugmentation. The isolates from long-term PCB-contaminated sediments *Ochrobactrum anthropi*, *Pseudomonas mandelii*, *Starkeya novella*, and *Achromobacter xylosoxidans* (Dudášová et al. 2012) were used for bioaugmentation as well. In strains *P. stutzeri* (Dercová et al. 2009) and *A. xylosoxidans* (Dudášová et al. 2013), *bphA* gene was detected.

13.2.3 Sediment Sampling

Sediment sampling protocol was in agreement with the technical norm ISO 5667–12:2001. Overview of all sampling sites is presented in Fig. 13.4. The sediment corer sampler (UWITEC Corp., Austria) was used as a sampling device. The sampler used is a transparent plastic tube that allows for visual examination and sample partition into several layers. Studied sediments were sampled from industrial Strážsky canal that flows into the water reservoir and represents abundant source of contamination.

13.2.4 Isolation of the Bacterial DNA

Isolation of the bacterial DNA of the used strains by thermolysis was described in our previous work (Dercová et al. 2009). For isolation, a culture from solid media was used and incubated in 50 μl sterile distilled water for 10 min at 95 °C in a thermoblock. After thermolysis, the mixture was stirred and centrifuged (1,300 rpm). QIAamp DNA Mini Kit (Qiagen, USA), a set for isolation of the total DNA from pure culture, was used.

For detection and amplification of the gene of bacterial DNA, a PCR technique with specific primers was used. Composition of the mixture was 34.5 μl sterile deionized water, 5 μl of polymerase buffer (Finnzymes, Finland), 1 μl dNTP (10 mM), 2 × 1 μl of primers for detection of gene bphA1 (primers F350 and R674 at a concentration of 10 pmol·ml^{-1}), 2 μl BSA, 0.5 μl of polymerase (Finnzymes, Finland), and 5 μl of DNA sample. PCR amplifications were performed in an automated thermal cycler Techne (Progene, USA) with an initial denaturing (94 °C for 5 min), followed by 35 cycles of denaturing (94 °C, 30 s), annealing (55 °C, 30 s), extension (72 °C, 1 min), and concluded by a single final extension (72 °C, 10 min). A suitable concentration of the agarose gel was used: from 1 % (for fragments with number of bp higher than 100) up to 1.5 % (for fragments with number of bp below 100).

Fig. 13.4 Sediment sampling device (UWITEC, Austria)

To determine fragment size, a marker of 100 bp DNA standard was used (Sambrook et al. 1989) (Fig. 13.5).

13.2.5 Bioavailability Assay: Supercritical Fluid Extraction (SFE) of Sediments

To homogenize a sample, 100 g of sediments was homogenized with the same volume of liquid CO_2 (Zougagh et al. 2004). Aliquot part of this mixture was put into SFE extraction vessel. The contaminated sediment was pre-extracted with selective SFE system. The extraction temperature was 40 °C, the pressure was set to 12 MPa, and extraction time was 1 h (10 min static and 50 min dynamic extraction), repeated 6 times. Extract of sample was trapped on solid phase of Florisil® sorbent (Method 3562). Extractions were done on 0.2 g of copper powder and 1 g sediment portions at a time with HP 7680 T SFE. After each extraction, the trap was eluted with dichloromethane solvent (DCM) to obtain PCBs and reactivate sorbent, each extract was concentrated to the final volume 0.5 ml. All extracts were fortified with the internal

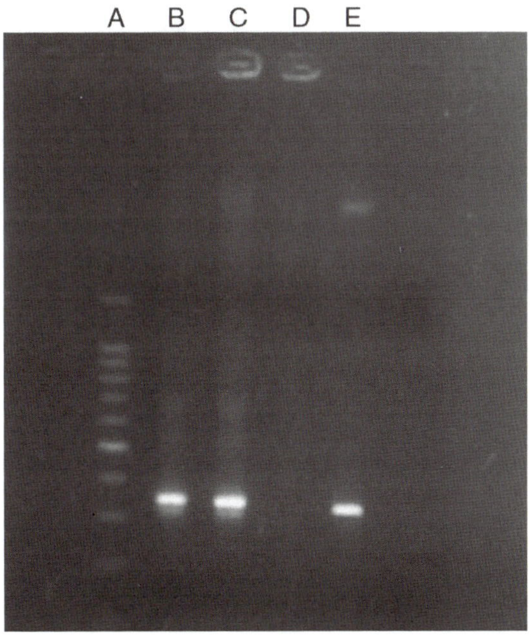

Fig. 13.5 Gel electrophoretic proof (1.5 % gel) of the presence of *bphA1* gene in the DNA isolated from strain *Pseudomonas stutzeri*

Gel description		*bphA1* gene
A line	Marker 100 bp	
B line	*P. stutzeri*	Yes
C line	*P. stutzeri*	Yes
D line	Negative isolate	No
E line	Positive control	Yes

standard 25 μl of PCB121 and then subjected to gas chromatography–mass spectrometry (GC-MS) analysis.

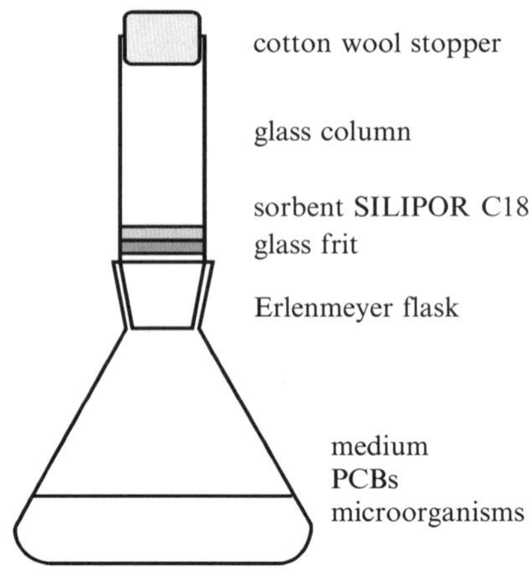

Fig. 13.6 Apparatus for simultaneous monitoring of evaporation and biodegradation of PCBs (Dercová et al. 1996)

13.2.6 Biodegradation Assay: Biostimulation/ Bioaugmentation Approach

Biodegradation was carried out in Erlenmeyer flasks equipped with sorbent Silipor (0.5 g) column that was closed with a cotton wool stopper to maintain sterile environment and allow for gas diffusion. This apparatus for simultaneous monitoring of evaporation and biodegradation of PCBs was described previously (Dercová et al. 1999a, b). The apparatus allows to determine mass balance of the tested compound. The evaporation rate constants of the individual PCB congeners presented in Delor 103 were published in our previous papers (Vrana et al. 1995, 1996a). To assess aerobic biodegradation of PCBs present in long-term contaminated sediments from Strážsky canal with autochthonous microflora under biostimulation conditions, 0.75 g of cut plant material containing terpenes (ivy leaves or pine needles) was added to 50 g of wet sediment. Biodegradation was carried out under stationary conditions with intermittent mixing at 28 °C in the dark. The control for the abiotic decrease of PCBs was represented by sterilized sediment. The samples were analyzed at the beginning of the experiment and on the 7th, 14th, and 21st days. Total flask contents were taken for PCB analysis. To evaluate aerobic biodegradation of PCBs in the bioaugmented sediments, 50 g of dry sediment and inoculum of *Pseudomonas stutzeri* with the final concentration 2 g l^{-1}in the minimal liquid media mixed with the sediment and replace into the apparatus (Fig. 13.6) were used. Other conditions were the same as in an experiment with non-bioaugmented sediment described previously.

13.2.7 Soxhlet Extraction

A sample of the sediment was desiccated at laboratory temperature and sieved using 71 μm diameter sieves. Two gram sediment and 0.5 cm layer of Na_2SO_4 were added into the Soxhlet apparatus. The sediment was covered with cotton wool moisturized in chloroform. A 100 ml extraction flask with 40 ml of n-hexane as extraction medium was attached to the apparatus. After 1 h, 20 ml of n-hexane was added and the extraction was continued for 3 h. The column was washed by n-hexane. The cleaning procedure was repeated to obtain transparent extract. Subsequently, after concentration at vacuum evaporator, 1 ml of n-hexane was added and content was transported to the test tube. The prepared sample was analyzed using GC. The extract was concentrated on vacuum evaporator to the final volume 1–2 ml, 1 g of copper powder was added to eliminate sulfur, and the mixture was ultrasonicated for 45 min. Subsequently, the mixture was filtered through 5 cm of Florisil layer.

13.2.8 PCB Extraction and GC ECD Analysis

The sediments were extracted for 4 h in the Soxhlet apparatus with n-hexane, filtered, and purified through a 5 cm layer of Florisil until the sample became transparent. One μl of the n-hexane extract was injected into the Hewlett–Packard gas chromatograph HP7890A. Separation of the PCBs was performed on a capillary nonpolar column HP-5 (30 m×0.32 mm inner diameter) coated with 0.25 μm (5 %-phenyl)-methylpolysiloxane phase. Electron capture detector (300 °C) was utilized, helium was used as carrier gas (flow rate 1.8 ml·min^{-1}) and nitrogen as makeup gas (25 ml·min^{-1}), and the pressure was 68.13 kPa. The temperature program was as follows: the initial temperature 80 °C for 1 min, followed by an increase at a rate of 30 °C min^{-1} until the temperature reached 160 °C and then kept at the temperature 1 min, followed by an increase at a rate of 4 °C·min^{-1} until the temperature reached 260 °C and then kept at the temperature 3 min. The total time of analyses was 32.7 min. The indicator PCB congeners recommended by EPA (US EPA Methods 8089/8081) and three selected PCB congeners were analyzed. Calibration solution was prepared for PCBs 8, 28, 52, 101, 118, 138, 153, 180, and 203. Quantitative evaluation of the results was performed based on comparison of the peak areas corresponding to the indicator congeners in the chromatograms of the samples and of the standards (ISO 6468).

13.3 Results and Discussion

13.3.1 Characterization of PCB-Contaminated Sediments

Extensive laboratory-scale research has been performed on spiked sediments suspended in artificial mineral media (Bedard 2003). In contrast to the conditions used in the cited papers, our experiments were carried out in the real PCB historically contaminated sediments used in dry form with the addition of minimal mineral medium or in naturally wet form. We expected that the experiments might allow collecting relevant information for predicting the actual potential of biological processes in the final in situ remediation of contaminated sediments.

In the used sediment samples, in addition to the presence of PCBs (commercial mixture Delor 103), presence of polyaromatic hydrocarbons (PAHs), pesticides, as well as of heavy and toxic metals has been established, too (not shown). In relation to the determined concentrations of all pollutants, it was demonstrated that PCBs represented the majority contaminant found in several times higher concentrations than other contaminants.

We observed that tri-CBs (CBs 18, 28, and 31) and hexa-CBs (CBs 138, 149, and 153) represented about 57 % of the total PCB content. It has been found that in Delor 103 and Delor 104, tri-CB congeners were the most abundant ones, whereas in the industrial mixtures Delor 105 and Delor 106, hexa-CBs were the predominant congeners (Taniyasu et al. 2003). Indeed, the PCB profile established by us for the sediments sampled from Strážsky canal resembles these types of Delors. Our data resembled very closely the statistically calculated data obtained by Hiller et al. (2010) for the sediment sampled from the same contaminated location from Zemplínska Šírava, a water reservoir, in which Strážsky canal is flowing through Laborec River (part 1.2.). It is recognized that the bottom sediments are substantially contaminated by the commercial mixtures of PCBs Delor 103 and Delor 106 (equivalent to Aroclor 1242 and 1260). Indeed, the profile established by us for sediments from Strážsky canal (not shown) and Zemplínska Šírava (Hiller et al. 2010, see Sect. 1.2) resembles these types of PCB commercial mixtures Delor 103 and Delor 106.

13.3.2 Congener-Specific Analysis of PCBs

In the experiments, PCB congener-specific analysis was performed: PCB 28 (3 Cl: 2,4,4′-), PCB 52 (4 Cl: 2,2′5,5′-), PCB 101 (5 Cl: 2,2′,4,5,5′-), PCB 118 (5 Cl: 2,3′,4,4′,5-), PCB 138 (6 Cl: 2,2′,3,4,4′,5-), and PCB 153 (6 Cl: 2,2′,4,4′,5,5′-). Peaks were identified according to Mills et al. (2007) (Fig. 13.7, Table 13.1).

Basic parameters of the sampled sediments are listed in Table 13.2. Concentrations of the individual PCB congeners determined in the untreated sediments sampled in 2008 and 2009

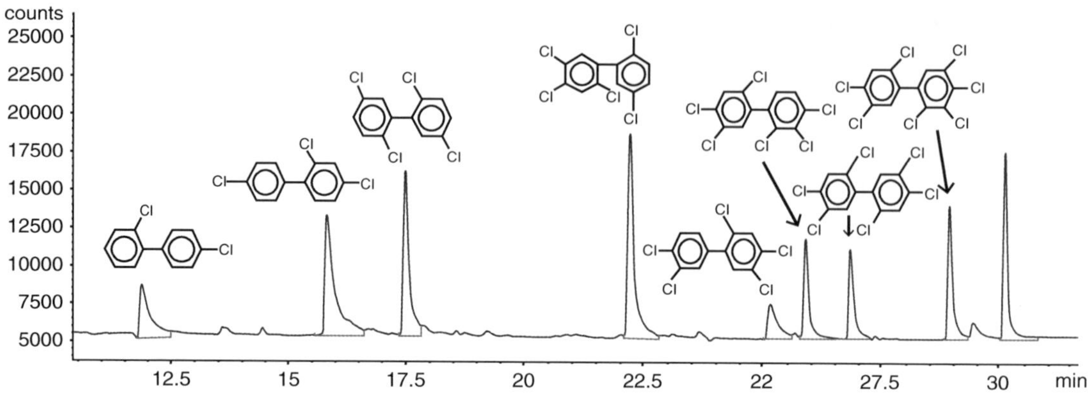

Fig. 13.7 GC ECD chromatogram of the six indicator PCB congeners

Table 13.1 PCB congeners evaluated in the experiments

IUPAC no.	Chlorine position
PCB 8	2, 3, 6; 2, 3′, 6
PCB 28[a]	2, 3′, 4, 5
PCB 52[a]	2, 2′, 3, 4′, 5, 5′
PCB 101[a]	2, 2′, 3, 4, 5, 5′
PCB 118	2, 3′, 4, 4′, 5
PCB 153[a]	2, 2′, 4, 4′, 5, 5′
PCB 138[a]	2, 2′, 3, 4, 4′, 5
PCB 180[a]	2, 2′, 3, 4, 4′, 5, 5′
PCB 203	2, 2′, 3, 4, 4′, 5, 5′, 6

[a]The six indicator PCB congeners

Table 13.2 Basic parameters of the sampled PCB-contaminated sediment

Parameter	Range
pH	6.94–7.24
ORP	−197–439 mV
Dry mass	20–63 %
Organic portion	1.05–7.06 %
TOC	21.9 g·kg^{-1} d.w.

are listed in Table 13.3. The sample from 2008 was taken for biodegradation experiments. Data shown in Table 3.3 prove that the contamination decrease annually very slowly. Decrease was probably caused by accumulation of PCBs in some aquatic organisms and depended also on the sampling time during the year and sampling place.

13.4 Bioavailability of PCB Congeners

To determine biologically relevant PCB fraction, we have applied the SFE method. SFE can discriminate PCBs available for degrading bacteria and simultaneously potential toxic fraction of the particular pollutant as well. In the earlier works (Nilsson et al. 2003, 2006; Hallgren et al. 2006), it was proven that SFE treatment preferentially removed those molecules of pollutants (PCBs) that were available for uptake by organisms.

For comparison, two sediment samples sampled from the long-term contaminated industrial Strážsky canal were analyzed, the first collected in 2008 and the second one in 2009. No noticeable differences have been observed in the obtained data with an exception of the two low chlorinated PCBs, easily degradable and vaporizable congeners PCB8 and PCB28. As can be seen, amount of bioavailable fractions changed from 91 to 6 % of the analyzed PCB congeners or from 58 to 7.6 %. Values and comparison of the bioavailable fractions of the indicator PCB congeners (ng·g^{-1}) determined in the sediments sampled from the S. canal in 2008 and 2009 expressed as SFE/Sox (%) ratio are introduced in Tables 13.4, 13.5 and Fig. 13.8. Bioavailability of the individual PCB congeners determined in 2009 reached slightly lower values when compared with the sampling of 2008.

Table 13.3 Concentration of the individual PCB congeners determined in the sampled sediments (mg PCB·kg^{-1} of dry sediment)

PCB congener (μg·g^{-1})									
Sampling year	8	28	52	101	118	138	153	180	203	Total
2008	2.541	8.164	5.336	5.033	1.900	6.342	8.585	6.392	1.867	46.161
2009	0.600	4.504	6.771	4.343	2.600	7.269	8.779	6.034	1.665	42.625

Table 13.4 Bioavailable fraction of the PCB congeners in the contaminated sediment sampled from Strážsky canal (sampling site: eastern part of Slovakia, surroundings of the former PCB producer) (sampling year 2008)

PCB congener	SFE (ng/g)	Soxhlet (ng/g)	SFE/Sox (%) – bioavailable fraction
c PCB28 (ng/g)	7,444.9	8,164.1	91.2
c PCB52 (ng/g)	4,300.7	5,335.9	80.6
c PCB101 (ng/g)	1,892.0	5,033.3	37.6
c PCB118 (ng/g)	736.2	12,237.0	6.0
c PCB138 (ng/g)	2,463.6	6,341.8	44.0
c PCB153 (ng/g)	3,778.8	8,585.1	38.8
c PCB180 (ng/g)	2,455.9	6,392.4	38.4

Table 13.5 Bioavailable fraction of the PCB congeners in the contaminated sediment sampled from Strážsky canal (sampling site: eastern part of Slovakia, surroundings of the former PCB producer) (sampling year 2009)

PCB congener	SFE (ng/g)	Soxhlet (ng/g)	SFE/Sox (%) – bioavailable fraction
PCB28	2,623.0	4,503.9	58.2
PCB52	3,894.4	6,771.3	57.5
PCB101	2,113.1	4,342.6	48.7
PCB118	873.3	11,539.4	7.6
PCB153	2,874.0	8,779.3	32.7
PCB138	2,920.0	7,269.2	40.2
PCB180	2,061.0	6,033.6	34.2

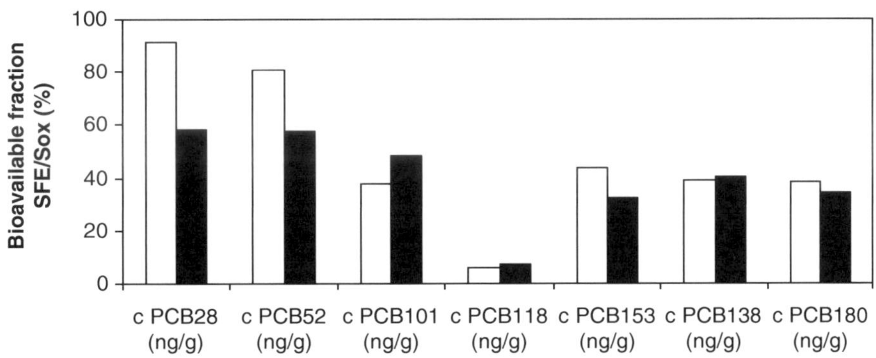

Fig. 13.8 Comparison of the bioavailable fraction of the indicator PCB congeners (ng·g^{-1}) determined in the sediments sampled from *S. canal* in 2008 and 2009 expressed as SFE/Sox (%) ratio. *SFE* supercritical fluid extraction, *Sox* Soxhlet extraction. □ sampling year 2008; ■ sampling year 2009

13.5 Evaluation of Ecotoxicity and Genotoxicity of PCB-Contaminated Sediments

Toxicity of sediments depends on the physicochemical properties of the particular contaminant (dilution in water, evaporation, polarity). These properties influence adsorption of PCBs on sediment particles or their arrangement in the particular phase (water, solid matrix, and air). Chemical analysis of contaminant concentration alone is not sufficient for determination of the environmental risk. Sometimes, the by-products of degradation are more toxic than the parent compound. Lovecká et al. (2004) and Vrana et al. (1996b) proved higher toxicity of chlorobenzoic acids, bacterial degradation products of PCBs produced by the "upper" degradation pathway. In case of several bioindicator models, tests with water solutions did not render unambiguous results due to the hydrophobic character of some toxic compounds, e.g., PCBs (Borja et al. 2005). Our previous experiments performed with bacteria *Vibrio fischeri* demonstrated different toxic effects of the water and isopropyl alcohol solutions of PCBs on the bioluminescence of the tested organism (Dercová et al. 2009). PCBs are hydrophobic compounds, and it is obvious that toxicity was expressed more markedly in the case of the less polar solvent. Solubility of PCBs in isopropyl alcohol is higher than in water, and this fact may cause higher bioavailability and increase of solution toxicity.

Ecotoxicity of the sediments was assessed applying aquatic plant toxicity test using the standard bioindicator *Lemna minor* and bioluminescent bacteria *Vibrio fischeri*, respectively. The sediments sampled from Strážsky canal were more toxic than those sampled from Zemplínska Šírava. Our results revealed that the toxicity had increasing tendency from the west to the east side of the water reservoir in spite of the fact that in the west part the contaminated Laborec River ran through an inflow canal into the reservoir. A possible explanation for this observation is that the outflow canal is situated directly on the opposite side and due to the movement of sediments during flood in spring and summer times. To summarize, both sediments (sampled in Strážsky canal and Zemplínska Šírava) revealed detrimental impact on biota and represented a source of adverse effects on life functions of the tested bioindicators. Thus, it can be assumed that these sediments pose potential ecological and human health risks (more details in Dercová et al. 2009).

Genotoxicity of PCB-contaminated sediments was determined in our experiments by studying their effect on histidine-dependent strains of *Salmonella typhimurium* (TA98 and TA100) using Ames test. The results show that using the sediments sampled from Strážsky canal, a statistically significant average increase of induced revertant colonies of *Salmonella typhimurium* TA98 has been established. Similar results were obtained at the evaluation of genotoxicity of the commercial mixture Delor 103 using *Salmonella typhimurium* TA100 (Dercová et al. 2009).

13.6 Isolation and Identification of PCB-Degrading Bacteria from Historically PCB-Contaminated Soil and Sediments

There are several approaches that allow for selection of cultivable microorganisms useful for bioaugmentation. We used enrichment method with PCBs and biphenyl for isolation and identification using 16SrRNA phylogenetic analysis and SIP (stable isotope probing) method for the PCB-degrading bacteria from the contaminated sediments (Dudášová et al. 2013).

Isolation of the two bacterial strains *Pseudomonas stutzeri* and *Alcaligenes xylosoxidans* from the long-term contaminated soil and of the two bacteria *Pseudomonas veronii* and *Ochrobactrum anthropi* from the contaminated sediments was described in our previous work (Dercová et al. 1996) and the latest one (Dudášová et al. 2013). The last four bacterial strains were identified in CCM (Czech Collection of Microorganisms) (Masaryk University, Brno, Czech Republic).

With *Pseudomonas stutzeri* and *Ochrobactrum xylosoxidans*, detection of *bphA* gene encoding biphenyldioxygenase, the important starting enzyme of PCB degradation, was performed (Dercová et al. 2008; Dudášová et al. 2013). Preadaptation of catabolic bacteria to the target contaminant, prior to inoculation, led to better survival in adverse environment and more efficient degradation of target pollutant. Isolated pure bacteria were added to the natural contaminated sediments using an approach called reinoculation. The novel approach to obtain present autochthonous strains with the particular catabolic activity is stable isotope probing (SIP) analyses. This method provides opportunities to link microbial diversity with function and to identify those cultivable and yet-to-be cultivated organisms which are involved in biodegradation in the field (Uhlík et al. 2009).

The subsequent step included microbiological analysis of the contaminated sediments and isolation of pure bacterial cultures capable of degrading PCBs. In order to determine the genetic potential for their biodegradability, the gene *bphA1* encoding the enzyme biphenyldioxygenase, responsible for the first step of PCB aerobic degradation, was identified using a PCR technique (Dercová et al. 2009). The ultimate goal of the work was to perform aerobic biodegradation of PCBs in the sediments. The bacteria present in both sediments are able to degrade certain low chlorinated congeners (Dercová et al. 2008). Our previous work deals also with biodegradation of PCBs in the two samples of the abovementioned contaminated sediments: (a) in the natural sediments by autochthonous microbial consortium and (b) in the bioaugmented sediments inoculated by allochthonous bacterial strains, two bacterial isolates from long-term PCB-contaminated soil *Pseudomonas stutzeri* and *Alcaligenes xylosoxidans*. Both approaches were applied under the biostimulation conditions, with addition of glucose or biphenyl as co-substrates. The highest PCB degradation was observed in the bioaugmented sediment inoculated with bacterial strain *P. stutzeri*. Addition of biphenyl, as the co-substrate and the inducer, positively affected biodegradation of PCBs. The *bphA1* gene was identified in genome of *P. stutzeri*, a potential PCB degrader isolated from long-term PCB-contaminated soil.

Our current study is focused on the characterization of 15 bacterial strains isolated from long-term PCB-contaminated sediment isolated from Strážsky canal. PCB-degrading strains were isolated and identified as *Microbacterium oleivorans*, *Stenotrophomonas maltophilia*, *Brevibacterium* sp., *Ochrobactrum anthropi*, *Pseudomonas mandelii*, *Rhodococcus* sp., *Achromobacter xylosoxidans*, *Stenotrophomonas* sp., *Ochrobactrum* sp., *Pseudomonas aeruginosa*, and *Starkeya novella* by the 16S rRNA gene sequence phylogenetic analysis. The study presents a newly isolated bacterial strain *S. novella* with PCB-degrading ability in minimal mineral medium as well as in PCB-contaminated sediment. For *Achromobacter xylosoxidans*, the *bphA* gene was identified. The best growth ability in the presence of both carbon sources (biphenyl and PCB vapor) was obtained for *Ochrobactrum* sp. and *Rhodococcus* sp. The highest potential for bioaugmentation of PCB-contaminated sediment showed *Achromobacter* sp. (in more detail, see Dudášová et al. 2013).

13.7 Plant Terpenes as Alternative Inducers of PCB Degradation Applicable During In Situ Decontamination

Microbial communities from different habitats usually have different response to same amendments. The ability to manipulate microbial activity by making organic carbon amendments can have important consequences for PCB remediation. Amendments with salicylic acid and biphenyl can improve PCB removal by inducing the synthesis of PCB-transforming enzyme systems and serving as carbon and energy sources for PCB degraders (Donnelly et al. 1994; Gilbert and Crowley 1997; Luo et al. 2008). Several researchers have successfully applied the approach to stimulate PCB removal in laboratory and field studies (Fava et al. 1999; Dzantor et al. 2002).

There have been attempts to find alternative natural nontoxic substrates that can induce the biphenyl/PCB pathway for the purpose of PCB decontamination (Focht 1995). Certain plant compounds, including flavonoids (Donnelly et al. 1994), lignin (Higson 1992; Furukawa 1994), and terpenes (Focht 1995; Hernandéz et al. 1997; Gilbert and Crowley 1998), can serve as natural substrates for the induction of the *bph* genes. Terpene addition to contaminated soils enhanced biodegradation of PCBs (Hernandéz et al. 1997; Gilbert and Crowley 1998; Dercová et al. 2003a, b). Terpenes were thus considered to be natural inducers of the biphenyldioxygenase, the first and the most important enzyme that starts biodegradation of PCBs. Enrichment of contaminated sediments or soil with terpenes might improve the bioremediation of PCBs (Hernandéz et al. 1997).

We have also studied the effects of four plant materials containing terpenes (natural compounds of plant essential oils) – orange peels, tangerine peels, pine needles, and ivy leaves – and two synthetic terpenes, carvone and limonene, on the biodegradation of Delor 103 by *Pseudomonas stutzeri*. The highest biodegradation of PCBs assayed in minimal mineral medium with the addition of mentioned plant materials was observed in the presence of ivy leaves (52.8 %) and pine needles (50.5 %). The enhancement of biodegradation was observed in some cases even without the increase of bacterial biomass. Addition of synthetic terpenes, limonene and carvone, resulted in an effect similar to that observed with addition of biphenyl, known as inducer of PCB degradation. It has been concluded that the role of different plant terpenes should be further studied in more detail to evaluate exactly if terpenes serve as inducers of biphenyldioxygenase or as biostimulators of bacterial growth representing thus an additional carbon source. It seems likely that when improvement of biodegradation was not followed by an increase of biomass, added terpenes served as inducers of the biphenyldioxygenase (Dudášová et al. 2013).

13.8 The Effects of PCBs as the Environmental Stress Factor on Bacterial Degraders: Adaptation Responses of the Cell Membrane to the Presence of Toxic Organic Compounds

Results of a study of fluidity of bacterial membrane of PCB degraders allow us to evaluate the adaptation of bacterial cell membrane (as the primary contact place of xenobiotics and bacteria) to environmental stress caused by the presence of toxic compounds in the environment. Such adaptation increases probability to survive and colonize in adverse environment by selection of resistant strains of bacterial degraders (Murínová et al. 2013).

The purpose of this part of our work was to analyze the effect of PCBs as potential stress factors on the cell membrane, especially on the profile of fatty acids in the two bacterial membrane lipids – phosphatidylcholine and phosphatidylethanolamine.

To evaluate the effects of chloroaromatic compounds as potential environmental stressors on bacterial cells, the effects of PCBs on biomass production of a PCB-degrading strain *Pseudomonas stutzeri* (an isolate from long-term contaminated soil, Dercová et al. 2008) and on the fatty acid profile of its major membrane lipids were examined. Growth based on biomass weight was stimulated when PCBs were added at the time of inoculation, but PCB addition 3 days after inoculation led to a significant decrease in biomass concentration. Simultaneous addition of PCBs plus biphenyl or PCBs plus carvone negatively affected biomass of *P. stutzeri* (addition of biphenyl or carvone at the time of inoculation and PCBs to 3-day-old culture). In the presence of PCBs alone, the amount of the prevalent fatty acids C16:0 and C17-cyclopropyl fatty acid (C17-CP) of *P. stutzeri* in total and neutral lipids was significantly reduced. When PCBs were added together with carvone (carvone at the time of inoculation and PCBs after 3 days), a significant reduction of the content of these fatty acids was observed, but, in addition, the content of oleic, cis-vaccenic, and cyclononadecanic (C19-CP) acids was increased. When PCBs were combined with biphenyl, the amount of prevalent fatty acids was reduced, while that of oleic, cis-vaccenic, and cyclononadecanic acids was increased in total and neutral lipids. Addition of 3-chlorobenzoic acid led to a significant growth inhibition and to the production of oleic and cis-vaccenic acids in the phosphatidylcholine membrane fraction (Zorádová et al. 2011).

The aim of the other work was to analyze the effect of PCBs in the presence of natural and synthetic terpenes as potential inducers of PCB degradation on the cell membrane, especially on the profile of fatty acids in the bacterial membrane lipids, *cis/trans* ratio of fatty acids, lipid accumulation, degree of saturation, and bacterial growth. The bacteria chosen for the study were the abovementioned strain of *P. stutzeri* previously shown to degrade PCBs and *Burkholderia xenovorans* LB400 (control strain possessing *bphA* gene). We supposed that addition of the inducers of PCB degradation would cause decrease of the toxic effects of PCBs and also lead to smaller adaptation changes in the cytoplasmic membrane.

According to the results obtained, it could be concluded that natural terpenes, mainly those contained in ivy leaves and pine needles, decreased membrane adaptation mechanism caused by PCBs in the studied strains. Growth inhibition effect decreased upon addition of these natural compounds to the cultivation medium or sediment mixed with mineral medium. The amount of fatty acids that can lead to elevated membrane fluidity increased in both strains after the addition of the two natural terpene sources. The cells adaptation changes were more prominent in the presence of carvone, limonene, and biphenyl than in the presence of natural terpenes, as indicated by growth inhibition, lipid accumulation, and *cis/trans* isomerization. Addition of biphenyl or carvone simultaneously with PCBs increased the *trans/cis* ratio of fatty acids in membrane fractions probably as a result of fluidizing effects of PCBs. The adaptation changes were more pronounced in the presence of PCBs as a sole carbon source. This could suggest that PCBs alone have stronger fluidizing effect on bacterial membrane and increase adaptation mechanisms than when added together with biphenyl or terpene inducers.

Our results reveal that degree of cell damage depends on adaptation mechanisms of the individual bacterial strain and from the time of addition of the particular contaminant to bacterial cells during growth phase. The lower adaptation changes of bacterial cell membrane in the presence of contaminants, the higher resistance of the strain, and the higher probability for its successful application in remediation technology could be achieved (Zorádová et al. 2011; Zorádová-Murínová et al. 2012).

Microbial growth, assessed on the basis of biomass weight, was stimulated when PCBs were added to bacteria suspended in minimal mineral media at the time of inoculation, but PCB addition 3 days after inoculation led to a significant decrease in biomass amount. Simultaneous addition of PCBs plus biphenyl or PCBs plus carvone negatively affected *P. stutzeri* biomass (addition of biphenyl or carvone at the time of inoculation and PCBs to 3-day-old culture). Thus, summarizing, our data demonstrated that PCBs inhibited bacterial growth and modified membrane fluidity by changing the fatty acids profile in two major membrane lipid fractions phosphatidylcholine and phosphatidylethanolamine and in all membrane lipids as well as in nonpolar lipids inside the cells. Increased membrane fluidity as a consequence of the presence of PCBs led to the enhanced growth of biomass that might be explained by the biodegradation ability of the used *P. stutzeri* strain. It is worth mentioning that addition of 3-CBA (Vrana et al. 1996b) led to a similar increase of membrane fluidity but resulted in a significant inhibition of bacterial growth.

13.9 Biostimulation and/or Bioaugmentation of PCB-Contaminated Sediment

The presence of active autochthonous adapted microflora in the sampled long-term PCB-contaminated sediments was confirmed in our previous work. Our results revealed that with the more contaminated sediment of Strážsky canal higher biodegradation was determined than in less contaminated sediment of Zemplínska Šírava water reservoir, both located in PCB-contaminated area (Dercová et al. 2009). Currently, we have isolated using 16S rRNA phylogenic analysis of 15 bacterial strains with PCB-degradative ability from the studied historically contaminated sediments. Five from them revealed PCB-degradative ability (see Sect. 13.4) (Dudášová et al. 2013).

Aerobic and anaerobic biodegradation of PCBs in the contaminated sediments occurred and is continuing but very slow during the least invasive bioremediation approach natural attenuation. Biodegradation can be accelerated by nutrient or inducer addition during the process of biostimulation, and/or by inoculum addition (autochthonous or allochthonous), using a bioaugmentation strategy to promote the degradation of target contaminant, PCBs.

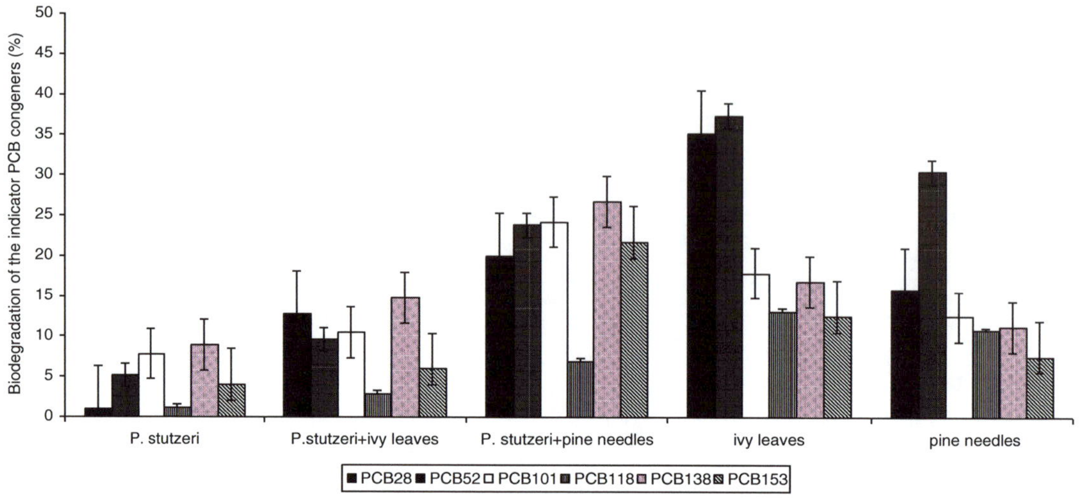

Fig. 13.9 Biodegradation of the indicator PCB congeners under bioaugmentation, biostimulation, and combination of bioaugmentation and biostimulation conditions in the PCB-contaminated sediments in the presence of *Pseudomonas stutzeri* inoculum and plant terpenes as the potential inducers (ivy leaves and pine needles, addition of 5 g cut plants). Cultivation conditions: 50 g of dry sediment with 100 ml of minimal mineral media, pH 6.5, initial concentration of inoculum 1 g·l^{-1}, 28 °C, in the dark, and 21-day stationary cultivation with intermittent mixing

Our attention is currently focused on biodegradation of PCBs in the natural sediments using biostimulation and/or bioaugmentation approaches of bioremediation. Biodegradation experiments were performed in the intact (untreated) sediments by autochthonous microbial consortium and in the bioaugmented sediments inoculated by the abovementioned bacterial isolate from long-term PCB-contaminated soil *Pseudomonas stutzeri* (Dercová et al. 1996). Biphenyl as the co-substrate and the inducer improved biodegradation of PCBs. The *bphA1* gene, encoding enzyme biphenyldioxygenase responsible for PCB degradation, was identified in a genome of the bacteria *P. stutzeri* in our previous work (Dercová et al. 2009). Presence of this important enzyme makes these bacteria attractive and prospective candidates for application as bacterial degraders suitable for the bioaugmentation of PCB-contaminated biotopes.

Figure 13.9 summarizes biodegradation of PCB congeners observed under different conditions: bioaugmentation (*P. stutzeri*), bioaugmentation combined with biostimulation (addition of *P. stutzeri* and ivy leaves or *P. stutzeri* and pine needles), and biostimulation (autochthonous microflora and ivy leaves or pine needles). Number of CFUs of the indigenous soil microorganisms present in sediment sampled from Strážsky canal was about 5×10^5 in 1 g of the dried sediment (used in bioaugmentation strategy) and about 7×10^6 in 1 g of the wet sediment (used in biostimulation strategy). Figure 3.4 illustrates the time course of biostimulation (addition of plant terpenes) during biodegradation of PCBs by indigenous bacteria in the sample of contaminated sediment. The results obtained during biostimulation with pine needles were comparable with those obtained under the conditions of biostimulation combined with bioaugmentation (the addition of *P. stutzeri* and ivy leaves). In the experiments with addition of plant terpenes as inducers in the form of pine needles, the lowest ecotoxicity was determined in the sample of wet sediment after biodegradation (after 21 days)

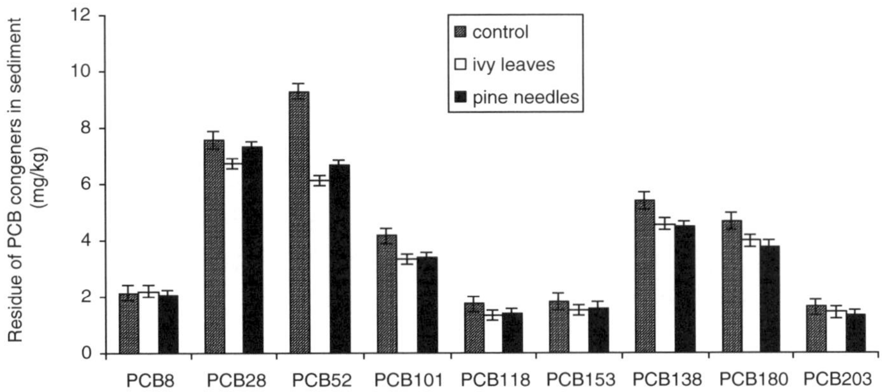

Fig. 13.10 Biostimulation time course after 21 days of stationary cultivation of PCB-contaminated sediment by naturally present microorganisms in the dark at 28 °C

using bioindicator *Lemna minor* (not shown). When the indigenous microbial population is present in the contaminated area, the addition of exogenous inoculum reveals very often not the expected synergistic but antagonistic effects due to competition of microorganisms towards substrate. The results indicated slightly antagonistic rather than synergistic effect of indigenous and exogenous bacteria (Fig. 13.10).

Under bioaugmentation conditions, degradative bacteria *P. stutzeri* was added as a biocatalyst to the sediment to strengthen biodegradative ability of present microflora. The results revealed that higher biodegradation of the indicator PCB congeners occurred under biostimulation than under bioaugmentation or combined bioaugmentation and biostimulation conditions. According to the results, biostimulation seems to be an efficient method to enhance biodegradation of PCBs using natural capacity of the long-term adapted indigenous microbial consortium. The method may allow for solving the problems with survival and colonization ability of the inoculated degraders. The biostimulation and bioaugmentation experiments using autochthonous bacteria isolated directly from PCB-contaminated sediments are currently running. It seems that *Achromobacter xylosoxidans* owning *bphA1* gene is a very promising PCB degrader (Dudášová et al. 2013).

13.10 Conclusions

We assume that our findings might be of high relevance for development of sediment-specific remediation technologies. Information obtained from microcosm studies performed with naturally PCB-contaminated sediment, plant terpene inducers, and autochthonous inoculum might be also useful in the preliminary design of a site-specific biostimulation/bioaugmentation strategy. Microbial degradation of PCBs is highly dependent on chlorine substitution and is highly strain dependent. In this area, much new knowledge is needed, in terms of commercially available nutrients, electron donors/acceptors, suitable bacterial degraders, individual strain or microbial consortium, and the technologies enabling their efficient incorporation into the contaminated sediments. In spite of still existing gaps, it is possible to say that bioremediation represents a potential ecological and economical alternative to decontamination of areas polluted with low concentrations of chlorinated aromatic pollutants.

Acknowledgements Financial support from the Scientific Grant Agency of the Ministry of Education, Science, Research, and Sports of Slovak Republic and Slovak Academy of Sciences (Grant No. 1/0734/12) is gratefully acknowledged. The authors are thankfull to Ing. Lucie Bielska (Masaryk University, Brno) for determination of bioavailability of PCBs in the contaminated sediments.

References

Alexander M (1994) Biodegradation and bioremediation. Academic Press, San Diego, p 320

Ang EL, Zhao H, Obbard JP (2005) Recent advances in the bioremediation of persistent organic pollutants via biomolecular engineering. Enzym Microb Technol 37:487–496

Apitz SE, Brils J, Marcomini A, Critto A, Agostini P, Micheletti CH, Pippa R, Scanferla P, Zuin S, Lánczos T, Dercová K, Kočan A, Petrík J (2006) Approaches and frameworks for managing contaminated sediments – a European perspective, Chapter 1. In: Reible D, Apitz S (eds) Assessment and remediation of contaminated sediments, NATO Science Series, Series IV: Earth and environmental sciences. Springer, Dordrecht, pp 5–82

Bedard D (2003) Polychlorinated biphenyls in aquatic sediments: environmental fate and outlook for biological treatment. In: Häggblom MM, Bossert ID (eds) Dehalogenation: microbial processes and environmental application. Kluwer Academic Publishers, Boston, pp 443–465

Björklund E, Nilsson T, Bøwadt S, Pilorz K, Mathiasson L, Hawthorne SB (2000) Introducing selective supercritical fluid extraction as a new tool for determining sorption/desorption behavior and bioavailability of persistent organic pollutants in sediment. J Biochem Biophys Method 43(1–3):295–311

Borja J, Taleon DM, Auresenia J, Gallardo S (2005) Polychlorinated biphenyls and their biodegradation. Process Biochem 40:1999–2013

Čertík M, Dercová K, Sejáková Z, Finďová M, Jakubík T (2003) Effect of polyaromatic hydrocarbons (PAHs) on the membrane lipids of bacterial cell. Biology 58:1111–1117

Chihib NE, Tierny Y, Mary P, Hornez JP (2005) Adaptational changes in cellular fatty acids branching and unsaturation of *Aeromonas* species as a response to growth temperature and salinity. Int J Food Microbiol 102:113–119

Cronan JE (2002) Phospholipid modifications in bacteria. Curr Opin Microbiol 5:202–205

Denich TJ, Beaudette LA, Lee H, Trevor SJT (2003) Effect of selected environmental and physicochemical factors on bacterial cytoplasmic membranes. J Microbiol Method 52:149–182

Dercová K, Baláž Š, Haluška Ľ, Horňák V, Holecová V (1995) Degradation of PCBs by bacteria isolated from long-time contaminated soil. Int J Environ Anal Chem 58:337–348

Dercová K, Vrana B, Baláž Š, Šándorová A (1996) Biodegradation and evaporation of polychlorinated biphenyls (PCBs) in liquid media. J Ind Microbiol 16:325–329

Dercová K, Vrana B, Baláž Š (1999a) A kinetic distribution model of evaporation, biosorption and biodegradation of polychlorinated biphenyls (PCBs) in the suspension of Pseudomonas stutzeri. Chemosphere 38:1391–1400

Dercová K, Vrana B, Tandlich R, Šubová Ľ (1999b) Fenton's type reaction and chemical pretreatment of PCBs. Chemosphere 39(15):2621–2628

Dercová K, Baláž S, Vrana B, Tandlich R (2003a) Aerobic biodegradation of polychlorinated biphenyls (PCBs): the fate, distribution, kinetics, and enhancement of PCB biodegradation efficacy in the bacterial cell suspension of Pseudomonas stutzeri. In: Sašek V, Baveye P, Glaser JA (eds) The utilization of bioremediation to reduce soil contamination: problems and solutions, vol 19, NATO Science Series, Series IV: Earth and environmental sciences. Kluwer Academic Publishers, Dordrecht, pp 95–113

Dercová K, Tandlich R, Brežná B (2003b) Application of terpenes as possible inducers of biodegradation of PCBs. Fresen Environ Bull 12(1):286–290

Dercová K, Čertík M, Maľová A, Sejáková Z (2004) Effect of chlorophenols on the membrane lipids of bacterial cells. Int Biodeter Biodegr 54:251–254

Dercová K, Čičmanová J, Lovecká P, Demnerová K, Macková M, Hucko P, Kušnír P (2008) Isolation and identification of PCB-degrading microorganisms from contaminated sediments. Int Biodeter Biodegr 62:219–225

Dercová K, Šeligová J, Dudášová H, Mikulášová M, Šilhárová K, Tóthová L, Hucko P (2009) Characterization of the bottom sediments contaminated with polychlorinated biphenyls: evaluation of ecotoxicity and biodegradability. Int Biodeter Biodegr 63:440–449

Dercová K, Dudášová H, Lukáčová L, Kočan A, Murín M, Pilváňová A (2010) Environmental and technical aspects of PCB disposal, properties, monitoring, destruction, and remediation of PCB-contaminated sites in Slovakia. Now we have to address the PCB mess PEN Magazine (PCBs Elimination Network Magazine). Stockholm Convention, Geneva, pp 47–48. http://www.pops.int

Dercová K, Dudášová H, Lukáčová L, Zorádová S, Hiller E, Hucko P, Tóthová L, Šilhárová K (2011) Evaluation of ecotoxicity, genotoxicity, and biodegradability of PCB-contaminated bed sediments. In: Kalogerakis N, Fava F (eds) CD ROM book of manuscripts, 5th European bioremediation conference, ID034, Chania, Crete, Greece, pp 1–6. Available also in: Abstract Book of 5th EBC, p 242, Chania, Crete, Greece

Diefenbach R, Keweloh H (1994) Synthesis of *trans* unsaturated fatty acids in *Pseudomonas putida* P8 by direct isomerization of the double bond of lipids. Arch Microbiol 162:120–125

Donnelly PK, Hedge RS, Fletcher JS (1994) Growth of PCB-degrading bacteria on compounds from photosynthetic plants. Chemosphere 28:981–988

Dudášová H, Lukáčová L, Zorádová S, Dercová K (2012) The effect of terpenes on biodegradation of PCBs. Int Biodeter Biodegr 69:23–27

Dudášová H, Lukáčová L, Murínová S, Puškárová A, Pangallo D, Dercová K (2013) Biodegradation ability of bacterial strains isolated from long-term PCB-contaminated sediment. J Basic Microbiol (in press). doi:10.1002/jobm.201200369

Duldhardt I, Gaebel J, Chrzanowski L, Nijenhuis I, Härtig C, Schauer F, Heipieper HJ (2010) Adaptation of anaerobically grown *Thauera aromatica, Geobacter sulfurreducens* and *Desulfococcus multivorans* to organic solvents on the level of membrane fatty acid composition. Microb Biotechnol 3(2):201–209

Dzantor EK, Woolston JE, Momen B (2002) PCB dissipation and microbial community analysis in rhizosphere soil under substrate amendment conditions. Int J Phytorem 4:283–295

Fava F, Bertini L, Pinelli D, Nocetini M (1999) Characterisation of the indigenous bacteria involved on the ex-situ aerobic bioremediation of polycyclic aromatic hydrocarbon-contaminated soil. Annali Chim 89:777–782

Field JA, Alvarez RS (2007) Microbial transformation and degradation of polychlorinated biphenyls. Environ Pollut 155:1–12

Focht DD (1995) Strategies for the improvement of aerobic metabolism of polychlorinated biphenyls. Curr Opin Biotechnol 6:341–346

Furukawa K (1994) Molecular genetics and evolutionary relationship of PCB-degrading bacteria. Biodegradation 5:289–300

Furukawa K (2003) Super bugs for bioremediation. Trends Biotechnol 21(5):187–190

Gilbert ES, Crowley DE (1997) Plant compounds that induce polychlorinated biphenyl biodegradation by Arthrobacter sp. strain B1B. Appl Environ Microbiol 5:1933–1938

Gilbert ES, Crowley DE (1998) Repeated application of carvone-induced bacteria to enhance biodegradation of polychlorinated biphenyls in soil. Appl Microbiol Biotechnol 50:489–494

Gutierrez JA, Nichols P, Couperwhite I (1999) Changes in whole cell-derived fatty acids induced by benzene and occurrence of the unusual 16:1 o6c in *Rhodococcus* sp. 33. FEMS Microbiol Lett 176:213–218

Hallgren P, Westbom R, Nilsson T, Sporring S, Björklund E (2006) Measuring bioavailability of polychlorinated biphenyls in soil to earthworms using selective supercritical fluid extraction. Chemosphere 63(9):1532–1538

Hazel JR, Williams EE (1990) The role of alterations in membrane lipid composition in enabling physiological adaptation of organisms to their physical environment. Prog Lipid Res 29:167–227

Heipieper HJ, de Bont JAM (1994) Adaptation of Pseudomonas putida S12 to ethanol and toluene at the level of the fatty acid composition of membranes. Appl Environ Microbiol 60:4440–4444

Heipieper HJ, Diefenbach R, Keweloh H (1992) Conversion of cis unsaturated fatty acids to trans, a possible mechanism for the protection of phenol-degrading *Pseudomonas putida* P8 from substrate toxicity. Appl Environ Microbiol 58(6):1847–1852

Heipieper HJ, Weber FJ, Sikkema J, Keweloh H, de Bont JAM (1994) Mechanisms of resistance of whole cells to toxic organic solvents. TIBTECH 12:409–415

Heipieper HJ, Meulenbeld G, van Oirschot Q, de Bont JAM (1996) Effect of environmental factors on the trans/cis ratio of unsaturated fatty acids in *Pseudomonas putida* S12. Appl Environ Microbiol 62:2773–2777

Heipieper HJ, Waard P, Meer P, Killian JA, Isken S, Bont JAM, Eggink G, Wolf FA (2001) Regiospecific effect of 1-octanol on cis-trans isomerization of unsaturated fatty acids in the solvent-tolerant strain *Pseudomonas putida* S12. Appl Microbiol Biotechnol 57:541–547

Heipieper HJ, Meinhardt F, Seggura A (2003) The cis-trans isomerase of unsaturated fatty acids in *Pseudomonas* and *Vibrio*: biochemistry, molecular biology and physiological function of a unique stress adaptive mechanism. FEMS Microbiol Lett 229:1–7

Heipieper HJ, Neumann G, Kabelitz N, Kastner M, Richnow HH (2004) Carbon isotope fractionation during cis-trans isomerization of unsaturated fatty acids in *Pseudomonas putida*. Appl Microbiol Biotechnol 66:285–290

Heipieper HJ, Neumann G, Cornelissen S, Meinhardt F (2007) Solvent-tolerant bacteria for biotransformations in two-phase fermentation systems. Appl Microbiol Biotechnol 74:961–973

Hernandez BS, Koh SC, Chial M, Focht DD (1997) Terpene utilizing isolates and their relevance to enhanced biotransformation of polychlorinated biphenyls in soil. Biodegradation 8:153–158

Higson FK (1992) Microbial degradation of biphenyl and its derivates. Adv Appl Microbiol 37:135–164

Hiller E, Zemanová L, Sirotiak M, Jurkovič Ľ (2010) Concentrations, distributions, and sources of polychlorinated biphenyls and polycyclic aromatic hydrocarbons in bed sediments of the water reservoirs in Slovakia. Environ Monit Assess 173:883–897

Junker F, Ramos JL (1999) Involvement of the cis/trans isomerase Cti in solvent resistance of *Pseudomonas putida* DOT-T1E. J Bacteriol 181:5693–5700

Kabelitz N, Santos PM, Heipieper HJ (2003) Effect of aliphatic alcohols on growth and degree of saturation of membrane lipids in *Acinetobacter calcoaceticus*. FEMS Microbiol Lett 220:223–227

Kaneda T (1991) Iso- and anteiso-fatty acids in bacteria: biosynthesis, function, and taxonomic significance. Microbiol Rev 55:288–302

Kočan A, Petrík J, Chovancová J, Drobná B (1994) Method for the group separation of non-ortho-, mono-ortho- and multi-ortho-substituted polychlorinated biphenyls and polychlorinated dibenzo-p-dioxins/polychlorinated dibenzofurans using activated carbon chromatography. J Chromatogr 665:139–153

Kočan A, Drobná B, Chovancová J, Kočan J, Petrík J, Szabová E (1998) Burden of the environment and human population in the area contaminated with PCBs. Interim report, Project No 105/98-24, IPCM Bratislava, November, 113 pp. (in Slovak)

Kočan A, Petrík J, Drobná B, Chovancová J, Jursa S, Pavúk M, Kovrižnych J, Langer P, Bohov P, Tajtáková M, Suchánek P (1999) Burden of the environment and human population in the area contaminated with PCBs. Final report, Project No 105/98-24, IPCM Bratislava, February, 217 pp. (in Slovak)

Kočan A, Petrík J, Jursa S, Chovancová J, Drobná B (2001) Environmental contamination with polychlorinated biphenyls in the area of their former manufacture in Slovakia. Chemosphere 43:596–600

Langer P, Tajtáková M, Kočan A, Petrík J, Koška J, Kšinantová L, Rádiková Ž, Ukropec J, Imrich R, Hučková M, Chovancová J, Drobná B, Jursa S, Vlček M, Bergman A, Athanasiadou M, Hovander L, Shishiba Y, Trnovec T, Šebökova E, Klimeš I (2006) Thyroid ultrasound volume, structure and function after long-term high exposure of large population to polychlorinated biphenyls, pesticides and dioxin. Chemosphere 69:118–127

Lovecká P, Melenová I, Kučerová P, Nováková H, Macková M, Ruml T, Pazlarová J, Demnerová K (2004) Products of biological PCB degradation and their ecotoxicity. Int Biodeter Biodegr 53:195–283

Luo W, D'Angelo EM, Coyne MS (2008) Organic carbon effects on aerobic polychlorinated biphenyl removal and bacterial community composition in soils and sediments. Chemosphere 70:364–373

Macková M, Uhlík O, Lovecká P, Viktorová J, Nováková M, Demnerová K, Sylvestre M, Macek T (2010) Bacterial degradation of polychlorinated biphenyls. In: Barton L, Mandl M, Loy A (eds) Geomicrobiology: molecular and environmental perspective. Springer Science, Dordrecht/New York, pp 347–366. ISBN 978-90-481-9203-8

McCullar MV, Koh S-C, Focht DD (2002) The use of mutants to discern the degradation pathway of 3,4 P-dichlorobiphenyl in Pseudomonas acidovorans M3GY. FEMS Microbiol Ecol 42:81–87

Megharaj M, Ramakrishnan B, Venkateswarlu K, Sethunathan N, Naidu R (2011) Bioremediation approaches for organic pollutants: a critical perspective. Environ Int 37:1362–1375

Mills SA, Thal DI, Barney J (2007) A summary of the 209 PCB congener nomenclature. Chemosphere 68:1603–1612

Mouhamadou B, Faure M, Sage L, Marcais J, Souard F, Geremia RA (2013) Potential of autochthonous fungal strains isolated from contaminated soils for degradation of polychlorinated biphenyls. Fungal Biol 117:268–274

Mrozik A, Piotrowska-Seget Z (2010) Bioaugmentation as a strategy for cleaning up of soils contaminated with aromatic compounds. Microbiol Res 165: 363–375

Mrozik A, Labuzek S, Piotrowska-Seget Z (2004) Changes in fatty acid composition in *Pseudomonas putida* and *Pseudomonas stutzeri* during naphthalene degradation. Microbiol Res 160:149–157

Mrozik A, Cycoń M, Piotrowska-Seget Z (2010) Changes of FAME profiles as a marker of phenol degradation in different soil inoculated with *Pseudomonas* sp. CF600. Int Biodeter Biodegr 64:86–96

Murínová S, Dudášová H, Lukáčová L, Lászlová K, Dercová K (2013) Adaptation responses of bacterial strains to environmental stress caused by the presence of toxic organic compounds. Chem Listy 107(5): 356–361

Nichols-Smith S, The SY, Kuhl TL (2004) Thermodynamic and mechanical properties of model mitochondrial membranes. Biochim Biophys Acta 1663:82–88

Nilsson T, Björklund E (2005) Selective supercritical fluid extraction as a tool for determining the PCB fraction accessible for uptake by chironomid larvae in a limnic sediment. Chemosphere 60(1):141–146

Nilsson T, Sporring S, Björklund E (2003) Selective supercritical fluid extraction to estimate the fraction of PCB that is bioavailable to a benthic organism in a naturally contaminated sediment. Chemosphere 53(8): 1049–1052

Nilsson T, Häkkinen J, Larsson P, Björklund E (2006) Selective supercritical fluid extraction to identify aged sediment-bound PCBs available for uptake by eel. Environ Pollut 140(1):87–94

Ohtsubo Y, Nagata Y, Kimbara K, Takagi M, Ohta A (2000) Expression of the bph genes involved in biphenyl/PCB degradation in Pseudomonas sp. KKS102 induced by the biphenyl degradation intermediate, 2-hydroxy-6-oxo-6-phenylhexa-2,4-dienoic acid. Gene 256:223–228

Pedrotta V, Witholt B (1999) Isolation and characterization of the cis–trans-unsaturated fatty acid isomerase of Pseudomonas oleovorans GPo12. J Bacteriol 181:3256–3261

Petrík J, Kočan A, Jursa S, Drobná B, Chovancová J, Pavúk M (2001) Polychlorinated biphenyls in sediments in Eastern Slovakia. Fresen Environ Bull 10:375–380

Pieters RJ, Lutje Spelberg JH, Kellogg RM, Janssen D (2001) The enantioselectivity of haloalkane dehalogenases. Tetrahedron Lett 42(3):469–471

Pritchard PH (1992) Use of inoculation in bioremediation. Curr Opin Biotechnol 3:232–243

Ramos JL, Duques E, Rodriguez-Hervas JJ, Godoy P, Haidours A, Reyes F, Fernandez-Barrero A (1997) Mechanisms for solvent tolerance in bacteria. J Biol Chem 272:3887–3890

Reineke W (1984) Microbial degradation of halogenated aromatic compounds. In: Gibson DT (ed) Microbial degradation of organic compounds. Marcel Dekker, New York, pp 319–360

Šajbidor J (1997) Effect of some environmental factors on the content and composition of microbial membrane lipids. Crit Rev Biotechnol 17(2):87–103

Sambrook J, Fritsch EF, Maniatis T (1989) Molecular cloning: a laboratory manual. Cold Spring Harbor Laboratory, Cold Spring Harbor, New York

Schlame M (2008) Cardiolipin synthesis for the assembly of bacterial and mitochondrial membranes. J Lipid Res 49:1609–1619

Sejáková Z, Dercová K, Tóthová L (2009) Biodegradation and ecotoxicity study of soil contaminated by pentachlorophenol (PCP) using bioaugmentation and addition of sorbents. World J Microb Biotechnol 25:243–252

SHMÚ (2004) Initial assistance to the Slovak Republic to meet its obligations under the Stockholm Convention on Persistent Organic Pollutants (POPs). Proposal of the national implementation plan under the Stockholm Convention on POPs in Slovakia Project SLO/01/G31. Technical report no 5 (Final version), SHMÚ, Bratislava, April, 131 pp

Sikkema J, de Bont JAM, Poolman B (1994) Interactions of cyclic hydrocarbons with biological membranes. J Biol Chem 269:8022–8026

Sikkema J, de Bont JA, Poolman B (1995) Mechanisms of membrane toxicity of hydrocarbons. Microbiol Rev 59:201–222

Svobodová K, Plačková M, Novotná V, Cajthaml T (2009) Estrogenic and androgenic activity of PCBs, their chlorinated metabolites and other endocrine disruptors estimated with two *in vitro* yeast assays. Sci Total Environ 407:5921–5925

Sylvestre M (2013) Prospects for using combined engineered bacterial enzymes and plant systems to rhizoremediate polychlorinated biphenyls. Environ Microbiol 15(3):907–915

Tandlich R, Brežná B, Dercová K (2001) The effect of terpenes on the biodegradation of PCBs by *Pseudomonas stutzeri*. Chemosphere 44:1547–1555

Tandlich R, Vrana B, Payne S, Dercová K, Baláž Š (2011) Biodegradation mechanism of biphenyl by a strain of Pseudomonas stutzeri. J Environ Sci Health Part A Toxic/Hazard Subst Environ Eng 46(4):337–344

Tandlich R, Martišková M, Dercová K, Baláž Š (2013) Characterization of the chlorobenzoate hydrophobicity using the 1-octanol/water partition system. Fres Environ Bull 22(1):22–29 (the same article but published after revision in another journal)

Taniyasu S, Kannan K, Holoubek I, Ansorgova A, Horii Y, Hanari N et al (2003) Isomer-specific analysis of chlorinated biphenyls, naphthalenes and dibenzofurans in Delor: polychlorinated biphenyl preparations from the former Czechoslovakia. Environ Pollut 126:169–178

Tříska J, Kuncová G, Macková M, Nováková H, Paasivirta J, Lahtiperä M, Vrchotová N (2004) Isolation and identification of intermediates from biodegradation of low chlorinated biphenyls (DELOR-103). Chemosphere 54:725–733

Uhlík O, Jecná K, Leigh MB, Macková M, Macek T (2009) Sci Tot Environ 407:3611–3619

Unterman R, DeFlaun M, Steffan R (2000) Advanced in situ bioremediation – a hierarchy of technology choices. In: Rehm HJ, Reed G (eds) Biotechnology, vol 11b, 2nd edn, Environmental processes II: soil decontamination. Klein J (ed). Wiley-Verlag, Weinheim, FRG

Vasilyeva GK, Strijakova ER (2007) Bioremediation of soils and sediments contaminated by polychlorinated biphenyls. Microbiology 76(6):639–653

Vasseur P, Bonnard M, Palais F, Eom IC, Morel JL (2008) Bioavailability of chemical pollutants in contaminated soils and pitfalls of chemical analyses in hazard assessment. Environ Toxicol 23(5):652–656

Vítková M, Dercová K, Molnárová J, Tóthová L, Polek B, Godočíková J (2011) The effect of lignite and *Comamonas testosteroni* on pentachlorophenol biodegradation and soil ecotoxicity. Water Air Soil Pollut 218:145–155

von Wallbrunn A, Richnow HH, Neumann G, Meinhardt F, Heipieper HJ (2003) Mechanism of cis–trans isomerization of unsaturated fatty acids in *Pseudomonas putida*. J Bacteriol 185:1730–1733

Vrana B, Dercová K, Baláž Š (1995) Monitoring evaporation polychlorinated biphenyls (PCBs) in long-term degradation experiments. Biotechnol Tech 9: 333–338

Vrana B, Dercová K, Baláž Š (1996a) Evaporation kinetics of polychlorinated biphenyls (PCBs) during biodegradation experiments. Biotechnol Tech 10:37–40

Vrana B, Dercová K, Baláž Š, Ševčíková A (1996b) Effect of chlorobenzoates on the degradation of polychlorinated biphenyls (PCBs) by *Pseudomonas stutzeri*. World J Microb Biotechnol 12:323–326

Vrana B, Tandlich R, Baláž Š, Dercová K (1998) The aerobic biodegradation of polychlorinated biphenyls by bacteria. Biologia 53:251–266

Weber FJ, Isken S, de Bont JAM (1994) Cis/trans isomerization of fatty acids as a defence mechanism of Pseudomonas putida strains to toxic concentrations of toluene. Microbiology 140:2013–2017

Zhiwei Y, Palkovičová Ľ, Drobná B, Petrík J, Kočan A, Trnovec T, Hertz-Picciotto I (2007) Comparison of organochlorine compound concentrations in colostrum and mature milk. Chemosphere 66:1012–1018

Zorádová S, Dudášová H, Lukáčová L, Dercová K, Čertík M (2011) The effect of polychlorinated biphenyls (PCBs) on the membrane lipids of *Pseudomonas stutzeri*. Int Biodeter Biodegr 65:1019–1023

Zorádová-Murínová S, Dudášová H, Lukáčová L, Čertík M, Šilharová K, Vrana B, Dercová K (2012) Adaptation mechanisms of bacteria during the degradation of polychlorinated biphenyls in the presence of natural and synthetic terpenes as potential degradation inducers. Appl Microbiol Biotechnol 94:1375–1385

Zougagh M, Valcárcer M, Ríos A (2004) Supercritical fluid extraction: a critical review of its analytical usefulness. Trends Anal Chem 23:399–406

Metal Organic Frameworks: New Smart Material for Biological Application

14

P. Bansal, L.M. Bharadwaj, A. Deep, S.K. Rohilla, and R.K. Salar

Abstract

Metal organic frameworks are class of hybrid porous material made up of inorganic clusters connected with organic linkers, which offers many opportunities for applications in gas storage, chemical engineering, sensing applications, adsorbent material, catalysis, optoelectronics, and biological and medical fields. The unique physical and chemical properties of MOF and its ability of chemical tailorability make them potential candidate for biological applications like drug delivery, diagnosis, and imaging, as biosensors and many more. A wide range of possible compositions allows toxicologically acceptable formulations of MOFs. Many families of MOFs are under consideration for this purpose, and improvement is still needed in terms of stability, toxicity, and biocompatibility. Many biological applications of MOFs have been investigated, and many are still to be explored. This chapter covers the abovementioned aspects of MOFs and their use as new smart material for biological applications.

14.1 Introduction

Synthesizing novel materials has long been cherished as a stepping-stone in technological advancement. It is not long back when zeolites and activated carbon, known to be the most versatile, are used primarily in wide-ranging industrial applications. But as the need grows for more efficient, economical, and highly specific functions, conventional adsorbents were not found effective enough to deal with such problems. Although improved synthesis and different posttreatment procedures of zeolites and activated carbon resulted into some of their derivatives, the need of the hour is to design and synthesize materials that could be more effective. In the quest for designing novel adsorbents, attention has been paid to develop hybrid structures involving both inorganic and organic components by employing novel synthetic routes. The general concept was to take advantage of both the metal coordination and functionalities of the organic components. The concept of reticular synthesis, which can be

P. Bansal • S.K. Rohilla • R.K. Salar (✉)
Department of Biotechnology, Chaudhary Devi Lal University, Sirsa 125 055, India
e-mail: rajsalar@rediffmail.com

L.M. Bharadwaj • A. Deep
Central Scientific Instruments Organisation, Sector 30-C, Chandigarh 160 030, India

R.K. Salar et al. (eds.), *Biotechnology: Prospects and Applications*,
DOI 10.1007/978-81-322-1683-4_14, © Springer India 2013

Table 14.1 Some commonly used organic ligands in MOFs

Common name	IUPAC name	Structural formula
Oxalic acid	Ethanedioic acid	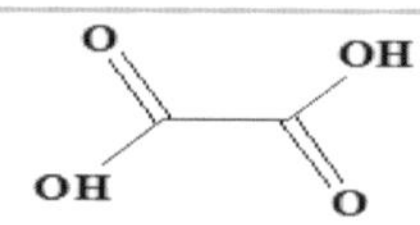
Terephthalic acid	Benzene-1,4-dicarboxylic acid *p*-Phthalic acid	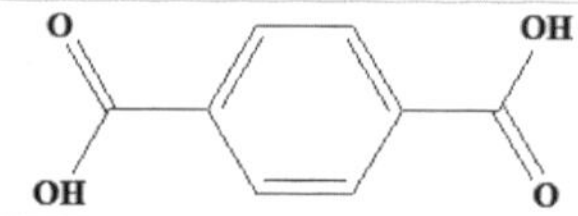
Succinic acid	Butanedioic acid	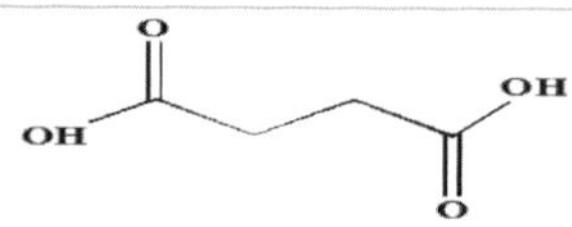
Trimesic acid	Benzene-1,3,5-tricarboxylic acid	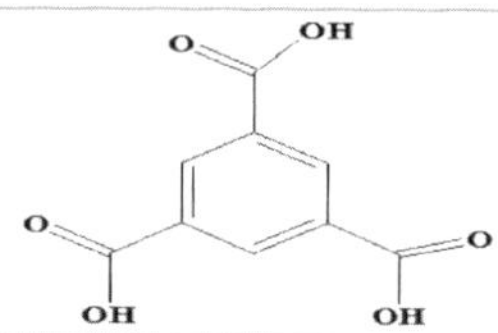
Pyrrodiazole	1*H*-1,2,4-triazole	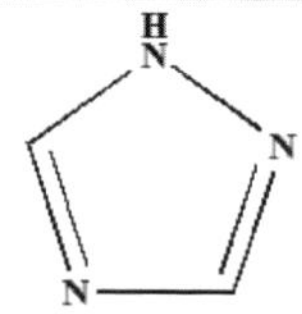

described as the process of assembling judiciously designed rigid molecular building blocks into predetermined ordered structures or networks, held together by strong bonding is found to be the key to the true design of novel solid-state materials. The building blocks of metal organic frameworks (MOFs) are carefully chosen so that their properties are retained and exhibited by product material. The choice of linker and metal significantly affects the structure and properties of MOF.

Metal organic frameworks are a special class of coordination polymers (CPs) which are crystalline and porous compounds having strong metal ligand interactions and generally belong to second and third generation of CPs. Early researches (Hoskins and Robson 1990; Yaghi et al. 1995) pointed out rich possibility of new material structure and properties offered by these coordination polymers. Normally the term metal organic framework is applied to those coordination polymers which exhibit porosity. These are a new class of crystalline hybrid materials whose crystal structure is made up of extended 3D network of metal ions or small discrete clusters connected through multidentate organic spacers. The inorganic part generally consists of first-row transition metals and cyanide, citrate, triazole, oxalate, carboxylate, etc. as organic linkers (Yaghi et al. 2003) (Table 14.1).

Metal organic frameworks are solid coordination polymers (CPs), formed by extended networking between metal ions and multidentate organic molecules. CPs include a large variety of materials containing metal ions and organic molecules which can combine to form crystalline or amorphous and porous or nonporous solids. CPs can be broadly divided into three generations of materials based on behavior of guest molecules and porosity of CPs.

First-generation CPs have a porous system sustained by guest molecules, which collapse irreversibly upon removal of guest molecules such as analyte and solvent molecules. Second-generation CPs have robust porous system which maintains permanent porosity even after removal of guest molecules. Third-generation CPs have flexible porous system, which changes reversibly depending on the presence and absence of guest molecules or in response to external stimuli such as light, temperature, and electric field. This generation of CP is also known as "breathing material" and "dynamic porous coordination polymers" (Comma et al. 2010).

Three important properties of MOFs which make them very interesting and useful in many applications are their crystallinity, porosity, and strong metal ligand interaction. The presence of strong metal ligand interaction leads to permanent porosity of the material. Thus, it is also possible to remove solvent molecules without structural collapse. Metal organic frameworks (MOFs) are a new development on the interface between molecular coordination chemistry and material science. MOFs are materials in which metal to organic ligand interactions yield porous coordination networks with record-setting surface areas surpassing activated carbons and zeolites. The crystal structure of MOFs is analogous to that of zeolites and other related inorganic porous materials, but there is a clear distinction between them concerning porosity and range of stability.

14.1.1 Zeolite vs. MOFs

Zeolites are microporous aluminosilicate minerals. It is a Greek word where *Zeo* means boil and *lithos* means stone. Zeolites are widely used in industry for water purification, catalysis, laundry detergents, medicine, and agriculture (Yichang et al. 2011). Zeolites have a porous structure and these are extraordinarily robust. Some of the more common mineral zeolites are analcime, clinoptilolite, natrolite, stibnite, chabazite, heulandites, and phillipsite. In other words, zeolite is an inorganic porous material having a highly regular structure of pores and chambers that allow some molecules to pass through and causes others to be either excluded or broken down. Their porosity yields internal surface areas which are relatively large, thereby facilitating a wide range of applications. Although crystalline MOFs share some common features with zeolites such as large internal surface areas and uniform pore and cavity sizes, albeit there are some important differences too. Compared to traditional inorganic zeolites whose pore sizes are normally fixed, the dynamic features of MOFs can produce materials with adjustable pore sizes. The presence of organic components in MOFs also make a noticeable difference. MOFs can be synthesized in much greater chemical variety than zeolites because of their organic component. MOFs are less stable than zeolites due to the presence of organic building blocks. Many MOFs exhibit permanent porosity, while others collapse when solvent is removed, but persistence of microporosity after solvent evacuation is essential in many applications (James 2003; Lee et al. 2009).

As stated earlier, metal organic framework (MOFs) is composed of two major components: a metal ion or cluster of metal ions and an organic molecule called linker. The organic units are mono-, di-, tri-, or tetravalent ligands. The choice of metal and linker has significant affects on the structure and properties of the MOFs. Since 1990s research into material with polymeric, sometimes porous structures based on metal ions and organic bridging ligands has increased tremendously. MOFs can be constructed from a limitless number of bridging ligands and metal ions, which allows the properties of the material to be tailored for a particular application. As shown in Fig. 14.1, depending on the vacant sites on the metal ion and the connectivity of the organic linker, 1-, 2-, and 3-dimensional polymers can be formed. The use of different organic ligands as the skeleton allows fine-tuning of the size and shape of the void which can also be functionalized with different functional groups (Basu et al. 2009).

14.1.2 Properties of MOFs

James (2003) elaborated important properties of MOFs as follows.

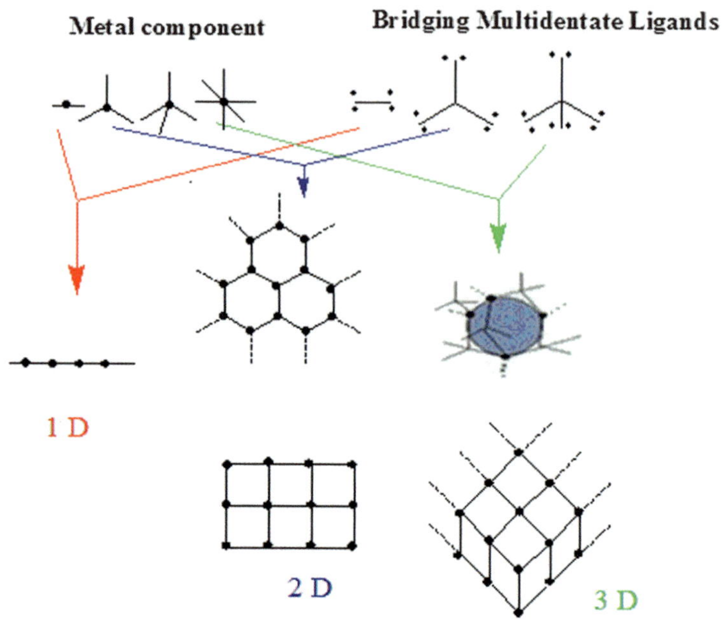

Fig. 14.1 *1D*, *2D*, and *3D* structure formation of MOFs by metal component and multidentate ligands

14.1.2.1 Very Large Surface Area

MOFs exhibit large surface area of up to 3,000 m^2/g (MOFs-177 has 4,500 m^2/g) and specific pore volume of up to 1 cm^3/g which are among the highest ever reported for any material.

14.1.2.2 Low Density

MOFs have an extremely wide-open structure in which the free space available may reach up to 90 % of the crystal volume. This results in the lowest density attained for any crystalline material (0.21 g/cm^3 for IRMOFs-16).

14.1.2.3 Stability

Due to lower strength of coordination bonds, MOFs are expected to be less thermally stable and decomposes at lower temperature. The loss of crystallinity and collapsing of network on loss of solvent guest molecule is a common observation. However, structure is regained on reexposure of guest molecule.

14.1.2.4 Chirality

MOFs are prepared in a modular or building block fashion, meaning thereby the chirality can be introduced simply by choosing chiral building blocks, most conventionally as the organic ligands.

14.1.2.5 Ion Exchange and Solubility

Many MOFs are found to undergo anion exchange to the crystals of silver-bipyridyl polymer [Ag (1) (NO_3)].

14.1.2.6 Luminescence and Nonlinear Optical Properties

Zhang et al. (2008) synthesized, characterized, and studied thermal and nonlinear optical properties of zinc/lead isophthalates luminescent metal organic frameworks.

An important feature of MOFs is that the pore size, shape, dimensionality, and chemical environment can be finely controlled by selecting building blocks, i.e., metal ion and organic linker and the networking between them. Though the wide choice of metal and infinite choice and design of ligands, a broad range of structural, magnetic, electrical, optical, and catalytic properties can be incorporated into such materials. MOFs also demonstrate molecular sieve and shape selective properties. It is also possible to modify and functionalize the organic ligands for specific applications. This unique combination of properties has triggered an extensive investigation into a number of applications.

14.2 Nomenclature of MOFs

There is no standardized nomenclature for MOFs (such as that used for zeolites). However, researchers adopt different ways for nomenclature.

- Some use descriptive name for MOFs prepared in their labs, consisting of an acronym followed by a number (which represents the chronological order of preparation), for example, MOFs-n (metal organic framework).
- Some name them on the type of structure, for example, ZIF series (zeolitic imidazole framework).
- Naming is also done on the basis of laboratory in which synthesis is done, for example, HKUST (Hong Kong University of Science and Technology).
- An alternative method is to use the empirical formula of the material, for example, $[Zn_4(O)(bdc)_3]$ (bdc- 1,4-benzenedicarboxylate).

14.3 Synthesis

MOFs synthesis is usually carried out in liquid phase either by using a pure solvent or a mixture of solvents. The commonly used method involves mixing the solutions containing metal and organic ligands. Alkaline, alkaline earth metals, transition metals, and rare earth metals have been successfully used for the synthesis of MOFs. Rigid organic molecules are generally preferred as ligands because they favor formation of crystalline, porous, stable MOFs. The nature of the solvent and ligand and the presence of other molecules affect the crystal structure of the material obtained. Thus a specific metal ligand combination can give rise to a number of different structures (polymorphism). For example, Tian et al. (2007) have prepared seven different zinc imidazolate frameworks with a general formula [Zn(im)2. xG] (im=imidazolate, G=guest molecule, x=0.2–1) by varying solvent only keeping all other parameters constant.

Synthesis of MOFs generally can be carried out by the following methods:

1. *Solvothermal* – It is basically a chemical reaction in which precursors are allowed to react in the presence of a solvent under moderate to high pressure (1–100 atm.) and temperature (100–1,000 °C) that facilitate the interaction of precursors during synthesis. Solvothermal allows to control shape, size, and crystallinity by controlling parameters such as time, temperature, solvent, and precursor. Some commonly used solvents are liquid NH_3, HF (hydrogen fluoride), HBr (hydrogen bromide), HCl (hydrogen chloride), etc. For example, $Mn_3\text{-}(C_8O_4H_4)_3(C_5H_{11}ON)_2$ was synthesized by using 1,4-dicarboxylic acid as ligand and $Mn(NO_3)\cdot 6H_2O$ as a source of metal ions (Poulsen et al. 2005).
2. *Hydrothermal* – Hydrothermal is also based on the same working principle as the solvothermal except that in hydrothermal synthesis only water can be used as a solvent. Temperature is maintained above 100 °C and below supercritical temperature of water, i.e., 374 °C.
3. *Layer-by-layer deposition* – The basic concept of this method is based on surface chemistry and solid-phase synthesis of complex polymers. This approach is based upon sequential immersion of functionalized surfaces into solution of the building blocks of the MOFs, i.e., the organic ligand and inorganic metal unit. A controlled and highly oriented growth of homogeneous films of MOFs can be obtained by this method. For the first time, layer by layer synthesis of HKUST-1 $[Cu_3(btc)_2\cdot n(H_2O)]$ was done by sequential dipping of -COOH-functionalized Au (gold) substrate in solution of $Cu_2(CH_3COOH)_2\cdot H_2O(Cu(Ac)_2)$ and 1,3,5-benzenetricarboxylic acid (H_3btc) (Zacher et al. 2009; Biemmi et al. 2007).
4. *Others* – Some other methods, e.g., microwave-assisted method, are used for fast synthesis of MOF, i.e., synthesis within minutes (Reiter et al. 2006; Taylor et al. 2008). Some other methods include microemulsion and reverse-phase microemulsion where oil in water or water in oil is used. Here, micelles formed act as nanoreactors (Reiter et al. 2006; Taylor et al. 2008). Ultrasound method (Qiu et al. 2008), electrochemical method (Mueller et al. 2006), and ionothermal method (Parnham and Morris 2007) are some other less commonly used methods for MOF synthesis.

Table 14.2 Some commonly used classes of MOF with their entities and important property

Name	Moieties	Property
MIL-101 (Materials Institute Lavoisier)	Chromium+3, terephthalate	Very high surface area
HKUST-1 (Hong Kong University of Science and Technology)	Cu+2, benzenetricarboxylate	Good thermal stability
ZIFs(ZIF-7,ZIF-8,ZIF-9) (Zeolite imidazolate framework)	Zn+2,Co+2, imidazolate	Highly stable
COF-101 (Covalent organic framework)	Carbon, oxygen, boron, hydrogen	Very low density
IRMOF-1 (Isoreticular metal organic framework)	Zn+2, benzenedicarboxylic acid	Good adsorption properties

Filtering and drying of MOFs in wet processing need to be carried out very carefully. Due to very high surface area and porosity, large amount of solvent get adsorbed. Mild conditions of temperature and pressure must be applied that may vary for individual case (Czaja et al. 2009). It is very important to control the pore size because if the pores are too large, numerous incidence of MOF lattices develop simultaneously, forming an interlocked network of numerous structures. It means network formation within large void occurs. This results in reduction in size of the individual voids. Some of the commonly used MOFs are summarized in Table 14.2.

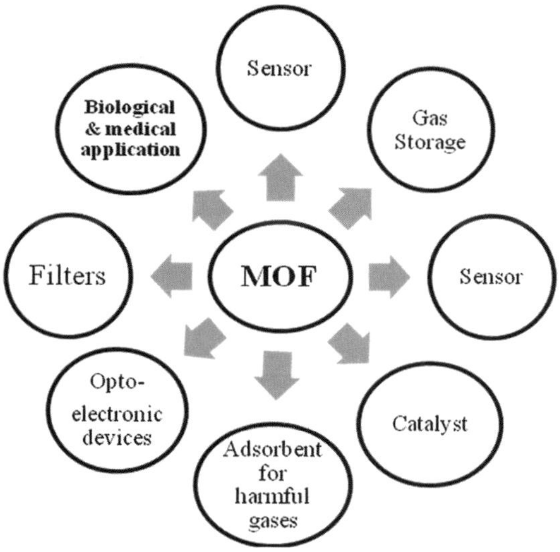

Fig. 14.2 Application of MOFs in various fields

14.4 Applications of MOFs

MOF is an interesting class of materials that finds its application in various fields (Fig. 14.2) due to their unique properties like very high surface area, nanoscale channels, and ability to functionalize according to requirement, diverse nature, and stability. As a large number of metals and ligands are available, different combination of MOFs can be obtained and can be characterized as per our needs.

14.4.1 Gas Storage

MOFs exhibit considerable potential for the adsorptive storage of both hydrogen and methane in energy applications as well as gas separation and purification for industrial applications. MOFs attract attention as materials for adsorptive gas storage because of the following properties: (1) MOFs have exceptionally high surface area and chemically tunable structures. (2) MOFs are free of dead-volume, so there is almost no loss of storage capacity as a result of space blocking by non-accessible volume. (3) MOFs have a fully reversible uptake-and-release behavior: since the storage mechanism is based primarily on the process of physiosorption. (4) There are no large activation barriers that otherwise have to be overcome when liberating the adsorbed gas.

14.4.1.1 Hydrogen Storage

Due to imminent depletion of petroleum reserves, there is considerable interest in the development of nonpetroleum energy carriers for use in transportation. Hydrogen has the potential to be an attractive option because it has a high energy content (120 MJ/kg compared to 44 MJ/kg for gasoline), produces clean exhaust product (water vapor without CO_2 or NO_x), and can be derived from a variety of primary energy sources. However, the specific energy of uncompressed hydrogen gas is very low, and considerable attention must be given to denser storage methods if hydrogen is to emerge as a serious option for the energy storage. Hydrogen molecules are stored in MOFs by adsorbing to its surface. MOF-filled gas cylinder can store more gas as compared to an empty cylinder because of the more adsorption that takes place on the surface of MOFs. The storage capacity of MOFs is limited by the liquid-phase density of hydrogen because the benefits provided by MOFs can be realized only if the hydrogen is in its gaseous state.

14.4.1.2 Methane Storage

Compressed natural gas (CNG) vehicles are already on the roads today, but methane pressure in the fuel tank of such vehicles can approach 3,600 psi (248 bars). To get economical energy output, safe storage, and transport, the DOE has set targets for methane storage under 35 bars. To achieve this pressure, a number of MOFs in the IRMOFs series, i.e., Zn-based MOFs with the same metal cluster but varying linear organic linkers, were tested for methane storage. IRMOFs-6 was found to be the highest in this series surpassing MOFs-5 and IRMOFs-3.

14.4.2 MOFs as Sensors

A wide range of topology and variable pore sizes of MOFs lead to its use for sensing specific analyte.

14.4.2.1 Recognition of Small Molecules

Many existing technologies such as mass spectroscopy and chemiluminescence lack the combination of selectivity, sensitivity, portability, and low cost. The tailorable nanoporosity and very high surface area of MOFs make them ideal candidate for recognizing analyte in sensing applications. MOFs contain functional pores to direct their specific and unique recognition of small molecules through several types of interactions, e.g., van der Waals interactions of the framework surface with the substrate, metal–substrate interactions, and hydrogen bonding of the framework surface with the substrate. We can functionalize the pores to direct specific recognition of small molecules, thus emerging porous MOFs serve as novel functional materials for gas storage, separation, and sensing. For example, MOF-76 is used for recognizing a variety of anions (Chen et al. 2008).

14.4.2.2 Chemical Vapors and Gases

MOF-based devices have been developed for sensing applications. Impedimetric humidity sensors have been designed for gas atmosphere of O_2, CO_2, C_3H_8, NO, and H_2 and hydrophilic chemicals like methanol and ethanol (Achmann et al. 2009). ZIF-8 has been successfully used for the sensing of gases, for example, sensing of hexane and cyclohexane (sterically more demanding) based on the pore size (Lu and Hupp 2010).

14.4.2.3 Pesticides

A new lanthanum-based MOF has been developed and characterized for the determination of range of pesticides such as malathion, methyl parathion, and pyrimicarb from lettuce (Barreto et al. 2010). A biphenyltetracarboxylic acid-based MOF has been designed for adsorption of MOF and detection via stripping voltammetric analysis (Wen et al. 2010). Thin film of doped Basolite has been found quite effective as sensor for commonly used pesticide Basolite (Kumar et al. 2012).

14.4.3 MOFs as Catalyst

Over the past 10 years, the notion of metal organic frameworks as catalysts has advanced from being a largely hypothetical one to a fledgling "real" application encompassing more than two dozen experimental examples. Catalytic MOFs can be considered as the analogs of enzymes due to their potential ability in multi-catalyst reactions and high reactivity due to strong metal ligand interaction. Tremendous research is being done in this area. For example, Shultz et al. (2009) successfully synthesized ZnPO-MOFs, a highly porous pillared paddlewheel MOFs, from $Zn(NO_3)\cdot6H_2O$, $DPyDPhF_5Por$ (5,15-dipyridyl-10,20-bis(pentafluorophenyl)porphyrin) and the ligand H_4TCPB (1,2,4,5-tetrakis(4-carboxyphenyl)benzene) and then used it for the catalysis of an intermolecular acyl-transfer reaction (Lee et al. 2009).

14.4.4 Adsorbent for Harmful Gases

IRMOFs have been used as selective adsorbent for harmful gases such as sulfur dioxide, ammonia, chlorine, benzene, and carbon monoxide. MOFs show much higher potential in the adsorption of harmful gases as compared to activated carbon (Britt et al. 2008). Carbon dioxide emission in the atmosphere always remains a growing concern. Currently, the largest single point sources of CO_2 emission are power plants that produce streams of flue gas, exhausted combustion smoke with CO_2 concentrations of about 15 % at 1 atm. Secondly, combustion also adds to the CO_2 concentration. While storage of CO_2 is not much of a challenge, the separation of CO_2 represents a significant problem that must be addressed by the development of high CO_2 affinity materials before any thought of geologic sequestration can be effectively implemented. The use of MOFs-74, a Zn-based material, and its Co-, Ni-, and Mg-based analogs in the uptake of CO_2 at low pressure has shown significant results. The Mg-based analog of MOFs-74 (Mg/DOBDC) was found to uptake *ca.* 35 wt% CO_2 at 1 atm. and room temperature. This value is significantly higher than that of any other physiosorptive material under the same conditions including zeolites (Britt et al. 2008).

14.4.5 MOFs as Optoelectronic Devices

MOFs have potential to be used as optoelectronic materials. It includes use of MOFs in photovoltaic solar cells and electroluminescence devices. The presence of inorganic semiconductor quantum entities in close contact with organic molecules offers their potential application in this field, as organic molecules can serve to activate the semiconductor quantum dots upon external stimulus (Xamena et al. 2007). The network formation in galena (PbS) by organic and inorganic entities increases both the band gap and defect reaction energies, thus favoring utility of hybrid material in solar cell applications as compared to galena (Walsh 2010).

14.4.6 MOFs as Filters

Conventionally, distillation is generally used for separation of mixtures, but these processes are energy intensive and sometimes not economical. Metal organic frameworks are regarded useful for the same purpose due to high stability in different environments and high selectivity. Nanofiltration membrane based on metal organic framework is being incorporated as fillers in polydimethylsiloxane (PDMS) membranes. These membranes are then used in solvent-resistant nanofiltration (SRNF). The MOF-filled membranes showed increased permeability but lower retention as compared to unfilled membranes. It had been successfully used in case of separating Rose Bengal from isopropanol (Basu et al. 2009).

14.4.7 Biological and Medical Applications

Important aspect for designing material for biological application is their stability and toxicity, and MOFs can be designed by maintaining balance between the two. The wide range of chemical composition and the possibility of tailoring morphology and surface chemistry of frameworks make it possible to formulate toxicologically acceptable formulations. Moreover, very high porosity and tunable pore sizes and pore volumes ensure high loading capacity. Many compounds with same chemical composition as MOFs have been approved for medical use as prescribed drugs. For example, iron fumarate having the same chemical composition as MIL-88A is used as oral iron supplement (Hinks et al. 2010). But the practical use of MOF in biomedical application is still in its infancy. One of the major challenge is to control size of MOFs for producing stable formulations preventing agglomeration and amenable to different administration routes. Toxicity, administration route, behavior of MOF in biological system, bioadhesion, etc. are some other issues. Some of the biological applications of MOFs are summarized here.

14.4.7.1 MOFs as Drug Delivery Carrier

To overcome the drawbacks such as low drug loading capacity found in inorganic delivery materials (zeolites) and too rapid delivery "burst effect" offered by organic routes, highly porous MOFs have been used as new drug carriers (Lee et al. 2009; Horcajada et al. 2009). Biocompatibility and toxicity are another major issues while considering novel drug delivery carriers. MOFs can be considered as optimal drug delivery material due to wide range of diversity, tunable pore sizes, and possibility of functionalization. Organic ligands provide biocompatibility and good drug loading capacity, whereas inorganic entity helps in controlled release. In tests involving ibuprofen as model drug system, MOFs showed very high drug capacity of 1.4 g of drug/g of MOF. MIL (Materials of Institute Lavoisier) family of MOF is one of the first successfully used MOF for drug delivery (Ferey et al. 2005). Sometimes nanoencapsulation is done to decrease toxicity and other side effects.

14.4.7.2 BioMOFs

BioMOFs also known as bioactive MOF are constructed by using biologically therapeutically active molecules as linkers. These are MOFs with permanent porosity constructed with rigid biomolecules such as peptides, drugs etc., which act as building blocks and are stable as well as crystalline in biological buffers for several weeks. In this approach, a bioactive molecule is used as a linker or a bioactive metal is used as an inorganic part to avoid toxicity concerns associated with nondegradable linkers and metal after the release of drug. Under simulated physiological conditions MOF degrade and release the drug. As many drugs exhibit ligands such as carboxylates, phosphates, and amines, this method can be used for development of fully biocompatible drug delivery systems (McKinlay et al. 2010).

14.4.7.3 MOFs in Imaging

Ferey et al. (2005) used MIL family of MOFs in magnetic resonance imaging due to their enhanced enormous porosity and very large pores size. Iron carboxylate modification of MIL-101 and loading of anticancer drug has been done to improve the capabilities of MOF as contrast agent and in drug delivery (Kathryn et al. 2009; McKinlay et al. 2010). Nanoscale Gd MOF nanoparticles with composition $Gd(1,4\text{-}BDC)15(H_2O)_2$ have been used as a contrast agent in MRI (Rowe et al. 2009; Rocca and Lin 2010). However, Gd-based MOF would offer high toxicology during clinical studies. Therefore, silica-coated Mn-based MOFs were tried and found to show reduced toxicity as compared to Gd-based MOF (Taylor et al. 2008). Achieving good MRI contrast enhancement by paramagnetic metal-based MOF is not a big deal, but suitable combination of biocompatibility and exceptional MRI contrast enhancement capabilities are still a challenge.

14.4.7.4 MOF in Sensing

MOFs which can monitor oxygen level, glucose level, iron, and other biomolecules along with presence of toxins are important in biological applications. Fe–BTC has shown significant sensing capacity for ethanol, methanol, and water (Achmann et al. 2009).

14.4.7.5 NO Storage

NO is a crucial gas involved in various biological processes such as neuronal, immune, and vascular system. It is also a neurotransmitter and neuromodulator (Bose et al. 2009). Supply of NO from a storage material is important for a number of applications including in vitro and in vivo antibacterial and antithrombotic applications. Different Co-, Ni-, and Cr-based MOFs were first studied for storing and delivering bioactive NO gas and were found highly efficient (McKinlay et al. 2008). Use of polymers and zeolites as a storage and supply material for NO has been limited due to the release of associated carcinogenic side products. HKUST-1 has shown adsorption capacity of 9 mmol NO/g MOF at 298 K and 1 bar and releases 1 μmol NO/g MOF when it comes in contact with water (Xiao et al. 2007). CPO-27-Co and CPO-27-Ni are found to adsorb twofold amount of NO and released 7,000 times NO as compared to HKUST-1 (McKinlay et al. 2008).

14.4.8 Important Aspects for Biological Applications

14.4.8.1 Toxicity

Besides the advantages of using MOFs, there are some issues such as toxicity which should be considered before implementing MOFs in biological applications. To control toxicity, nontoxic or less toxic metal ion and organic linkers should be used while formulating MOFs. Iron seems to be most promising candidate for designing MOFs for biological applications with high LD_{50} value of 30 g/kg (Eines 2011). Toxicity for the same metal ion may vary depending upon organic ligand used. Carboxylate and phosphate linkers are expected to have suitable toxicity with optimal LD_{50} values for biological applications. The toxicity studies of MOFs have attracted much attention in recent years and need to be explored further.

14.4.8.2 Biodegradation

Degradation of MOF under physiological conditions should be taken into consideration (Horcajada et al. 2010). For biological applications, stability of MOF is an important criterion as we do not want its intrinsic accumulation as well as too quick degradation. MOF should remain intact until it is needed and functioning, and after that it should be metabolized. Degradation capability may depend upon metal and organic ligand used, its size, physiological conditions, etc. Coating of MOF with some gel or silica can enhance its life (Taylor et al. 2008).

14.4.8.3 Size Requirement

Whenever it comes to biological applications, bulky MOFs in size range of micro- to millimeters are not required (Spokoyny et al. 2009). Depending upon the mode of administration, size needs may vary, but for most of the applications, nanosized MOFs commonly called as nano-MOFs (NMOF) are found to be suitable. Because of their small size, there will be less damage due to impact; in addition to that small size enhances stability by forming more homogenous solutions inside the body.

14.5 Functionalization/Surface Modification

Surface modification is an important aspect of working with metal organic framework. Surface of MOF can be functionalized with organic molecules for specific application purpose. PEG (polyethylene glycol) is a commonly used hydrophilic polymer used for coating MOFs to control interaction of MOF with biological media and to increase blood circulation for minutes up to several hours (Horcajada et al. 2010). For sensing of DPA (dipicolinic acid), coating of NMOFs having general composition $Ln(BDC)1.5(H_2O)2$ where Ln = Eu3+, Gd3+ Tb3+ and BDC = 1,4benzenedicarbgoxylate with silica shell has enhanced water dispersibility and biocompatibility of MOF (Rieter et al. 2007). α-Hopeite (poly-hydrate zinc phosphate) microparticles act as nucleation seeds (in both liquid and solid state) and also functionalize MOF core and not the outer surface as in conventional methods for functionalization of MOFs (Falcaro et al. 2011). Surface functionalization can also be done for imaging purpose by modifying surface of silica-coated MOF with an in vitro imaging agent (1,3,5,7-tetramethyl-4,-4-difluoro-8-bromomethyl-4-bora-3a,4a-diaza-s-indacene) (Taylor et al. 2009) (Fig. 14.3).

14.6 Conclusion

Metal organic frameworks are new class of smart functional material having crystalline nature and are generally porous coordination polymers. The presence of unbelievable level of porosity, surface area, micro-/nano-range pore size, and wide chemical organic inorganic composition and

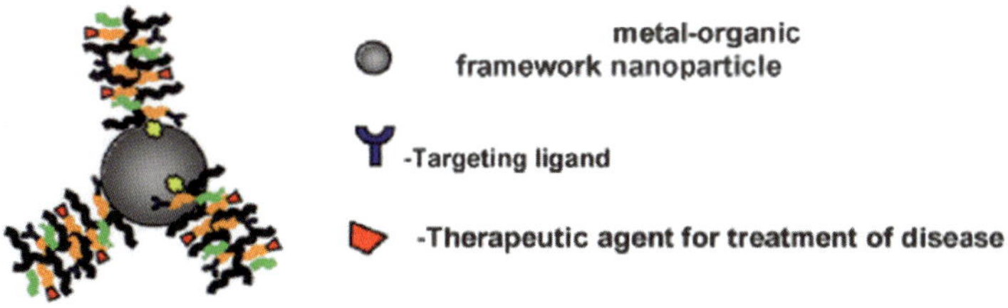

Fig. 14.3 Functionalization and loading of MOF by therapeutic agent (Rowe et al. 2009)

ability to functionalize brought these materials to the attention of researchers. In applications involving adsorptive separation, MOF offers much advantage over zeolites and activated carbon as process utilizing MOFs becomes more efficient, energy saving, and environmentally benign. It not only offers very high surface area but also provides size/shape and selectivity which are very important for separation and catalysis purposes. Biocompatible MOF can be synthesized for medical application. Porous, nonporous, luminescent, and nonluminescent MOFs can be synthesized by varying metal component and organic ligand for different applications. In applications like drug delivery and in vivo imaging, MOFs are far better than carbon nanotubes (CNTs) and zeolites as till today there is dispute about the toxic effects of CNTs in biological systems. Due to wide range of available metal entity and organic ligand, a large number of MOFs can be designed according to our needs having different properties. Research on biomedical applications of MOFs is gaining momentum day by day, and this new class of porous material is likely to replace other traditional materials for drug/therapeutic delivery and storage applicants in biological systems. When we talk about biological applications, the main concern is toxicity and stability, and more research has to be done in this area.

The major limitation of MOFs is about its stability as the organic component may degrade at very high temperature and in adverse conditions of pH, chemicals, and pressure. But still ZIFs (zeolite imidazolate frameworks) are found to be quite stable at high temperature range (550 °C for ZIF-8) and chemically resistant to boiling alkaline water and organic solvent. In the past two decades, much of the research work has been done, and much is still to be conceded. Furthermore, scores of the applications and properties of MOFs are still to be explored.

References

Achmann S, Hagen G, Kita J, Malkowsky IM, Kiener C, Moos R (2009) Metal organic frameworks for sensing applications in gas phase. Sensors 9:1574–1589

Barreto AS, Silva RL, Silva SCGS, Rodrigues MO, Simone CA, Sa GF, Junior SA, Navickiene S, Mesquita ME (2010) Potential of a metal organic framework as a new material for solid phase extraction of pesticides from lettuce (*Lactuca Sativa*) with analysis by gas chromatography mass spectrometry. J Sep Sci 33:3811–3816

Basu S, Michael M, Odena AC, Alaerts L, Vos DED, Vankelecom IFJ (2009) Solvent resistant nanofiltration (SRNF) membranes based on metal organic frameworks. J Membr Sci 344:190–198

Biemmi E, Scherb C, Bein T (2007) Oriented growth of the metal organic framework $Cu_3(BTC)_2(H2O)_3$.xH2O tunable with functionalized self-assembled monolayers. J Am Chem Soc 129:8054–8055

Bose M, Olivan B, Laferrere B (2009) Stress and obesity: the role of the hypothalamic-pituitary-adrenal axis in metabolic disease. Curr Opin Endocrinol Diabetes Obes 16:340–346

Britt D, Tranchemontagne D, Yaghi OM (2008) Metal-organic framework with high capacity and selectivity for harmful gases. PNAS 105:11623–11627

Chen B, Wang L, Zapata F, Qian G, Lobkovsky EB (2008) A luminescent metal organic framework for the recognition and sensing of anions. J Am Chem Soc 130:6718–6719

Comma A, Garcia H, Laore FX (2010) Engineering metal organic frameworks for heterogeneous catalysis. Chem Rev 110:4506–4655

Czaja AU, Trukhan N, Muller U (2009) Industrial applications of metal organic frameworks. Chem Soc Rev 38:1284–1293

Eines EG (2011) Preparation and characterization of metal organic framework for biological applications. Dissertation, RUHR University, Bochum

Falcaro P, Hill AJ, Nairn KM, Jasieniak J, Mardel JI, Bastow TJ, Mayo SC, Gimona M, Gomez D, Whitfield

HJ, Ricco R, Patelli A, Marmiroli B, Amenitsch H, Colson T, Villanova L, Buso D (2011) A new method to position and functionalize metal-organic framework crystals. Nat Commun 237(2):1–8

Ferey G, Mellot-Draznieks C, Serre C, Millange F, Dutour J, Surble S, Margiolaki I (2005) A chromium terephthalate based solid with unusually large pore volume and surface area. Science 309:2040–2042

Hinks NJ, McKinlay AC, Xiao B, Wheatley PS, Morris RE (2010) Metal organic frameworks as NO delivery materials for biological applications. Microporous Mesoporous Mater 129:330–334

Horcajada P, Chalati T, Serre C, Gillet B, Baati T (2009) Porous metal–organic-framework nanoscale carriers as a potential platform for drug delivery and imaging. Nat Mater 9:172–178

Horcajada P, Chalati T, Serre C, Gillet B, Sebrie C, Baati T, Eubank JF, Heurtaux D, Clayette P, Kreuz C, Chang JS, Hwang YK, Marsaud V, Veronique B, Phuong NC, Cynober L, Gil S, Ferey G, Couvreur P, Gref R (2010) Porous metal-organic-framework nanoscale carriers as a potential platform for drug delivery and imaging. Nat Mater 9(2):172–178

Hoskins BF, Robson R (1990) Design and construction of a new class of scaffolding like materials comprising infinite polymeric frameworks of 3-D linked molecular rods. J Am Chem Soc 112:1546–1554

James S (2003) Metal organic framework. Chem Soc Rev 32:276–288

Kathryn ML, Taylor P, Rocca JD, Zhigang X, Sylvie T, Wenbin L (2009) Post synthetic modifications of iron carboxylate nanoscale metal organic frameworks for imaging and drug delivery. J Am Chem Soc 131:14261–14263

Kumar P, Kumar P, Deep A, Bharadwaj LM (2012) Doped zinc organic frameworks for sensing of pesticide. Adv Mater Res 1543:488–489

Lee JY, Farha OK, Roberts J, Scheidt KA, Nguyen ST, Hupp JT (2009) Metal–organic framework materials as catalysts. Chem Soc Rev 38:1250–1259

Lu G, Hupp JT (2010) Metal-organic frameworks as sensors: a ZIF-8 based Fabry-Perot device as a selective sensor for chemical vapors and gases. J Am Chem Soc 132:7832–7833

McKinlay AC, Xiao B, Wragg DS, Wheatley PS, Meegson LL, Morris RE (2008) Exceptional behavior over the whole- adsorption- delivery cycle for NO for porous metal organic frameworks. J Am Chem Soc 130:10440–10444

McKinlay AC, Morris E, Horcajada P, Ferey G, Gref R, Couvreur P, Serre C (2010) BioMOFs: metal – organic frameworks for biological and medical applications. Angew Chem 49:6260–6266

Mueller U, Schubert M, Teich F, Puetter H, Arndt KS, Pastre J (2006) Metal organic frameworks prospective industrial applications. J Mater Chem 16:626–636

Parnham ER, Morris RE (2007) Ionothermal synthesis of zeolite, metal organic framework and inorganic organic hybrids. Acc Chem Res 40(10):1005–1013

Poulsen D, Bentien A, Chevalier M, Iversen B (2005) Synthesis, physical properties, multitemperature crystal structure, and 20 K synchrotron X-ray charge density of a magnetic metal organic framework structure, $Mn_3(C_8O_4H_4)_3(C_5H_{11}ON)_2$. J Am Chem Soc 127:9156–9166

Qiu LG, Li ZQ, Wu Y, Wang W, Xu T, Jiang X (2008) Facile synthesis of nanocrystals of microporous metal-organic framework by an ultrasonic method and selective sensing of organoamines. Chem Commun 31:3642–3644

Reiter WJ, Taylor KML, An H, Lin W, Lin W (2006) Nanoscale metal organic frameworks as potential multimodal contrast enhancing agents. J Am Chem Soc 128(8):9024–9025

Rieter WJ, Taylor KML, Lin W (2007) Surface modification and functionalization of nanoscale metal organic frameworks for controlled release and luminescence sensing. J Am Chem Soc 129:9852–9853

Rocca JD, Lin W (2010) Nanoscale metal–organic frameworks: magnetic resonance imaging contrast agents and beyond. Eur J Inorg Chem 24:3725–3734

Rowe MD, Thamm DH, Kraft SL, Boyes SG (2009) Polymer modified gadolinium metal organic framework nanoparticles used as multifunctional nanomedicines for the targeted imaging and treatment of cancer. Biomacromolecules 10:983–993

Shultz AM, Farha OK, Hupp JT, Nguyen ST (2009) A catalytically active, permanently microporous MOF with metalloporphyrin struts. J Am Chem Soc 131:4204–4205

Spokoyny AM, Kim D, Sumrein A, Mirkin CA (2009) Infinite coordination polymer nano- and microparticle structures. Chem Soc Rev 38:1218–1227

Taylor KML, Rieter WJ, Lin W (2008) Manganese based nanoscale metal organic framework for magnetic resonance imaging. J Am Chem Soc 130: 14358–14359

Taylor KML, Rocca JD, Xie Z, Tran S, Lin W (2009) Postsynthetic modifications of iron-carboxylate nanoscale metal organic frameworks for imaging and drug delivery. J Am Chem Soc 131:14261–14263

Tian YQ, Zhao YM, Chen ZX, Zhang GN, Weng LH, Zhao DY (2007) Design and generation of extended zeolitic metal-organic frameworks (ZMOFs): synthesis and crystal structures of zinc(II) imidazolate polymers with zeolitic topologies. Chemistry 13: 4146–4154

Walsh A (2010) Defect process in a PbS metal organic framework: a quantum confined hybrid semiconductor. J Phys Chem Lett 1:1284–1287

Wen LL, Wang F, Leng XK, Wang CG, Wang LY, Gong JM, Li DF (2010) Efficient detection of organophosphate pesticide based on a metal-organic framework derived from biphenyltetracarboxylic acid. Cryst Growth Des Commun 10:2835–2838

Xamena FXL, Corma A, Garcia H (2007) Applications for metal organic frameworks (MOFs) as quantum dot semiconductors. J Phys Chem 111:80–85

Xiao B, Wheatley PS, Zhao XB, Fletcher AJ, Fox S, Rossi AG, Megson IL, Bordiga S, Regli L, Thomas KM, Morris RE (2007) High capacity hydrogen and nitric oxide adsorption and storage in a metal organic framework. J Am Chem Soc 129:1203–1209

Yaghi OM, Richardson DA, Li G, Davis CE, Groy TL (1995) Open-framework solids with diamond-like structures prepared from clusters and metal-organic building blocks. Mater Res Soc Symp Proc 371:15–19

Yaghi OM, O'Keeffe M, Ockwig NW, Chae HK, Eddaoudi M, Kim J (2003) Reticulate synthesis and design of new materials. Nature 423:705–714

Yichang P, Yunyang L, Gaofeng Z, Lan Z, Zhiping L (2011) Rapid synthesis of imidazolate framework-8 (ZIF-8) nanocrystals in an aqueous system. Chem Commun 47:2071–2073

Zacher D, Shekhah O, Woll C, Fisher RA (2009) Thin films of metal-organic frameworks. Chem Soc Rev 38:1418–1429. Guest editor Jeffery Long and Omar yaghi

Zhang L, Qln YY, Li ZJ, Lin QP, Cheng JK, Zhang J (2008) Topology analysis and non-linear-optical-active properties of luminescent metal organic framework materials based on zinc/lead isophthalates. Inorg Chem 47:8286

Potential Bioherbicides: Indian Perspectives

15

K.R. Aneja, V. Kumar, P. Jiloha, M. Kaur, C. Sharma, P. Surain, R. Dhiman, and A. Aneja

Abstract

Weeds, one of the major kinds of pests, continue to cause major problems in agriculture throughout the world, reducing yield and quality of crops by competing for water, nutrients, and sunlight, essential for vigorous crop growth. Due to the recent trends in environmental awareness concerning the side effects of herbicides, public pressure is mounting to force industry to develop safer, more environmental friendly approaches for controlling weeds. Microbial-based pesticides, referred to as bioherbicides, for the management of weeds offer such an approach. In majority of the cases, the bioherbicides include fungal organisms as the active ingredients; therefore, the term mycoherbicide has often been used interchangeably with bioherbicide. Considerable progress has been made during the past four decades in the use of fungi as biocontrol agent of weeds. There has been a great number of naturally occurring fungal strains researched for possible use as mycoherbicides, but only a small proportion have been developed to commercial products. Currently, a total of 17 mycoherbicides (8 in the USA, 4 in Canada, 2 in South Africa, and 1 each in the Netherlands, Japan, and China) have been registered around the globe. The advancement of formulation techniques is of paramount importance to the continued development of mycoherbicides. It is also essential to continue intensive screening programs for the selection of fungal pathogens, especially hemibiotrophs, if mycoherbicides are to become a viable component of

K.R. Aneja (✉)
Vaidyanath Research, Training and Diagnostic Centre, Vaidyanath Research and Healthcare Complex, Urban Estate, Shopping centre, Sector-3, Kurukshetra-136 118, Haryana (India)
e-mail: anejakr@yahoo.ca

V. Kumar • P. Jiloha • M. Kaur • P. Surain • R. Dhiman
Department of Microbiology, Kurukshetra University, Kurukshetra, India

C. Sharma
Guru Nanak Khalsa College, Yamuna Nagar, Haryana, India

A. Aneja
University Health Centre, Kurukshetra University, Kurukshetra, India

R.K. Salar et al. (eds.), *Biotechnology: Prospects and Applications*,
DOI 10.1007/978-81-322-1683-4_15, © Springer India 2013

integrated weed management in the future. Recent trend is the application of several host-specific fungal pathogens in a bioherbicide mixture as a multicomponent bioherbicide system for simultaneous, broad-spectrum weed biocontrol. Many microbes, although they rarely have an effect under natural conditions, can be developed as effective bioherbicides. At Kurukshetra, during the last 30 years, searches for fungal BCAs have been made on 26 weeds (7 aquatic and 19 terrestrial), and a number of them have been evaluated for their biocontrol potential against the notorious weeds of this region. The mycoherbicides which have been commercialized and are in the process of commercialization have been discussed in detail in this chapter.

15.1 Introduction

A weed is "a plant that has the potential to have a detrimental effect on economic, social, and conservation values." Weeds, one of the major kinds of pests, continue to cause major problems in agriculture throughout the world, reducing yield and quality of crops by competing with indigenous plants for space, moisture, nutrients, and light, but some weeds have little impact on the overall health and function of the community, while others have a dramatic impact, with the ability to spread rapidly and cause damage such as congress grass (Aneja 2009). Li et al. (2003) reported that out of about 30,000 species of widely distributed weeds, 1,800 species cause yield loss by about 9.7 % of total crop production every year in the world. Weeds can reduce crop yield by as much as 12 % which results to $32 billion losses as a whole (Chutia et al. 2007).

Management of weeds is a necessary but expensive challenge. Herbicides, chemicals that kill or suppress plants, annually account for over $14 billion (Kiely et al. 2004). Chemical weed control is not an ideal option in organic cropping systems. Recent trends in environmental awareness concerning the side effects of chemical pesticides in general have prompted researchers, plant pathologists, and microbiologists to develop novel, sustainable, and protective weed control strategies. Weed management in organic vegetable production systems must involve the use of different strategies, all with the goal of achieving economically acceptable weed control and crop yield. Biological weed control using biological control agents (BCAs), particularly fungal pathogens, offers us a tremendous opportunity to provide agriculture with effective tools for abundant crop production while minimizing impacts on health and the environment (Panetta 1992; Aneja 2009). The term bioherbicide is used to refer to herbicide based on natural biological agents such as fungi, bacteria, viruses, protozoans, and nematodes. When the organism used is a fungus, the product is termed a mycoherbicide. However, the use of pathogens other than fungi as bioherbicides is limited. Therefore, the term "mycoherbicide" has often been used interchangeably with "bioherbicide" (Watson 1989). Hemibiotrophic fungi combine both biotrophic and necrotrophic strategies. During the initial biotrophic phase host's immune system and cell death are actively suppressed allowing invasive hyphae to spread throughout the infected plant tissue. This phase is followed by a necrotrophic phase during which toxins are secreted by the pathogen to induce host cell death (Koeck et al. 2011). *Colletotrichum lindemuthianum* and certain varieties of *C. gloeosporioides* are well-known examples of hemibiotrophs.

Despite many economic, social, and environmental benefits ascribed to biological control, it is reasonable to ask why more mycoherbicide products are yet to become widely available in the market and acceptable by the farmers. This review will provide an update on the status of mycoherbicides that have been commercialized and are in the pipeline for commercialization.

15.2 Strategies for Biological Control

The strategies used to control weeds with plant pathogens often fall into two categories:

the classical or inoculative strategy and the mycoherbicidal or inundative strategy.

15.2.1 The Classical or Inoculative Strategy

The classical strategy involves the release of a coevolved pathogen into an exotic environment where the target is an alien or non-indigenous species. Usually the pathogen is released over a small part of the target weed population using non-formulated inoculums, and the control of the target weed is dependent upon self-perpetuation and natural dispersal of the pathogen. The first successful example of classical strategy is the introduction of *Puccinia chondrillina* from Mediterranean (South Europe) to control *Chondrilla juncea* (Skeleton weed) in Australia (Quimby 1982; Aneja and Mehrotra 1996). Other notable examples are given in Table 15.1.

15.2.2 The Mycoherbicidal or Inundative Strategy

Where an agent that occurs as a pest in its native range is cultured, mass-produced, registered, marketed, and applied inundatively like chemical pesticides come in the inundative strategies. Mycoherbicides are developed from fungal pathogens that normally incite diseases at endemic levels in specific weed population, increased in fermentation tanks, and applied in the form of spray that uniformly kill or suppress weed growth without harm to crop or any nontarget species in the environment (Aneja 2009). Strategic framework for evaluation and development of mycoherbicides is represented below:

Table 15.1 Successful examples of control of weeds through classical approach

Weed	Bio-agent	Kind of bio-agent	Reporting country
Chondrilla juncea	*Puccinia chondrillina*	Plant pathogen	Australia
Cyperus rotundus	*Bactra verutana*	Shoot-boring moth	India, Pakistan, USA
Eupatorium riparium	*Entyloma compositarum*	Plant pathogen	USA
Hydrilla verticillata	*Hydrellia pakistanae*	Shoot fly	USA
Orobanche cernua	*Sclerotinia* spp.	Plant pathogen	USA
Parthenium hysterophorus	*Puccinia abrupta* var. *partheniicola*	Plant pathogen	Mexico
Parthenium hysterophorus	1. *Zygogramma bicolorata*	Leaf-eating beetle	Mexico
	2. *Epiblema strenuana*	Stem-galling insect	Australia
	3. *Conotrachelus* sp.	Stem-galling insect	Australia
Rumex spp.	1. *Uromyces rumicis*	Plant pathogen	USA
	2. *Gastrophysa viridula*	Beetle	USA
Tribulus terrestris	*Microlarinus lareynii* and *M. lypriformis*	Pod weevil	USA

Initiation of field Initiation of field surveys to search for suitable fungal pathogens

↓

Identify highly pathogenic (or virulent) isolates

↓

Develop an efficient way of mass production of inoculums followed by technology for drying and formulation of drying spores ensuring their stability and shelf life.

↓

Study the disease cycle to determine the optimum conditions for infection and disease development

↓

Check the host specificity and toxicity tests for human safety

↓

Develop an appropriate formulation and application technology

↓

Screening trials for biocontrol potential in laboratory and green house

↓

Small scale field tests on local experiment station

↓

Large scale test at different locations in cooperation with the farmers

↓

Obtain registration of the product

The reasons why bioherbicides are being preferred over chemical herbicides are as follows:

- They are inherently less harmful than conventional pesticides.
- Effective in very small quantities and often decompose very quickly.
- They are nontoxic to humans, domestic animals, and plants.
- They are economically feasible, safe, and non-pathogenic to nontarget organisms.
- When used as a component of IPM programs, biopesticides can greatly decrease the use of conventional pesticides, while crop yields remain high.
- Fungi used as mycoherbicides are easy to isolate, mass produce, identify, grow, and manipulate.

Mycoherbicide research to control agricultural and environmental weeds began in the 1940s. The earliest experiments simply involved moving indigenous fungi between populations of target weeds (e.g., the fungus *Fusarium oxysporum* used against prickly pear cactus and *Opuntia ficus-indica* in Hawaii, before the release of the *Cactoblastis cactorum* moth). A total of 17 mycoherbicides have been registered, of these, 8 registered in the USA, 4 in Canada, 2 in South Africa, and 1 each in the Netherlands, Japan, and China (Table 15.2) (Aneja 2009; Dagno et al. 2012), as discussed below.

15.2.2.1 *Acremonium diospyri*

Acremonium diospyri (Fig. 15.1), commonly called the persimmon wilt fungus, is a hyphomycetous

Table 15.2 Examples of commercial bioherbicides and type of formulation used

S. No.	Target weed	Fungus	Product name	Year of registration	Formulation type
Liquid formulations					
1.	Persimmon (*Diospyros virginiana*) trees in rangelands	*Acremonium diospyri*	*Acremonium diospyri*	1960	Conidial suspension
2.	Dodder (*Cuscuta chinensis* and *C. australis*) in soybeans	*Colletotrichum gloeosporioides* f. sp. *cuscutae*	Lubao	1963	Conidial suspension
3.	Milkweed vine (*Morrenia odorata*)	*Phytophthora palmivora* (*P. citrophthora*)	DeVine[R]	1981	Liquid spores suspension
4.	Yellow nutsedge (*Cyperus esculentus*)	*Puccinia canaliculata*	Dr. BioSedge	1987	Emulsified suspension
5.	Turfgrass (*Poa annua*) in golf courses, *Acacia* sp.	*Cylindrobasidium laeve*	Stumpout™	1997	Liquid (oil) suspension
6.	Woody plants blackberry weed (*Prunus serotina*)	*Chondrostereum purpureum*	BioChon™	1997	Mycelial suspension in water
7.	*Hakea gummosis* and *H. sericea* in native vegetation	*Colletotrichum acutatum*	Hakatak	1999	Conidial suspension
8.	Deciduous tree species	*Chondrostereum purpureum*	MycoTech™ Paste	2004	Paste
9.	Alder, aspen, and other hardwoods	*Chondrostereum purpureum*	Chontrol™ (Ecoclear™)	2004	Spray emulsion and paste
10.	Dodder species	*Alternaria destruens*	Smolder[R]	2008	Conidial suspension
11.	Soda apple (*Solanum viarum*)	Tobacco mild green mosaic virus	SolviNix™	2009	Foliar spray suspension
Solid formulations					
1.	Northern jointvetch (*Aeschynomene virginica*)	*Colletotrichum gloeosporioides* f. sp. *aeschynomene*	Collego™ LockDown™	1982	Wettable powder
2.	Sicklepod and coffee senna (*Cassia* spp.)	*Alternaria cassiae*	Casst™	1983	Solid
3.	Water hyacinth (*Eichhornia crassipes*)	*Cercospora rodmanii*	ABG-5003	1984	Wettable powder
4.	Velvetleaf (*Abutilon theophrasti*)	*Colletotrichum coccodes*	Velgo[R]	1987	Wettable powder
5.	Round-leaved mallow (*Malva pusilla*)	*Colletotrichum gloeosporioides* f. sp. *malvae*	BioMal[R]	1992	Mallet wettable powder
6.	*Hakea gummosis* and *H. sericea* in native vegetation	*Colletotrichum acutatum*	Hakatak	1999	Granular (dry conidia)
7.	Dyers wood (*Isatis tinctoria*) in farms, rangeland, waste areas, and roadsides	*Puccinia thlaspeos*	Woad Warrior	2002	Powder
8.	Dandelion (*Taraxacum officinale*) in lawns/turf	*Sclerotinia minor*	Sarritor	2007	Granular

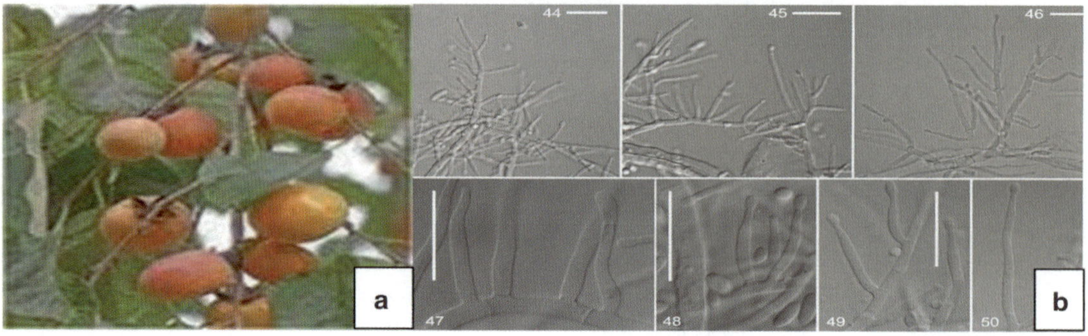

Fig. 15.1 (**a**) Persimmon tree; (**b**) *Acremonium diospyri*, unicellular hyaline conidia borne on conidiogenous cells arising from hyphae

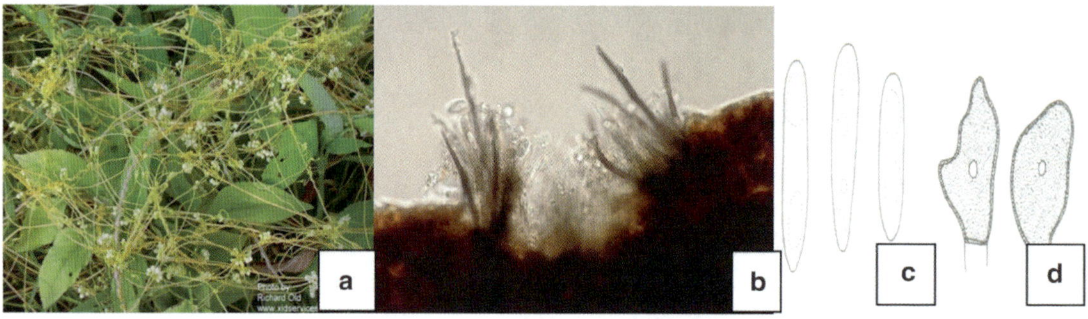

Fig. 15.2 (**a**) Dodder *(Cuscuta chinensis*); (**b–d**) *Colletotrichum gloeosporioides* f. sp. *cuscutae*; (**b**) acervulus with conidiophores, conidia, and setae; (**c**) unicellular hyaline conidia; (**d**) appressoria

fungus which had been used as a mycoherbicide since 1960 to control persimmon trees (*Diospyros virginiana*) (Fig. 15.1) in rangeland of south central Oklahoma. All trees in a grove may be killed within 3 years following hand inoculation of 80 % or more of the trees with a conidial suspension of the fungus. Spread is rapid and quick death of plant occurs. First notice of the disease was in Tennessee in 1933. By 1938, only 5 % of the persimmons in the infected stand were alive. However, it is not commercially available now (Schroers et al. 2005).

15.2.2.2 Lubao

Lubao is a formulation of a selected strain of *Colletotrichum gloeosporioides* f. sp. *cuscutae* (Fig. 15.2) used as a mycoherbicide in People's Republic of China to control dodder (*Cuscuta chinensis* and *C. australis*) parasitic on broadcast-planted soybeans. It has been used since its discovery in 1963 for practical control of this parasitic weed. Spore concentrations of 2×10^7 spores/ml are applied with a hand sprayer until runoff and provide greater than 80 % control of dodder, resulting in yield loss reductions of 30–80 % (Templeton 1992). An improved formulation of Lubao 2 is still in use today (Wan and Wang 2001).

15.2.2.3 DeVine[R]

DeVine is a liquid suspension of chlamydospores of *Phytophthora palmivora* (*Phytophthora citrophthora*) (Fig. 15.3). This mycoherbicide was registered in 1981 in the USA and is the first registered mycoherbicide in the world. It was marketed by Abbott Laboratories, the USA. It is being used as a postemergent directed spray to control

Fig. 15.3 (**a**) *Morrenia odorata* weed; (**b**) Sporangia of *Phytophthora palmivora*; (**c**) DeVine product

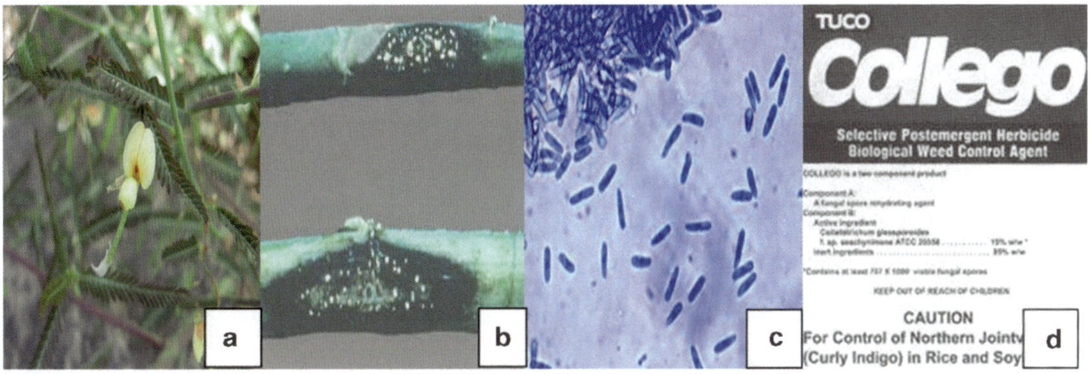

Fig. 15.4 (**a**) Northern jointvetch (*Aeschynomene indica*); (**b**) jointvetch stems infected with *Colletotrichum gloeosporioides* f. sp. *aeschynomene*; (**c**) unicellular hyaline conidia; (**d**) Collego product

milkweed vine (*Morrenia odorata*) (Fig. 15.3) in Florida citrus groves. The product is diluted with water and applied around citrus trunks by conventional sprayers to control vines that are adjacent to the trees. It attacks strangled vine without harming citrus tree roots, fruits, or foliage. Care is taken not to expose *P. palmivora* to wetting agent, fertilizers, or chemical pesticides that are detrimental to the viability of the fungal spores. It has a shelf life of 6 weeks in refrigerated storage (TeBeest and Templeton 1985).

15.2.2.4 Collego™

Collego is a dry powdered formulation of the fungus *Colletotrichum gloeosporioides* f. sp. *aeschynomene* (Fig. 15.4). The fungus is characterized by sickle-shaped, unicellular hyaline conidia produced in acervuli. The product was registered in 1982 for the selective control of northern jointvetch (*Aeschynomene virginica*) (Fig. 15.4), a leguminous weed, whose black seeds contaminate harvests of rice and soybeans in the USA. When Collego is applied to northern jointvetch, conidia of the fungus germinate and infect the plant within 48 h. Visual symptoms appear within 7–10 days in the form of disease lesions on the plant. The lesions are elongated black spots that develop gray centered with scattered light orange pustules. Over the next few weeks, lesions elongate, enlarge, and grow together until stems are girdled. Plant parts above girdled area collapse and die. The formulation consists of 15 % viable spores and 85 % inert materials. It is applied postemergence, aerially, and with land-based sprayers. Most of the plants die within 5 weeks after application. The formulation is stored in

Fig. 15.5 (**a**) Black cherry (*Prunus serotina*); (**b**) application of fungus on red alder; (**c**) application on bigleaf maple; (**d**) sporophores of *Chondrostereum purpureum* (growth on the cut stump); (**e**) basidiocarps of *C. purpureum* allowing complete wood decay; (**f**) scanning electron micrographs of the basidium and basidiospores; (**g**) formulated product of Chontrol™; (**h**) MycoTech™ product; (**i**) BioChon product

air-conditioned area at a temperature of 40–80 °F. Spore germination reduces by exposure to temperature below 32 °F or above 105 °F. Collego is safe to use and has no hazards to human or environment, and it is eco-friendly (Bowers 1986).

15.2.2.5 Chontrol™ (Ecoclear™), MycoTech™, and BioChon™

Chondrostereum purpureum (Fig. 15.5) is a cosmopolitan basidiomycetous fungus that is found in temperate regions of the northern and southern hemispheres. It is ubiquitous in Canada and common in the USA south to Virginia in the east and to northern California in the west. *C. purpureum* causes silver leaf disease that attacks deciduous trees and shrubs. The fungus blocks xylem vessels which provide structural support to the plant and also transport sap containing nutrients up to the leaves resulting in wilting of plants (Wall 1986, 1990).

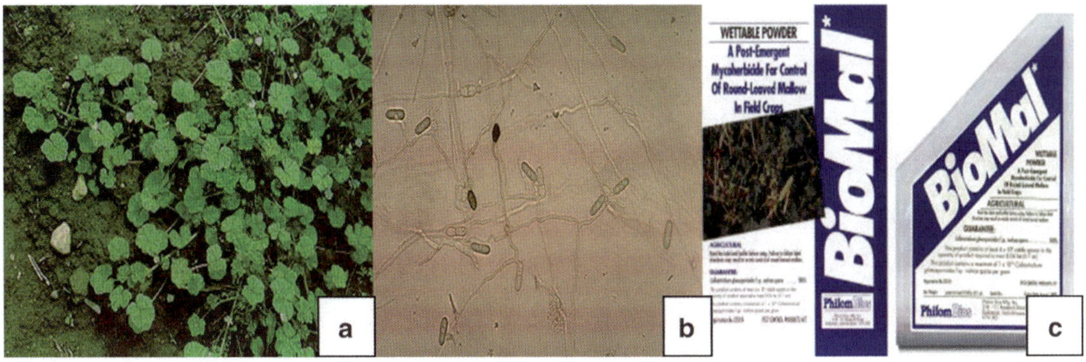

Fig. 15.6 (**a**) Round-leaved mallow (*Malva pusilla*); (**b**) germinating conidia of *Colletotrichum gloeosporioides* f. sp. *malvae*; (**c**) BioMal product

Chontrol™ or Ecoclear™ formulations of *C. purpureum* were registered as mycoherbicides in 2004 in Canada to control alder, aspen, and other hardwoods deciduous trees and shrubs. Two formulations of the fungus are Ecoclear™ paste, a wet paste formulation, and Ecoclear™ EC, a spray emulsion formulation, for application with a back-pack sprayer (Barton 2005).

MycoTech™ gel is another formulated product containing viable mycelium but not spores of *C. purpureum* (Fig. 15.5). It was developed as a mycoherbicide also in 2004 in Canada. It is applied as a thin layer over the surface area of freshly cut stumps (Fig. 15.5) of deciduous trees within 30 min of cutting. Extensive spread of the hyphae takes place throughout the xylem after application of MycoTech that inhibits resprouting and regrowth and causes brown staining and wood decay (Barton 2005).

BioChon™ is the fourth formulation of the fungus *C. purpureum* that is used for controlling broad-leaved weedy trees like black/wild cherry (*Prunus serotina*) in the forests. It was developed in 1997 in the Netherlands and Canada.

15.2.2.6 BioMal®

BioMal as a mycoherbicide was registered in 1992 in Canada. It is formulated as a wettable powder in which hydrophilic conidia of *Colletotrichum gloeosporioides* f. sp. *malvae* are absorbed on silica gel. The fungus causes anthracnose (stem and leaf lesions). It is native to western Canada and is used against an annual weed, *Malva pusilla,* which is round-leaved mallow (Fig. 15.6) belonging to *Malvaceae* family. Conidia germinate under suitable conditions within 20 h to form black appressoria with pegs that penetrate the epidermal cells. Secondary hyphae are then formed which colonize and destroy the plant tissue. Hyphal stroma forms beneath the cuticle of necrotic tissue, and conidia are produced in a gelatinous matrix in acervuli in the lesions. These can cause a number of secondary disease cycles depending on weather conditions. The first symptoms of the disease are dark sunken stem lesions. These turn grayish in the center with an almost black margin. Lesions often coalesce and girdle the stems or petioles resulting in wilting above the lesions. Mallow leaves and flowers may also become infected and develop wilted and necrotic zones, and ultimately the plant dies. To be effective as a bioherbicide, *C. gloeosporioides* f sp. *malvae* requires 12–15 h dew after application, or >6 mm rain for the first 48 h, and relatively cool (20 °C) temperature. Delaying application to meet these conditions enhances control. All stages of round-leaved mallow are susceptible. BioMal is compatible with many broadleaf herbicides and is synergistic when applied in sequence with them, but it is strongly inhibited by selective graminicides (Grass herbicides), fungicides, and many surfactants (Makowski and Mortensen 1992; Morin et al. 1996).

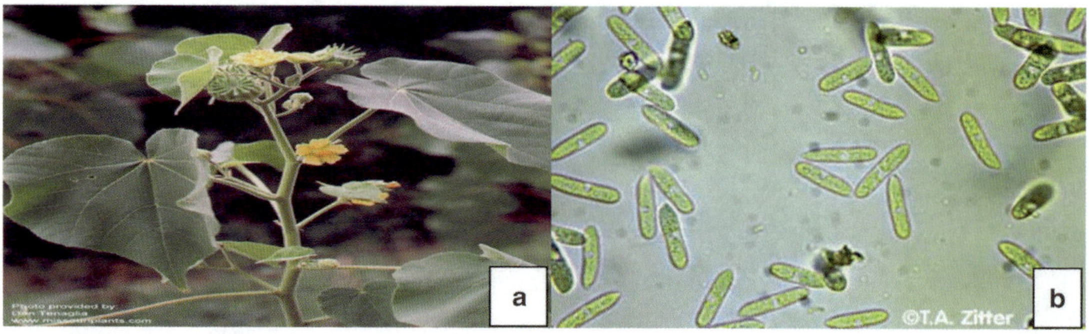

Fig. 15.7 (**a**) Velvetleaf (*Abutilon theophrasti*); (**b**) Conidia of *Colletotrichum coccodes*

Fig. 15.8 (**a**) Dandelion weed (*Taraxacum officinale*); (**b**) Sclerotia of *Sclerotinia minor*; (**c**) Sarritor product

15.2.2.7 Velgo^R

Velgo^R was developed as a mycoherbicide of the fungus *Colletotrichum coccodes* (Fig. 15.7) in 1987 for velvetleaf (*Abutilon theophrasti*) control in corn (*Zea mays*) and soybeans (*Glycine max*) in the US Corn Belt and southern Ontario, Canada. When inoculated at the two to three leaf stages, the fungus alone at the rate of 1×10^9 spores m^{-2} kills 40 % of plants (Mortensen 1988; Dauch et al. 2002).

15.2.2.8 Sarritor™

Sarritor™, a granular formulation of the fungus, *Sclerotinia minor*, was registered in 2007 in Canada to control the Dandelion (*Taraxacum officinale*) (Fig. 15.8) weed in lawns/turf. It was registered as a biological herbicide by Dr. Alan Watson of McGill University. It is a naturally occurring fungus that is normally associated with rotting lettuce. It was isolated from diseased lettuce plants in Quebec. When applied preemergent, the fungus reduced dandelion emergence by 78 %; postemergent application 10 days after sowing dandelions led to a 97 % reduction. After application to dandelion weed in lawns/turf, the fungus grows into the weed and absorbs the plant tissues until the weed is completely gone. Once the weed is gone, the fungus disappears (Abu-Dieyeh and Watson 2007).

15.2.2.9 Smolder®

Smolder^R, the formulation of *Alternaria destruens* strain 059 (Fig. 15.9), was registered to control *Cuscuta* spp. in the USA on May 5, 2005. *A. destruens* is a ubiquitous, naturally occurring fungus which was initially isolated from swamp dodder (*Cuscuta gronovii*) (Fig. 15.9) in 1986. The pathogen is indigenous to the USA and has been found to be parasitic to several *Cuscuta* species, including dodder, large seed dodder, field

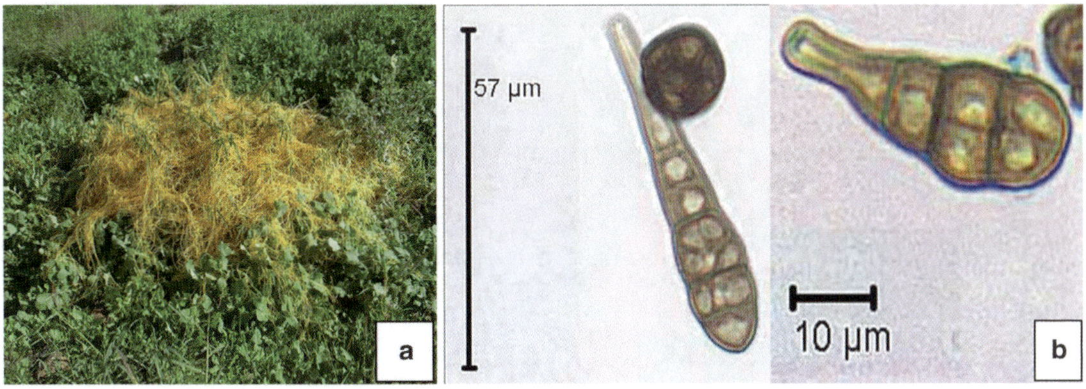

Fig. 15.9 (**a**) Swamp dodder (*Cuscuta gronovii*); (**b**) muriform, beaked, dark-colored conidia of *Alternaria destruens*

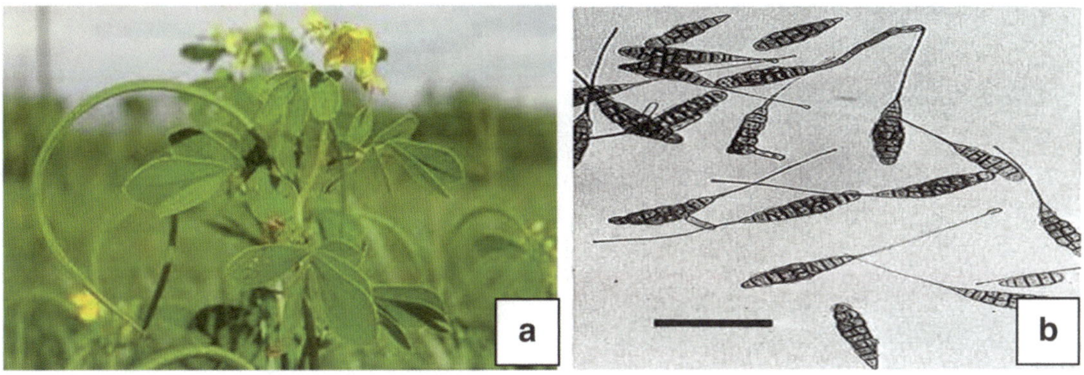

Fig. 15.10 (**a**) Sicklepod weed (*Cassia obtusifolia*); (**b**) Conidia of *Alternaria cassiae* with long thin beak

dodder, and small seed dodder. The fungus acts by infecting and suppressing dodder at early and late stages of growth. Smolder is available in two forms. *Smolder G* (a soil-applied granule product contains 4.4 % *A. destruens*) is used as herbicidal agent against dodder species in agriculture, dry bogs, and ornamental nurseries. It has been manufactured by Loveland Products Inc., USA. Smolder G (a soil applied granular product contains 4.4 % *A. destruens*) is used as herbicidal agent against dodder species in agriculture, dry bogs and ornamental nurseries. It has been manufactured by Loveland Products Inc USA. Smolder G granules which is applied to a moist surface at a rate of 50 pounds (one bag) per acre at, or immediately prior to dodder emergence. *Smolder WP*, a liquid formulation, consists of two components A and B: Component A is a water-soluble product of the fungus and component B is a liquid adjuvant, and the two products are combined with sufficient water to bring the final mix volume of 30 gal and applied when dodder are to reach the top of crop canopy (Chutia et al. 2007; Aneja and Mehrotra 2011).

15.2.2.10 Casst™

Casst™ mycoherbicide, formulation of the fungus *Alternaria cassiae* (Fig. 15.10), has been shown to be an effective biocontrol agent for sicklepod (*Cassia obtusifolia*), coffee senna (*Cassia occidentalis*), and showy crotalaria (*Crotalaria spectabilis*), resulting in leaf-blight symptoms. The pathogen has a narrow host-range and causes stunting and death of inoculated sicklepod seedling. It was registered in 1983 in the USA. This mycoherbicide has shown potential for controlling weed alone as well as with integration of herbicides (Charudattan et al. 1986).

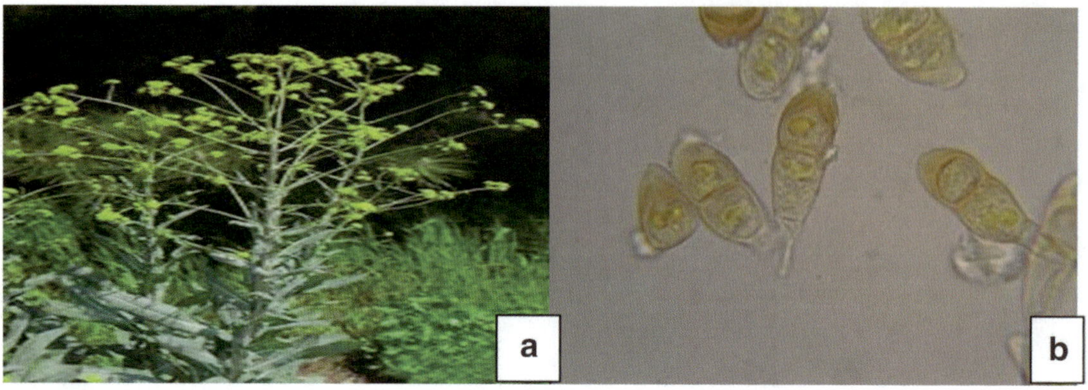

Fig. 15.11 (**a**) Woad (*Isatis indigotica*) weed; (**b**) Teleutospores of *Puccinia thlaspeos*

Fig. 15.12 (**a**) Yellow nutsedge (*Cyperus esculentus*); (**b**) Urediniospores of *Puccinia canaliculata*

15.2.2.11 Woad Warrior

Woad warrior, a mycoherbicide of the fungus *Puccinia thlaspeos* (Fig. 15.11), was registered in 2002 in the USA. It contains "strain woad," which is known as dyer's woad rust that was used to control dyer's woad (*Isatis tinctoria*), a noxious weed in eight western states. *Puccinia thlaspeos,* an autoecious, microcyclic rust, infects dyer's woad rosettes in the first year of growth. Healthy plants begin to show infection in the second spring and produce a flower stalk with yellow flowers. However, infected plants carry infection in the second year and leaves are chlorotic and malformed, with numerous spermatia and telia that are produced on young leaves of the bolting plant. Flowers and seeds are rarely produced on infected plants. Hence, reduce the reproductive capacity of the population and reduce the quantity of seeds produced in direct proportion to the incidence of infected plants (Stirk et al. 2006).

15.2.2.12 Dr. BioSedge®

Dr. BioSedge® was registered in 1987 in the USA of the rust fungus *Puccinia canaliculata* (Fig. 15.12) for the control of yellow nutsedge (*Cyperus esculentus*). It is endemic rust that completely parasitizes yellow nutsedge. When applied in early spring, *P. canaliculata* inhibited flowering and reduced yellow nutsedge stand and new tuber formation by 46 % and 66 %, respectively. But this product has failed due to uneconomic production system and resistance in some weed biotypes, and it is not commercially available now (Phatak et al. 1983).

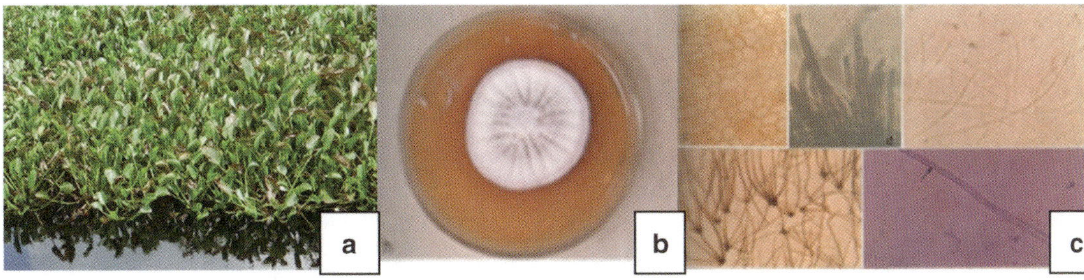

Fig. 15.13 (**a**) Infected water hyacinth plants with *Cercospora rodmanii*; (**b**) Colony on an agar medium (**c**) conidia and conidiophores

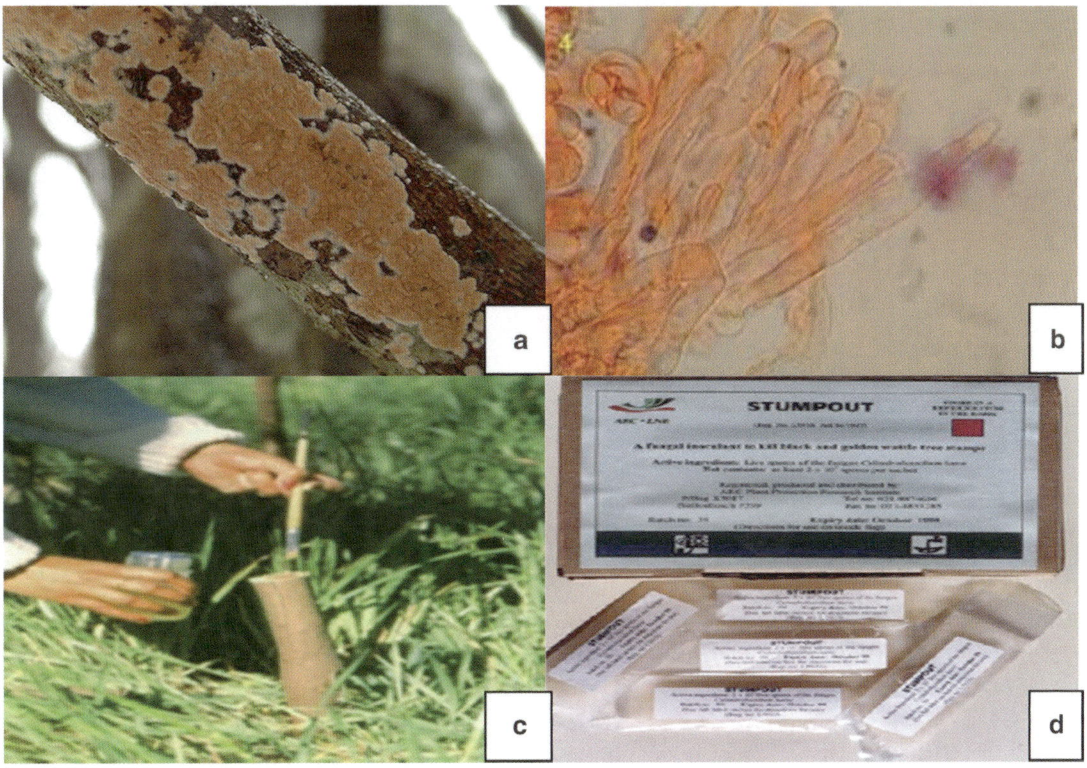

Fig. 15.14 (**a**) *Cylindrobasidium laeve* growing on the bark on tree stem; (**b**) hyphae of the fungus; (**c**) formulation applied on cut stump like paint; (**d**) Stumpout product

15.2.2.13 ABG-5003

ABG-5003 is the formulation of the fungus *Cercospora rodmanii* (Fig. 15.13) for controlling water hyacinth. This fungus was first isolated in 1973 by Conway from diseased water hyacinth plants found in the Rodman reservoir, Florida, USA. The experimental formulation of ABG-5003 was developed in 1984 by Abbott laboratories which consisted mycelium and spores and applied as wettable powder; however, this formulation was not found to be very efficient when applied in natural conditions. The notable advantage of this fungus is that it is compatible with various chemical herbicides and insects for controlling water hyacinth (Conway et al. 1978; Charudattan 2001).

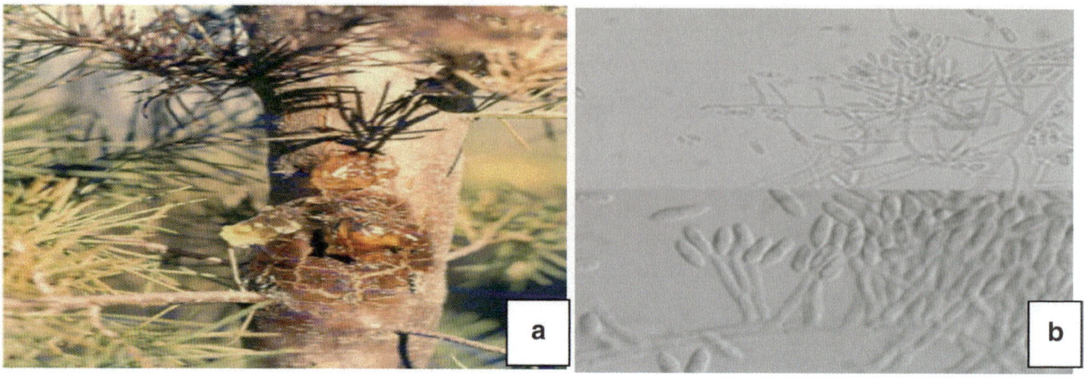

Fig. 15.15 (**a**) Silky hakea weed (*Hakea sericea*); (**b**) conidiophores and conidia of *Colletotrichum acutatum*

Fig. 15.16 *Gibbago trianthemae* on the *T. portulacastrum*: (**a**) weed in the field; (**b**) leaf spots; (**c**) germinating phaeodictyoconidia

15.2.2.14 Stumpout™

The product Stumpout™ (Fig. 15.14) containing the basidiospores of the rust fungus *Cylindrobasidium laeve* in oil was registered in 1997 in South Africa for controlling wattle stumps, turfgrass (*Poa annua*) in golf courses, and *Acacia* spp., i.e., both the black (*A. mearnsii*) and golden (*A. pycnantha*) wattle. At the time of application, the product is added to sunflower oil, and 1–2 ml is painted onto the fresh cut surface of the tree stump. The stumps die within a year of treatment. There is a regular demand for the product from conservation organizations and land owners. Stumpout™ is produced in a small factory on the premises of PPRI Weed Pathology Unit, Stellenbosch (Morris et al. 1999).

15.2.2.15 Hakatak®

Hakatak product has been developed in 1990 as a biological herbicide in South Africa using the fungus *Colletotrichum acutatum* (Fig. 15.15) for controlling *Hakea gummosis* and *H. sericea* in native vegetation. It can be applied as aerially broadcasted pellets or by wound inoculation containing liquid spore suspension. Preparation consisting of dry spores of the fungus is available to interested users (Morris et al. 1999).

15.2.2.16 *Gibbago trianthemae*: A Possible Mycoherbicide for *Trianthema portulacastrum*

T. portulacastrum (*Aizoaceae*) (Fig. 15.16), commonly known as horse purslane, is commonly found in Haryana, Punjab, Rajasthan, Uttar Pradesh, and Delhi, infesting important agricultural crops such as mustard, maize, sorghum, sugarcane, cotton, mung bean, potato, soybean, and onion (Aneja 2010). *Gibbago trianthemae* Simmon, a phaeodictyoconidial hyphomycetes (Hemibiotrophic fungus) (Fig. 15.17)

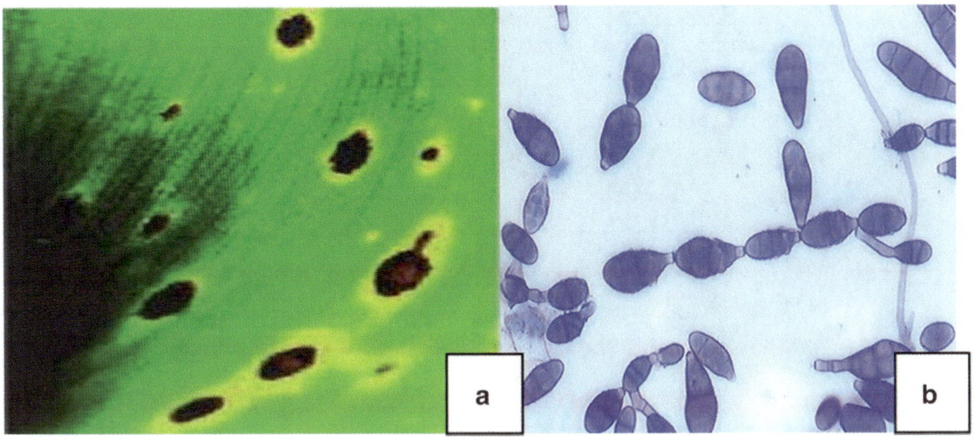

Fig. 15.17 *Alternaria alternata* on water hyacinth: (**a**) leaf spots; (**b**) beak-shaped conidia arising in chains

fungus, causing leaf spots has been isolated from the USA and Kurukshetra, India, from the diseased plants (Aneja and Kaushal 1998). *G. trianthemae* has shown high potential to reduce this weed. In experiment pots, defoliation started after 20 days of inoculum spraying. Percent infection on leaves ranged between 72 and 84 % after 30 days postinoculation with conidial suspension at concentration of 2.2×10^5 conidia/ml of the pathogen. The reduction in number of leaves was 36–37 % in both covered and uncovered pots; reduction in plant height in covered pot was 53 % and uncovered pots 60 %. The biomass/plant was also reduced to 65 % in covered pots and 85 % in uncovered pots (Aneja et al. 2000). *G. trianthemae* has been found to be highly aggressive toward horse purslane and is a desirable candidate for development as a mycoherbicide.

15.2.3 Biological Control of *Eichhornia crassipes*

A leaf spot disease in both young and mature leaves caused by *Alternaria alternata* was isolated from Kurukshetra in 1989. *A. alternata* (Fig. 15.17) is a type of cosmopolitan fungus that induces spots and lesions mainly on leaves and less severely on stolons and finally leads to complete death of the plant. When growing on an agar media, this fungus produces black to brownish colony. This fungus produces beak-shaped conidia arising in simple chains. During pathogenicity, when applied to 7-day-old mycelial culture disk to the leaves surface, disease started as small necrotic spots and developed into a leaf blight that entirely covered the whole leaf. The toxin produced by *A. alternata* (Fr.) Keissler is known to play an important role in the pathogenesis of the blight disease of water hyacinth. In 1995, Abbas et al. reported AAL toxin which is effective herbicide at low concentration against a number of weeds including water hyacinth. This pathogen on evaluation for its biocontrol potential on the experimental pits showed host specificity and killing of the host alone and in combination with insect *Neochetina eichhorniae* (Aneja and Singh 1989).

Another leaf spot disease caused by *Fusarium chlamydosporum* (Fig. 15.18) was recorded from Kurukshetra (Haryana). Macroconidia of this fungus are sickle shaped with a narrowly rounded to pointed apex. The infection of water hyacinth leaves after 1 month postinoculation with conidia of *F. chlamydosporum* ranged between 25 and 54 %. In covered pits, infection was lower than the uncovered pits. Green house and laboratory experiment results revealed that large-sized leaves exhibited more infection than the small- and medium-sized leaves. This pathogen has also the potential to be developed as a future mycoherbicide against this weed (Aneja et al. 1990).

A large-size leaf spot disease caused by *Alternaria eichhorniae* (Fig. 15.19) was firstly

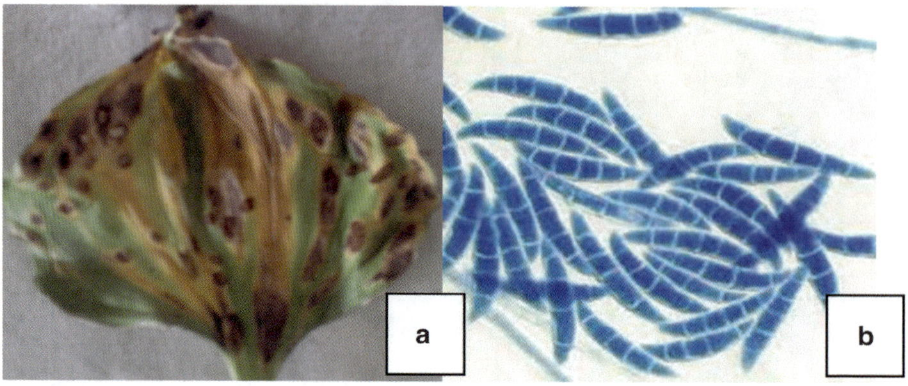

Fig. 15.18 *Fusarium chlamydosporum* on water hyacinth: (**a**) leaf spots; (**b**) sickle-shaped macroconidia

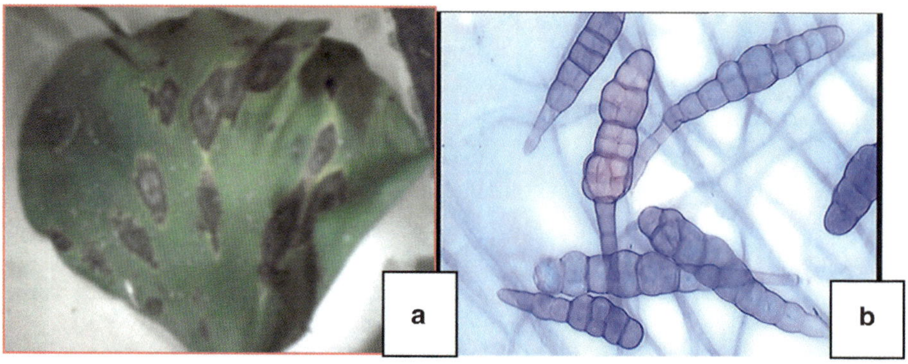

Fig. 15.19 *Alternaria eichhorniae* on water hyacinth: (**a**) large-sized leaf spots; (**b**) longed beaked muriform conidia

reported on water hyacinth from India by NagRaj and Ponappa in 1970. The fungus was later discovered on water hyacinth in Australia, Bangladesh, Indonesia, and South Africa. This fungus causes necrotic leaf spots with dark centers and brownish black margins on water hyacinth leaves. When growing the fungus in nutrient media, it produces red-colored phytotoxic compound. The fungus produced two nonspecific phytotoxins, i.e., Bostrycin and 4-deoxybostrycin (Charudattan and Rao 1982). These two toxins are phytotoxic to the leaves of water hyacinth. This fungus has shown the presence of all the desirable features of a potential mycoherbicide such as conidial production on simple agar media, biocontrol efficacy, host specificity, and nontoxicity to human beings and hence the possibility for its development into a commercial mycoherbicide in the near future.

15.2.4 Biological Control of *Parthenium hysterophorus*

Parthenium hysterophorus L. is an aggressive invasive alien weed species, native to the USA (Fig. 15.20) but now widely spread in Asia, Africa, and Australia (Kohli et al. 2006). It has entered India before 1910 and achieved the status of world's worst weed since 1956. *P. hysterophorus* is commonly called as congress grass, white top, carrot weed, chatak chandani, bitter weed, ramphool, and gajar grass in India (Madan et al.

Fig. 15.20 (**a**) A flowering plant; (**b**) leaf spots due to *Alternaria alternata*; (**c**) conidia in chain

2011). It has become one of the world's seven most devastating and hazardous weed and covered 14.25 million ha of farm land during 2001–2007, compared to two million hectares in 1991–2000. It is spreading at an alarming rate and not only compete with cultivated crops but also deplete the nutrient pool of the soil in which have they grown.

To control this weed, bioherbicidal activity of culture filtrates of nine phytopathogenic fungi, namely, *Alternaria alternata* (Fig.15.20), *Drechslera australiensis*, *D. hawaiiensis*, *D. biseptata*, *D. rostrata*, *Fusarium oxysporum, F. solani*, *Monilia sitophila*, and *Cladosporium* sp., was evaluated against *Parthenium* weed. In laboratory bioassays, the effect of original (100 %) as well as lower concentrations (75, 50, and 25 %) of these cultural filtrates was studied on germination and early seedling growth of *Parthenium.* Cultural filtrates of different concentrations of *A. alternata*, *Cladosporium* sp., and *D. rostrata* significantly suppressed the germination of *Parthenium* seeds by 70–90 %, 13–73 %, and 27–50 %, respectively. Cultural filtrates of these fungi also exhibited pronounced adverse effects on the seedling root and shoot growth. Among other fungal species, cultural filtrates of *D. australiensis*, *D. hawaiiensis*, *F. oxysporum,* and *F. solani* significantly reduced the root and shoot length of *Parthenium* seedlings. Foliar spray bioassay was performed using cultural filtrates of three fungal species, namely, *A. alternata*, *F. solani,* and *D. rostrata*. In this bioassay, three sprays of fungal cultural filtrates, with 4-day intervals each, were carried out on 1- and 2-week-old pot-grown seedlings of *Parthenium.* Root and shoot growth of *Parthenium* weed is markedly suppressed by the cultural filtrates of all the three fungal species. *Sclerotium rolfsii* (teleomorph: *Athelia rolfsii),* cause a severe collar rot disease in *Parthenium*, and has been suggested as a potential mycoherbicidal agent (Shukla and Pandey 2006).

15.3 Conclusions

Agriculture has had to face the destructive activities of numerous pests from time immemorial, leading to radical decrease in yields of crops. Weeds cause maximum reduction/loss in the yields of crops than other pests and diseases. The level of activity in mycoherbicide research has increased tremendously since the early 1970s. Despite gashing significant knowledge in many fields such as pathogenicity, host specificity, formulation, and mass production, the concept of bioherbicide is suffering from limited success, and it is reasonable to ask why more bioherbicide products are yet to become widely available and acceptable in the marketplace. The need of the hour is better understanding of the mode of action of mycoherbicides involved in the host-pathogen interactions, which consequently leads to enhance virulence of a pathogen and/or suppression of the host plant's defense. Toxins and/or enzymes produced by fungal pathogens may play an important role in host-pathogen interactions and could represent important tools in future for improving, directly or indirectly, the efficacy of mycoherbicides. Although the use of mycoherbicides is the only safe, cost-effective, truly successful, and environmentally sustainable method of weed control, however, much remains to be done in the use of fungi for weed control, especially in the developing countries where the coordination among the private companies and the universities, Institutes is leading.

References

Abbas HK, Duke SO, Paul RN, Riley RT, Tanaka T (1995) AAL-toxin, a potent natural herbicide which disrupts sphingolipid metabolism of plants. Pestic Sci 43(3):181–187

Abu-Dieyeh MH, Watson AK (2007) Grass over seeding and a fungus combine to control *Taraxacum officinale*. J Appl Ecol 44:115–124

Aneja KR (2009) Biotechnology: an alternative novel strategy in agriculture to control weeds resistant to conventional herbicides. In: Lawrence R, Gulati AK, Abraham G (eds) Antimicrobial resistance from emerging threats to reality. Narosa Publishing House, New Delhi, pp 160–173

Aneja KR (2010) Biological control of horse purslane (*Trianthema portulacastrum* L.) by fungal pathogens. J Mycopathol Res 48(2):181–185

Aneja KR, Kaushal S (1998) Occurrence of *Gibbago trianthemae* on horse purslane in India. J Biol Control 12(2):157–159

Aneja KR, Mehrotra RS (1996) Strategies in biological control of weeds with plant pathogens. In: Agnihotri VP, Prakash O, Kishan R, Misra AK (eds) Disease scenario in crop plants. International Books and Periodicals Supply Service, Delhi, pp 243–252

Aneja KR, Mehrotra RS (2011) Fungal diversity and biotechnology. New Age International Publishers, New Delhi, pp 621–622

Aneja KR, Singh K (1989) *Alternaria alternata* (Fr.) Keissler a pathogen of water hyacinth with biocontrol potential. Trop Pest Manag 35:354–356

Aneja KR, Srinivas B, Singh K (1990) Three new pathogenic fungi of water hyacinth from India. Trop Pest Manag 36(1):76

Aneja KR, Khan SA, Kaushal S (2000) Management of horse purslane (*Trianthema portulacastrum* L.) with *Gibbago trianthemae* Simmons in India. In: Spencer NR (ed) Proceedings of the X international symposium on biological control of weeds, Montana State University, Bozeman, MT, pp 27–33

Barton J (2005) Bioherbicides: all in a day's for a superhero. In: What's new in biological control of weeds? Manaaki Whenua Landcare Research, New Zealand Ltd., Lincoln, pp 4–6

Bowers RC (1986) Commercialization of Collego™ -an industrialist's view. Weed Sci (Suppl) 34:24–25

Charudattan R (2001) Biological control of weeds by means of plant pathogens: significance for integrated weed management in modern agro ecology. BioControl 46:229–260

Charudattan R, Rao KV (1982) Bostrycin and 4-deoxybostrycin: two nonspecific phytotoxins produced by *Alternaria eichhorniae*. Appl Environ Microbiol 43:846–849

Charudattan R, Walker HL, Boyette CD, Ridings WH, Beest Te DO, Van Dyke CG, Worsham AD (1986) Evaluation of *Alternaria cassiae* as a mycoherbicide for sicklepod (*Cassia obtusifolia*) in regional field tests. Southern Cooperative Service Bulletin Alabama: Alabama Agricultural Experiment Station, Auburn University, pp 1–19

Chutia M, Mahanta JJ, Bhattacharyya N, Bhuyan M, Boruah P, Sharma TC (2007) Microbial herbicides for weed management: prospects, progress and constraints. Plant Pathol J 6:200–218

Conway KE, Freeman TE, Charudattan R (1978) Development of *Cercospora rodmanii* as a biological control for *Eichhornia crassipes*. In: Proceedings 5th international symposium on aquatic weeds, Amsterdam, pp 225–230

Dagno K, Lahlali R, Diourte M, Jijakli MH (2012) Present status of the development of mycoherbicides against water hyacinth: successes and challenges-a review. Biotechnol Agron Soc Environ 16(3):360–368

Dauch AL, Watson AK, Jabaji SH (2002) Detection of the mycoherbicide Velgo in velvetleaf field soil using strain-specific primers in the 168 Canadian phytopathological society annual meeting. Can J Plant Pathol, Waterton Lakes National Park, Alberta, Canada, vol 24, pp 383–384

Kiely T, Donaldson D, Grube A (2004) Pesticides industry sales and usage, 2000 and 2001 market estimates. US Environmental Protection Agency Office of Pesticide Programs, Washington, DC

Koeck M, Hardham AR, Dodds PN (2011) The role of effectors of biotrophic and hemibiotrophic fungi in infection. Cell Microbial 13(12):1849–1857

Kohli RK, Batish DR, Singh HP, Dogra K (2006) Status, invasiveness and environmental threats of three tropical American invasive weeds (*Parthenium hysterophorus* L., *Ageratum conyzoides* L., *Lantana camara* L.). Biol Invasions 8:1501–1510

Li yongguan, Sun Z, Zhuang X, Xu L, Chen S, Li M (2003) Research progress on microbial herbicides. Crop Prot 47:252

Madan H, Gogia S, Sharma S (2011) Antimicrobial and spermicidal activities of *Parthenium hysterophorus* Linn. and *Alstonia scholaris* Linn. Indian J Nat Prod Resour 2(4):458–463

Makowski RMD, Mortensen K (1992) The first mycoherbicide in Canada: *C. gloeosporioides* f. sp. *malvae* for round-leaved mallow control. In: Richardson RG (ed) Proceeding of the 1st international weed control congress, Monash University, Melbourne, Australia, pp 298–300

Morin L, Derby JAL, Kokko EG (1996) Infection process of *Colletotrichum gloeosporioides* f. sp. *malvae* on Malvaceae weeds. Mycol Res 100(2):165–172

Morris MJ, Wood AR, den Breeÿen A (1999) Plant pathogens and biological control of weeds in South

Africa: a review of projects and progress during the last decade. Afr Entomol Mem 1:129–137

Mortensen K (1988) The potential of an endemic fungus, *Colletotrichum gloeosporioides* for biological control of round-leaved mallow (*Malva pusilla*) and Velvet leaf (*Abutilon theophrasti*). Weed Sci 36:473–478

NagRaj TR, Ponappa KM (1970) Blight of water hyacinth caused by *Alternaria eichhorniae*. Trans Br Mycol Soc 55(1):123–130

Panetta FD (1992) The role of prediction in determining the quarantine significance of weeds. In: Richardson RG (ed) Proceedings of the first international weed control congress. Weed Science Society of Victoria Inc., Melbourne, pp 381–383

Phatak SC, Summer DR, Wells HD, Bell DK, Glaze NC (1983) Biological control of yellow nutsedge with the indigenous rust fungus *Puccinia canaliculata*. Science 219:1446–1447

Quimby PC (1982) Impact of diseases on plant populations. In: Biological control of weed with plant pathogens. Wiley, New York, pp 47–60

Schroers HJ, Geldenhuis MM, Wingfield MJ, Schoeman MH, Yen YF, Shen WC, Wingfield BD (2005) Classification of the guava wilt fungus *Myxosporium psidii,* the palm pathogen *Gliocladium vermoesenii* and the persimmon wilt fungus *Acremonium diospyri* in Nalanthamala. Mycologia 97(2):375–395

Shukla R, Pandey AK (2006) Maximization of production of oxalic acid from *Sclerotium rolfsii,* a mycoherbicidal agent against *Parthenium*. Ann Pl Protect Sci 14(1):202–205

Stirk WA, Thomson SV, Staden JV (2006) Effect of rust-causing pathogen (*Puccinia thlaspeos*) on auxin-like and cytokinin- activity in dyer's woad (*Isatis tinctoria*). Weed Sci 54:815–820

TeBeest DO, Templeton GE (1985) Mycoherbicides: progress in the biological control of weeds. Plant Dis 69(1):6–10

Templeton GE (1992) Use of *Colletotrichum* strains as mycoherbicides. In: Bailey JA, Jeger MJ (eds) *Colletotrichum*: biology, pathology and control. CAB International, Wallingford, pp 358–380

Wall RE (1986) Pathogenicity of *Chondrostereum purpureum* to yellow birch. Plant Dis 70:158–160

Wall RE (1990) The fungus *Chondrostereum purpureum* as a silvicide to control stump sprouting in hardwoods. North J Appl For 7:17–19

Wan FH, Wang R (2001) Biological weed control in China: an update report on Alien invasive species. Workshop on alien invasive species, IUCN Regional biodiversity programme, Asia. Colombo, Sri Lanka, pp 8–19

Watson AK (1989) Current advances in bioherbicide research. Brighton Crop Protection Conference – Weeds 3:987–996

Nitrite Biosensors: Analytical Tools for Determination of Toxicity Due to Presence of Nitrite Ions

16

V. Dhull, A. Gahlaut, A. Gothwal, J.S. Duhan, and V. Hooda

Abstract

The chapter reviews the current state of art in the field of nitrite sensors based on electrochemical transducers with their salient features and wide application in healthcare, food industry, environmental monitoring, etc. Increased anthropogenic activities, rased the concentration of nitrite to alarming situation, directly putting an adverse effect on environment and natural habitat, and is of serious health concern. The issue of nitrites toxicity led to the implementation of rules to restrict their level in drinking waters and foodstuffs. From the last 20 years, the growing need of portable tool for onsite nitrite analysis leads to outburst of numerous approaches for development of efficient nitrate biosensors. The present review provides the global perspective in regard to nitrate biosensors with diverse fabrication strategies and materials adopted. Use of different fabrication strategies, leading to improved performance of biosensors, is also discussed.

16.1 Introduction

Over the past three decades, the world has witnessed major revolutions in agriculture and industrial sectors. The practices involved have direct or indirect relation with the increase of nitrate and nitrite ion concentration in the environment and food chain. The gradual increase of nitrites to the environment occurs through the chemical conversion of atmospheric nitrogen oxides (NO_x) generated by combustion processes (industrial, domestic and automobile) (Galloway 2003; Victorin 1994). Applications of nitrites containing products in industries like textile, metal, petroleum, and pharmaceutical industries are also recognised (Dutton 2004). Continuous pressure for more agriculture production leads to extensive use of N-fertilisers in farming activities that through anaerobic process converted into nitrites (Dutt and Davis 2002). The monitoring and continuous evaluation of nitrite concentration in food, water,

V. Dhull
Department of Bio & Nanotechnology, Guru Jambeshwar University of Science & Technology, Hisar 125001, India

A. Gahlaut • A. Gothwal • V. Hooda (✉)
Centre for Biotechnology, M.D. University, Rohatk 124001, India
e-mail: advance.biotech@gmail.com

J.S. Duhan
Department of Biotechnology, Chaudhary Devi Lal University, Sirsa 125055, India

R.K. Salar et al. (eds.), *Biotechnology: Prospects and Applications*, DOI 10.1007/978-81-322-1683-4_16, © Springer India 2013

soil and biological fluids is vulnerable for the society due to its proven physiological and clinical impacts. Levels of nitrate/nitrite have now been under regulation with the promulgation of rules to restrict their levels in drinking and food products. Nitrite concentration has now been considered as a major physiological parameter for clinical studies.

Frequent exposure to high concentration of nitrate/nitrite leads to adverse impact on public health and ecological systems. Nitric oxide has critical role in cellular signalling and physiological processes, and methaemoglobinemia is the principal adverse health effect caused by excessive nitrites intake particularly in infants, often referred to as the blue baby syndrome (Dutt and Davis 2002; Hord et al. 2009). Concern has been raised on the potential role of nitrite in forming carcinogenic N-nitroso compounds (NOCs) via reaction with secondary amines (Lijinsky 1999; Larsson et al. 2006). The information generated from the analytical surveillance of nitrite in food products is fundamental for the management of health risks. The issue of nitrite toxicity led to the implementation of rules to restrict their level in drinking waters and foodstuffs. In order to manage environmental and health risks caused by nitrites exposure, governmental agencies have promulgated rules and directives to restrict the level of these ions in drinking waters and food products. European Community has established the maximum admissible levels of nitrite in drinking water at 0.1 mg/l. Likewise, the World Health Organization has set these limits at 3 mg/l (NO_2^-). More recently, following the European Food Safety Authority recommendations, 2006/52/EC directive has reduced the authorised levels for nitrites in meat and other food products, which should be controlled on the basis of added rather than residual amounts (e.g. 150 mg/kg of nitrites in meat products).

There are several analytical techniques for the nitrite detection which include gas chromatography (Jain et al. 1997), ion exchange LC with spectrophotometer (Gapper et al. 2004), automated methods (Pasquali et al. 2010), photometrical method (Tu et al. 2010), HPLC (Kodamatani et al. 2009) and enzymatic methods (Chen et al. 2008). However, these centralised and sophisticated analytical techniques are less preferred due to tedious sample preparation steps for removal of interfering compounds (Washko et al. 1989), some have low sensitivity, overestimation of nitrite level due to the oxidation of several species other than nitrite present in the sample, low minimum detection limit and finally costly. Above-mentioned instruments are representing delays in solving problems in emergency situations where the rapid diagnostic of disease, food quality control, or environmental pollution monitoring are critical issues. Enzyme-based methods have been preferred over other methods because they are simple, easy to use, highly specific and sensitive. Nitrite reductase is the enzyme used in nitrite determination methods. The use of free enzyme increases the cost besides its cumbersome handling. This can be overcome by use of immobilised enzymes for development of biosensor. With the increasing knowledge of nitrate role in physiology and related clinical implications, the awareness leads to development of nitrite determination assays and tools. There is now exponential potential of nitrite analytical tools in food industry, pollution control and clinical diagnostics. From the last 20 years, the growing need of portable tool for onsite nitrite analysis leads to outburst of numerous approaches for development of efficient nitrate biosensors. The present chapter provides the global perspective in regard to nitrate biosensors with diverse fabrication strategies and materials adopted. The outcome and significant approaches were categorically summarised in Table 16.1.

16.2 Enzyme Used for Electrochemical Determination of Nitrite

The nitrite concentration can be detected by using the nitrite reductase. Nitrite reductase is an important class of enzymes that catalyse the reduction of nitrite (Kiang et al. 1975). There are several types of enzymes such as multiheme,

Table 16.1 Different sensors for nitrite detection

Sr. No.	Target analyte	Detection technique	Enzyme immobilisation technique	Electrode/transducer	Linearity range	Detection limit	References
I. Horseradish peroxidase (HRP)/catalase-based biosensor							
1	Nitrite	Amperometric	Entrapment	[Zn_2CrABTS] LDH	–	4 μM	Chen et al. (2008)
2	Nitrite	Conductometric	Cross-linking	Gold interdigitated thin film electrodes	0.3–446 μM	0.3 μM	Zazoua et al. (2009)
II. Cytochrome c-based biosensors							
3	Nitrite	Cyclic voltammetry	–	Multiwalled carbon nanotube	0.1–29 μM 29–254 μM	0.01 μM	Chen et al. (2009)
4	Nitrite	Cyclic voltammetry	Electrochemical polymerisation	Nano diamond (PANI-g-ND)	0.5–3 μM	1.16 μM	Gopalan et al. (2010)
5	Nitrite	Electrochemical	Ionic bonding	P3MT and MWCNT	10–100 μM	0.05 μM	Eguílaz et al. (2010)
6	Nitrite	Atomic Force microscopy, X-ray photoelectron spectroscopy	–	Platinum-implanted indium tin oxide electrode	5–800 μM	0.4 μM	Liang et al. (2012)
III. Superoxide dismutase-based biosensor							
7	Nitrite	Cyclic voltammetry	Matrix confinement	Pt electrode	0.5–2,000 μM	0.5 μM	Rajesh et al. (2010)
IV. Nitrite reductase-based biosensors							
8	Nitrite	Potentiometric	–	–	10^{-4} M–5×10^{-2} M	–	Kiang et al. (1975)
9	Nitrite	Amperometric	Gel immobilised	Glassy carbon electrode	–	–	Scharf et al. (1995)
10	Nitrite	Amperometric	Surface immobilisation	Gold electrode	–	0.01 g/l	Sasaki et al. (1998)
11	Nitrite	Amperometric	Entrapment	Glassy carbon electrode	5.4–43.4 μM	5.4 μM	Silva et al. (2004)
12	Nitrite	Amperometric	Polymer entrapment	Glassy carbon electrode	1.5–260 μM	1.5 μM	Quan et al. (2006)
13	Nitrite	Amperometric	Entrapment	Glassy carbon electrode	0.015–2.35 μM	4 μM	Chen et al. (2007)
14	Nitrite	Potentiometric	Matrix entrapment	Glassy carbon electrode	75–800 μM	60 μM	Almeida et al. (2007)
15	Nitrite	Conductometric	Cross-linking	Planer interdigitated electrode	0.2–120 μM	0.05 μM	Zhang et al. (2009)
16	Nitrite	Amperometric	Entrapment	Porous silica glass	0.25–50 μM	120 nM	Silveira et al. (2010)
17	Nitrite	Amperometric	Co-immobilisation	Carbon paste screen-printed electrode	10–200 μM	7 μM	Serra et al. (2011)

(continued)

Table 16.1 (continued)

Sr. No.	Target analyte	Detection technique	Enzyme immobilisation technique	Electrode/transducer	Linearity range	Detection limit	References
V. Nonenzymatic sensor							
18	Nitrite	Amperometric	Pt electrode		–	–	Badea et al. (2001)
19	Nitrite	Cyclic voltammetry	Pt electrode		10–1,000 μM	2 μM	Biagiotti et al. (2007)
20	Nitrite	Cyclic voltammetry	PPy–Pt nanocomposite on glassy carbon electrode		5.0×10^{-7}–1.0×10^{-3} M	1.5×10^{-7} M	Li and Lin (2007)
21	Nitrite	Amperiometric	Glassy carbon electrode		10–1,000 μM	2.3 μM	Huang et al. (2008)
22	Nitrite	Electrochemical	CNT modified with screen-printed carbon electrode (SPCES)		–	0.96 μM	Lin et al. (2009)
23	Nitrite	Anodic voltammetry	Glassy carbon electrode		0.5–7.5 μmol l^{-1}	0.1 μmol l^{-1}	Santos et al. (2009)
24	Nitrite	X-ray diffraction	MnO_2/MQPOE composite electrode		–	Low	Xia et al. (2009)
25	Nitrite	Electrochemical	Choline chloride modified glassy carbon electrode		4.0×10^{-7}–7.5×10^{-4} M	1.0×10^{-7} M	Wang et al. (2009)
26	Nitrite	Absorption spectrophotometry	Triacetyl cellulose membrane		–	1.76 nmol l^{-1}	Ensafi and Amini (2010)
27	Nitrite	SEM	Glass carbon electrode		1.0×10^{-7}–2.5×10^{-3} M	2.4×10^{-8} M	Yang et al. (2010)
28	Nitrite	Cyclic voltammetry	$CuMe_2bpaCl_2$/collagen film-coated electrode		–	–	Isoda et al. (2010)
29	Nitrite	Amperiometric	Glassy carbon electrode		5.0×10^{-3}–2.5×10^{-1} M	3.18×10^{-4} M	Valentini et al. (2012)
30	Nitrite	Cyclic voltammetry	MWCNT		5–6.75 μM	1.65 μM	Zhang et al. (2011)
31	Nitrite	SEM, TEM, ray diffraction	Hollow hematite nano polyhedrons		0.004–3 mM	–	Xia et al. (2012)
32	Nitrite	Optical	Triacetyl cellulose membrane		10.12–1,012 ng mol l^{-1}	8.3 ng mol l^{-1}	Ensafi and Amini (2012)
33	Nitrite	Electrochemical	Micro-sensor		–	50 μg N l^{-1}	Larsen et al. (2000)

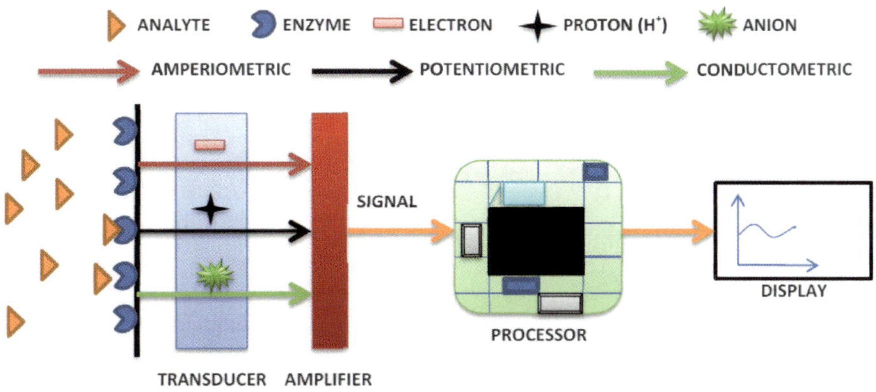

Fig. 16.1 Systematic diagram showing types of electrochemical biosensors

copper-containing enzymes (Strehlitz et al. 1996) which breakdown nitrite into different groups including nitric oxides and ammonium ions. Cytochrome c nitrite reductase (ccNIR) is a multiheme enzyme that converts nitrite to ammonia on each active site (Scharf et al. 1995). In active site, iron is bound to a protoporphyrin IX ring that is covalently linked to the enzyme's proteins. Horse reddish peroxidase/catalase and cytochrome c were also used for the determination of nitrite concentration (Chen et al. 2008).

16.3 Principle of Electrochemical Biosensor

The working of electrochemical biosensors is mainly based on the use of a biological component/bio-receptor element which is in direct contact with an active transducer (electrode) to obtain an analytical useful signal by coupling biochemical and electrochemical interactions (Thévenot et al. 1999). The principle of electrochemical sensors is that when an electroactive analyte is exposed to a fixed or varying potential of some predefined patterns which causes oxidation or reduction of analyte on the working electrode surface, this leads to the generation of electrochemical signal which can be measured by fluctuation on electron flux. This signal can be detected electrochemically (Fig. 16.1). These can be further categorised as amperometric, potentiometric and conductometric on the bases of transducer types.

16.3.1 Amperometric Biosensors

These biosensors are based on the movement of electrons as a result of enzyme-catalysed redox reactions. The substrate or product can transfer electrons. This results in change in current flow that can be measured. The magnitude of current is proportional to the substrate concentration.

16.3.2 Potentiometric Biosensor

In these biosensors, changes in the ionic concentration are determined by the use of ion-selective electrodes, since many enzyme reactions involve the release or absorption of the hydrogen ions which generate biological signals and converted into electronic signal by potentiometric transducers. The most commonly used electrodes are pH metre glass electrodes (for cations), glass pH electrodes coated with a gas selective membrane (for CO_2, NH_3 or H_2S) or solid state electrodes. The potential difference between potentiometric electrode and reference electrode can be measured which is proportional to the substrate concentration.

16.3.3 Conductometric Biosensor

Many reactions in biological system involve change in the ionic concentration. This change alters the electrical conductivity which can be measured through conductometric biosensor (Fig. 16.1).

16.4 Use of Horseradish Peroxidase (HRP)/Catalase for the Development of Nitrite Biosensor

As reported by Titov and Petrenko (2003), the concentration of nitrite in micromolar induced a considerable decrease in catalase activity, in which nitrite acts as a competitive inhibitor. The specific catalase inhibition by nitrite was first determined by UV-visible spectroscopy. The presence of 1 mM $NaNO_2$ in the enzyme solution (0.66 μg/ml) caused a decrease in catalase activity of 30 %, whereas nitrite had no effect on HRP activity even increasing the concentration ten times.

Inactivated enzymes have lower affinity for the substrate (nitrite) called enzyme inhibition, and the degree of inhibition is proportional to the concentration of nitrite in the sample. The inhibition effect of nitrite was also tested amperometrically at 0.0 V with the bi-enzyme configuration (60/60 μg catalase/LDH), the H_2O_2 concentration being fixed at 10 μM. As expected, the addition of nitrite induced an increase in the current intensity related to the enzymatic reduction of H_2O_2. This biosensor presented a rapid and stable response to nitrite within 15–30 s, depending on the nitrite concentration range (Chen et al. 2008). Conductometric biosensor was also developed for the determination of nitrite in sample (Zazoua et al. 2009). Biosensors based on the HRP/catalase for nitrite detection were listed in Table 16.1(I).

Although sensitive biosensors based on enzyme inhibition have few limitations (1) since it is inhibited by many other compounds, these analytical tools are not selective and cannot be used for quantitation of either an individual or a class of pesticides which may be required to monitor detoxification processes, for example, detoxification of nitrite. (2) These protocols involve multiple steps requiring measurement of the uninhibited activity of enzymes, followed by incubation of the sensor with the analyte sample for 10–15 min (and even longer for good sensitivity) and the measurement of the enzyme again to determine the degree of inhibition.

16.5 Cytochrome c-Based Nitrite Biosensor

The first oxidative nitrite biosensor based on Cyt *c* was reported recently and a possible mechanism-biocatalytic oxidative property to nitrite was proposed (Geng et al. 2008), which could be divided into three steps as shown in the following reactions:

$$Fe^{2+} - \text{Cyt c} \rightarrow Fe^{3+} - \text{Cyt c} + e- \quad (16.1)$$

$$Fe^{3+} - \text{Cyt c} \rightarrow \left[Fe^{4+} - \text{Cyt c}\right]^{\bullet} + 2e- \quad (16.2)$$

$$\left[Fe^{4+} - \text{Cyt c}\right]\bullet + NO_2^- + H_2O \rightarrow Fe^{3+} - \text{Cyt c} + NO_3^- + 2H^+ \quad (16.3)$$

Firstly, ferrous Cyt *c* was oxidised to ferric form at low potential, and then the ferric Cyt *c* was further oxidised into $[Fe^{4+}$–Cyt $c]^{\bullet}$ at higher potential. $[Fe^{4+}$–Cyt $c]^{\bullet}$ is a highly reactive Cyt *c* π-cation which could oxidise NO_2^- into NO_3^- in the solution. Cyt *c* was immobilised onto L-cysteine (L-Cys) self-assembled monolayer-modified P3MT/MWCNT/GCE at neutral pH, where positively charged Cyt *c* strongly interacts with L-Cys negatively charged sites. The developed Cyt *c*/L-Cys/P3MT/MWCNT/GCE biosensor was applied for the biocatalytic determination of nitrite through the oxidation of NO_2^- to NO_3^- by electrogenerated $[Fe^{4+}$–Cyt *c*] species (Eguílaz et al. 2010). Liang et al. (2012) developed a biosensor based on immobilised cytochrome c which exhibited a rapid response to H_2O_2. Chen et al. (2009) fabricated a novel nitrite biosensor by immobilising cytochrome *c* (Cyt *c*) onto the multiwalled carbon nanotubes–poly (amidoamine) (PAMAM)–chitosan (MWNT–PAMAM–Chit) nanocomposite modified glass carbon electrode (GCE). The nitrite biosensor showed a fast response to nitrite (about 5 s) in two concentration intervals, one was from 0.1 to 29 μM and the other from 29 to 254 μM. Gopalan et al. (2010) grafted polyaniline chains onto nanodiamond (PANI-g-ND) through electrochemical polymerisation of aniline in the presence of amine functionalised ND.

Several electrochemical biosensors based on the cytochrome c activity have been reported (as shown in Table 16.1(II)).

16.6 Superoxide Dismutase (SOD)-Based Nitrite Biosensor

Use of SOD for the development of nitrite biosensor has been reported. Rajesh et al. (2010) developed a highly sensitive biosensor for the direct and simultaneous determination of superoxide anion radical ($O_2^{\cdot -}$) and nitrite (NO_2^-) by incorporation of carbon nanotube (CNT) solubilised in nafion in polypyrrole (PPy) matrix on Pt electrode followed by immobilisation of Cu, ZnSOD (SOD) on it. The biosensor exhibited a linear response over the concentration range from 0.1 to 750 μM, with a detection limit of 0.1 ± 0.03 μM for $O_2^{\cdot -}$ and a corresponding linear range of 0.5–2,000 μM, with a detection limit of 0.5 ± 0.025 μM for NO_2^- (Table 16.1(III)).

16.7 Nonspecific Proteins for Nitrite Detection

Myoglobin and haemoglobin are the nonspecific proteins which are used for the nitrite determination by the development of the electrochemical biosensors such as the amperiometric and the voltammetry.

16.8 Nitrite Reductase-Based Biosensor for Determination of Nitrite

Nitrite reducing enzymes (NiRs) are innate candidates for playing the role of bio-recognition elements in nitrite bio-sensing devices. Two different NiRs are able to catalyse the six electron reduction of nitrite to ammonia (NH_4^+) and nitric oxide (NO), according to the following Eqs. 16.4 and 16.5:

$$NO_2^- + 8H^+ + 6e^- \rightarrow NH_4^+ + 2H_2O \quad (16.4)$$

$$NO_2^- + 2H^+ + e^- \rightarrow NO + H_2O \quad (16.5)$$

Various biosensors based on nitrite reductase (Table 16.1(IV)) by immobilising the nitrite reductase onto several different insoluble supports provided better solution for several problems which include loss of enzyme, storage life of biosensor, time of enzymatic response and enzyme stability maintenance. It offers easy use of disposable devices in stationary or in flow systems. There are several insoluble supports which can be used for immobilisation of nitrite reductase as follows: glassy carbon electrode (Quan et al. 2006), (ZnCr-AQS) layer (Chen et al. 2007), poly (pyrrole–viologen) matrix (Silva et al. 2004), graphite electrode (Silveira et al. 2010), carbon paste screen-printed electrodes (Serra et al. 2011), gel-immobilised glassy carbon electrode (Scharf et al. 1995), gold electrode (Sasaki et al. 1998), glassy carbon electrode (Almeida et al. 2007) and gold interdigitated electrodes (Zhang et al. 2009).

16.9 Nonenzymatic Sensor

The sensing technologies which do not use the enzyme for the detection of analyte are generally referred as nonenzymatic sensors (Table 16.1(V)). Microbial biosensor for the detection of nitrite was developed by Larsen et al. (2000). It is composed of a biodegradation chamber for denitrifying bacteria converting nitrite to nitrous oxide and an inside nitrous oxide micro-sensor to produce electrochemical signals. Sensors other than microbial biosensor were also being reported.

16.10 Role of Nanoparticles in the Fabrication of Nitrite Biosensor

In recent years, nanotechnological approaches can be used to introduce the nanoparticles in the electrode modification and electrochemical biosensors development. These nanostructures show extraordinary properties like high mechanical strength, large surface area, good conductivity and extremely miniaturised size. Role of nanomaterials and their applications were also discussed in literature (Guo and Wang 2007). Use of nanoscale materials in the development of electrochemical biosensors gives rise to advancement in detection techniques and methods (Pumera et al. 2007). Nowadays, different types of nanomaterials are used for the development of new sensing techniques. Among all of them, carbon nanotubes (CNTs) showed remarkable application in the sensor technology because of its excellent electrical conductivity, good stability and high mechanical strength. Carbon nanotubes play a major role in preventing the electrode surface fouling when incorporated onto the electrode surface which causes enhanced rate of electron transfer during the process (Kachoosangi et al. 2009). Vairavapandian et al. (2008) has discussed the various strategies through which CNTs are prepared and modified. Incorporation of metal nanoparticles into the carbon nanotubes matrices enhances its catalytic behaviour (Vairavapandian et al. 2008). By exploiting the properties of nanomaterials, they can be used as a good support for immobilisation of enzyme.

16.11 Conclusion

High concentration of nitrites cause adverse effect on environment as well as on human health. To control the continuous increase in concentration of nitrite in the environment there is need to limit anthropogenic activities. During the 1970s, spectrophotometry, capillary electrophoresis, polarographic analysis, chromatography, etc. were the only methods to detect the nitrite concentration. As nitrite quantification becomes more important, new methodologies were developed to overcome the drawbacks associated with conventional methods. The diagnostic industries have been challenged to develop new analytical tools which are fast, accurate, selective and cost-effective that can be exploited commercially. Sensor technology has been developed in the past years as a competitive approach that includes nonenzymatic and enzyme-based sensors. These sensors directly catalyse the reaction involved during the estimation of the nitrite concentration. These methods have some drawbacks as they are not specific, have low detection limit, give false-positive results due to the generation of other electroactive species and are not accurate. Also the reusability of these sensors is limited. Various approaches have been channelled by reactions based on different types of transducers that include amperometric, potentiometric and conductometric. Apart from these sensors, some biological molecules or the agents especially enzymes and bacteria are used for the development of analytical tools. These bio-sensing techniques exploit the properties of the biological component which specifically interact with a target analyte. Bacterial biosensors are less preferred over enzymatic biosensors as they are difficult to store, require extra precautions during handling, etc. The enzymes which are used for the fabrication of enzymatic biosensors include nitrite reductase, HRP/catalase, superoxide dismutase and cytochrome c. Among all these enzymes, nitrite reductase is the most frequently used enzyme for the detection purpose because of its specificity to nitrite. Nitrite reductase catalyse the breakdown of nitrite into ammonium ions or the nitric oxide depending upon the pH of the medium, and these products can be sensed by the selective electrodes which are referred as the working electrodes. Immobilisation of enzyme on the different supports improves the reusability of the enzymes. The electrodes made up of the nanoparticles can also be used for the detection purpose. Transducer modification by nanomaterials and genetic engineering of the bio-components improve the working ability of transducers. The electro-catalytical properties of the nanostructures like their action as electron transfer mediators or electrical wires, large surface to volume ratio, structural robustness and biocompatibility bring revolutionary changes in electrochemical biosensor development. Therefore, overall studies on the nitrite determination pointed towards the development of advanced analytical tool which are able to detect nitrite at low concentration with more accuracy in the sample.

Acknowledgement Authors are thankful to Department of Science and Technology (DST), New Delhi, for providing JRF fellowship (INSPIRE) to first author. The authors are also thankful to UGC and DST New Delhi for providing financial assistance to Centre for Biotechnology, M D University, Rohtak (India).

References

Almeida MG, Silveira CM, Moura JJG (2007) Biosensing nitrite using the system nitrite reductase/Nafion/methyl viologen-a voltammetric study. Biosens Bioelectron 22:2485–2492

Badea M, Amine A, Palleschi G, Moscone D, Volpe G, Curulli A (2001) New electrochemical sensors for detection of nitrites and nitrates. J Electroanal Chem 509:66–72

Biagiotti V, Valentini F, Tamburri E, Terranova ML, Moscone D, Palleschi G (2007) Synthesis and characterization of polymeric films and nanotubule nets used to assemble selective sensors for nitrite detection in drinking water. Sens Actuators B Chem 122:236–242

Chen H, Mousty C, Cosnier S, Silveira C, Moura JJG, Almeida MG (2007) Highly sensitive nitrite biosensor based on the electrical wiring of nitrite reductase by [ZnCr-AQS] LDH. Electrochem Commun 9:2240–2245

Chen H, Mousty C, Chen L, Cosnier S (2008) A new approach for nitrite determination based on a HRP/catalase biosensor. Mater Sci Eng 28:726–730

Chen Q, Ai S, Zhu X, Yin H, Ma Q, Qiu Y (2009) A nitrite biosensor based on the immobilization of Cytochrome *c* on multi-walled carbon nanotubes-PAMAM-chitosan nanocomposite modified glass carbon electrode. Biosens Bioelectron 24:2991–2996

Dutt J, Davis J (2002) Current strategies in nitrite detection and their application to field analysis. J Environ Monit 4:465–471

Dutton R (2004) Problems with volatile corrosion inhibitors in the metal finishing industry. Met Finish 102:12–15

Eguílaz M, Agüí L, Sedeño PY, Pingarrón JM (2010) A biosensor based on cytochrome *c* immobilization on a poly-3-methylthiophene/multi-walled carbon nanotubes hybrid-modified electrode. Application to the electrochemical determination of nitrite. J Electroanal Chem 644:30–35

Ensafi AA, Amini M (2010) A highly selective optical sensor for catalytic determination of ultra-trace amounts of nitrite in water and foods based on brilliant cresyl blue as a sensing reagent. Sens Actuators B Chem 147:61–66

Ensafi AA, Amini M (2012) Highly selective optical nitrite sensor for food analysis based on Lauth's violet–triacetyl cellulose membrane film. J Food Chem 132:1600–1606

Galloway JN (2003) The global nitrogen cycle of the chapter. In: Heinrich DH, Karl KT (eds) Treatise on geochemistry. Pergamon, Oxford, pp 557–583

Gapper LW, Fong BY, Otter DE, Indyk HE, Woollard DC (2004) Determination of nitrite and nitrate in dairy products by ion exchange LC with spectrophotometric detection. Int Dairy J 14:881–887

Geng R, Zhao G, Liu M, Li M (2008) A sandwich structured SiO_2/cytochrome *c*/SiO_2 on a boron-doped diamond film electrode as an electrochemical nitrite biosensor. Biomaterials 29:2794–2801

Gopalan AI, Lee KP, Komathi S (2010) Bioelectrocatalytic determination of nitrite ions based on polyaniline grafted nanodiamond. Biosens Bioelectron 26:1638–1643

Guo S, Wang E (2007) Synthesis and electrochemical applications of gold nanoparticles. Anal Chim Acta 598:181–192

Hord NG, Tang Y, Bryan NS (2009) Food sources of nitrates and nitrites: the physiologic context for potential health benefits. Am J Clin Nutr 90:1–10

Huang X, Li Y, Chen Y, Wang L (2008) Electrochemical determination of nitrite and iodate by use of gold nanoparticles/poly(3-methylthiophene) composites coated glassy carbon electrode. Sens Actuators B Chem 134:780–786

Isoda N, Yokoyama H, Nojiri M, Suzuki S, Yamaguchi K (2010) Electroreduction of nitrite to nitrogen oxide by a copper-containing nitrite reductase model complex incorporated into collagen film. Bioelectrochemistry 77:82–88

Jain A, Smith RM, Verma KK (1997) Gas chromatographic determination of nitrite in water by precolumn formation of 2-phenylphenol with flame ionization detection. J Chromatogr A 760:319–325

Kachoosangi RT, Musameh MM, Yousef IA, Yousef JM, Kanan SM, Xiao L, Davies SG, Russell A, Compton RG (2009) Carbon nanotube-ionic liquid composite sensors and biosensors. Anal Chem 81:435–442

Kiang CH, Kuan SS, Guilbault GG (1975) A novel enzyme electrode method for the determination of nitrite based on nitrite reductase. Anal Chim Acta 80:209–214

Kodamatani H, Yamazaki S, Saito K, Tomiyasu T, Komatsu Y (2009) Selective determination method for measurement of nitrite and nitrate in water samples using high-performance liquid chromatography with post-column photochemical reaction and chemiluminescence detection. J Chromatogr A 1216:3163–3167

Larsen LH, Damgaard LR, Kjær T, Stenstrom T, Jensen AL (2000) Fast responding biosensor for on-line determination of nitrate/nitrite in activated sludge. Water Res 34:2463–2468

Larsson SC, Orsini N, Wolk A (2006) Processed meat consumption and stomach cancer risk: a meta-analysis. J Natl Cancer Inst 98:1078–1087

Li J, Lin X (2007) Electrocatalytic reduction of nitrite at polypyrrole nanowire–platinum nanocluster modified glassy carbon electrode. Microchem J 87:41–46

Liang F, Jia M, Hu J (2012) Pt-implanted indium tin oxide electrodes and their amperometric sensor applications for nitrite and hydrogen peroxide. Electrochim Acta 75:414–419

Lijinsky WN (1999) Nitroso compounds in the diet. Mutat Res 443:129–138

Lin CY, Vasantha VS, Ho KC (2009) Detection of nitrite using poly (3, 4 ethylenedioxythiophene) modified SPCEs. Sens Actuator B Chem 140:51–57

Pasquali CEL, Gallego PA, Hernando PF, Velasco M, Alegria JSD (2010) Two rapid and sensitive automated methods for the determination of nitrite and nitrate in soil samples. Microchem J 94:79–82

Pumera M, Sanchez S, Ichinose I, Tang J (2007) Electrochemical nanobiosensors. Sens Actuat B 123:1195–1205

Quan D, Min DG, Cha GS, Nam H (2006) Electrochemical characterization of biosensor based on nitrite reductase and methyl viologen co-immobilized glassy carbon electrode. Bioelectrochemistry 69:267–275

Rajesh S, Kanugula AK, Bhargava K, Ilavazhagan G, Kotamraju S, Karunakaran C (2010) Simultaneous electrochemical determination of superoxide anion radical and nitrite using Cu, ZnSOD immobilized on carbon nanotube in polypyrrole matrix. J Bios 26:689–695

Santos WJR, Lima PR, Tanaka AA, Tanaka SM, Kubota LT (2009) Determination of nitrite in food samples by anodic voltammetry using a modified electrode. J Food Chem 113:1206–1211

Sasaki S, Karube I, Hirota N, Arikawa Y, Nishiyama M, Kukimoto M, Horinouchi S, Beppu T (1998) Application of nitrite reductase from *Alcaligenes faecalis* S-6 for nitrite measurement. Biosens Bioelectron 13:1–5

Scharf M, Moreno C, Costa C, Vandijk C, Payne WJ, Legall J, Moura I, Moura JJG (1995) Electrochemical studies on nitrite reductase toward a biosensor. Biochem Biophys Res Commun 209:1018–1025

Serra AS, Jorge SR, Silveira CM, Moura JJG, Jubete E, Ochoteco E, Cabañero G, Grande H, Almeida MG (2011) Cooperative use of cytochrome cd_1 nitrite reductase and its redox partner cytochrome c_{552} to improve the selectivity of nitrite biosensing. J Anal Chim Acta 693:41–46

Silva SD, Cosnier S, Almeida MG, Moura JJG (2004) An efficient poly (pyrrole–viologen)- nitrite reductase biosensor for the mediated detection of nitrite. Electrochem Commun 6:404–408

Silveira CM, Gomes SP, Araújo AN, Conceição M, Montenegro BSM, Todorovic S, Viana AS, Silva RJC, Moura JJG, Almeida MG (2010) An efficient non-mediated amperometric biosensor for nitrite determination. Biosens Bioelectron 25:2026–2032

Strehlitz B, Grundig B, Schumacher W, Kroneck PMH, Vorlop K, Kotte H (1996) A nitrite sensor based on a highly sensitive nitrite reductase mediator-coupled amperometric detection. Anal Chem 68:807–816

Thévenot DR, Tóth K, Durst RA, Wilson GS (1999) Electrochemical biosensors: recommended definitions and classification. Pure Appl Chem 71:2333–2348

Titov VY, Petrenko YM (2003) Nitrite catalase interaction as an important element of nitrite toxicity. Biochemistry 68:627

Tu X, Xiao B, Xiong J, Chen X (2010) A simple miniaturised photometrical method for rapid determination of nitrate and nitrite in freshwater. Talanta 82:976–983

Vairavapandian D, Vichchulada P, Lay MD (2008) Preparation and modification of carbon nanotubes: review of recent advances and applications in catalysis and sensing. Anal Chim Acta 62:6119–6129

Valentini F, Cristofanelli L, Carbone M, Palleschi G (2012) Glassy carbon electrodes modified with hemin-carbon nanomaterial films for amperometric H_2O_2 and NO_2^- detection. J Electrochim Acta 63:37–46

Victorin K (1994) Review of the genotoxicity of nitrogen oxides. Mutat Res 55:31743–31755

Wang P, Mai Z, Dai Z, Li Y, Zou X (2009) Construction of Au nanoparticles on choline chloride modified glassy carbon electrode for sensitive detection of nitrite. Biosens Bioelectron 24:3242–3247

Washko PW, Hartzell WO, Levine M (1989) Ascorbic acid analysis using high performance liquid chromatography with coulometric electrochemical detection. Anal Biochem 181:276–282

Xia C, Ning W, Lin G (2009) Facile synthesis of novel MnO_2 hierarchical nanostructures and their application to nitrite sensing. Sens Actuator B Chem 137:710–714

Xia C, Yanjun X, Ning W (2012) Hollow Fe_2O_3 polyhedrons: one-pot synthesis and their use as electrochemical material for nitrite sensing. Electrochim Acta 59:81–85

Yang S, Liu X, Zeng X, Xia B, Gu J, Luo S, Mai N, Wei W (2010) Fabrication of nano copper/carbon nanotubes/chitosan film by one-step electrodeposition and its sensitive determination of nitrite. Sens Actuator B Chem 145:762–768

Zazoua A, Hnaien M, Cosnier S, Renault NJ, Kherrat R (2009) A new HRP/catalase biosensor based on microconductometric transduction for nitrite determination. Mater Sci Eng 29:1919–1922

Zhang Z, Xia S, Leonard D, Renault NJ, Zhang J, Bessueille F, Goepfert Y, Wang X, Chen L, Zhu Z, Zhao J, Almeida MG, Silveira CM (2009) A novel nitrite biosensor based on conductometric electrode modified with cytochrome *c* nitrite reductase composite membrane. Biosens Bioelectron 24:1574–1579

Zhang Y, Yuan R, Chai Y, Li W, Zhong X, Zhong H (2011) Simultaneous voltammetric determination for DA, AA and NO_2^-based on graphene/poly-cyclodextrin/MWCNTs nanocomposite platform. Biosens Bioelectron 26:3977–3980

Evaluation of Low-Cost Bio-technology for Community-Based Domestic Wastewater Treatment

17

Rani Devi

Abstract

Domestic wastewater is the primary source of pathogens (disease-causing microorganisms) and putrescible organic substances. Pathogenic substances are excreted in faeces and sewage, potentially causing direct threat to public health. Putrescible organic matter causes another type of threat to water quality. So, wastewater must be treated before its discharge into other water bodies or onto the land. There are a large number of technologies used for wastewater treatment, and these have their own advantages or disadvantages. On the basis of advantages of technology, an important area of wastewater treatment is biological treatment and recently known as biotechnology. Microorganisms are isolated then selected, mutated and genetically engineered for effective bioremediation of wastewater. It can be applied for both in situ and ex situ treatment of contaminated wastewater. Biotechnology-based wastewater treatment involves bioaugmentation, bioremediation, bioreactors and phytoremediation. Wastewater can be treated by aerobic or anaerobic or blending of both the techniques. Considering the fact that environmental biotechnology can be operated in an inexpensive, flexible and also environmentally friendly way, there is a long list of areas where biotechnology-based approaches can and will expand in the near future.

17.1 Introduction

The debatable approach over humanity's impact on the environment has come of age. This chapter is focused about the wastewater generated due to human activities and its treatment by adopting the biotechnology-based approach. Wastewater may include sewage, storm water and water that have been used for various purposes around the community. Unless properly treated, it can harm health and the environment. Most communities generate wastewater from residential and nonresidential sources. Residential wastewater includes sewage that may be black water (wastewater from toilets) or grey water

R. Devi (✉)
Department of Energy and Environmental Sciences, Ch. Devi Lal University Sirsa, Haryana, India
e-mail: rani_sahu@yahoo.com

R.K. Salar et al. (eds.), *Biotechnology: Prospects and Applications*,
DOI 10.1007/978-81-322-1683-4_17, © Springer India 2013

(wastewater from all sources except toilets). Nonresidential wastewater includes wastewater generated from offices, businesses, department stores, restaurants, schools, hospitals, farms, manufacturers and other commercial, industrial and institutional entities. Storm water is a nonresidential source and carries trash and other pollutants from streets, as well as pesticides and fertilisers from yards and fields.

Domestic wastewater/domestic sewage is the primary source of pathogens (disease-causing microorganisms) and putrescible organic substances. Because pathogens are excreted in faeces, all sewage from cities and towns is likely to contain pathogens of some type, potentially presenting a direct threat to public health. Putrescible organic matter presents a different sort of threat to water quality. So treatment of domestic wastewater is very important.

There is an increasing interest in environmental biotechnology owing to a worldwide need to maintain clean soil, air and water. The major technological developments are in plant and microbial biology. Plants can be more readily engineered for resistances that enhance yield or produce new products, whereas microorganisms are exploited for their catalytic diversity and ease of genetic engineering.

There are a large number of techniques/technologies used for wastewater treatment, and most of them have their own merits and demerits. On the basis of merits of technology, an important area of wastewater treatment is biological treatment, popularly reclassified in recent years as biotechnology. Biotechnology has its origins from an old science where we find applications in the antiquities and nowadays is classified as a new technology due to transgenecity and undergoing resurgence in a wide range of applications, including past/present/future applications for the pollution engineer. Biotechnology has been increasingly applied to wastewater treatment for decades. This technology involves the manipulation of whole organisms; populations of organisms and nucleic components hold much promise for improving the quality of water and wastewater. Biotechnology offers various means of manipulating the fundamental processes of wastewater treatment, e.g. designing techniques for treating specific components of wastewater treatment. Microorganisms and enzymes have been successfully used in diverse bioremediation applications (Pletsch et al. 1999; Macek et al. 2000; Gavrilescu 2004; Jördening and Winter 2004). Effective and controlled bioremoval of nitrate and phosphate contamination from wastewater has become possible (Khin and Annachhatre 2004; Liu and Tay 2004).

Biotechnology and its subdiscipline of genetic engineering are currently in more practice for the improvement of wastewater characteristics and helping for control of water quality and quantity stresses. Ultimately the improvement of wastewater quality, in principle, should also create environmental advantage.

17.2 Wastewater

Wastewater is 99.9 % water. The other 0.1 % is what causes concern. That 0.1 % includes other materials that make up only a small portion of wastewater but can be present in large enough quantities of water to endanger public health and the environment. Because practically anything that can be flushed down a toilet, drain or sewer can be found in wastewater, even household sewage contains many potential pollutants. The wastewater components that most wastewater facilities are designed to remove are suspended solids, biodegradable organics and pathogenic organisms. Thus, the impurities present in wastewater may be classified as *inorganic, organic, nutrients, oil and grease and pathogens* as shown in Fig. 17.1.

Inorganic impurities include minerals, metals and their compounds, such as sodium, potassium, calcium, magnesium, cadmium, copper, lead, nickel and zinc, and are common in wastewater from both residential and nonresidential sources. They can originate from a variety of sources in the community including industrial and commercial sources, storm water and inflow and infiltration from cracked pipes and leaky manhole covers. Most inorganic substances are relatively stable

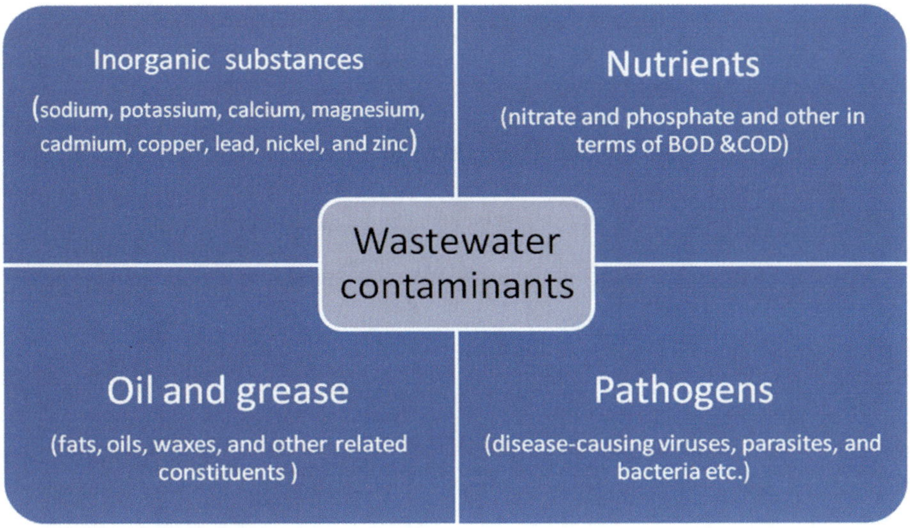

Fig. 17.1 Contaminants of domestic wastewater

and cannot be broken down easily by organisms in wastewater.

Nutrients in wastewater often include large amounts of nitrogen and phosphorus in the form of nitrate and phosphate, respectively, and that promote plant growth. Organisms only require small amounts of nutrients in biological treatment, so there normally is an excess of available nutrients in treated wastewater. In severe cases, excess nitrogen and phosphorous can result in *eutrophication*, the nutrient enrichment of water bodies causing excessive growth of aquatic plants like algae, cyanobacteria, rooted aquatic vegetation, and duckweed, which leads to the depletion of dissolved oxygen in the water body, and the aquatic plants die, fall to the bottom and are decomposed by aerobic bacteria. The oxygen depletion can reduce the populations of indigenous fish and other oxygen-consuming organisms.

Oil and *grease* includes the combination of fats, oils, waxes and other related constituents found in wastewater. Fatty organic materials from animals, vegetables and petroleum are not quickly broken down by bacteria and can cause pollution in receiving environments. When large quantity of oils and greases are discharged to receiving water, this increases BOD level, causing aesthetically unpleasing conditions and foul odours and attracting flies and mosquitoes and other disease vectors. In some cases, too much oil and grease causes septic conditions in ponds and lakes by preventing oxygen from the atmosphere from reaching the water. Organic impurities are composed of the carbon-based chemicals that are the building blocks of most living things. Organic impurities include materials in wastewater that originate from plants, animals or synthetic organic compounds and enter wastewater as human wastes, paper products, detergents, cosmetics and foods and from agricultural, commercial and industrial sources. Organic compounds normally are some combination of carbon, hydrogen, oxygen, nitrogen and other elements. Many organics are proteins, carbohydrates or fats and are biodegradable, which means they can be consumed and broken down by organisms.

However, even biodegradable materials can cause pollution. Some organic compounds are more stable than others and cannot be quickly broken down by organisms, posing an additional challenge for treatment. This is the case of many synthetic organic compounds developed for agriculture and industry. In addition, certain synthetic organics are highly toxic. Insecticides and herbicides are toxic to humans, fish and aquatic plants and often are disposed of improperly in drains or carried in storm water. In receiving waters, they kill or contaminate fish, making them unfit to eat. They also can damage processes in treatment plants. Benzene and toluene are two toxic organic compounds found in some solvents, pesticides and other products. New synthetic organic compounds are being developed all the time, which can complicate treatment efforts.

Pathogens including many disease-causing viruses, parasites and bacteria also are present in wastewater and enter from anywhere in the community. These pathogens often originate from people and animals that are infected with or are carriers of a disease, e.g. grey water and black water from typical homes contain enough pathogens to pose a risk to public health. Other sources in communities include hospitals, schools, farms and food processing plants.

17.3 Wastewater Treatment

Wastewater contains a large number of impurities of diverse nature including inorganic contaminants, organic contaminants, nutrients, oil and grease and pathogens. These impurities are very dangerous for the environment. Before its disposal into running water, in underground sources or on land, it should be treated.

Wastewater treatment consists of applying known technology to improve or upgrade the quality of a wastewater. Wastewater treatment is done in successive steps like collecting the wastewater in a central, segregated location, i.e. the Wastewater Treatment Plant, and subjecting the wastewater to various treatment processes in continuously flowing system (continuous flow or 'open' systems) rather than as 'batch' or a series of periodic treatment processes in which treatment is carried out in batches of wastewaters. A complete treatment system may consist of the application of a number of physical, chemical and biological processes to the wastewater as shown in Fig. 17.2.

Physical treatment involves sedimentation (clarification), screening, aeration, filtration, flotation and skimming, degasification and equalisation. It includes processes where no gross chemical or biological changes are carried out and strictly physical phenomena are used to improve or treat the wastewater.

Chemical treatment involves chlorination, ozonation, neutralisation, coagulation, adsorption and ion exchange. It consists of using some chemical reaction or reactions to improve the water quality. Probably the most commonly used chemical process is chlorination. Chlorine, a strong oxidising chemical, is used to kill bacteria and to slow down the rate of decomposition of the wastewater.

Biological treatment includes *aerobic* (activated sludge treatment methods, trickling filtration, oxidation ponds, lagoons, aerobic digestion) and *anaerobic* (anaerobic digestion, septic tanks, lagoons). These uses microorganisms, mostly bacteria, in the biochemical decomposition of wastewaters to stabilise end products. More microorganisms or sludges are formed, and a portion of the waste is converted to carbon dioxide, water and other end products.

17.4 Biotechnology for Domestic Wastewater Treatment

Biotechnology-based or biological treatment process is one of the most widely used removal methods as well as for partial or complete stabilisation of biologically degradable substances in wastewaters. The degree of impurities removal is depending upon the nature of the wastewater. The characteristics of wastewaters are measured in terms of chemical oxygen demand (COD), biochemical oxygen demand (BOD) and volatile suspended solids *(VSS)*, and a typical characteristic of domestic wastewater is as shown in Table 17.1.

Most biological waste and wastewater treatment processes employ bacteria as primary microorganisms; certain other microorganisms may play an important role. Degradation of organic matter is affected by its use as food by microorganisms

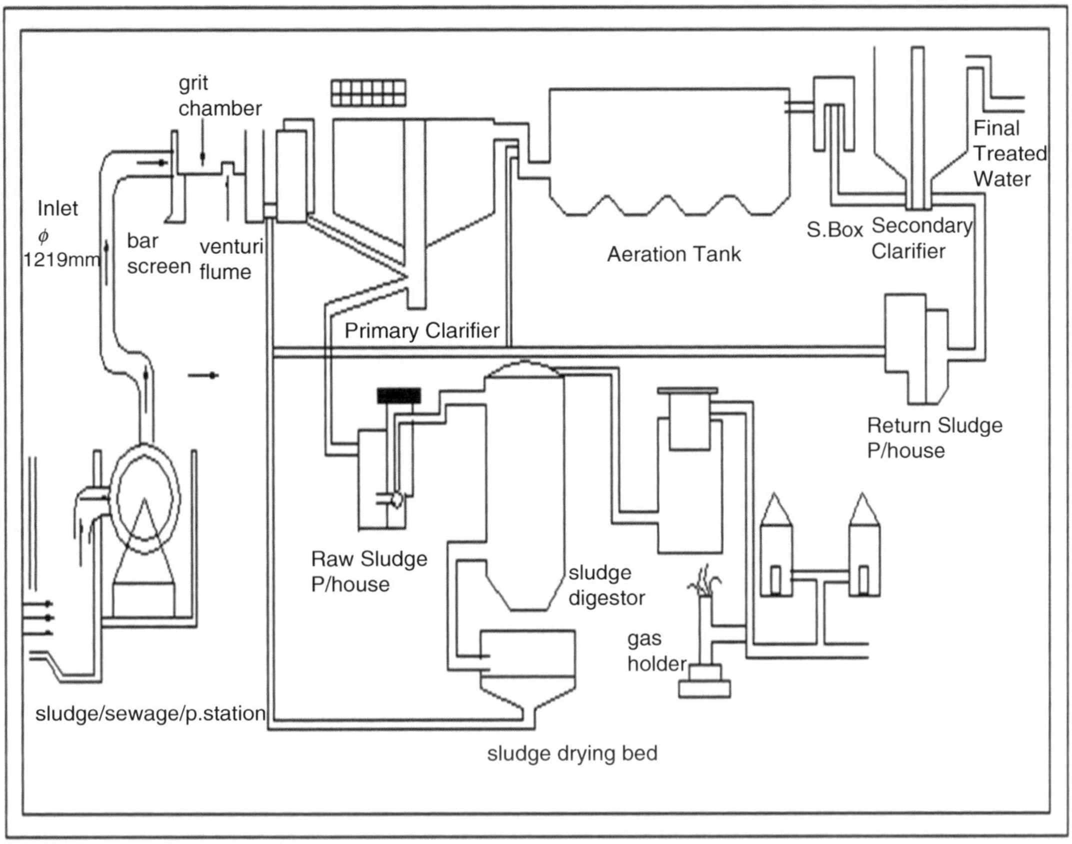

Fig. 17.2 Main units for the domestic wastewater treatment

Table 17.1 Physico-chemical analysis of domestic wastewater and CPCB permissible limits

Parameters	Values	Maximum permissible limits (CPCB norms)
pH	7.2	6.8–8.5
Electrical conductance (mho/cm)	3.1×10^{-4}	1×10^{-3}
Temperature (°C)	20.5	16–32
Turbidity (nephalometer turbidity unit)	320	5–10
Total solids (mg/l)	1,031	500
Total suspended solids (mg/l)	341	10–50
Total dissolved solids (mg/l)	690	450
Chemical oxygen demand (mg/l)	1,080	100
Biochemical oxygen demand (mg/l)	505	20–30
MPN (coliform cells/100 ml)	2×10^{9}	1.5×10^{7}

Source: Devi and Dahiya (2006)

to produce protoplasm for new cells during the growth process.

First of all microorganisms are isolated then selected, mutated and genetically engineered for effective bioremediation capabilities (Renner 1997; Pieper and Reineke 2000) even to degrade recalcitrant pollutants, to achieve enhanced rates of degradation of target compounds and to assure better survival and colonisation in target polluted sites. Microorganisms have been the main focus of the effort for improving bioremediation capabilities, but use of higher plants in phytoremediation is also a significant developing area (Macek et al. 2000; Glick 2003). Ecologically integrated mixed bioremediation systems are towards the more popular options for bioremediation process. Bioremediation processes can be applied for both in situ and ex situ treatment of contaminated wastewater.

These techniques can offer significant cost and environmental benefits in comparison with

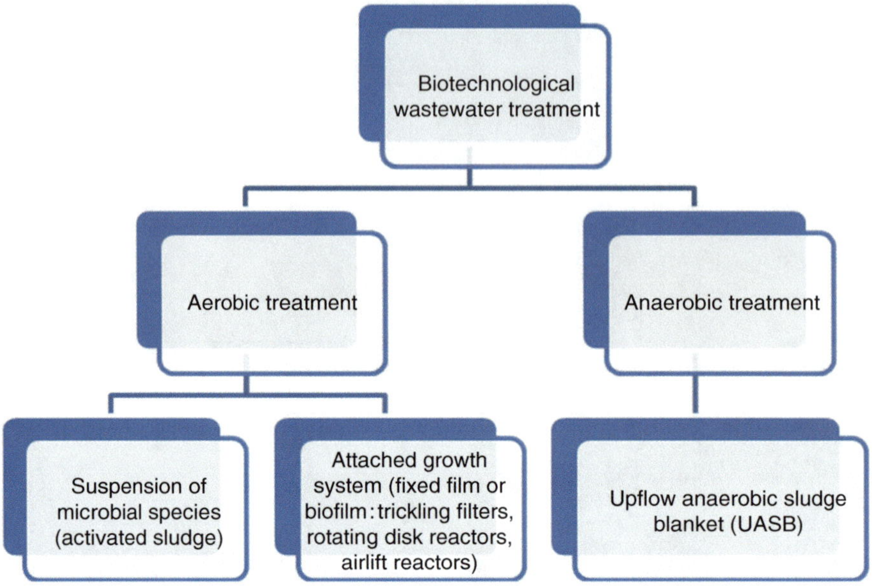

Fig. 17.3 A layout of biotechnological wastewater treatment process

alternative technologies (Lee and de Mora 1999; Jördening and Winter 2004; Khan et al. 2004) and provide the industry significant new tools for enhancing profitability and sustainability.

The important microbial species for wastewater treatment are Acinetobacter calcoaceticus, Acinetobacter radioresistens, Aeromonas sp., Alcaligenes latus, Alteromonas sp., Bacillus subtilis, Corynebacterium petrophilum, Kingella denitrificans, Micrococcus sp., Nocardia amarae, Pseudomonas aeruginosa, Pseudomonas carboxydohydrogena, Rhodococcus aurantiacus, Rhodococcus globerulus, Rhodococcus rubropertinctus, Sphingobacterium thalpophilum, Torulopsis bombicola, Deinococcus radiodurans (genetically modified and radiation-resistant species), etc.

17.4.1 Process of Wastewater Treatment

Biotechnology-based wastewater treatment involves the following techniques:

Bioaugmentation: The process of increasing the efficiency of the naturally occurring microbial population to concentrate or accumulate specific compounds. It is achieved by adding oxygen, water and nutrients.

Bioremediation: The use of naturally occurring or genetically modified microorganisms to break down or degrade hazardous substances into less or non-toxic substances.

Bioreactors: Closed container in which microorganisms are maintained under controlled conditions for the purpose of creating or destroying specific species.

Phytoremediation: The use of vegetative species for the remediation of polluted sites/water bodies.

17.4.1.1 Wastewater Treatment

A simple flow diagram of domestic wastewater treatment based on biotechnological approach as shown in Fig. 17.3:

17.4.1.1.1 Aerobic Wastewater Treatment

In this process sufficient oxygen is required, and easily biodegradable organic material is treated. The basic reaction in aerobic treatment is:

1. $\text{Organic material} + O_2 \xrightarrow[\textit{others nutrients}]{\textit{cells}} CO_2 + H_2O + \text{new cells}$
2. Microbial cells undergo progressive auto-oxidation of the cell mass:

$$\text{Cells} + O_2 \rightarrow CO_2 + H_2O + NH_3$$

Aerobic treatment process may involve suspension of microbial species (activated sludge) or attached growth system (fixed film or biofilm: trickling filters, rotating disc reactors, airlift reactors). In this biofilter system, the wastewater passes down through a floating filter, and filter material is continuously removed from the top of the reactor, cleaned and added back at the bottom of the reactor, avoiding the need for backwashing. This system achieved a high chemical oxygen demand removal capacity per unit area, but the removal capacity was limited by biofilm processes, and so research is now aimed at changing the filter material to increase the biofilm surface area.

The facts about the current aerobic technology to treat wastewater are that it generally occurs in a centralised way that is effective at first sight. When observing carefully it reveals that such treatment is still poorly understood: it relies on microbial communities which come to existence but, as yet, are hard to control by, for example, the addition of certain. Moreover, the conventional treatment is costly, i.e. 100 Euro per inhabitant per year. Another major drawback is the fact that it is not holistic at all: half of the organic matter (COD) becomes sludge which has to be disposed of either by depositing it onto the soil or by incineration.

17.4.1.1.2 Anaerobic Wastewater Treatment

In these processes oxygen supply is not required, and the energy intake is very slow. These processes are not very frequently used for domestic wastewater but used only to remove polychlorinated biphenyls (PCBs) and chloroform and dechlorination of solvent trichloroethylene (TCE) with high rate anaerobic wastewater treatment technology, i.e. sulphate reduction and bioremediation. In sulphur reduction, sulphates react with organic matter and convert it into disulfide and carbon dioxide. Then disulfide reacts with oxygen and produce elemental sulphur and water. Finally, in most instances, anaerobic wastewater treatment is mainly focused on organic carbon removal and needs to be completed by an adequate after treatment for N, P and infectious propagules. Upflow anaerobic sludge blanket (UASB) reactors can be used to treat sulphur-rich wastewater (Tuppurainen et al. 2002; Lens et al. 2004). This process produces less quantities of biological excess sludge with high treatment efficiency, low capital costs, no oxygen requirements, methane production and low nutrient requirements. Anaerobic digestion performs poorly at low temperatures and therefore needs to be linked to low-value heat recovery processes.

17.5 Biotechnology-Based Low-Cost Domestic Wastewater Technique and Their Future Prospects

Biotechnology-based domestic wastewater treatment system is very economical due to the following reasons:

1. Greatly reduced dependence on nonrenewable fuels and other resources
2. Reduced potential for pollution of industrial processes and products
3. Ability to safely destroy accumulated pollutants for remediation of the environment
4. Improved economics of production
5. Sustainable production of existing and novel products

The fact that microbial biocatalysis is relatively very cheap is the predominant reason why so many environmental technologies prefer to rely on microorganisms. As a comparison: 1 kg waste (dry matter) costs in terms of incineration of the order of Rs. 30; biological mineralisation costs generally a factor 10 less. The second major advantage of microbiological processes is that they are flexible; they adapt to variable conditions (self-regulation) and also to new molecules or combinations of new molecules.

A good example is the fact that in recent years the recalcitrant herbicide atrazine has been found to be degraded by microorganisms (Evy et al. 2012; Zhang et al. 2010). Microbial species and communities continue to 'learn' to metabolise or co-metabolise various xenobiotics. Another fact is that environmental biotech is perceived as 'green'. We have all experienced that conventional sewage treatment installations with their predominance of concrete and electromechanics are not very much liked by the public, whilst reed beds and wetlands are considered environmentally useful. The message that the latter are quite expensive and can often not perform the treatment in a sufficient way is difficult to convey to the public. In our experience, the above-mentioned features govern the continuous growing interest in environmental biotechnology.

Considering the fact that environmental biotechnology can be operated in an inexpensive, flexible and also environmentally friendly way, there are a series of areas where microbial biocatalysis can and will expand in the near future.

17.6 Conclusion

Globally environmental biotechnology has increased importance and emphasis on the treatment wastewater. Environmental biotechnology is a boon for the society, and by adopting this biotechnology, it is a powerful technology for treating the domestic wastewater and achieving clean development mechanism for water sustainability.

It has advantages and disadvantages; some are scientific and ecological whilst others relate to its organisation on a global basis. Instead of resulting in polarisation between the developed and developing world, biotechnology could lead to mutual benefits. It led to greatly reduced dependence on nonrenewable fuels and other resources or a lower by-product production and its disposal. Thus, biotechnology has a key role to play in the novel approaches to design wastewater treatment based on decentralised sanitation and reuse.

The biotechnology, however, has some disadvantages. The certain species of organisms that prove invasive and/or toxic is a major concern from the environmental point of view. Culturally, developing countries may be disadvantaged by the near monopoly of biotechnology by transnational companies based in the developed world. These issues reflect the close relationships between wastewater treatment, the environment and society. Achieving greater involvement of biotechnology for environmental purposes will require joint R&D efforts by government and industry.

References

Devi R, Dahiya RP (2006) Chemical oxygen demand (COD) reduction in domestic wastewater by fly ash and brick kiln ash. Int J Water Air Soil Pollut 174(1–4):33–46

Evy A, Abigail M, Nilanjana D (2012) Microbial degradation of atrazine, commonly used herbicide. Int J Adv Biol Res 2(1):16–23

Gavrilescu M (2004) Removal of heavy metals from the environment by biosorption. Life Sci Eng 4(3): 219–232

Glick BR (2003) Phytoremediation: synergistic use of plants and bacteria to clean up the environment. Biotechnol Adv 21:383–393

Jördening HJ, Winter J (eds) (2004) Environmental biotechnology: concepts and applications. Wiley-VCH, Weinheim, pp 39–48

Khan FI, Husain T, Hejazi R (2004) An overview and analysis of site remediation technologies. J Environ Manag 71:95–122

Khin T, Annachhatre AP (2004) Novel microbial nitrogen removal processes. Biotechnol Adv 22:519–532

Lee K, de Mora S (1999) In situ bioremediation strategies for oiled shoreline environments. Environ Technol 20:783–794

Lens P, Von der Mass P, Zandvoort M, Vallero M (2004) New development in anaerobic environmental biotechnology. In: Ujang Z, Henze M (eds) Environmental biotechnology: advancement in water and wastewater application in tropics. IWA Publishing, London, pp 13–18

Liu Y, Tay JH (2004) State of the art of biogranulation technology for wastewater treatment. Biotechnol Adv 22(5):33–63

Macek T, Macková M, Káš J (2000) Exploitation of plants for the removal of organics in environmental remediation. Biotechnol Adv 18:23–34

Pieper DH, Reineke W (2000) Engineering bacteria for bioremediation. Curr Opin Biotechnol 11:262–270

Pletsch M, Santos de Araujo B, Charlwood BV (1999) Novel biotechnological approaches in environmental remediation research. Biotechnol Adv 17:679–687

Renner R (1997) On the trail of bioremediating microbes. Environ Sci Technol 31:188–189

Tuppurainen KO, Vaisanen O, Rintal JA (2002) Sulphate reducing laboratory scale high rate anaerobic reactors for treatment of metal and sulphate containing mine wastewater. Environ Technol 23:599–608

Zhang CZ, Zhang ZY, Liu XJ, Jiang W, Wu Y (2010) Dissipation and environmental fate of herbicide H-9201 in carrot plantings under field conditions. Food Chem 119:874–879

Part IV

Animal Biotechnology

18 Whole Genome Sequencing Strategies and Development of *Orbivirus* Sequence Database: Implications for Novel dsRNA Virus Detection

Sushila Maan, Manjunatha N. Belaganahalli, Narender S. Maan, and Peter P.C. Mertens

Abstract

The genus *Orbivirus* is the largest of the genera within the family *Reoviridae*, containing 22 recognised virus species as well as 15 unclassified 'orbiviruses', which could potentially represent further new species. The orbiviruses are transmitted by both ticks and/or haematophagous insect vectors. They have a wide host range that includes domestic and wild ruminants, equines, marsupials, sloths, bats, birds and humans. Low-level serological cross-reactions between different species of orbiviruses and lack of reference strains/antisera for existing *Orbivirus* species make serological identification of new virus isolates difficult. Recently, whole genome sequence data (WGS) has become an important tool for the detection, classification and epidemiological investigations of different pathogens. This study presents full genome sequence database of all known 22 *Orbivirus* species (including 5 unclassified viruses). Development of novel sequencing strategies and phylogenetic analysis of the orbiviruses using this database has identified five novel *Orbivirus* species and has facilitated development of a pan-orbivirus RT-PCR assay that can be used to identify the RNA of any *Orbivirus* species. These techniques will support *Orbivirus* discovery with greater accuracy than before and can be used for definitive diagnosis of suspected *Orbivirus* infection.

S. Maan (✉) • N.S. Maan
LLR University of Veterinary and Animal Sciences, Hisar, Haryana 125004, India
e-mail: sushilamaan105@gmail.com

M.N. Belaganahalli
Arbovirus Molecular Research Group, Institute for Animal Health, Pirbright Laboratory, Ash Road, Woking, Surrey GU24 0NF, UK

P.P.C. Mertens
Vector-borne Viral Diseases (VVD) Programme, Institute for Animal Health, Pirbright Laboratory, Ash Road, Woking, Surrey GU24 0NF, UK

18.1 Introduction

The genus *Orbivirus* is the largest of the genera within the family *Reoviridae*, containing 22 recognised virus species as well as 15 unclassified 'orbiviruses'. The orbiviruses are transmitted by both ticks and haematophagous insect vectors. The orbiviruses have a wide host range that includes domestic and wild ruminants, equines, marsupials, sloths, bats, birds and humans. Bluetongue virus

R.K. Salar et al. (eds.), *Biotechnology: Prospects and Applications*,
DOI 10.1007/978-81-322-1683-4_18, © Springer India 2013

(BTV), African horse sickness virus (AHSV) and epizootic haemorrhagic disease virus (EHDV) are the most economically important orbiviruses.

In this study, full genomes of 22 orbiviruses (including 5 unclassified viruses) were sequenced to generate a complete sequence database representing the entire genus *Orbivirus*. Sequencing and phylogenetic analysis of the orbiviruses using this database has identified five novel *Orbivirus* species (Mitchell River virus, Pata virus, Andasibe virus, Japanaut virus and Matucare virus) and grouped four unclassified viruses (TRV, TMEV, CMPV and SLOV) into existing species (*Changuinola virus, Lebombo virus, Corriparta virus* and *Umatilla virus*, respectively). This study provides the most comprehensive phylogenetic study of genus *Orbivirus* to date, using genetic sequences for all of the genome segments.

The phylogenetic trees show characteristic branching patterns that reflect vector grouping, supporting the hypothesis that orbiviruses are evolving through 'co-speciation' with their vectors. This dataset has also helped in the development of novel pan-orbivirus diagnostic assays, group- and type-specific RT-PCR assays for BTV, AHSV, EHDV and equine encephalosis virus (EEV), which can be successfully used for *Orbivirus* discovery.

This chapter will outline and describe the unique features of *Orbivirus* genome, presents a complete reference database for the entire genus *Orbivirus* and summarises sequencing strategies for these viruses, which can be used to further understand their taxonomic relationships, epidemiology and evolutionary processes and facilitate the development of molecular diagnostic assays.

18.2 History of the Orbiviruses

The Book of Exodus describes the fifth plague of Egypt, potentially the first record of a true epizootic (a disease that infects animals but not humans) caused by the orbiviruses, possibly involving not only bluetongue virus (BTV) but also African horse sickness virus (AHSV) and epizootic haemorrhagic disease viruses (EHDV) (Marr and Malloy 1996).

Bluetongue (BT) was first described as a disease in cattle and sheep in the late eighteenth century by a French biologist Francois de Vaillant, who identified it as 'tong-sikte' during his travels in the Cape of Good Hope between 1781 and 1784 (Gutsche 1979). However, the first detailed scientific descriptions of the disease in sheep were given by Hutcheon (1902) and Spreull (1902) who named it as 'malarial catarrhal fever'. Later, Spreull suggested, the name of the disease should be changed to 'bluetongue' (BT) based on the distinctive although infrequent cyanotic appearance of the mouth and tongue in some severely affected animals (Spreull 1905). The first confirmed outbreak of BT outside Africa occurred in Cyprus in 1943 (Gambles 1949). However, it is believed that periodic although unconfirmed outbreaks may have previously occurred in Cyprus since 1924 (Polydorou 1985).

African horse sickness (AHS) is another ancient viral disease that may have caused an epidemic in the Yemen during 1327 (Moule 1896). However, the virus almost certainly originated in Africa, with early observations made by Father Monclaro in 1569 (Theal 1899). AHSV has probably been present in southern Africa from distant times, although the disease AHS was not recognised until 1719, when over 1,700 animals died (Henning 1956). Until relatively recently AHSV has been confined primarily to sub-Saharan Africa, which may reflect the distribution of zebra, regarded as its primary wildlife host. However, occasional incursions have also been recorded in North Africa and the Arabian Peninsula. During the period 1959–1961, AHSV spread across Saudi Arabia, Syria, Lebanon, Jordan, Iraq, Turkey, Cyprus, Iran, Afghanistan, Pakistan and India killing many hundred thousand animals (Mellor and Hamblin 2004; Rafyi 1961).

Epizootic haemorrhagic disease (EHD) was first formally identified in 1955 by Richard Shope and colleagues in New Jersey (USA), following reports of a fatal epizootic in white-tailed deer (*Odocoileus virginianus*) (Shope et al. 1955). However, retrospective studies suggest that the disease had existed in wild ruminants throughout the south-eastern USA since 1890, where it was known among woodsmen and hunters as 'blacktongue' (Shope 1955).

After 1950, many orbiviruses have been isolated either from vertebrate hosts or from insect vectors, often named after the location of their first isolation (Karabatsos 1985).

18.3 The Family *Reoviridae*

The ‘reoviruses’ (used here to include any member of the family *Reoviridae*) are non-enveloped, icosahedral viruses, which have genomes composed of 9–12 linear segments of dsRNA, with a total genome size of 19–32 kbp (Attoui et al. 2005a; Mertens et al. 2005b). The family *Reoviridae* contains a total of 15 recognised genera, distributed across a wide variety of geological ‘econiches’, infecting terrestrial and non-terrestrial vertebrates, terrestrial and non-terrestrial invertebrates, plants, fungi and protozoa (Attoui et al. 2009a; Mertens 2004; Mertens et al. 2005a). Some of these genera contain viruses that have icosahedrally arranged surface projections, known as ‘turrets’ or ‘spikes’, which are included in the subfamily *Spinareovirinae*, while others have a more bristly or smooth appearance and are included in the subfamily *Sedoreovirinae* (Attoui et al. 2006, 2009a; Mertens et al. 2005a).

18.4 Genus *Orbivirus*

The genus *Orbivirus* is the largest of the genera within the family *Reoviridae*, containing 22 recognised virus species as well as 15 unclassified ‘orbiviruses’ (Borden et al. 1971; Murphy et al. 1971). The orbiviruses are transmitted by both ticks and haematophagous insect vectors (phlebotomine flies, mosquitoes or *Culicoides*). The orbiviruses have a wide host range that includes domestic and wild ruminants, equines, marsupials, sloths, bats, birds and humans. *Bluetongue virus* is the *Orbivirus* type species and has been most extensively studied, while *African horse sickness virus* and *epizootic haemorrhagic disease virus* represent other virus species within the genus *Orbivirus*, family *Reoviridae* (Table 18.1) (Brown et al. 1992; Cowled et al. 2009; King et al. 2011; Martins et al. 2007; Mertens et al. 2005a; Vieira Cde et al. 2009).

18.5 Classification and Differentiation of the *Orbivirus* Species

Viruses within each *Orbivirus* species share common antigens that are detectable by complement-fixation (CF), agar gel immunodiffusion (AGID), fluorescent antibody tests (FAT) or enzyme-linked immunosorbent assays (ELISA) (Gorman et al. 1983; Mertens et al. 2005a). Consequently the different *Orbivirus* species match the previously recognised ‘serogroups’ of these viruses (Mertens et al. 2005a). However, in some cases low-level or ‘one-way’ cross-reactions have been detected in some serological assays and in cross-hybridisation studies, between different species, e.g. between the BTV and EHDV serogroups (Della-Porta et al. 1985; Huismans et al. 1979; Moore 1974), potentially making virus identification, diagnosis and classification more difficult (Borden et al. 1971; Della-Porta et al. 1985; Moore 1974; Moore and Lee 1972).

Although virus ‘species’ is the most fundamental taxonomic category in all biological classifications, it was only in 1991 that the ICTV agreed that the concept of virus species should be uniformly applied in virus classification and defined virus species as a ‘polythetic class’ rather than as a traditional ‘universal class’.

In common with the other genera within the family *Reoviridae*, ‘reassortment of genome segments during co-infection of the same cell, by closely related orbiviruses, resulting in an exchange of genetic information and creation of viable progeny virus strains’, is considered to be the prime determinant for inclusion of different viruses within a single *Orbivirus* species (Mertens et al. 2005a). However, in the absence of data concerning the ability of specific reovirus strains to reassort (particularly for new virus isolates), other parameters can be used (either singly or in combination) to identify members of the same virus species. This provides a polythetic definition of virus species as recommended by ICTV.

A polythetic class consists of members which all have a number of properties in common but which may not all have been shown to share any

Table 18.1 List of Orbivirus species in the genus *Orbivirus*

S. No	Species	Abbreviation	No. of serotypes	Year[a]	Vertebrate hosts	Vectors
1	*African Horse sickness virus*	AHSV	9	1899	Equines and dog	*Culicoides* spp.
2	*Bluetongue virus*	BTV	26	1905	Domestic and wild ruminants	*Culicoides* spp.
3	*Changuinola virus*	CGLV	12	1960	Sloth, armadillo, anteater, rodents and human (mosquito catcher)	Phlebotomine flies
4	*Chenuda virus*	CNUV	7	1954		
5	*Chobar Gorge virus*	CGV	2	1970	Bats (isolation) and domestic animals (serology)	Ticks
6	*Corriparta virus*	CORV	6	1960	Shorebird, human and domestic fowls (serology)	Culex mosquitoes
7	*Epizootic haemorrhagic disease virus*	EHDV	7	1955	White-tailed deer, cattle	*Culicoides* spp.
8	*Equine encephalosis virus*	EEV	7	1967	Equines	*Culicoides* spp.
9	*Eubenangee virus*	EUBV	4	1963	Kangaroos, wallabies and cattle	Mosquitoes and *Culicoides* spp.
10	*Great island virus*	GIV	36	1971	Seabirds, domestic animals, horses and humans	Ticks
11	*Ieri virus*	IERIV	3	1955	Birds	Mosquitoes
12	*Lebombo virus*	LEBV	1	1956	Man, rodent	Mosquitoes
13	*Orungo virus*	ORUV	4	1959	Man (isolation), sheep, cattle, goat and camels (serology)	Mosquitoes
14	*Palyam virus*	PALV	13	1956	Domestic and wild ruminants	*Culicoides*, mosquitoes and ticks
15	*Peruvian horse sickness virus*	PHSV	1	1997	Horse	Mosquitoes
16	*St Croix river virus*	SCRV	1	1994	Unknown	Ticks
17	*Umatilla virus*	UMAV	4	1969	Birds	Mosquitoes
18	*Warrego virus*	WARV	3	1969	Marsupials (isolation) and cattle (serology)	*Culicoides* spp., mosquitoes
19	*Wallal virus*	WALV	3	1970	Marsupials	*Culicoides* spp.
20	*Wad Medani virus*	WMV	2	1952	Domestic animals and rodents	Ticks
21	*Wongorr virus*	WGRV	8	1970	Cattle and marsupials	*Culicoides* spp., Mosquitoes
22	*Yunnan orbivirus*	YUOV	2	1997	Cattle, sheep and donkeys	Mosquitoes

Tentative species					
1	Andasibe virus	ANDV	1979	Unknown hosts	Mosquitoes
2	Breu Branco virus		1988	Unknown	Mosquitoes
3	Codajas virus	COV	1984	Unknown hosts	Mosquitoes
4	Ife virus	IFEV	1971	Bats, rodents, birds, ruminants	Mosquitoes
5	Itupiranga virus	ITUV	1976	Unknown hosts	Mosquitoes
6	Japanaut virus	JAPV	1969	Bats and some unknown hosts	Mosquitoes
7	Kammavanpettai virus	KMPV	1963	Birds	Unknown
8	Lake Clarendon virus	LCV	1981	Birds, cattle (serology)	Ticks
9	Matucare virus	MATV	1963	Unknown hosts	Ticks
10	Minacu virus		1996	Unknown hosts	Mosquitoes
11	Tembe virus	TMEV	1963	Unknown hosts	Mosquitoes
12	Tracambe virus	TRV	1985	Unknown hosts	Mosquitoes
13	Stretch Lagoon orbivirus	SLOV	2002	Equines and goats	Mosquitoes
14	California mosquito pool virus	CMPV	1974	Unknown hosts	Mosquitoes
15	Golok virus		1981	Unknown hosts	Mosquitoes

[a]Year of first isolation of prototype serotype in the species

single specific characteristic that could be used as a defining and discriminating property of the species (van Regenmortel and Mahy 2004).

Based on these criteria, 22 species of orbiviruses are currently recognised together with 15 unassigned viruses that may represent further species (Cowled et al. 2009; King et al. 2011; Martins et al. 2007; Mertens et al. 2005a; Vieira Cde et al. 2009). Each of the different *Orbivirus* species contains a number of distinct virus isolates and distinct serotypes, the identity of which is determined by the specificity of reactions between neutralising antibodies generated during infection of the vertebrate host and protein components of the outer capsid (Gould and Eaton 1990; Huismans and Erasmus 1981; Mertens and Attoui 2009; Mertens et al. 1989). In recent years, genetic sequence data has steadily become more important in the identification of different orbiviruses (Anthony et al. 2007; Aradaib 2009; Aradaib et al. 2009; Attoui et al. 2005b; Cowled et al. 2007; Maan et al. 2011b, 2012; Shaw et al. 2007) providing additional information concerning the classification of virus species, 'types' and even different virus lineages from different geographic regions (topotypes), which cannot be generated by earlier serological assays.

Bluetongue virus (BTV) that causes an infectious but non-contagious disease of domestic and wild ruminants (bluetongue, BT) has been studied in greatest depth providing a useful paradigm for the structural properties, organisation and replication strategies of other orbiviruses. BTV is the prototype species of the genus *Orbivirus* and is transmitted between its mammalian hosts primarily via the bite of infected midges of the genus *Culicoides*. There are 26 distinct serotypes of BTV, which exist as a complex mixture of different topotypes (geographic variants) and nucleotypes (closely related groups of serotypes) and strains that have the potential to exchange genome segments (Maan et al. 2009). None of these (serotypes, topotypes or nucleotypes) has any formal taxonomic status, as classification of the reoviruses does not currently extend beyond virus species (Maan et al. 2011b, c) (Maan et al. 2007a; Mertens et al. 2005a).

18.5.1 Properties of the Orbiviruses

The orbivirus particle molecular weight (Mr) is $\sim 10.8 \times 10^7$ and the core Mr is $\sim 6.7 \times 10^7$. The 10 linear dsRNA segments of the orbivirus genome range in size from 3.9 kb (segment 1) down to 0.8 kb (segment 10) and are packaged in exactly equimolar ratios, one of each segment per particle (Huismans et al. 1979; Mertens et al. 2005a; Verwoerd et al. 1970). The RNA segments are numbered segment 1 to segment 10 (Seg-1 to Seg-10) in the order of decreasing molecular weight and in the order of their migration during agarose gel electrophoresis (AGE) (Mertens et al. 2005a).

The genome segments encode 7 structural proteins (VP1-VP7) and four nonstructural proteins (NS1, NS2, NS3/3a and NS4) (Belhouchet et al. 2011a; Mertens et al. 1984, 2005a). In common with other members of the family *Reoviridae*, the majority of the orbivirus genome segments are thought to be monocistronic. However, two related proteins can be produced from the same ORF, starting at different in-frame initiation codons on Seg-9 and Seg-10 – generating VP6 and VP6a, or NS3 and NS3a, respectively (French et al. 1989; Mertens et al. 1984, 2005a; Wade-Evans 1990; Wade-Evans et al. 1992; Wu et al. 1992). Bioinformatic analyses have also identified a second out-of-frame (+1) ORF in Seg-9 of BTV, which is translated in infected mammalian cells into a further nonstructural protein (Firth 2008). An alternate out-of-frame ORF was also identified in Great island virus (GIV) Seg-9 (Belhouchet et al. 2010). The first, 'ORF-1', encodes the helicase,VP6(Hel), thought to be involved in RNA replication, while the second, 'ORF-2', encodes a novel protein identified as VP6(dBP) or NS4, BTV (in GIV and BTV, respectively) which contains a dsRNA-binding domain similar to those located in dsRNA-binding proteins of other dsRNA viruses (Attoui et al. 2000a; Belhouchet et al. 2010, 2011b; Mohd Jaafar et al. 2005; Ratinier et al. 2011). Full details of each protein are available on the dsRNA website, http://www.iah.bbsrc.ac.uk/dsRNA_virus_proteins/Orbivirus.htm.

The orbivirus genome segments have hexanucleotide termini, which are often fully conserved and usually at least partially conserved (between the different genome segments) within isolates of each *Orbivirus* species and to a lesser extent between different *Orbivirus* species (Mertens and Sangar 1985; Mertens and Diprose 2004; Mertens et al. 2005a; Rao et al. 1983) (http://www.reoviridae.org/dsRNA_virus_proteins/CPV-RNA-Termin.htm). The noncoding region (NCR) of all the orbivirus genome segments that have been sequenced contains two conserved base pair (bp) at either terminus (+ve 5′-GU.AC-3′) (Mertens et al. 2005a). The first and last two nucleotides of each BTV genome segment are inverted complements and are identical to those found in other orbiviruses. These inverted repeats may be involved in controlling the efficiency of viral mRNA translation (Roy 1989). For BTV, the 5′-NCRs range from 8 to 34 bp, while the 3′-NCRs are 24–116 bp in length (Maan et al. 2008; Mertens et al. 2005a).

In most mammalian cell lines (e.g. BHK, BSR, Vero, PS-EC), replication of orbiviruses usually results in cell lysis or cytopathic effect (CPE) and the release of virus particles (Karabatsos 1985; Mertens et al. 1987). However, in a variety of insect cell cultures (C6/36 and KC cells), it replicates with no extensive cell lysis or cytopathic effect (CPE) (Mertens et al. 1996, 2005a).

The orbiviruses are 'true' arboviruses that are transmitted between their vertebrate hosts by ticks and/or haematophagous insect vectors (phlebotomine flies, mosquitoes or *Culicoides*) with a wide host range that collectively includes domestic and wild ruminants, equines, marsupials, sloths, bats, rodents, birds, and humans (Attoui et al. 2009a; Karabatsos 1985; King et al. 2011; Mertens et al. 2005a). Based primarily on studies of BTV, it is clear that the orbiviruses can replicate in many insect and mammalian cell types. In their mammalian hosts, they principally infect vascular endothelial cells, mononuclear phagocytic cells in the lungs/lymphoid tissues and skin fibroblasts (Darpel 2007). *Orbivirus* infection of insects cells (e.g. KC cells) results in persistent infection, with little detectable cytopathology. The virus is released more slowly than from infected mammalian cells and there is no evidence for shut-off of host cell protein synthesis. Some mammalian cell types (such as γδ T cells and dendritic cells) may also become persistently infected, playing an important role in the movement and dissemination of the virus in the infected animal (Hemati et al. 2009; Mertens and Attoui 2009; Takamatsu et al. 2003).

18.5.2 Disease and Epidemiology of Orbiviruses

Although orbiviruses can infect and replicate in both their mammalian and arthropod hosts, they appear to have little adverse effect on the vector insects, which become persistently infected for life (Fu et al. 1999). However, the disease can vary from fatal to silent/inapparent in the mammalian host, influenced by many factors including the level of stress (e.g. UV light), virus strain/attenuation/adaptation for the host, the species/breed/age of the host itself and the level of acquired immunity (immune response) by the host (Anthony 2007; Attoui et al. 2009a).

BTV, EHDV and AHSV are currently the most economically important orbiviruses. However, viruses within several other species are also potentially important, either regionally or globally (including EEV, PALV and PHSV). BTV and EHDV can both infect domestic and wild ruminants, while AHSV infects equines. All three diseases (BT, EHD and AHS) are listed by OIE as notifiable diseases (Maclachlan and Guthrie 2010; OIE 2010). Currently, BTV is distributed around the world with and expanding the list of new serotypes (http://www.reoviridae.org/dsRNA_virus_proteins/btv-serotype-distribution.htm). It is also likely that there are several other undiscovered arboviruses (including other orbiviruses) in regions such as South America and India.

During the last three decades, the incidence of a number of diseases that were previously unknown or thought to be controlled has increased dramatically. Bats have recently been recognised as a significant reservoir of different pathogens, with several different viruses attributed to them, including Hendra and Nipah viruses. The orbiviruses Ife, Japanaut, Fomede and Chobar Gorge viruses have also been isolated from bats and may have potential to 'emerge' as more significant threats to human and domesticated animal (Calisher et al. 2006).

Some novel orbiviruses have been discovered as the result of disease outbreaks in vertebrate hosts, for example, 'Peruvian horse sickness virus' (PHSV), which was first isolated in horses from Peru in 1997. Subsequently, a second isolate of PHSV, Elsey station virus, was isolated from horses in Australia (Attoui et al. 2009b). However it is also possible for the characterised viruses, classified within established *Orbivirus* species, to emerge causing major outbreaks of disease, as seen with BTV in northern Europe (2006–2010).

18.5.3 *Orbivirus* Diagnosis

Orbivirus diagnosis has traditionally been based primarily on virus isolation and subsequent characterisation of the virus isolate using serological assays. However, these assays are time-consuming and can take several weeks to confirm initial clinical diagnoses (Akita et al. 1992; Tabachnick et al. 1996; Wade-Evans et al. 1990).

Nucleic acid hybridisation techniques (both radioactive and nonradioactive methods) and reverse-transcription polymerase chain reaction (RT-PCR) were initially proposed as rapid, highly specific, sensitive tools for detection of orbiviruses in clinical specimens (Brown et al. 1988a, b; de Mattos et al. 1989; Gonzalez and Knudson 1987, 1988; Gould 1988; Huismans and Cloete 1987; Koekemoer and Dijk 2004; Mohammed et al. 1996; Roy et al. 1985b; Squire et al. 1985a, b; Venter et al. 1991; Zientara et al. 1998). Several further modifications of these techniques were developed in the 1990s, including nested RT-PCR, multiplex RT-PCR, immuno-PCR, random primer PCR, in situ PCR and PCR-enzyme-linked oligonucleotide-sorbent assay (PCR-ELOSA) (Katz et al. 1993; MacLachlan et al. 1994).

More rapid and more sensitive diagnostic assays based on RT-PCR (both conventional and real time) were developed at the beginning of the twenty-first century (Anthony et al. 2007; Aradaib et al. 2003, 2005, 2009; Billinis et al. 2001; Maan et al. 2010a; Zientara et al. 2002; Batten et al. 2010; Hoffmann et al. 2009; Jimenez-Clavero et al. 2006; Monaco et al. 2011; Orru et al. 2006; Quan et al. 2010; Rodriguez-Sanchez et al. 2008; Shaw et al. 2007; Toussaint et al. 2007; Wilson et al. 2009a, b). Recently, a rapid molecular strategy based on RT-PCR has been developed for the rapid detection of all orbiviruses using the partial sequences available in the GenBank (Palacios et al. 2011). However, the test has not been validated for all of *Orbivirus* species.

Recent developments in sequencing technologies and the establishment of databases containing sequence data for reference strains have also turned sequencing and phylogenetic comparisons into major tools for the identification and differentiation of even closely related orbiviruses.

18.6 Development of Sequencing Strategies for dsRNA Viruses

Recent advances in RT-PCR and high-throughput DNA sequencing methods have contributed to the emergence of new disciplines, such as genomics, proteomics and bioinformatics. With the development of more rapid and technically less demanding sequencing methods, cross-hybridisation analyses have been superseded by comparative sequencing studies for the dsRNA viruses (Attoui et al. 2000b; Maan et al. 2007a, b; Potgieter et al. 2002).

The earliest approach to orbivirus sequencing was the analyses of cloned fragments by Maxam and Gilbert method (Ghiasi et al. 1985; Purdy et al. 1985; Roy et al. 1985a). Later Sanger's dideoxy method was used to sequence cloned BTV genome segments (Kowalik and Li 1989). However, the majority of other orbiviruses (partial segments) were analysed by shotgun sequencing using random hexamers (Gould 1987; Parkes and Gould 1996). The procedures known as single-primer amplification technique (SPAT) and full-length amplification of cDNA (FLAC) were subsequently developed to further increase the efficiency and accuracy of sequencing for segmented dsRNA viruses (Attoui et al. 2000b; Maan et al. 2007b). In these techniques a single primer is used to amplify full-length cDNA copies of individual dsRNA segments. Consequently full-length segments can be cloned and sequenced using vector-specific primers (Attoui et al. 2001, 2005b; Maan et al. 2007a). Recent technical breakthroughs have also introduced additional powerful alternatives to traditional Sanger sequencing, known as high-throughput sequencing (HTS), or 'next-generation' sequencing methods, that include the Roche 454 FLX system and the Illumina/Solexa Genome Analyzer.

The generation of viral genome sequence data has provided a basis for virus taxonomy, phylogenetics, diagnostic test development (for surveillance, diagnosis and molecular epidemiology) and vaccine development (via epitope prediction and recombinant protein synthesis) to address fundamental questions of virus biology through mutation and reverse genetics studies, pathogenesis, virulence and evolution (Attoui et al. 2001; Carpi et al. 2010; Cowled et al. 2007; Maan et al. 2008; Palacios et al. 2011). Although important data can be generated from an analysis of individual orbivirus genome segments, sequence analysis of the entire genome provides further information concerning reassortment and virus strain evolution. However, most orbivirus sequence data in GenBank is incomplete and lacks terminal sequences, due to technical difficulties in amplifying and sequencing the ends of the viral RNA, compared with sequencing internal regions of genes, and many sequences are derived from poorly identified strains or isolates.

In recent years we have developed improved and optimised methods for the synthesis, amplification and sequencing of full-length cDNAs from dsRNA genes and associated sequencing strategies described as 'full-length amplification of cDNAs' (FLAC) (Maan et al. 2007b). They also include the development of direct sequencing methods (without cloning) for the resulting full-length cDNAs. This is an improved version of the 'single-primer amplification technique' (SPAT – (Attoui et al. 2000b)).

Individual dsRNA genome segments need to be separated, or selected in some way prior to sequencing, to avoid generating data from several different RNAs segments simultaneously, making the data unreadable. Separation can be achieved by electrophoresis, by bacterial cloning of cDNAs or by synthesis of cDNAs using primers that are specific for a single genome segment. However, the design of such specific primers is difficult for any segment that has not already been sequenced. Even then, mis-priming (especially during reverse transcription), RNA secondary structure and high GC content can result in a failure to generate exclusively full-length cDNA products in the RT-PCR reactions. Inefficient transfection and cloning of vectors in host bacteria are also factors that can cause problems, particularly with larger dsRNA segments. To describe FLAC briefly, a 35-base single-stranded self-priming oligonucleotide 'anchor primer' (5′p-GAC CTC TGA GGA TTC TAA AC/iSp9/T CCA GTT TAG AAT CC-OH 39) was ligated to the 3′ (uncapped) termini of unfractionated BTV genomic dsRNA. The 'ligated' dsRNA was fractionated on gel and purified. The purified ligated dsRNA was denaturated and then used for the synthesis of cDNA copies, primed by the base-paired 'hairpin' structure of the anchor primer. Amplification of cDNAs was then performed using primer 5-15-1 (5′GAGGGATCC AGTTTAGAATCCTCAGAGGTC 3′) containing a Bam HI restriction site (underlined). FLAC, which prevents mis-priming during cDNA synthesis by reverse transcriptase and is not dependent on previous sequence data for primer design, can therefore be used to amplify any large and previously uncharacterised dsRNA virus gene.

A new strategy is also described for direct Sanger sequencing of the terminal regions of the full-length cDNAs generated by FLAC. Both termini of the PCR products generated by FLAC are identical. Previous reports of single-primer amplification methods suggest that a cloning step would therefore be essential to avoid generating sequence data from both termini simultaneously (Attoui et al. 2000b; Shapiro et al. 2005). However, the 5′ and 3′ termini of each BTV genome segment end in six conserved base pairs. These are conventionally represented as (5′-GUUAAA—— ACUUAC-3′) for the positive strands (Mertens and Sangar 1985). These sequences were exploited in the design of universal sequencing primers (phased primers). The phased primers consist of the sequence of primer 5-15-1 minus the Bam HI restriction site, plus six 3′ terminal nucleotides equivalent to the conserved 5′ end of the +ve (forward primer) or −ve (reverse primer). These primers differ by only three bases at their 3′ ends. However, this confers sufficient specificity to ensure that they only prime DNA synthesis for sequence generation from the relevant end of the genome segment.

The conserved nucleotides selected in the design of these universal sequencing primers are common to all genome segments of BTV (Mertens et al. 2005a). Other reoviruses also have terminal sequences that are conserved, although different from those of BTV (http://www.reoviridae.org/dsRNA_virus_proteins/CPV-RNA-Termin.htm). A similar approach was therefore adopted for the design of phased primers for sequence analysis of the 22 *Orbivirus* species (Belaganahalli et al. 2011, 2012). The sequences generated using the phased primers do not include the conserved termini which are part of the priming site. However, these sequences can be obtained using near-terminal outward-facing primers with the full-length cDNAs as a sequencing template.

This method was used to analyse whole genome sequences (WGS) of several multiple BTV isolates, without a cloning step. It is also applicable to other uncharacterised dsRNA virus genes that have conserved RNA termini (e.g. those of the other reoviruses), as demonstrated by its successful use with Indian sheep orthoreovirus (ISOV) (Maan et al. 2007b). Sequencing of the central portion of the viral genes is then possible, using the initial data generated to design sequencing primers. Because these methods do not involve a cloning step, they generate a consensus sequence for the original RNA gene population, avoiding cloning 'artefacts' and increasing the speed of analysis. Using these methods, sequence data can be obtained from a novel virus isolate within 48 h. These novel sequencing methods developed for dsRNA viruses have been used in the creation of a sequence database for the genome segments of many different global BTV isolates (>300 WGS) (see http://www.reoviridae.org/dsRNA_virus_proteins/ReoID/viruses-at-iah.htm).

These techniques have/are being used to generate a sequence database for the individual genome segments and whole genomes of different strains of BTV, EHDV, AHSV and other orbiviruses from around the world. These techniques, which are applicable to any viruses with segmented dsRNA genomes and conserved RNA termini, make it possible to generate sequence data rapidly from multiple isolates for molecular epidemiology studies. The sequences are particularly useful as they are linked to specific isolates of each virus in the orbivirus reference collection (ORC) (see www.iah.bbsrc.ac.uk/dsRNA_virus_proteins/ReoID/orbiviruses.htm) at IAH Pirbright, which can be used for further studies.

18.7 Next-Generation DNA Sequencing Platforms for BTV and Other Orbiviruses

The next-generation sequencing technologies offer novel and rapid ways for genome-wide characterisation. DNA sequencing is in the process of an enormous technological shift marked by dramatic throughput increases, a precipitously dropping per-base cost of raw sequence. As the users and developers of the DNA sequencing techniques realised, the great limitations of the Sanger sequencing protocols for even larger sequence output were the need for gels or polymers used as sieving separation media for the fluorescently labelled DNA fragments, the relatively low number of samples which could be analysed in parallel and the difficulty of total automation of the sample preparation methods. These limitations initiated efforts to develop techniques without gels, which would allow sequence determination on very large numbers of samples in parallel (meaning many strands of DNA are sequenced at once, hence far faster and cheaper than Sanger sequencing). Novel DNA sequencing techniques provide high speed and throughput, such that genome sequencing projects that took several years with the Sanger technique can now be completed in a matter of weeks (Niedringhaus et al. 2011). The advantage of these platforms is the determination of the sequence data from amplified single DNA fragments, avoiding the need for cloning of DNA fragments.

18.7.1 Second-Generation Sequencing

Several reviews of what were first called 'next-generation' or, more precisely, second-generation sequencing technologies have appeared (Mardis 2008; Niedringhaus et al. 2011). It is proposed to classify the second-generation technologies as a combination of a synchronised reagent wash of nucleoside triphosphates (NTPs) with a synchronised optical detection method. However, this definition is not rigid, as several real-time synthesis strategies, which comprise third-generation technologies, also rely on optical detection, with Pacific Biosciences' single DNA polymerase sequencing method being a prime example. Second-generation technologies rely upon sequencing by ligation or sequencing by synthesis, including pyrosequencing and reversible chain termination. Commercially available instruments from Roche, Illumina, Helicos and Life Technologies deliver several Gbp of DNA sequence per week in the form of short contiguous fragments or reads.

18.7.2 Third-Generation Sequencing Technologies

These include alternative approaches to improve second-generation technologies, as well as novel approaches to sequencing that include the use of scanning tunnelling electron microscope (TEM), fluorescence resonance energy transfer (FRET), single-molecule detection and protein nanopores. Two of the leading third-generation sequencing technologies (Pacific Biosciences and Complete Genomics) still rely on optical detection of fluorescent events but aim to increase sequencing speed and throughput. Ion Torrent technology, on the other hand, uses ion-sensitive field-effect transistors (ISFETs) to eliminate the need for optical detection of sequencing events. Nanopore technologies, such as Oxford Nanopore, also aim to remove optics as well as the need for DNA amplification in their sequencing design by measuring changes in conductivity across a nanopore. Nonoptical TEM approaches used by Halcyon Molecular and ZS Genetics require million-dollar capital equipment and, to date, have limited throughput yet, in principle, could give the sequence of thousands of contiguous bases. Finally, new methods involving optical methods are being developed that allow for previously unattainable levels of long-range mapping, which is essential for accurate assembly of certain genomes.

Although second- and third-generation sequencing technologies are being used currently within Arbovirus Molecular Research Group (AMRG), IAH Pirbright for some orbiviruses, e.g. BTV, African horse sickness virus (AHSV), Epizootic haemorrhagic disease virus (EHDV) and equine encephalosis virus (EEV) for which we already have a reference sequence, which is a prerequisite to align short read lengths. However, for the remaining *Orbivirus* species, they are in limited use in the absence of reference sequence data. Shorter read lengths are one of the major disadvantages that lie in these technologies. The number of DNA bases per sequenced fragment is much lower than Sanger reads. Another limiting factor of the new technology remains the overall high cost for generating the sequence with very high throughput, even though compared with Sanger sequencing the cost per base is lower by several orders of magnitude. Reduction of sequencing errors is another factor; in this respect the Sanger sequencing technique remains competitive in the immediate future. Other limitations in some applications are short read lengths, nonuniform confidence in base calling in sequence reads, particularly deteriorating 30-sequence quality in technologies with short read lengths and generally lower reading accuracy in homopolar stretches of identical bases. The huge amount of data generated by these systems (over a gigabase per run) in the form of short reads presents another challenge to developers of software and more efficient computer algorithms.

18.8 Development of Diagnostic Assays for Orbiviruses

Recent developments in sequencing technologies and the establishment of databases containing sequence data for reference strains have also turned sequencing and phylogenetic comparisons into major tools for the identification and differentiation of even closely related orbiviruses. Recent sequencing studies show that sequence variations in Seg-2 correlate with BTV serotype (Maan et al. 2007a). This has made it possible to develop RT-PCR assays for typing BTV (Mertens et al. 2007). Sequence analyses of Seg-2 and comparisons to data for reference strains of existing serotypes have helped in identifying two new BTV types that were recently been detected in Europe and the Middle East (Hofmann et al. 2008; Maan et al. 2011c).

In addition to providing a basis for studies of the relationships, evolution and reassortment of different orbiviruses, sequencing and phylogenetic studies can also be used to track the movement and origins of specific virus isolates (Anthony et al. 2009, 2010; Batten et al. 2008; Maan et al. 2008, 2010b; Nomikou et al. 2009; Oura 2011; Ozkul et al. 2009). Further advancement of sequencing methodologies involving 'deep sequencing' has facilitated full genome analyses and helped to identify reassortment events that occur during the co-circulation of different virus strains. Although full genomes from multiple isolates of BTV, EHDV, AHSV and EEV have been sequenced, there are few available sequence data for other *Orbivirus* species.

As the techniques used for virus discovery continue to advance (e.g. more advanced sequencing and array technologies), more and more viruses are being identified from new or existing geographical locations (Cowled et al. 2007, 2009; Jansen et al. 2009; Maan et al. 2011c; Palacios et al. 2011; Victoria et al. 2008; Vieira Cde et al. 2009). Novel arboviruses are also emerging, or known pathogens are re-emerging, in different ecosystems due to changes in human socio-economic activities, increased trade and other anthropological activities (Arankalle et al. 2007; Baylis 2002; Gibbens 2012; Gould and Higgs 2009; Maclachlan 2010; Maclachlan and Guthrie 2010; Weaver and Reisen 2010).

The early identification of an emerging pathogen is important, making it possible to design and implement suitable surveillance and control measures before it causes a major disease epidemic. Virus isolation in cell culture and/or RT-PCR assays using consensus or degenerate primers to detect members of a large virus group or genus, followed by more specific PCRs and/or sequencing of the cDNA products, is a common strategy of virus discovery.

The emergence and re-emergence of different orbiviruses in North Africa, Europe, America and Australia are well documented (Allison et al. 2010, 2012; Baylis 2002; Hooper et al. 1999; Kirkland 2005; Maclachlan and Guthrie 2010; Rose et al. 2000; Wilson et al. 2007). Current advanced sequencing technologies have allowed us to generate sequence data for many orbiviruses, which in turn has helped in the development of fast, reliable, sensitive and specific, conventional and real-time RT-PCR diagnostic assays for the differential detection and identification of these viruses (Maan et al. 2010a; Mertens et al. 2007; Orru et al. 2006; Shaw et al. 2007). This has reduced our reliance on conventional virus isolation techniques and serological assays, which can take several weeks to confirm an initial clinical diagnosis and identify the virus type responsible. Recently, these molecular techniques have helped in identifying two completely novel BTV serotypes (types 25 and 26) from different geographical areas (Switzerland and Kuwait) (Hofmann et al. 2008; Maan et al. 2011c). Orbiviruses within the same virus species show a high level of identity in the nucleotide sequence of their cognate genes, providing the possibility for reassortment of genome segments within each species. Serogroup and serotype-specific orbivirus assays targeting conserved or type-specific regions of the virus genome were developed for BTV, EHDV, AHSV and PALV (Anthony et al. 2007; Aradaib 2009; Aradaib et al. 2009; Clavijo et al. 2010; Maan et al. 2010a, 2011a, 2012; Mertens et al. 2007; Monaco et al. 2011; Rodriguez-Sanchez et al. 2008; Shaw et al. 2007; Stone-Marschat et al. 1994; Wade-Evans et al. 1990; Wilson et al. 2009b; Zientara et al. 1993).

The higher levels of diversity between members of different *Orbivirus* species and a lack of comprehensive sequence data for all *Orbivirus* species have made it difficult to design genus-specific 'consensus' primers for a pan-orbivirus assay. Recently, attempts were made to develop consensus PCR for all orbiviruses based on Seg-1 using the limited sequence data available in GenBank (Palacios et al. 2011). However, the size of the RT-PCR product was very small (~190 bp) providing a sequence for only 132 bp, and it did not include any tick-borne orbiviruses.

In this study, we have generated a complete sequence database for representative isolates of all of the existing and formally recognised species within the genus *Orbivirus*. Sequencing and phylogenetic analysis of the orbiviruses using this database has identified five novel *Orbivirus* species (Mitchell River virus, Pata virus, Andasibe virus, Japanaut virus and Matucare virus) and grouped four unclassified viruses (TRV, TMEV, CMPV and SLOV) into existing species (*Changuinola virus*, *Lebombo virus*, *Corriparta virus* and *Umatilla virus*, respectively). This study provides the most comprehensive phylogenetic study of genus *Orbivirus* to date, using genetic sequences for all of the genome segments.

Based on these data we have designed primers for pan-orbivirus conventional RT-PCR diagnostic assay targeting Seg-1 (Belaganahalli 2012). *Orbivirus* consensus primers were designed from a multiple sequence alignment of all available Seg-1 sequences for the orbiviruses (including viruses described in Table 18.1). Twenty-one viruses that belong to 18 *Orbivirus* species along with 4 unclassified orbiviruses were successfully detected by a pan-orbivirus RT-PCR assay using this primer Pair A. These assays could detect orbivirus RNA at very low levels in mixed samples. This illustrates the value of the assay for diagnosis with more specificity and sensitivity, particularly when linked with sequence analyses. The assay detected both eastern topotype and western topotype of orbivirus isolates.

RT-PCR methods have potential to replace or reduce the need for experimental animals to make antisera for diagnostic assays. More than 161 serologically distinct orbiviruses are described in the Ninth Report of the International Committee on Taxonomy of Viruses (Attoui et al. 2011; King et al. 2011), and this number is likely to grow as advanced research tools helping to screen more arboviruses in new ecosystems or in old collections. The lack of sequence data for all *Orbivirus* species has previously hampered the development of comprehensive or pan-genus-specific diagnostic tests, as well as evolutionary studies on the orbiviruses.

Zoonotic orbiviruses are present in certain geographical areas (Belhouchet et al. 2010; Brown et al. 1991; Dilcher et al. 2012; Tomori and Fabiyi 1976) that could be identified using these assays on inactivated samples, reducing any potential threat to the health of investigators. The assay may therefore have 'One Health' significance in screening viruses from those geographic areas.

Acknowledgements This paper is based on a presentation made at the International conference on Biotechnology: Emerging Trends (ICB-2012) organized by Department of Biotechnology, Chaudhary Devi Lal University, Sirsa, Haryana, India from 18-20th September 2012.

I would like to thank the organisers for inviting me to the meeting.

We are extremely grateful to the many colleagues at IAH Pirbright, who had their direct or indirect contributions to these studies.

References

Akita GY, Chinsangaram J, Osburn BI, Ianconescu M, Kaufman R (1992) Detection of bluetongue virus serogroup by polymerase chain reaction. J Vet Diagn Invest 4:400–405

Allison AB, Goekjian VH, Potgieter AC, Wilson WC, Johnson DJ, Mertens PP, Stallknecht DE (2010) Detection of a novel reassortant epizootic hemorrhagic disease virus (EHDV) in the USA containing RNA segments derived from both exotic (EHDV-6) and endemic (EHDV-2) serotypes. J Gen Virol 91:430–439

Allison AB, Holmes EC, Potgieter AC, Wright IM, Sailleau C, Breard E, Ruder MG, Stallknecht DE (2012) Segmental configuration and putative origin of the reassortant orbivirus, epizootic hemorrhagic disease virus serotype 6, strain Indiana. Virology 424:67–75

Anthony SJ (2007) Genetic studies of epizootic haemorrhagic disease virus (EHDV). Linacre College, University of Oxford, Oxford

Anthony S, Jones H, Darpel KE, Elliott H, Maan S, Samuel A, Mellor PS, Mertens PP (2007) A duplex

RT-PCR assay for detection of genome segment 7 (VP7 gene) from 24 BTV serotypes. J Virol Methods 141:188–197

Anthony SJ, Maan N, Maan S, Sutton G, Attoui H, Mertens PP (2009) Genetic and phylogenetic analysis of the core proteins VP1, VP3, VP4, VP6 and VP7 of epizootic haemorrhagic disease virus (EHDV). Virus Res 145:187–199

Anthony SJ, Darpel KE, Maan S, Sutton G, Attoui H, Mertens PP (2010) The evolution of two homologues of the core protein VP6 of epizootic haemorrhagic disease virus (EHDV), which correspond to the geographical origin of the virus. Virus Genes 40:67–75

Aradaib IE (2009) PCR detection of African horse sickness virus serogroup based on genome segment three sequence analysis. J Virol Methods 159:1–5

Aradaib IE, Smith WL, Osburn BI, Cullor JS (2003) A multiplex PCR for simultaneous detection and differentiation of North American serotypes of bluetongue and epizootic hemorrhagic disease viruses. Comp Immunol Microbiol Infect Dis 26:77–87

Aradaib IE, Mohamed ME, Abdalla TM, Sarr J, Abdalla MA, Yousof MA, Hassan YA, Karrar AR (2005) Serogrouping of United States and some African serotypes of bluetongue virus using RT-PCR. Vet Microbiol 111:145–150

Aradaib IE, Mohamed ME, Abdalla MA (2009) A single-tube RT-PCR for rapid detection and differentiation of some African isolates of palyam serogroup orbiviruses. J Virol Methods 161:70–74

Arankalle VA, Shrivastava S, Cherian S, Gunjikar RS, Walimbe AM, Jadhav SM, Sudeep AB, Mishra AC (2007) Genetic divergence of Chikungunya viruses in India (1963–2006) with special reference to the 2005–2006 explosive epidemic. J Gen Virol 88:1967–1976

Attoui H, Billoir F, Biagini P, de Micco P, de Lamballerie X (2000a) Complete sequence determination and genetic analysis of Banna virus and Kadipiro virus: proposal for assignment to a new genus (Seadornavirus) within the family Reoviridae. J Gen Virol 81:1507–1515

Attoui H, Billoir F, Cantaloube JF, Biagini P, de Micco P, de Lamballerie X (2000b) Strategies for the sequence determination of viral dsRNA genomes. J Virol Methods 89:147–158

Attoui H, Stirling JM, Munderloh UG, Billoir F, Brookes SM, Burroughs JN, de Micco P, Mertens PP, de Lamballerie X (2001) Complete sequence characterization of the genome of the St Croix River virus, a new orbivirus isolated from cells of Ixodes scapularis. J Gen Virol 82:795–804

Attoui H, Mohd Jaafar F, Belhouchet M, Biagini P, Cantaloube JF, de Micco P, de Lamballerie X (2005a) Expansion of family Reoviridae to include nine-segmented dsRNA viruses: isolation and characterization of a new virus designated Aedes pseudoscutellaris reovirus assigned to a proposed genus (Dinovernavirus). Virology 343:212–223

Attoui H, Mohd Jaafar F, Belhouchet M, Aldrovandi N, Tao S, Chen B, Liang G, Tesh RB, de Micco P, de Lamballerie X (2005b) Yunnan orbivirus, a new orbivirus species isolated from Culex tritaeniorhynchus mosquitoes in China. J Gen Virol 86:3409–3417

Attoui H, Jaafar FM, Belhouchet M, de Micco P, de Lamballerie X, Brussaard CP (2006) Micromonas pusilla reovirus: a new member of the family Reoviridae assigned to a novel proposed genus (Mimoreovirus). J Gen Virol 87:1375–1383

Attoui H, Maan S, Anthony SJ, Mertens PPC (2009a) Bluetongue virus, other orbiviruses and other reoviruses: their relationships and taxonomy. In: Mellor PS, Baylis M, Mertens PPC (eds) Bluetongue monograph. Elsevier/Academic Press, London, pp 23–552

Attoui H et al (2009b) Peruvian horse sickness virus and Yunnan orbivirus, isolated from vertebrates and mosquitoes in Peru and Australia. Virology 394:298–310

Attoui H et al (eds) (2011) Reoviridae. Elsevier/Academic Press, London

Batten CA, Maan S, Shaw AE, Maan NS, Mertens PP (2008) A European field strain of bluetongue virus derived from two parental vaccine strains by genome segment reassortment. Virus Res 137:56–63

Batten CA, van Rijn PA, Oura CA (2010) Detection of the European 'field' strain of bluetongue virus serotype 6 by real-time RT-PCR. Vet Microbiol 141:186–188

Baylis M (2002) The re-emergence of bluetongue. Vet J 164:5–6

Belaganahalli MN (2012) Full genome sequencing and phylogenetic analyses of different *Orbivirus* species. University of London, London

Belaganahalli MN, Maan S, Maan NS, Tesh R, Attoui H, Mertens PP (2011) Umatilla virus genome sequencing and phylogenetic analysis: identification of stretch lagoon orbivirus as a new member of the Umatilla virus species. PLoS One 6:e23605

Belaganahalli MN, Maan S, Maan NS, Nomikou K, Pritchard I, Lunt R, Kirkland PD, Attoui H, Brownlie J, Mertens PP (2012) Full genome sequencing and genetic characterization of Eubenangee viruses identify Pata virus as a distinct species within the genus Orbivirus. PLoS One 7:e31911

Belhouchet M, Mohd Jaafar F, Tesh R, Grimes J, Maan S, Mertens PP, Attoui H (2010) Complete sequence of Great Island virus and comparison with the T2 and outer-capsid proteins of Kemerovo, Lipovnik and Tribec viruses (genus Orbivirus, family Reoviridae). J Gen Virol 91:2985–2993

Belhouchet M, Mohd Jaafar F, Firth AE, Grimes JM, Mertens PPC, Attoui H (2011a) Detection of a fourth orbivirus non-structural protein. PLoS One 6(10):e25697. doi:10.1371/journal.pone.0025697

Belhouchet M, Mohd Jaafar F, Firth AE, Grimes JM, Mertens PP, Attoui H (2011b) Detection of a fourth orbivirus non-structural protein. PLoS One 6:e25697

Billinis C, Koumbati M, Spyrou V, Nomikou K, Mangana O, Panagiotidis CA, Papadopoulos O (2001) Bluetongue virus diagnosis of clinical cases by a duplex reverse transcription-PCR: a comparison with conventional methods. J Virol Methods 98:77–89

Borden EC, Shope RE, Murphy FA (1971) Physicochemical and morphological relationships of some arthropod-borne viruses to bluetongue virus – a new taxonomic group. Physicochemical and serological studies. J Gen Virol 13:261–271

Brown SE, Gonzalez HA, Bodkin DK, Tesh RB, Knudson DL (1988a) Intra- and inter-serogroup genetic relatedness of orbiviruses. II. Blot hybridization and reassortment in vitro of epizootic haemorrhagic disease serogroup, bluetongue type 10 and Pata viruses. J Gen Virol 69:135–147

Brown SE, Morrison HG, Buckley SM, Shope RE, Knudson DL (1988b) Genetic relatedness of the Kemerovo serogroup viruses: I. RNA-RNA blot hybridization and gene reassortment in vitro of the Kemerovo serocomplex. Acta Virol 32:369–378

Brown SE, Morrison HG, Karabatsos N, Knudson DL (1991) Genetic relatedness of two new orbivirus serogroups: orungo and lebombo. J Gen Virol 72:1065–1072

Brown SE, Gorman BM, Tesh RB, Knudson DL (1992) Isolation of bluetongue and epizootic hemorrhagic disease viruses from mosquitoes collected in Indonesia. Vet Microbiol 32:241–252

Calisher CH, Childs JE, Field HE, Holmes KV, Schountz T (2006) Bats: important reservoir hosts of emerging viruses. Clin Microbiol Rev 19:531–545

Carpi G, Holmes EC, Kitchen A (2010) The evolutionary dynamics of bluetongue virus. J Mol Evol 70:583–592

Clavijo A, Sun F, Lester T, Jasperson DC, Wilson WC (2010) An improved real-time polymerase chain reaction for the simultaneous detection of all serotypes of Epizootic hemorrhagic disease virus. J Vet Diagn Invest 22:588–593

Cowled C, Melville L, Weir R, Walsh S, Hyatt A, Van Driel R, Davis S, Gubala A, Boyle D (2007) Genetic and epidemiological characterization of Middle Point orbivirus, a novel virus isolated from sentinel cattle in northern Australia. J Gen Virol 88:3413–3422

Cowled C, Palacios G, Melville L, Weir R, Walsh S, Davis S, Gubala A, Lipkin WI, Briese T, Boyle D (2009) Genetic and epidemiological characterization of Stretch Lagoon orbivirus, a novel orbivirus isolated from Culex and Aedes mosquitoes in northern Australia. J Gen Virol 90:1433–1439

Darpel KE (2007) The bluetongue virus 'ruminant host – insect vector' transmission cycle; the role of Culicoides saliva proteins in infection. University of London, London

de Mattos CC, de Mattos CA, Osburn BI, Dangler CA, Chuang RY, Doi RH (1989) Recombinant DNA probe for serotype-specific identification of bluetongue virus 17. Am J Vet Res 50:536–541

Della-Porta AJ, Parsonson IM, McPhee DA (1985) Problems in the interpretation of diagnostic tests due to cross-reactions between orbiviruses and broad serological responses in animals. Prog Clin Biol Res 178:445–453

Dilcher M et al (2012) Genetic characterization of Tribec virus and Kemerovo virus, two tick-transmitted human-pathogenic Orbiviruses. Virology 423:68–76

Firth AE (2008) Bioinformatic analysis suggests that the Orbivirus VP6 cistron encodes an overlapping gene. Virol J 5:48

French TJ, Inumaru S, Roy P (1989) Expression of two related nonstructural proteins of bluetongue virus (BTV) type 10 in insect cells by a recombinant baculovirus: production of polyclonal ascitic fluid and characterization of the gene product in BTV-infected BHK cells. J Virol 63:3270–3278

Fu H, Leake CJ, Mertens PP, Mellor PS (1999) The barriers to bluetongue virus infection, dissemination and transmission in the vector, Culicoides variipennis (Diptera: Ceratopogonidae). Arch Virol 144:747–761

Gambles RM (1949) Bluetongue of sheep on Cyprus. J Comp Pathol 59:176–190

Ghiasi H, Purdy MA, Roy P (1985) The complete sequence of bluetongue virus serotype 10 segment 3 and its predicted VP3 polypeptide compared with those of BTV serotype 17. Virus Res 3:181–190

Gibbens N (2012) Schmallenberg virus: a novel viral disease in northern Europe. Vet Rec 170:58

Gonzalez HA, Knudson DL (1987) Genetic relatedness of corriparta serogroup viruses. J Gen Virol 68:661–672

Gonzalez HA, Knudson DL (1988) Intra- and inter-serogroup genetic relatedness of orbiviruses. I. Blot hybridization of viruses of Australian serogroups. J Gen Virol 69:125–134

Gorman BM, Taylor J, Walker PJ (1983) Orbiviruses. In: Joklik WK (ed) The reoviridae. Plenum Press, New York/London, pp 287–357

Gould AR (1987) The complete nucleotide sequence of bluetongue virus serotype 1 RNA3 and a comparison with other geographic serotypes from Australia, South Africa and the United States of America, and with other orbivirus isolates. Virus Res 7:169–183

Gould AR (1988) The use of recombinant DNA probes to group and type orbiviruses. A comparison of Australian and South African isolates. Arch Virol 99:205–220

Gould AR, Eaton BT (1990) The amino acid sequence of the outer coat protein VP2 of neutralizing monoclonal antibody-resistant, virulent and attenuated bluetongue viruses. Virus Res 17:161–172

Gould EA, Higgs S (2009) Impact of climate change and other factors on emerging arbovirus diseases. Trans R Soc Trop Med Hyg 103:109–121

Gutsche T (1979) There was a man. Timmins, Cape Town, p 4

Hemati B, Contreras V, Urien C, Bonneau M, Takamatsu HH, Mertens PP, Breard E, Sailleau C, Zientara S, Schwartz-Cornil I (2009) Bluetongue virus targets conventional dendritic cells in skin lymph. J Virol 83:8789–8799

Henning MM (ed) (1956) African horse sickness, perdesiekte, Pestis equorum. In: Animal diseases in South Africa, 3rd edn. Central News Agency, Pretoria

Hoffmann B, Eschbaumer M, Beer M (2009) Real-time quantitative reverse transcription-PCR assays specifically detecting bluetongue virus serotypes 1, 6, and 8. J Clin Microbiol 47:2992–2994

Hofmann MA, Renzullo S, Mader M, Chaignat V, Worwa G, Thuer B (2008) Genetic characterization of toggenburg orbivirus, a new bluetongue virus, from goats, Switzerland. Emerg Infect Dis 14:1855–1861

Hooper PT et al (1999) Epidemic of blindness in kangaroos–evidence of a viral aetiology. Aust Vet J 77:529–536

Huismans H, Cloete M (1987) A comparison of different cloned bluetongue virus genome segments as probes for the detection of virus-specified RNA. Virology 158:373–380

Huismans H, Erasmus BJ (1981) Identification of the serotype-specific and group-specific antigens of bluetongue virus. Onderstepoort J Vet Res 48:51–58

Huismans H, Bremer CW, Barber TL (1979) The nucleic acid and proteins of epizootic haemorrhagic disease virus. Onderstepoort J Vet Res 46:95–104

Hutcheon D (1902) Malarial catarrhal fever of sheep. Vet Rec 14:629–633

Jansen CC et al (2009) Arboviruses isolated from mosquitoes collected from urban and peri-urban areas of eastern Australia. J Am Mosq Control Assoc 25:272–278

Jimenez-Clavero MA et al (2006) High throughput detection of bluetongue virus by a new real-time fluorogenic reverse transcription-polymerase chain reaction: application on clinical samples from current Mediterranean outbreaks. J Vet Diagn Invest 18:7–17

Karabatsos N (1985) International catalogue of arboviruses including certain other viruses of vertebrates. American Society of Tropical Medicine and Hygiene, San Antonio

Katz JB, Alstad AD, Gustafson GA, Moser KM (1993) Sensitive identification of bluetongue virus serogroup by a colorimetric dual oligonucleotide sorbent assay of amplified viral nucleic acid. J Clin Microbiol 31:3028–3030

King A, Lefkowitz E, Adams MJ, Carstens EB (2011) Virus taxonomy: ninth report of the international committee on taxonomy of viruses. Elsevier, London.

Kirkland P (2005) Epidemic viral diseases of wildlife – sudden death in tammar wallabies, blind kangaroos, herpesviruses in pilchards-what next? Microbiol Aust 26(2):82–84

Koekemoer JJ, Dijk AA (2004) African horsesickness virus serotyping and identification of multiple co-infecting serotypes with a single genome segment 2 RT-PCR amplification and reverse line blot hybridization. J Virol Methods 122:49–56

Kowalik TF, Li JK (1989) Sequence analyses and structural predictions of double-stranded RNA segment S1 and VP7 from United States prototype bluetongue virus serotypes 13 and 10. Virology 172:189–195

Maan S, Maan NS, Samuel AR, Rao S, Attoui H, Mertens PP (2007a) Analysis and phylogenetic comparisons of full-length VP2 genes of the 24 bluetongue virus serotypes. J Gen Virol 88:621–630

Maan S, Rao S, Maan NS, Anthony SJ, Attoui H, Samuel AR, Mertens PP (2007b) Rapid cDNA synthesis and sequencing techniques for the genetic study of bluetongue and other dsRNA viruses. J Virol Methods 143:132–139

Maan S et al (2008) Sequence analysis of bluetongue virus serotype 8 from the Netherlands 2006 and comparison to other European strains. Virology 377:308–318

Maan S, Maan NS, Nomikou K, Anthony SJ, Ross-smith N, Singh KP, Samuel AR, Shaw AE, Mertens PPC (2009) Molecular epidemiology studies of bluetongue virus. In: Mellor PS, Baylis M, Mertens PPC (eds) Bluetongue. Elsevier/Academic Press, London, pp 135–166

Maan NS et al (2010a) RT-PCR assays for seven serotypes of epizootic haemorrhagic disease virus & their use to type strains from the Mediterranean region and North America. PLoS One 5:e12782

Maan S et al (2010b) Full genome characterisation of bluetongue virus serotype 6 from the Netherlands 2008 and comparison to other field and vaccine strains. PLoS One 5:e10323

Maan NS, Maan S, Nomikou K, Belaganahalli MN, Bachanek-Bankowska K, Mertens PP (2011a) Serotype specific primers and gel-based rt-PCR assays for 'typing' African horse sickness virus: identification of strains from Africa. PLoS One 6:e25686

Maan S, Maan NS, Nomikou K, Veronesi E, Bachanek-Bankowska K, Belaganahalli MN, Attoui H, Mertens PP (2011b) Complete genome characterisation of a novel 26th bluetongue virus serotype from Kuwait. PLoS One 6:e26147

Maan S et al (2011c) Novel bluetongue virus serotype from Kuwait. Emerg Infect Dis 17:886–889

Maan NS, Maan S, Belaganahalli MN, Ostlund EN, Johnson DJ, Nomikou K, Mertens PPC (2012) Identification and differentiation of the twenty six bluetongue virus serotypes by RT–PCR amplification of the serotype-specific genome segment 2. PLoS One 7:e32601

Maclachlan NJ (2010) Global implications of the recent emergence of bluetongue virus in Europe. Vet Clin North Am Food Anim Pract 26:163–171

Maclachlan NJ, Guthrie AJ (2010) Re-emergence of bluetongue, African horse sickness, and other orbivirus diseases. Vet Res 41:35

MacLachlan NJ, Nunamaker RA, Katz JB, Sawyer MM, Akita GY, Osburn BI, Tabachnick WJ (1994) Detection of bluetongue virus in the blood of inoculated calves: comparison of virus isolation, PCR assay, and in vitro feeding of Culicoides variipennis. Arch Virol 136:1–8

Mardis ER (2008) Next-generation DNA sequencing methods. Annu Rev Genomics Hum Genet 9:387–402

Marr JS, Malloy CD (1996) An epidemiologic analysis of the ten plagues of Egypt. Caduceus 12:7–24

Martins LC, Diniz JA, Silva EV, Barros VL, Monteiro HA, Azevedo RS, Quaresma JA, Vasconcelos PF (2007) Characterization of Minacu virus (Reoviridae: Orbivirus) and pathological changes in experimentally infected newborn mice. Int J Exp Pathol 88:63–73

Mellor PS, Hamblin C (2004) African horse sickness. Vet Res 35:445–466

Mertens P (2004) The dsRNA viruses. Virus Res 101:3–13

Mertens P, Attoui H (2009) Orbiviruses and bluetongue virus. In: Encyclopedia of life sciences (ELS). Wiley, Chichester: John Wiley & Sons, Ltd

Mertens PP, Diprose J (2004) The bluetongue virus core: a nano-scale transcription machine. Virus Res 101:29–43

Mertens PP, Sangar DV (1985) Analysis of the terminal sequences of the genome segments of four orbiviruses. Virology 140:55–67

Mertens P, Attoui H (2009) Orbiviruses and bluetongue virus. Encyclopedia of life sciences (ELS). Wiley, Chichester

Mertens PP, Brown F, Sangar DV (1984) Assignment of the genome segments of bluetongue virus type 1 to the proteins which they encode. Virology 135:207–217

Mertens PP, Burroughs JN, Anderson J (1987) Purification and properties of virus particles, infectious subviral particles, and cores of bluetongue virus serotypes 1 and 4. Virology 157:375–386

Mertens PP, Pedley S, Cowley J, Burroughs JN, Corteyn AH, Jeggo MH, Jennings DM, Gorman BM (1989) Analysis of the roles of bluetongue virus outer capsid proteins VP2 and VP5 in determination of virus serotype. Virology 170:561–565

Mertens PP, Burroughs JN, Walton A, Wellby MP, Fu H, O'Hara RS, Brookes SM, Mellor PS (1996) Enhanced infectivity of modified bluetongue virus particles for two insect cell lines and for two Culicoides vector species. Virology 217:582–593

Mertens PPC, Maan S, Samuel A, Attoui H (2005a) Orbiviruses, reoviridae. In: Fauquet CM, Mayo MA, Maniloff J, Desselberger U, Ball LA (ed) Virus taxonomy. Eighth report of the international committee on taxonomy of viruses. Elsevier/Academic Press, London, pp 466–483

Mertens PPC, Attoui H, Duncan R, Dermody TS (2005b) Reoviridae. In: Fauquet CM, Mayo MA, Maniloff J, Desselberger U, Ball LA (ed) Virus taxonomy. Eighth report of the international committee on taxonomy of viruses. Elsevier/Academic Press, London

Mertens PP, Maan NS, Prasad G, Samuel AR, Shaw AE, Potgieter AC, Anthony SJ, Maan S (2007) Design of primers and use of RT-PCR assays for typing European bluetongue virus isolates: differentiation of field and vaccine strains. J Gen Virol 88:2811–2823

Mohammed ME, Aradaib IE, Mukhtar MM, Ghalib HW, Riemann HP, Oyejide A, Osburn BI (1996) Application of molecular biological techniques for detection of epizootic hemorrhagic disease virus (EHDV-318) recovered from a sentinel calf in central Sudan. Vet Microbiol 52:201–208

Mohd Jaafar F, Attoui H, Bahar MW, Siebold C, Sutton G, Mertens PP, De Micco P, Stuart DI, Grimes JM, De Lamballerie X (2005) The structure and function of the outer coat protein VP9 of Banna virus. Structure 13:17–28

Monaco F, Polci A, Lelli R, Pinoni C, Di Mattia T, Mbulu RS, Scacchia M, Savini G (2011) A new duplex real-time RT-PCR assay for sensitive and specific detection of African horse sickness virus. Mol Cell Probes 25:87–93

Moore DL (1974) Bluetongue and related viruses in Ibadan, Nigeria: serologic comparison of bluetongue, epizootic hemorrhagic disease of deer, and Abadina (Palyam) viral isolates. Am J Vet Res 35:1109–1113

Moore DL, Lee VH (1972) Antigenic relationship between the virus of epizootic haemorrhagic disease of deer and bluetongue virus. Brief report. Arch Gesamte Virusforsch 37:282–284

Moule L (1896) Histoire de la Médecine Vétérinaire. Presses universitaires de France, Maulde, Paris, p 38

Murphy FA, Borden EC, Shope RE, Harrison A (1971) Physicochemical and morphological relationships of some arthropod-borne viruses to bluetongue virus – a new taxonomic group. Electron microscopic studies. J Gen Virol 13:273–288

Niedringhaus TP, Milanova D, Kerby MB, Snyder MP, Barron AE (2011) Landscape of next-generation sequencing technologies. Anal Chem 83:4327–4341

Nomikou K, Dovas CI, Maan S, Anthony SJ, Samuel AR, Papanastassopoulou M, Maan NS, Mangana O, Mertens PP (2009) Evolution and phylogenetic analysis of full-length VP3 genes of Eastern Mediterranean bluetongue virus isolates. PLoS One 4:e6437

OIE (2010) Chapter 1.2. Criteria for listing diseases. Terrestrial Animal Health Code. http://www.oie.int/index.php?id=169&L=0&htmfile=chapitre_1.1.2.htm. Accessed 15 Nov 2013

Orru G, Ferrando ML, Meloni M, Liciardi M, Savini G, De Santis P (2006) Rapid detection and quantitation of Bluetongue virus (BTV) using a Molecular Beacon fluorescent probe assay. J Virol Methods 137:34–42

Oura C (2011) Bluetongue: new insights and lessons learnt. Vet Rec 168:375–376

Ozkul A, Erturk A, Caliskan E, Sarac F, Ceylan C, Mertens P, Kabakli O, Dincer E, Cizmeci SG (2009) Segment 10 based molecular epidemiology of bluetongue virus (BTV) isolates from Turkey: 1999–2001. Virus Res 142:134–139

Palacios G et al (2011) Rapid molecular strategy for orbivirus detection and characterization. J Clin Microbiol 49:2314–2317

Parkes H, Gould AR (1996) Characterisation of Wongorr virus, an Australian orbivirus. Virus Res 44:111–122

Polydorou K (ed) (1985) Bluetongue in Cyprus. Alan R Liss Inc., New York

Potgieter AC, Steele AD, van Dijk AA (2002) Cloning of complete genome sets of six dsRNA viruses using an improved cloning method for large dsRNA genes. J Gen Virol 83:2215–2223

Purdy MA, Ghiasi H, Rao CD, Roy P (1985) Complete sequence of bluetongue virus L2 RNA that codes for the antigen recognized by neutralizing antibodies. J Virol 55:826–830

Quan M, Lourens CW, MacLachlan NJ, Gardner IA, Guthrie AJ (2010) Development and optimisation of a duplex real-time reverse transcription quantitative PCR assay targeting the VP7 and NS2 genes of African horse sickness virus. J Virol Methods 167:45–52

Rafyi A (1961) Horse sickness. Bull Off Int Epizoot 56:216–250

Rao CD, Kiuchi A, Roy P (1983) Homologous terminal sequences of the genome double-stranded RNAs of bluetongue virus. J Virol 46:378–383

Ratinier M et al (2011) Identification and characterization of a novel non-structural protein of bluetongue virus. PLoS Pathog 7:e1002477

Rodriguez-Sanchez B, Fernandez-Pinero J, Sailleau C, Zientara S, Belak S, Arias M, Sanchez-Vizcaino JM (2008) Novel gel-based and real-time PCR assays for the improved detection of African horse sickness virus. J Virol Methods 151:87–94

Rose KA, Kirkland PD, Davis RJ, Cooper DW, Blumstein DT (2000) An epizootic of sudden death in tammar wallabies (Macropus eugenii). American Association of Zoo Veterinarians and International Association for Aquatic Animal Medicine Annual Conference Proceeding, pp 166–170

Roy P (1989) Bluetongue virus genetics and genome structure. Virus Res 13:179–206

Roy P, Purdy MA, Petre J, Rao CD (1985a) Cloning and nucleotide sequencing of bluetongue virus genomes. Prog Clin Biol Res 178:363–370

Roy P, Ritter GD Jr, Akashi H, Collisson E, Inaba Y (1985b) A genetic probe for identifying bluetongue virus infections in vivo and in vitro. J Gen Virol 66:1613–1619

Shapiro A, Green T, Rao S, White S, Carner G, Mertens PP, Becnel JJ (2005) Morphological and molecular characterization of a Cypovirus (Reoviridae) from the mosquito Uranotaenia sapphirina (Diptera: Culicidae). J Virol 79:9430–9438

Shaw AE et al (2007) Development and initial evaluation of a real-time RT-PCR assay to detect bluetongue virus genome segment 1. J Virol Methods 145:115–126

Shope RE (1955) An infectious fibroma of deer. Proc Soc Exp Biol Med 88:533–535

Shope RE, MacNamara LG, Mangold R (1955) Epizootic hemorrhagic disease of deer. New Jersey Outdoors (November). Report. pp 16–21

Spreull J (1902) Report from veterinary surgeon Spreull on the result of his experiments with the malarial catarrhal fever of sheep. Agric J Cape Good Hope 20:469–477

Spreull J (1905) Malarial catarrhal fever (bluetongue) of sheep in South Africa. J Comp Pathol 18:321–337

Squire KR, Chuang RY, Chuang LC, Doi RH, Osburn BI (1985a) Molecular cloning and hybridization studies on bluetongue virus serotype 17. Prog Clin Biol Res 178:355–361

Squire KR, Chuang RY, Chuang LF, Doi RH, Osburn BI (1985b) Detecting bluetongue virus RNA in cell culture by dot hybridization with a cloned genetic probe. J Virol Methods 10:59–68

Stone-Marschat M, Carville A, Skowronek A, Laegreid WW (1994) Detection of African horse sickness virus by reverse transcription-PCR. J Clin Microbiol 32:697–700

Tabachnick WJ, MacLachlan NJ, Thompson LH, Hunt GJ, Patton JF (1996) Susceptibility of Culicoides variipennis sonorensis to infection by polymerase chain reaction-detectable bluetongue virus in cattle blood. Am J Trop Med Hyg 54:481–485

Takamatsu H, Mellor PS, Mertens PP, Kirkham PA, Burroughs JN, Parkhouse RM (2003) A possible overwintering mechanism for bluetongue virus in the absence of the insect vector. J Gen Virol 84:227–235

Theal GM (1899) Records of South-Eastern Africa collected in various libraries and archive departments in Europe. Government of Cape Colony, Cape Town, South Africa Report no. 3, p 224

Tomori O, Fabiyi A (1976) Neutralizing antibodies to Orungo virus in man and animals in Nigeria. Trop Geogr Med 28:233–238

Toussaint JF, Sailleau C, Breard E, Zientara S, De Clercq K (2007) Bluetongue virus detection by two real-time RT-qPCRs targeting two different genomic segments. J Virol Methods 140:115–123

van Regenmortel MH, Mahy BW (2004) Emerging issues in virus taxonomy. Emerg Infect Dis 10:8–13

Venter EH, Viljoen GJ, Nel LH, Huismans H, van Dijk AA (1991) A comparison of different genomic probes in the detection of virus-specified RNA in Orbivirus-infected cells. J Virol Methods 32:171–180

Verwoerd DW, Louw H, Oellermann RA (1970) Characterization of bluetongue virus ribonucleic acid. J Virol 5:1–7

Victoria JG, Kapoor A, Dupuis K, Schnurr DP, Delwart EL (2008) Rapid identification of known and new RNA viruses from animal tissues. PLoS Pathog 4:e1000163

Vieira Cde M, Nunes MR, da Silva EV, Carvalho VL, Nunes Neto JP, Cruz AC, Casseb SM, Vasconcelos HB, Quaresma JA, Vasconcelos PF (2009) Full-length sequencing and genetic characterization of Breu Branco virus (Reoviridae, Orbivirus) and two related strains isolated from Anopheles mosquitoes. J Gen Virol 90:2183–2190

Wade-Evans AM (1990) The complete nucleotide sequence of genome segment 7 of bluetongue virus, serotype 1 from South Africa. Nucleic Acids Res 18:4919

Wade-Evans AM, Mertens PP, Bostock CJ (1990) Development of the polymerase chain reaction for the detection of bluetongue virus in tissue samples. J Virol Methods 30:15–24

Wade-Evans AM, Mertens PP, Belsham GJ (1992) Sequence of genome segment 9 of bluetongue virus (serotype 1, South Africa) and expression analysis demonstrating that different forms of VP6 are derived from initiation of protein synthesis at two distinct sites. J Gen Virol 73:3023–3026

Weaver SC, Reisen WK (2010) Present and future arboviral threats. Antiviral Res 85:328–345

Wilson A, Carpenter S, Gloster J, Mellor P (2007) Re-emergence of bluetongue in northern Europe in 2007. Vet Rec 161:487–489

Wilson WC, O'Hearn ES, Tellgren-Roth C, Stallknecht DE, Mead DG, Mecham JO (2009a) Detection of all eight serotypes of Epizootic hemorrhagic disease virus by real-time reverse transcription polymerase chain reaction. J Vet Diagn Invest 21:220–225

Wilson WC, Hindson BJ, O'Hearn ES, Hall S, Tellgren-Roth C, Torres C, Naraghi-Arani P, Mecham JO, Lenhoff RJ (2009b) A multiplex real-time reverse transcription polymerase chain reaction assay for detection and differentiation of Bluetongue virus and Epizootic hemorrhagic disease virus serogroups. J Vet Diagn Invest 21:760–770

Wu X, Chen SY, Iwata H, Compans RW, Roy P (1992) Multiple glycoproteins synthesized by the smallest

RNA segment (S10) of bluetongue virus. J Virol 66: 7104–7112

Zientara S, Sailleau C, Moulay S, Plateau E, Cruciere C (1993) Diagnosis and molecular epidemiology of the African horse sickness virus by the polymerase chain reaction and restriction patterns. Vet Res 24:385–395

Zientara S, Sailleau C, Moulay S, Cruciere C, El-Harrak M, Laegreid WW, Hamblin C (1998) Use of reverse transcriptase-polymerase chain reaction (RT-PCR) and dot-blot hybridisation for the detection and identification of African horse sickness virus nucleic acids. Arch Virol Suppl 14:317–327

Zientara S et al (2002) Identification of bluetongue virus serotype 2 (Corsican strain) by reverse-transcriptase PCR reaction analysis of segment 2 of the genome. Vet Rec 150:598–601

Fetal Mesenchymal Stem Cells in Farm Animals: Applications in Health and Production

19

P.S. Yadav and B.R. Gulati

Abstract

Fetal stem cells derived from amniotic fluid, amnion, umbilical cord, and fetal fibroblast have provided new insights regarding the nature and potential of these cells and can be an alternative source of stem cells in livestock, as scientifically validated embryonic stem (ES) cell lines are yet not available in these species. In this review we shall state the art, envision, and the prospects of fetal mesenchymal stem cells in livestock and their application in assisted reproduction and in veterinary medicine. Fetal stem cells express pluripotency markers and share similar growth kinetics providing strong support to the notion that these cells may be biologically closer to embryonic stem cells. The placenta and fetal adnexa, such as umbilical cord blood, umbilical cord matrix, amnion, amniotic fluid (AF) and fetal fibroblasts are immensely valuable and easily accessible sources of pluripotent and progenitor cells. These cells represent the intermediate between ES and adult stem cells regarding proliferation rates and plasticity features. These cells can be used in assisted reproduction and veterinary health applications. General applications include veterinary regenerative medicine and as donor cells for cloning and transgenic and iPS cell production. In equines, these cells can be used for orthopedic injuries, including repair of damaged ligaments and tendons and for laminitis. In domestic livestock, these can be used for producing environment-friendly transgenic animals for less methane or phosphorus production and in vitro meat production. Specific application of stem cells in canine can be in spinal injuries and diabetes.

P.S. Yadav (✉)
Central Institute for Research on Buffaloes,
Hisar 125001, India
e-mail: psyadav@scientist.com

B.R. Gulati
National Research Center on Equines,
Hisar 125001, India

R.K. Salar et al. (eds.), *Biotechnology: Prospects and Applications*,
DOI 10.1007/978-81-322-1683-4_19, © Springer India 2013

19.1 Introduction

In twenty-first century, the most exciting and emerging area of medicinal research is stem cell research. The fetal adnexa such as umbilical cord, amnion, and amniotic fluid have been proposed as an ideal source of stem cells in livestock due to their noninvasive isolation procedure, availability of large tissue mass for harvesting cells, ease of access without any ethical reservation, and multilineage differentiation potential. In view of technical and ethical concerns to establish ES (Yadav et al. 2005) cells and the restricted differentiation potential of adult stem cells, establishing the fetal stem cells could be a promising alternative approach. In recent years, fetal stem cells (FSCs) have emerged as a potential "half way house" between ES cells and adult stem cells (Abdulrazzak et al. 2010). The adnexal tissue cells preserve some of the characteristics of primitive embryonic layers from which they originate (Pappa and Anagnou 2009). Non-embryonic fetal-derived stem cells open new perspective for regenerative medicine in livestock species. Recent studies demonstrated the isolation of stem cells from sources such as umbilical cord blood, Wharton's jelly, and amniotic fluid and membrane from various livestock and domestic species like equine, canine, bovine, buffalo, feline, caprine, ovine, and swine (Cremonesi et al. 2011; Yadav et al. 2012a, b). Histopaque and Ficoll density centrifugation were the most commonly used methods for isolation of mononuclear cells from umbilical cord blood and amniotic fluid. Mechanical chopping or enzymatic digestion of adnexal tissues or both were used for primary culture initiation in various studies. DMEM or MEM basal media supplemented with varying concentrations of FBS and other growth factors were used for cell culture by different workers. The cells cultured for prolonged periods were characterized for stemness properties through biochemical and molecular markers. High alkaline phosphatase activity and expression of key pluripotency transcription factors like Oct-4, Nanog, and Sox2 and immunostaining for expression of stage-specific embryonic antigens, tumor rejection antigen, etc., were employed in different studies for validation of stemness characteristics. Trans-differentiation of these cells into different lineages after induction was confirmed through lineage-specific staining or lineage-specific gene expression studies. Some reports demonstrated their ability to cryopreserve well and maintain their phenotypic and growth characteristics upon thawing which is of utmost importance for their therapeutic use.

19.2 Transplantation Studies

Porcine umbilical cord matrix cells engrafted and proliferated successfully in the rodent model of Parkinson's disease without requiring immune suppression (Medicetty et al. 2004). Porcine amniotic fluid-derived stem cells injected in immunodeficient mice (Balb/c-Nu) developed no teratoma (Chen et al. 2011). Effect of transplanted mesenchymal stem cells from Wharton's jelly (WJ) of caprine umbilical cord on wound healing revealed that reepithelialization was complete in 7 days in treated wounds whereas in control the wounds showed incomplete epithelization even after 12 days of wounding, elucidating the beneficial effect of caprine WJ cells on cutaneous wound healing (Azari et al. 2011). Canine umbilical cord blood (UCB)-derived MSCs transplanted in spinal cord-injured dogs resulted in recovery of nerve function without any inflammatory response even after direct transplantation of allogenic UCB-MSCs (Lim et al. 2007). Tendon injuries are common clinical problem in race horses. Lange-Consiligio et al. (2011) observed hyperintense population of equine umbilical cord MSCs in isolated tendons with artificial lesions indicating their possible role in equine orthopedics. These findings indicate that stem cell derived from fetal adnexa holds promise for therapeutic applications and may accelerate the field of veterinary medicine.

Application of the stem cells in understanding fundamental events in embryonic development, therapeutic delivery system, and animal model testing of pharmaceutical research has made it the fastest growing area of research in biological sciences. With decoding of the genome sequences, stem cell promises to resolve many mysteries of the developmental biology. Stem cells provide efficient means to investigate the expression, regulation, and functions of genes involved in mammalian differentiation and development. What has really attracted the attention of general public as well as scientists is the potential use of stem cells in transplantation and cell replacement therapy. There is considerable interest in obtaining stem cell lines from the livestock species of economic importance. A stem cell is a special kind of cell that has a unique capacity to renew itself and to give rise to specialized cell types. A stem cell can also be defined as an uncommitted cell that remains uncommitted unless it receives signal(s) to develop into a specialized cell. There are two types of stem cells: adult stem cells and embryonic stem cells. While the adult stem cells are either unipotent or multipotent and can be available in any tissue, embryonic stem cells can only be derived from the inner cell mass of the blastocysts and are the pluripotent cells. Now from the last few years, a third type of cells from extraembryonic tissues named as fetal stem cells are gaining importance particularly in livestock species where true embryonic stem cells are yet not available. Here livestock fetal stem cells from amniotic fluid, amniotic membrane, umbilical cord blood and cord matrix, fetal fibroblast, and placental cells are discussed.

19.3 Amniotic Fluid (AF) Stem Cells

The AF, being a safe, reliable, and simple screening tool, is widely used for numerous basic and biomedical applications. The AF comprises of normal embryonic or fetal chipping cells derived from the three germ layers (ectoderm, endoderm, and mesoderm); it possesses the natural precursors of all differentiation lineages.

The isolation of AF cells with stem cell features opens many new venues including regeneration of tissues and ultimately even organs. AF stem cells of early gestation seem to express higher levels of endoderm- and mesoderm-specific markers compared with those of later gestation, while ectodermal markers show no difference. These are clonally expanded in mesenchymal stem cells that exhibit a series of stem cell-specific markers including Oct-4, Nanog, and SSEA-4 and have a wide range in differentiation potential. Recent reports suggest that AF stem cells might have additional benefits. The deployment of AF stem cells for tissue regeneration offers advantages over the use of ES or adult stem cells, namely, (1) AF represents a convenient and non-contested source for obtaining stem cells; (2) isolating them is relatively simpler and rapid; (3) no feeder layers are required for their culturing; (4) they display no spontaneous differentiation in culture; and (5) their stem cell phenotype is not affected by long-term storage. The AF cells in water buffaloes collected from young fetuses (50–70 days) maintained a characteristic round shape and then (80–100 days gestation) changed to cells with different morphologies including spherical cells with nucleus, spherical cells without nucleus, polygonal cells, and freely floating cells (Yadav et al. 2011). The AF cells were found to expand without feeder layer over a period of 100 days (up to passages 20) and overexpress the AP, Oct-4, Nanog and Sox2, GAPDH, and β-actin. These features are also noted in human AF cells. It has been noted that bubaline AF cells could be cultured and maintained in vitro for a prolonged period and offer a potential source of multipotent cells for applications like therapeutic assisted reproduction in animals (Yadav et al. 2011).

19.4 Amniotic Membrane Cells

The amniotic membrane is a tissue of fetal origin and is composed of three major layers: a single epithelial layer, a thick basement membrane, and an avascular mesenchyme. There are no nerves, muscles, or lymphatics in the amnion. It can be easily separated from the underlying chorion, with which it never truly fuses at the cellular level. The amnion obtains its nutrition and oxygen from the surrounding chorionic fluid, the amniotic fluid, and the fetal surface vessels. One of the basic functions of the amniotic membrane is protection against desiccation to the developing embryo and provides an environment for suspension in which the embryo can grow without distortion by pressure from surrounding structures. The amnion also plays an important role during parturition through maintenance of uterine contraction and secretion of prostaglandins, especially prostaglandin E2. Amnion also expresses prostaglandin-biosynthesis enzymes such as phospholipase, prostaglandin synthase, and cyclooxygenase which are regulated by chorionic gonadotropin, and their receptors are found on the amniotic epithelium. Amniotic epithelium is metabolically highly active throughout gestation, and it is also responsible for regulating the pH of the amniotic fluid. Reports in human demonstrated that both amniotic epithelium cells and amniotic mesenchymal cells express stem cell markers such as OCT-4, which is specifically expressed in ES and germ cells; GATA-4, a marker of definitive embryonic and visceral endoderm; hepatocyte nuclear factor-3β, a marker of definitive endoderm; nestin, which is an intermediate protein and a neural stem cell-specific marker; and Nanog. These facts suggest that not only amniotic epithelium cells but also amniotic mesenchymal cells possess pluripotency.

The bubaline AM cells have been found to exhibit polygonal shape and completed 21 passages in 75 days of continuous culture and expressed Oct-4 (Yadav et al. 2011). The AM and cells derived from it are positive for AP, Nanog, Oct-4, and Sox2 overexpression (Mann et al. 2012). In equine, the differentiation induction for pancreatic and osteogenic cells yielded oil red O-positive cells for pancreatic and alizarin red positive for osteogenic cells. The technique for sampling, isolation, and expansion of equine AM stem cells forms a basis for establishment of a database for collecting and preserving stem cells used in the field of equine regenerative medicine. Human AM are used as healing accelerator and bone induction in dogs. The AM decreased fibrinoleukocytic exudates, inflammation, and suitable experimental cover for different injuries, and further acellular AM has the potential for rapid improvement and bone induction. The AM contains collagen, laminin, and fibronectin which provide an appropriate substrate for bone induction. This substrate has been reported to promote bone induction and might contribute to induction of progenitor cells and/or stem cells in the area where it had been undertaken and has also differentiated into bone.

The AM-derived cells have potential for osteogenic, adipogenic, chondrogenic, and myogenic differentiation in vitro and thus can have potential for use as engraftment material. As fetal tissue is usually discarded after birth, it involves no ethical concerns if it is used as source of MS stem cells. Equine amnion has further advantages for use in stem cell work as it is a rich and easily accessible source of MS cells, as equine amnion is easy to sample immediately after birth and is not attached with the rest of the placenta.

19.5 Fetal Fibroblasts and Stem Cell-Like Cells

Stem cell-like multipotent stem cells have been noted in the abattoir-derived fetuses of water buffaloes as reports indicate that the pluripotency genes expressed by ES cell are also expressed by fetal fibroblast. In this direction, the expression of AP, Oct-4, Sox2A and Sox2B, β-actin GAPDH, and Nanog has been detected in bubaline fetal fibroblasts (Yadav et al. 2011). The murine and porcine fetal explant cells have also been found to express the pluripotency genes. It was shown that fetal somatic explants contain a subpopulation of somatic stem cells, which can be induced to display features of lineage-uncommitted stem cells. After injection into blastocysts, fetal fibroblast cells differentiated into a variety of cell types including those of the mesodermal lineage; they even migrated into the genital ridge. In vitro, the fetal stem cells exhibit characteristics of ES cells, including extended self-renewal; expression of Oct-4, Stat3, and Akp2 (Tnap); and growth as multicellular aggregates (Kues et al. 2005). This indicates that fetal tissue contains stem cells with greater potency than previously thought, hence, might serve as a new source of animal stem cells.

19.6 Umbilical Cord Blood Stem Cells

The umbilical cord is a noncontroversial source of mesenchymal-like stem cells. The umbilical cord blood (UCB), which is normally discarded, can be easily collected at the time of delivery. Collection can be accomplished by venipuncture of the umbilical vein of the placenta still in utero or after the expulsion of the placenta itself. Advantages of UCB stem cells are their high proliferative capacity, low risk of viral contamination response to alloantigen, their availability, and donor safety. However, there are limited reports on the cord blood from livestock; main limitation in collection of cord blood in livestock including equines is breakage of the umbilical cord during parturition.

In normal calving, after expulsion of the placenta, small quantity of cord blood (5–6 ml) can be collected. Morphology of buffalo umbilical cord blood, newborn calf blood, and adult buffalo blood revealed that cord blood parameters differed significantly from newborn calf and adult buffalo blood (Singh et al. 2012a, b).

In equine, UCB had significantly lower total erythrocyte (RBC) count, total leucocyte (WBC), lymphocyte, and granulocyte count than those of mare and foal. While RBC count was the highest in foals, WBC, lymphocyte, and granulocyte counts were highest in mares followed by foal and UCB, respectively. Hemoglobin, percent hematocrit values, and platelet count were the least in UCB, followed by mare and foal blood. Unlike buffalo, equine UCB is more intimate to mare blood, as there was significant difference in 8 parameters in UCB and foal blood out of 18 parameters. In man, horse, cattle, and dogs, umbilical cord blood-MS cells have multipotent abilities. UCB-derived stem cells in domestic animals are capable of differentiation in vitro, not only toward mesenchymal cell lineage (osteogenic, chondrogenic, myogenic, and adipogenic) but also toward endodermal (hepatogenic) and ectodermal (neurogenic) lineages under appropriate culturing conditions.

19.7 Wharton's Jelly Stem Cells

The umbilical cord provides stem cells in two compartments: umbilical cord blood (UCB) and umbilical cord matrix, also known as Wharton's jelly (WJ). Besides humans, the stromal cells with certain pluripotency markers have also been reported from WJ of pig and buffalo (Yadav et al. 2012a, b).

The WJ surrounding the two arteries and single vein of the umbilical cord has been observed in buffaloes. The salient features of buffalo cells from WJ include sticky jelly-type tissue, slow growing primary colony (8–10 days embedded cells, spikes formed but not as clear as fibroblasts, and passage time is 3–4 days) (Yadav et al. 2011). Bubaline cord matrix cells could be cultured for more than 100 days in continuous cell culture which expressed Oct-4, Nanog, and Sox (Yadav et al. 2008). The identification of WJ as an alternative source of MS cells provides significant clinical benefits, namely, harvesting, reduction of risks associated with transmitting infections, and acceptable level of HLA mismatch.

19.8 Placenta Stem Cells

In transition from morula to blastocyst, first differentiation takes place in the ICM cells and the trophectoderm. ICM forms the epiblast and the fetus in early development as well as the source of ES cells. The trophectoderm forms the placenta which supports and protects the developing fetus. The stem cell populations derived from human placenta tissues are chorionic mesenchymal stromal cells and the chorionic trophoblastic cells, both demonstrating variable plasticity. The cells from the placental tissues exhibit the markers of pluripotency (SSEA-4, Oct-4, Stro-1, and TRA 1–81) which are typical mesenchymal markers and have capacity of a wide range of differentiation. They are capable of in vivo differentiation into various types of cell lineages if seeded in scaffolds. The pretreated placenta-derived human MS cells with a hyaluronan mixed ester of butyric and retinoic acid could cure infracted pig hearts. The treated pigs in comparison to untreated infracted pigs had 40 % smaller infarct scar size and a significant improvement of end-systolic wall thickening and circumferential shortening of the infarct border zone. Scarce reports on animal placenta cells suggest that further work on this source is required.

19.9 Stem Cells in Poultry

Poultry production is a well-organized industry. The chicken eggs have been used in the manufacture of vaccines for more than three decades. One of the remarkable advantages of the avian embryo is its accessibility and availability in plenty. Since the avian embryo is self-contained in a calcified eggshell, it lends itself to direct manipulation. However, compared to other livestock species, the overall progress toward establishing avian stem cells is slow. Advances in culturing avian embryos have also led to developments in avian stem cells. A modern layer lays approximately 300 eggs a year, and the egg white alone contains 4 g of protein. The hen takes much less time to reach sexual maturity than any other livestock species (http://www.poulvet.com/poultry/articles/3.ph, Oct., 1, 2011). Nevertheless, only a few laboratories were involved in research and development in avian ES as well as EG cells (Petitte et al. 2004). The main impetus for the isolation and culture of avian stem cells is the hope that they could be used to generate transgenic birds.

Avian ES cells differentiate into embryoid bodies (EB) as well as various somatic cell lineages and can be used for a variety of applications including production of chimera or transgenic birds for biopharming. With adaptations to high-throughput in ovo vaccination technology, it could be possible to generate high-grade avian somatic chimeras, thereby shortening the time needed for conventional poultry breeding programs to generate superior stock. Therefore, transgenic chicken holds a tremendous potential to revolutionize the biotechnology industry and would contribute significantly to the national economy. At present, many private research companies have initiated research in the field of avian transgenics to harness this powerful emerging technology (http://www.poulvet.com/poultry/articles/3.ph). However, development of stem cell lines and robust methods for production of transgenic birds has been more of a challenge.

19.10 Fetal Stem Cells in Animal Health and Production

The last two decades has seen a surge of interest in the research and clinical availability of stem cell treatments. Innovative techniques of cell and tissue processing, based on tissue engineering, have been developed. Cell expansion and tissue reconstruction through *ex vivo* cultures are core processes used to produce engineered tissues with sufficient structural integrity and functionality.

Fetal stem cells expressing pluripotency markers provide strong support to the notion that these cells may be biologically homologous to pluripotent ES cells. There are some reports describing the derivation of putative fetal stem cells in domestic ungulates and pet animals with varying results. Commercial companies are emerging with strategies to enhance stem cell research and applications in veterinary health and livestock production. Formed in 2002, Vet-Stem Inc. (www.vet-stem.com) is the first company to offer regenerative medicine to veterinarians. In 2003, the company introduced the first veterinary stem cell service in the United States. Its technology has been used to treat tendon, ligament, and joint injuries in more than 3,000 horses and more than 1,000 dogs.

The use of fetal and adult stem cells in veterinary medicine is of great promise and is likely to show rapid uptake, as commercially available safe treatments with adipose and bone marrow-derived cells become more widespread. The features and the potential therapeutic properties of fetal stem cells from various sources described here are expected to be studied in more depth and then implemented at the clinical level following international guidelines.

Cell-based therapies with embryonic, fetal, adult, or iPS cells are thought to have great potential for augmenting assisted reproduction and treatment of several degenerative diseases, which currently are without effective therapy. For instance, bone fracture, as well as damaged cartilage, tendons, and ligaments, heal poorly in horses. Equine iPS cells bring new therapeutic potential to the veterinary field and open up the opportunity to validate stem cell-based therapies before clinical studies in humans. As well, stem cell-based studies using the horse as a model more closely replicate human illnesses, when compared with studies in mice. Therefore, the use of reprogrammed cells in these animals may help enhance long-term tissue repair.

19.11 Conclusions

Science and medicine place a lot of hopes in the development of stem cell research and regenerative medicine. Worldwide media reports about stem cell therapies are becoming common as stem cell applications are being pursued in diverse areas including cardiology, orthopedics, oncology, internal medicine, and assisted reproduction. Till date, significant progress has been made in stem cell research, and new strategies for somatic cell reprogramming have been developed. In view of the ethical problems in deriving ES cells, the researchers are interested in discovering alternative ways of deriving pluripotent stem cells. Non-embryonic fetal stem cells can be isolated during gestation from many different tissues as well as from a variety of extraembryonic tissues such as the AF and placenta. In domestic ungulates as well as poultry, the ES and EG cells and adult stem cell technology are still at initial stages of development. Cells isolated from livestock species might have important application in studies in developmental biology and especially in unraveling appropriate culture conditions and markers for stem cells. Several potential applications of transgenic chickens have been identified, and the two main areas include agriculture and health care. In addition, although many of the non-embryonic stem cells have demonstrated ES cell-like morphology and expression of some ES cell-specific markers, concerns should be raised over the lack of data regarding long-term culture and maintenance of pluripotency. Ethical concerns in use of animals for production of therapeutic concerns have to be followed strictly. Stem cells obtained from transgenic livestock having desired traits hold the key for harvesting the maximum potential from this promising technology.

References

Abdulrazzak H, Moschidou D, Jones G, Guillot PV (2010) Biological characteristics of stem cells from foetal, cord blood and extraembryonic tissues. J R Soc Interface 6:689–706

Azari O, Babaei H, Derakshanfar A, Nematollahi-Mahani SN, Poursahebi R, Moshrerfi M (2011) Effect of transplanted mesenchymal stem cells isolated from Wharton's jelly of caprine umbilical cord on cutaneous wound healing: histopathological evaluation. Vet Res Commun 35:211–222

Chen J, Lu Z, Cheng De, Peng S, Wang H (2011) Isolation and characterization of porcine amniotic fluid-derived multipotent stem cells. PLoS One 6(5):e19964. doi:10.1371//journal.pone.0019964

Cremonesi F, Corradetti B, Lange CA (2011) Fetal adnexa derived stem cells from domestic animal: progress and perspectives. Theriogenology 75:1400–1415

Kues WA, Petersen B, Mysegades W, Carnwath JW, Niemann H (2005) Isolation of murine and porcine fetal stem cells from somatic tissue. Biol Reprod 72:1020–1028

Lange-Consiligio A, Corradetii B, Rutigliano L, Cremonesi F, Bizzaro D (2011) *In vitro* studies of horse umbilical cord matrix-derived cells: from characterization to labeling for magnetic resonance imaging. Open Tissue Eng Regen Med J 4:120–133

Lim JH, Byeon YE, Ryu HH, Jeong YH, Lee YW, Kim WH, Kang KS, Kweon OK (2007) Transplantation of canine umbilical cord blood-derived mesenchymal stem cells in experimentally induced spinal cord injured dogs. J Vet Sci 8:275–282

Mann A, Yadav RP, Singh J, Kumar D, Singh B, Yadav PS (2012) Culture, characterization and differentiation of cells from buffalo amnion. Cytotechnology. doi:10.1007/s10616-012-9464-z

Medicetty S, Bledsoe AR, Fahrenholtz CB, Troyer D, Weiss ML (2004) Transplantation of pig stem cells into rat brain: proliferation during the first 8 weeks. Exp Neurol 1:32–41

Pappa KI, Anagnou NP (2009) Novel sources of fetal stem cells: where do they fit on the developmental continuum? Regen Med 4(3):423–433, http://www.ncbi.nlm.nih.gov/pubmed/19438317

Petitte JN, Liu G, Yang Z (2004) Avian pluripotent stem cells. Mech Dev 121:1159–1168

Singh J, Kumar P, Mann A, Singh V, Duhan JS, Yadav PS (2012a) Cell morphology and mucopolysaccharides in

early gestation buffalo umbilical cord. Indian Vet J 89(11):36–38
Singh J, Singh V, Mann A, Singh JK, Jerome A, Sarkar SK, Duhan JS, Yadav PS (2012b) Collection of cord blood and comparison of its hematological parameters with newborn calf and adult buffalo. Indian J Anim Sci 82(8):865–867
Yadav PS, Kues WA, Herrmann D, Carnwath JW, Niemann H (2005) Bovine ICM derived cells express the OCT-4 ortholog. Mol Reprod Dev 72(2): 182–190
Yadav PS, Jayanti Tokas, Sharma RK, Inderjeet Singh, Sethi RK (2008) Buffalo amniotic fluid, umbilical cord matrix and early fetal explants as possible source of adult stem cells. In: IX annual conference of Indian society of animal genetics and breeding, NASC complex Delhi, 3–4 July 2008, p 118
Yadav PS, Mann A, Singh V, Yashveer S, Sharma RK, Singh I (2011) Expression of pluripotency genes in buffalo (*Bubalus bubalis*) amniotic fluid cells. Reprod Domest Anim 46:705–711
Yadav PS, Mann A, Singh J, Kumar D, Sharma RK, Singh I (2012a) Buffalo (*Bubalus bubalis*) fetal skin derived fibroblast exhibited characteristics of stem cells. Agric Res. doi:10.1007/s 40003-012-0013
Yadav PS, Singh RK, Singh B (2012b) Fetal stem cells in farm animals: applications in health and production. Agric Res 1:67–77

Animal Forensics and Applications 20

P. Minakshi, M. Shafiq, A. Alisha, K. Ranjan, P. Kumar, B. Anshul, R. Singh, Basanti, Supriya, and G. Prasad

Abstract

Forensic science is the application of a broad spectrum of sciences and technologies to investigate situations after getting the facts and to establish what occurred based on collected evidence. This is especially important in law enforcement where forensics is done in relation to criminal or civil law. In civil actions, forensics can help resolve a broad spectrum of legal issues through the identification, analysis, and evaluation of physical evidence. The field of forensic science covers document examination, DNA analysis using electronic or digital media, fingerprinting, autopsy techniques, forensic engineering, forensic anthropology, pathology, economics, accounting, biology, entomology, toxicology, and much more. In this chapter we have described different materials such as hair, blood, bone, teeth, saliva, nails, feathers, skin, leather, sperm, feces, and urine and different methods for extracting DNA from different sources. The applications of animal forensics can be broadly viewed in the following four categories such as animals can be the victim, can be the perpetrator, and can be the witness and wildlife forensics. Molecular animal forensics provides different genetic tools such as DNA sequencing, single nucleotide polymorphism (SNP), PCR-RFLP, and microsatellite analysis for species identification and for characterization or identification of a sample recovered from a crime scene or illegal wildlife traders and black markets involved in wildlife trade. The genetic identification can be done as species identification, identification of geographic origin, individual identification, etc. Mitochondrial and nuclear markers can be used for

P. Minakshi (✉) • M. Shafiq
A. Alisha • K. Ranjan • P. Kumar • B. Anshul
R. Singh • Basanti • Supriya • G. Prasad
Department of Animal Biotechnology,
LLR University of Veterinary and Animal Sciences, Hisar, India
e-mail: minakshi.abt@gmail.com

R.K. Salar et al. (eds.), *Biotechnology: Prospects and Applications*,
DOI 10.1007/978-81-322-1683-4_20, © Springer India 2013

genetic identification of the species. Identification of geographic origin is done by phylogeography and population assignment methods. To summarize, various techniques of individual identification, sexing, and parentage can be used. These techniques involve the microsatellite genotyping, DNA nucleotide sequencing, SNP typing, RAPD, and AFLP.

20.1 Introduction

Forensic science is the application of a broad spectrum of sciences and technologies to investigate and establish facts of interest about criminal or civil law. It is a specialism that aims to help judges solve legal issues, not only in criminal law but also in civil cases. It is a very broad field, crossing the boundaries between biology, physics, chemistry, and mathematics and including disciplines as varied as ballistics and botany. It also includes disciplines like veterinary sciences, genetics, pathology, and morphology. Forensic science finds applications in both humans and animals. In case of humans, it is more often used in human identity testing which has applications in crime solving, paternity testing, identifying accident victims and soldiers in war, solving inheritance claims, etc.

Animal forensics can be described as scientific tests or techniques used in the investigation of crimes against animals (domestic animals, agricultural animals, wildlife, etc.). In case of animals forensic science has much more varied applications. It helps in solving cases related to crime against animals, like animal poaching and blood sport. It also helps in preventing illegal trafficking of regulated species in the form of parts and products; settling legal claims related to animals; investigating unknown causes of death to endangered, threatened, and protected animal species; etc. Since animals also form a part of human diet, and there are many religious issues concerned with the diet, animal forensics thus help in protecting social and religious values of a society. Determining the species origin of animal tissues such as in cases of illegal commercialization and poaching of game animals is also one of the applications of animal forensics.

Conventional methods used in animal forensics include microscopic examination, protein-based assays, isoelectric focusing, and enzyme-linked immunoassays (ELISA). Before 1975, the application of forensic science protocols to animal and wildlife-related evidence was mostly limited to tentative family, genus, and species identifications of blood stains and loose hairs found at human crime scenes. These methods were usually based upon immunosorbent assays tests using relatively nonspecific antisera and microscopic comparisons. However, the results were not much useful in trying to link suspect, victim, and crime scene. Furthermore, in order to enforce the CITES (Convention on International Trade in Endangered Species Fauna and Flora) (www.cites.org) regulations in courts of law, the need for species-specific identifications was felt to address an underlying forensic issue: that illegal trafficking of regulated species would be in the form of parts and products. Thus, the need for wildlife forensics on an international scale was born.

There are other challenges to animal forensics as well. These include scant availability of samples (pieces of bones, flesh etc.), very old and degraded samples, samples obtained from closely related species, etc. Over the past 20 years, however, one particular biological tool has managed to address all these concerns and, in fact, revolutionized forensic investigations—the analysis of DNA. Since all living things contain DNA, and all DNA exhibit variability both among and within species, any biological material associated with a legal case carries in it information about its source. DNA analysis has evolved to become an indispensable tool of modern forensic science including animal forensics, employing extremely sensitive PCR-based techniques to analyze biological materials. Suspects can be linked to crime scenes using DNA evidence from as little as the saliva on a cigarette butt or skin cells on a steering wheel. Similarly, cases can now be solved decades after investigations were begun by analyzing degraded DNA from stored swabs or by analyzing DNA from degraded samples.

DNA-based forensic methods are much more efficient than the conventional methods because of so many obvious reasons. DNA is a stable molecule and contains identical genetic information irrespective of the origin of sample as far as the sample is taken from a same individual. The information content in DNA is species and/or individual specific and is more in DNA molecule as compared to that in protein molecules. Moreover, DNA-based forensic methods make it possible to deduce phylogenetic/evolutionary relations. In tropical countries, like India, due to high ambient temperature sample degradation occurs quickly. Moreover, particularly in India, there are fewer diagnostic laboratories, and so the samples have to be transported over considerably larger distances before reaching the laboratories. Also the field personnel are inadequately trained for collection, preservation, and transport of samples from collection sites to the laboratories. All these factors highlight the importance of DNA-based forensic methods.

This chapter mainly focuses on animal forensics based on DNA analysis and is divided into the following sections: materials used in forensics and protocols for extracting DNA from such materials, applications of animal forensics, and forensic genetic identification methods.

20.2 Material Used in Forensics

20.2.1 Hair

Comparative morphology, microscopy, and histology represent the classical methods in the field of forensic hair analysis. Type, number, and state of preservation of seized hair effect its value as trace evidence. Each mammalian species has hair with characteristic length, color, and root structure and various morphological characteristics. A typical hair consists of a root and hair shaft, which is basically composed of mark, cortex, and cuticle. The structure of the mark and of the hair cuticle is strictly species specific. The structure of the mark cells, the thickness of the marks and its continuity, cuticular pattern, medulla type, and medullary index also allow species differentiation. Microscopic analysis of hair roots allows not only the determination of growth phase but also a distinction between pulled out and naturally shed hairs.

20.2.1.1 DNA Extraction from Human Hairs

Hairs contain extremely small quantities of DNA (Higuchi et al. 1988). Since the content of nuclear DNA is too small for amplification, particularly those from naturally shed hairs or hair shafts rather than hair root, many studies have employed relatively abundant mtDNA (Sullivan et al. 1992; Baker et al. 2001; Pfeiffer et al. 1999; Vigilant 1999). Moreover, DNA extracted from hair is not always successfully amplified by PCR, suggesting the presence of PCR inhibitors in the extracted samples. It has already been revealed that the hair pigment melanin is a strong inhibitor of the PCR process (Yoshii et al. 1992, 1993; Wilson et al. 1995). More specifically, hair-dyeing was found to have a strong influence on PCR. In a study three different methods of DNA extraction were evaluated, and Chelex method was recommended for PCR experiments in view of its simplicity and cost-effectiveness (Suenaga and Nakamura 2005). The protocol for genomic DNA extraction recommended by them is described below.

Take two hair root segments (1 cm in length) and wash with 500 μl of 100 % ethanol in a small polypropylene test tube. After air-drying, place the hair in a 1.5-ml micro-centrifuge tube. Add 200 μl of 5 % Chelex® 100 and 10 μl of 10 mg/ml Proteinase K to the two pieces of hair placed in a 1.5-ml micro-centrifuge tube and mix well. The solution is incubated at 55 °C for at least 6–8 h or, alternatively, overnight. The mixture is vortexed and incubated in a boiling-water bath for 8 min. After centrifugation at 10,000–15,000 × g for 2–3 min, the supernatant is transferred to another 1.5-ml micro-centrifuge tube and is used for PCR amplification.

Chelex method is eco-friendly, cost effective, simpler, quick, and easy to perform and can also be used for DNA extraction from blood (Fig. 20.1), skin, feather, bone, muscle tissue, fecal sample, etc.

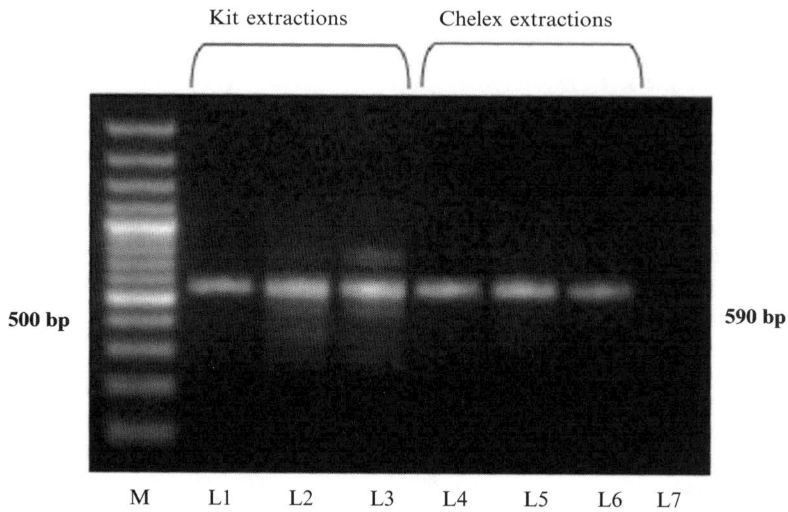

Fig. 20.1 PCR-amplified product of tiger blood DNA extractions (using kit and Chelex method) in 1.5 % agarose *M*: 100-bp ladder (MBI Fermentas); *Lanes 2–4*: Kit genomic DNA extractions; *Lanes 5–7*: Chelex genomic DNA extractions

20.2.1.2 Mitochondrial DNA (mtDNA) Extraction from Compromised Human Hairs

Since mtDNA is present in hundreds to thousands of copies per cell, it may therefore be better suited than nuclear DNA for ascertaining genetic information in cases where DNA amount may be limited, degraded, or both. The most polymorphic region of the human mtDNA genome is concentrated in two hypervariable segments within the noncoding, D-loop region (Greenberg et al. 1983). Analysis of DNA from ancient and historical anthropological samples has employed a silica-based method for DNA extraction (Hoss and Paabo 1993). In a study successful application of silica-based DNA extraction to compromised (water decomposed, incinerated, putrified) forensic hair shaft was carried out (Baker et al. 2001). The method described by them is given below.

Take 2 cm of a single strand of hair (cut hair shafts without the hair root), and wash with detergent in an ultrasonic water bath (Wilson et al. 1995). Place it in a pretreated 0.2 ml glass tissue homogenizer (Kontes Glass, Vineland, NJ) and grind in 100 μl of extraction buffer [10 M guanidine thiocyanate (GuSCN), 0.1 M Tris–HCl (pH 6.4), 0.2 M EDTA (pH 8.0), 1.3 % Triton X-100]. Transfer the homogenate to a UV-irradiated micro-centrifuge tube and pool with another 100 μl of extraction buffer used to rinse the homogenizer. Incubate the samples at 60 °C with slight agitation overnight (10–24 h). Isolate the DNA using the GeneClean II kit (Bio 101, Vista, CA). Add three volumes of sodium iodide and 5 μl of GlassMilk® to the sample, and incubate at 57 °C for 15 min with slight agitation. Following incubation, centrifuge the sample for 5 min at 12,000 × g and wash twice with the Bio 101 New Wash® solution. Remove any remaining New Wash® by centrifugation for an additional 3 min. Elute the sample in 30 μl of 10 mM Tris/1 mM EDTA buffer (pH 7.6) (TE) and incubate at 56 °C for 10 min. Centrifuge the sample again for 5 min and remove the supernatant and store at −20 °C until amplification.

20.2.2 Blood

The classical analytical methods of blood include blood group serology, the determination of serum proteins and isoenzymes, as well as the characterization of MHC antigens. However, current analysis possibilities of blood include the whole spectrum of molecular methods. The DNA is extracted from the nucleated white blood cells. The choice of the appropriate method of analysis depends on the quantity and quality of the available sample.

20.2.2.1 Extraction of Genomic DNA from Dried Blood Samples

Due to its ease and convenience in collection and transportation, even from geographically isolated populations, dried blood samples have become very popular for forensic as well as other public health purposes. It has been shown that any biological markers that can be measured from whole blood, serum, or plasma can be determined from dried blood specimens (Mei et al. 2001). In a study, Nguyen et al. (2012) developed a two-step lysis method for genomic DNA extraction from dried blood samples. The method was developed based on the Sumota Chaisomchit one-step lysis method (Chaisomchit et al. 2003). The two-step lysis method was shown to be more efficient and cost effective as compared to the one-step lysis method and commercial QIAamp® DNA Mini Kit-based method. The protocol for the two-step lysis method is described below.

Punch out three pieces of 3-mm-diameter circle from the dried blood spot samples into a 1.5-ml micro-centrifuge tube. Vortex with 200 μl of lysis buffer I (Tris–HCl 10 mM, MgCl2 5 mM, Triton X100 (1 %v/v), SDS 1 % w/v, EDTA 10 mM, and adjusted with pH 8.0) for 30 s and then incubate at 85 °C for 20 min. Cool down the lysate at room temperature for 10 min. Add 0.01 mg of Proteinase K and vortex again for 30 s. Add 100 μl of lysis buffer II (Tris–HCl 30 mM, EDTA 20 mM, SDS 3 %, and adjusted with pH 8.0). Vortex again for 30 s. Incubate at 65 °C for 1 h. Add the same amount of buffer phenol:chloroform:isoamyl alcohol (25:24:1) and mix well for 30 s. Centrifuge at 10,000 rpm for 4 min at room temperature and transfer the upper phase into a fresh tube. Treat with sodium acetate (3 M, pH 5.2) and isopropanol. Mix and centrifuge at 10,000 rpm for 4 min at room temperature. Remove supernatant and wash the pellet with 70 % ethanol. For long-term use, the pellet can be stored in 70 % ethanol at −20 °C. Prior to use, however, the pellet is centrifuged to remove the washing and dried at room temperature until there is no trace of ethanol. The genomic DNA is then resuspended in 50 μl of TE buffer [10 mM Tris, 1 mM EDTA].

20.2.3 Saliva

Fresh whole blood or blood-stained material is the primary source of an individual's DNA, however, in a study, Walsh et al. (1992) presented several methods for isolating DNA from saliva and saliva stains and showed that saliva can serve as an alternative source of DNA for known standards. They isolated DNA from fresh saliva and various saliva-stained materials (stored under different conditions), such as envelopes, buccal swabs, gags, and cigarettes (Walsh et al. 1992). The various methods described by them are described below.

20.2.3.1 DNA Extraction from Saliva

Take fresh saliva and centrifuge for 1 min. Resuspend the pellet in 0.7 ml of lysis buffer [10 mM tris(hydroxymethyl)aminomethane (Tris) (pH 8.0), 10 mM ethylenediaminetetraacetate (EDTA), 0.1 M sodium chloride (NaCl), and 2.0 % sodium dodecyl sulfate (SDS)] (Gill et al. 1985), and add 35 μl of 20 mg/ml Proteinase K. The saliva cell pellets can also be resuspended in 305 μl of 10 mM Tris (pH 7.6), 10 mM NaCl, 1 mM EDTA, 1 % SDS, and 0.65 μg/ml Proteinase K in the presence or absence of 39 mM dithiothreitol (DTT). If saliva is frozen at −20 °C, thaw it at 20 °C and use 1 ml aliquot for pellet formation and then treat with lysis buffer as above.

In case of buccal swabbing, remove cotton from the swab stem, and transfer to a micro-centrifuge tube. Add 0.7 ml of lysis buffer and 35 μl of Proteinase K (20 mg/ml). To obtain DNA from the cigarette butts, remove the filter ends of the cigarettes, cut into small pieces, and incubate in lysis buffer with Proteinase K (1 mg/ml final concentration) in a micro-centrifuge tube as above. Similarly, in case of stamps and the gummed edges of envelopes, cut into small pieces, and then transfer to micro-centrifuge tubes or 15 ml polypropylene tubes. Add lysis buffer to cover the samples, and make Proteinase K to a final concentration of 1 mg/ml.

The rest of the protocol is the same for all the above sample types. Incubate the samples overnight at 56 °C. Extract DNA with an equal volume of phenol/chloroform (1:1) and with an equal volume of chloroform. Add 1 μl of glycogen (20 mg/ml), and precipitate the DNA with an equal volume of isopropanol at −20 °C overnight. Pellet the DNA, wash with 80 % ethanol, air-dry, and resuspend in restriction buffer [60 mM Tris (pH 7.5), 10 mM magnesium chloride (MgCl), 100 mM NaC1, 35 mM 2-mercaptoethanol, and 1 mg/ml bovine serum albumin].

20.2.4 Bone and Teeth

The protein-mineral matrix of bone poses an effective physical and chemical barrier to environmental deterioration and biological attack. Therefore, bone and teeth samples are often the only, and almost always the best, biological material available for DNA typing.

20.2.4.1 DNA Extraction from Blood and Teeth

In a study, Loreille et al. (2007) presented a highly efficient protocol for the recovery of DNA from bone and teeth by full demineralization, resulting in full physical dissolution of the bone powder and quantitative recovery of all DNA released by the complete demineralization procedure (Loreille et al. 2007). They showed that the protocol significantly enhances the quantity of DNA that can be extracted and amplified from degraded skeletal remains. The protocol is described below.

Using an aluminum oxide sanding stone, extensively sand the entire surface of each bone or tooth sample to remove potential exogenous DNA. Sonicate the samples in 20 % bleach for 5 min. Following the bleach wash, rinse the bones in UV-irradiated water, and then sonicate again for 5 additional minutes in UV-irradiated water. Perform a final sonication wash step again using 100 % ethanol, and place the bone in a sterilized fume hood to air-dry overnight.

The next day, powder the samples in a sterilized process. Take 0.6–1.21 g of finely ground bone powder, and incubate overnight in 9–18 ml of extraction buffer (EDTA 0.5 M, 1 % lauryl sarcosinate) and 200 ml of 20 mg/ml Proteinase K, in a rotary shaker at 56 °C. Extract DNA with an equivalent volume of phenol:chloroform:isoamyl alcohol (25:24:1). Concentrate the supernatant to a volume slightly less than 2 ml using centrifugal filter units (e.g., by using Centriplus from Millipore). Transfer the remaining 2 ml of DNA extract into a Centricon 30 centrifugal filter unit (Millipore) and wash three times with irradiated water (UltraPure™ DNase/RNase-Free Distilled Water, Invitrogen).

This extraction protocol includes complete demineralization of the bone/tooth powder and significantly increases DNA yields, therefore, improving DNA typing results from degraded skeletal elements.

20.2.5 Nails

Nail clippings can be used as an alternative source of genomic DNA. They are often superior to other biological specimens because nail clippings offer long-term stability, even at room temperature, and are easily transportable. Moreover, as a biological specimen, nail clippings have low infectivity and can easily be obtained from subjects of any age, at any time, and in any place.

20.2.5.1 DNA Extraction from Nails

In a study Cline et al. (2003) developed a simple and efficient method for discrete isolation and purification of nail DNA and DNA from exogenous sources (exogenous material potentially originating from an attacker during self-defense) (Cline et al. 2003). The protocol is described below.

20.2.5.1.1 Exogenous DNA Isolation

1. Soak nail in 200 μl sterile 25 mM EDTA (in water) at room temperature for 1 h. Gently vortex periodically. Transfer liquid to a new tube.

2a. To this solution add 20 μl 10 % SDS and 1 μl 20 mg/ml Proteinase K. Incubate at 50–60 °C overnight. Extract DNA using phenol/chloroform as described in step 6 below.

20.2.5.1.2 Nail Preparation and DNA Isolation

2b. To the nail add 200 μl 1 % SDS/25 mM EDTA and 1 μl 20 mg/ml Proteinase K. Vortex and incubate for 1 h at room temperature.

3. Pipet or pour off liquid, and rinse nail 5–10 times with high-quality (18.3 M-ohm) sterile water. Following a final rinse with sterile water, centrifuge nail briefly and pipet off all remaining liquid.
4. Add 200 μl 2N NaOH to nail. Incubate overnight at room temperature. Vortex periodically if desired.
5. Following incubation, vortex nail to ensure it is completely solubilized. Neutralize solution (to pH 6–8, checked by spotting 1 μl onto pH paper), by adding 100 μl of 200 mM Tris (pH 7–8) and 34.5 μl concentrated HCl. Vortex immediately. If the pH is too low a precipitate will form. Adjust pH with dilute NaOH as needed to redissolve precipitate. (Note: the 34.5 μl of concentrated HCl (11.6N) is equal-normal with the 200 μl of 2N NaOH. If the HCl is more dilute, the volume added should be increased accordingly. Old HCl may work poorly.) Continue to step 6 or 7.

20.2.5.1.3 Organic Extraction

6a. Add an equal volume of phenol/chloroform (or PCI) to neutralized sample, vortex, spin at high speed for 5 min in a micro-centrifuge, and transfer aqueous (top) layer to a clean tube. Repeat this extraction if the aqueous layer is not clear.

6b. Precipitate DNA by adding 1/10 volume 3 M sodium acetate and 2 volumes of 95 % ethanol. Incubate at −20 °C for 1 h or longer.

6c. Centrifuge at high speed for 15–30 min. Note location of DNA pellet, which may or may not be visible. Carefully pipet off all liquid. Dry pellet and resuspend in TE (10 mM Tris/1 mM EDTA) at 10 μl/mg nail.

20.2.5.1.4 Microcon-100 Purification

7a. Add an equal volume of TE to the neutralized sample and place on column. Centrifuge at 500 × g for approximately 20 min or until liquid is pushed through. Discard flow-through.

7b. Add 200 μl TE to top portion of column and centrifuge at 500 × g as in step 7a. Repeat step 7b once.

7c. Collect retentate containing DNA (generally 10–20 μl remaining on top of column; add TE if needed to 10 μl/mg nail).

20.2.5.1.5 Removal of Nail Polish

Following step 1 above, add 100 μl acetone to nail, vortex, and soak at room temperature for 10 min. Draw off acetone, discard, and repeat. Allow nail to dry and continue to step 2.

20.2.6 Feathers

The use of feathers simplifies the sampling of avian genomic DNA, especially when blood extraction is difficult because of the age or the size of the bird. It also minimizes the stress on the bird. Moreover, it is a noninvasive method and is useful particularly in case of juvenile birds and small parrots in which blood extraction is very difficult because of the small size of the blood vessels.

20.2.6.1 DNA Extraction from Feathers

In a study Volo et al. (2008) presented an improved and modified protocol for extracting DNA from feathers (Volo et al. 2008). They used the protocol to successfully isolate DNA from molted feathers. The protocol is described below.

20.2.6.1.1 Feather Preparation

Wipe all surfaces with 10 % bleach. Run UV lights (if available) for 20 min. Heat water bath to 55 °C. Use of nitrile gloves helps to reduce static electricity that causes difficulty in handling feather material. Separate calamus tip from the rest of the feather and place in a tube. Make sure to cut above the superior umbilicus, so you can include it later. Fill tube with 70 % ethanol, and soak for 30 min. Set out more 1.5 ml centrifuge tubes and fill with double-distilled water; soak feathers for 30 min. Label 1.5 ml centrifuge tubes which will now serve as digestion tubes.

Use sterile petri dish as a catching surface and sterile scissors and sterile forceps for each sample. Cut feather rachis longitudinally along its length. Stop before superior umbilicus. Then cut horizontally ~5 cm of the feather tip into 2–3 mm pieces and place into tube. Cut out the superior umbilicus and place in tube. Cut it out as a small square around the "blood-dot." Pull out the papery material from inside the calamus and place in tube.

20.2.6.1.2 DNA Extraction

Prepare the following reagents:

1X TNE: Mix 100 mM NaCl, 50 mM Tris, and 25 mM EDTA. Bring solution to pH 7.5 with HCl, and then autoclave.

Digestion mix: To prepare 795 μl (volume required for each extraction) of digestion mix, mix 600 μl 1X TNE, 60 μl 1 M Tris–HCl, 45 μl Proteinase K (20 mg/ml), 10 μl of 25 % SDS (weight/volume), and 80 μl 1 M DTT (newly mixed).

Pipet 700 μl digestion mix to each tube. Incubate at 55 °C until most of the material dissolves, overnight to 1 week, usually 3–4 days. If material is not completely digested in 2 days, add more Proteinase K (20 μl of 25 mg/μl). Cool to room temperature after digestion is complete. Pipet 233 μl protein precipitation mix (7.5 M ammonium acetate) to each tube. Pulse-vortex to mix. Place in a freezer (−20 °C) for 30 min. Centrifuge (4 °C) at 13,000–16,000 rpm for 30 min. If some debris is still floating, recentrifuge for 10 min. Pour supernatant into a new 1.5 ml tube already containing 600 μl 100 % isopropanol (DNA grade). Add 1 μl glycogen (DNA carrier) to each tube. Mix samples by inverting tubes 50 times. Place in freezer (−20 °C) overnight. Centrifuge (4 °C) at 13,000–16,000 rpm for 30 min. Look for pellet at bottom of tube. Pour off supernatant and drain tube on clean, absorbent paper. Add 600 μl 70 % ethanol (DNA grade). Centrifuge (4 °C) at 13,000–16,000 rpm for 2 min. Carefully pour off ethanol. Air-dry tubes to remove all excess ethanol. It may be necessary to leave overnight to dry. When dried add 20–50 μl TE (10 mM Tris, pH 8.0, and 1 M EDTA pH 8.0) to rehydrate pellet, flicking tube to mix. Let it rehydrate overnight before use, or warm (35 °C) in a heating block to facilitate resuspension. Use 1–2 μl for PCR.

20.2.7 Skin and Leather

Once tanned it becomes extremely difficult to differentiate skin/leather of different species or skin of a wild animal from that of a domestic animal. Taking advantage of this fact, commercial products obtained from wild animals are sometimes sold as if they were from domestic animals and vice versa. In such cases, DNA analysis is the only solution for differentiating between processed skin samples of different species.

20.2.7.1 DNA Extraction from Skin and Leather

In a study, Ojeda et al. (2012) introduced a novel DNA extraction protocol for leather samples (Ojeda et al. 2012). DNA was successfully isolated from hides in various states of preservation such as raw hide and dry salted hide and skin. The protocol which is a modification of Sambrook et al. (1989) is described below.

Place a small tissue sample of 25 mg in a 1.5-ml tube. Add 500 μl lysis buffer (50 mM Tris–HCl, pH 8.0, 50 mM EDTA, 1 % SDS, and 50 mM NaCl) and 5 μl Proteinase K (20.1 mg/μl). Incubate at 55 °C in a water bath for 8 h with stirring. Centrifuge the samples for 20 min at 13,000 rpm. Remove 500 μl supernatant (avoiding the pellet as well as the superficial oily layer if any), and place in a new tube. Add 300 μl 5 M NaCl, and centrifuge the samples for 15 min at 13,000 rpm. Finally, recover 500 μl of the supernatant, and precipitate DNA by the addition of an equal volume of absolute isopropanol. Shake samples briefly, and then vortex thoroughly. Centrifuge tubes at 13,000 rpm for 15 min, and discard the liquid with care not to disturb the pellet. Wash pellet with 750 μl 70 % ethanol, and centrifuge for 5 min at 13,000 rpm to remove the alcohol. Dissolve DNA in 25 μl 1X TE buffer.

20.2.8 Sperm, Feces, and Urine

In contrast to human forensics, the molecular analysis of semen of animals plays hardly any role. The investigation of urine is useful in doping cases. The examination of feces has practical role in tracking wildlife.

20.3 Applications of Animal Forensics

The applications of animal forensics can be broadly viewed in the following four categories:

20.3.1 Animals Can Be the Victim

In a case report from Argentina, ranch cattle were stolen and slaughtered. The later found carcasses could be uniquely identified due to their brand (a specific farm mark on the animal coat for identification of individual animals) was used as reference sample. Their DNA profiles were compared with seized chunks of meat from a butcher's shop. The evidence (meat from butcher shop) agreed with the reference samples, so the meat could be assigned to the unambiguously killed cattle (Giovambattista et al. 2001). There are also numerous cases of cruelty to animals which include acts like bestiality and killing of animals for skin, bone, etc. In such cases a forensic scientist is expected to link traces of animal tissues available at the crime scene to the animals being killed or in case of bestiality, linking traces of sperm found in the vagina of an animal with the perpetrator or linking hairs or blood spots recovered from the perpetrator's body with the animals being assaulted.

20.3.2 Animals Can Be the Perpetrator

Animals can be perpetrators as well as victims and can be involved in an attack on a person or other animal. Thus, identification of such animals is needed as well as the identification of an unrestrained animal causing an accident or being responsible for property damage. Thus, again a forensic scientist is expected to link traces of animal material recovered from the crime scene to the animal involved in the act. DNA analysis was successfully used to identify a dog that had a miniature horse killed and another seriously injured. The offender animal was successfully identified with the help of traces of horse blood found at the edge of the dog's water bowl. They were consistent with the genetic profile of the dead horse (Agronis).

20.3.3 Animals Can Be the Witness

Animal DNA can link a suspect with a crime scene or victim. Transfer of DNA from hair, saliva, blood, urine, or feces can occur during the commission of a crime—from the victim's pet to the suspect or crime scene and from the suspect's pet to the victim or crime scene. In a case, the conviction of a man for the murder of a 7-year-old girl in California was supported by mitochondrial DNA analysis of dog hairs that matched a pet belonging to the victim.

20.3.4 Wildlife Forensics

Illegal trade of animals and their by-products is a growing global black market commerce. Since weight-for-weight wildlife trade is equally or more profitable than drugs or arms, and has less risk associated with it, illicit wildlife trade is attractive to criminals. It poses serious threats, both direct and indirect, to global biodiversity. A chain of events occurs which leads to the extinction of the species involved. Species sought for trade are directly affected by overexploitation. Overexploitation is fueled by the exaggerated values placed on rarer species by the black market. As a species becomes rarer from exploitation, its value on the black market escalates making it even more desirable despite the greater effort required to collect individuals from declining populations (Courchamp et al. 2006). Overexploitation of wild populations can cause global extinction. Wildlife trade also provides avenues for the introduction of exotics with the potential to spread disease to native species (Lips et al.

2006; Pedersen et al. 2007; Skerratt et al. 2007; Smith et al. 2006; Spinks and Shaffer 2007) or to become invasive (Keller and Lodge 2007; Normile 2004; Reed 2005; Weigle et al. 2005). A specialized field of animal forensics called wildlife forensics deals with such cases. Wildlife DNA forensics is an applied field that has emerged from a synthesis of conservation genetic research and forensic genetic practice to meet the increasing need for investigative tools in wildlife law enforcement. The ultimate objective of such investigative tools is identification of evidence items in order to determine the species, population, relationship, or individual identity of a sample.

At the heart of all this lies the challenge of establishing a relationship or individual identity of a degraded, compromised, and trace sample recovered from the crime scene. Since all living things contain DNA and all DNA exhibit variability both among and within species, any biological material associated with a legal case carries in it information about its source. Moreover, since techniques have been developed for extracting DNA even from traces of almost all sorts of samples, DNA forensics has evolved to become an indispensable tool of modern animal forensic science.

20.4 Forensic Genetic Identification Methods

Molecular animal forensics employs various genetic tools, for example, DNA sequencing, single nucleotide polymorphism (SNP), PCR-RFLP, and microsatellite analysis, to characterize or identify a sample recovered from crime scene or illegal wildlife traders and black markets involved in wildlife trade. The genetic identification can be done at three levels: species identification, identification of geographic origin, and individual identification.

20.4.1 Species Identification

To identify the species of an evidence sample using genetic analysis is the most common application in wildlife DNA forensics and is of lesser use in human forensics. Species identification is used in cases of illegal poaching in order to identify trace evidence in the field or from a suspect's possessions (Gupta et al. 2005). It has a particular use in the identification of traded products that have lost identifying morphological characters, such as processed wood (Deguilloux et al. 2002), traditional medicines (TMs) (Hsieh et al. 2003; Wetton et al. 2004; Peppin et al. 2008), and shark fins (Shivji et al. 2002; Chapman et al. 2003).

Genetic species identification relies on the isolation and analysis of various DNA markers present in eukaryotic genome such that they show variation among species but are generally conserved within species. There are two types of genetic markers: mitochondrial and nuclear markers.

20.4.1.1 Mitochondrial Markers

These include:

20.4.1.1.1 Mitochondrial Ribosomal RNA Markers

There are two ribosomal RNA (rRNA) genes, 12s rDNA and 16s rDNA, present inside animal mitochondria. Out of these 12s rDNA is highly conserved and is used to understand the genetic diversity of higher categorical levels, phyla and subphyla, while 16s rDNA is often used for studies at middle categorical levels, families or genera (Gerber et al. 2001). For molecular analysis, conserved primers are used to amplify these markers, and the amplicons are then sequenced. Sequencing data are then aligned and compared using appropriate bioinformatics tools. In a study, Turan (2008) used mitochondrial 16s rRNA to elucidate the pattern of relationships and systematic status of 4 genera, including nine species of skates living in the Mediterranean and Black Seas (Turan 2008).

20.4.1.1.2 Mitochondrial Protein-Coding Gene Markers

There are 13 protein-coding genes present in animal mitochondria. Compared to ribosomal RNA genes, mitochondrial protein-coding genes have faster evolutionary rates and are relatively conserved. Therefore, they are regarded as powerful markers for genetic diversity analysis at lower categorical levels, including families, genera, and species. Mitochondrial protein-coding genes that have been used for molecular analysis include cytochrome b (cyt b), NADH dehydrogenase subunit 5 (318 bp), and cytochrome oxidase I (COI) gene. However, cytochrome b is one of the most extensively used protein-coding genes of the mitochondrial genome. Its sequences have been used to understand the genetic diversity for better conservation management of Tibetan gazelle (*Procapra picticaudata*), a threatened species on the Qinghai-Tibet Plateau of China (Zhang and Jiang 2006). Similarly, NADH dehydrogenase subunit 5 (318 bp) has been used for phylogenetic analysis of multiple individuals of different species from the Felidae family, where it successfully differentiated eight clades reflecting separate monophyletic evolutionary radiations (Johnson and O'Brien 1997). Mitochondrial cytochrome oxidase I (COI) gene is most commonly used for developing DNA barcodes for species identification and biodiversity analysis. It has been used to reveal unrecognized species in several animal groups. DNA barcoding is a process in which one or a few genes are selected such that they are shared by most, if not all, organisms on earth and which show large interspecific but small intraspecific levels of variation. Such gene sequences can then be regarded as equivalent of species-specific barcodes.

A species identification method for 13 animal species, 9 of which were wild (wildebeest, zebra, Thomson's gazelle, impala, reedbuck, kongoni, oryx, warthog, and hippopotamus) and four were domestic (bovine, caprine, ovine, and porcine), was developed using a mitochondrial DNA marker in polymerase chain reaction coupled with restriction fragment length polymorphism (PCR-RFLP) analyses using *Rsa*I enzyme. A 700-bp region of mitochondrial cytochrome b gene/control (D-loop) was successfully amplified from all the samples, and a unique RFLP profile specific to each species was generated using *Rsa*I enzyme (Malisa et al. 2005).

20.4.1.1.3 Mitochondrial Control Region Markers

There is a noncoding region termed the control region (CR or D-loop) present in mitochondrial DNA. It has a role in replication and transcription of mitochondrial DNA and is approximately 1 kb in size. The D-loop segment exhibits comparatively higher level of variation than protein-coding sequences and is therefore used for identification of species and subspecies.

The decreasing order of conserved sequences in various mitochondrial DNA markers is 12S rDNA > 16S rDNA > cytochrome b > control region (CR or D-loop).

20.4.1.2 Nuclear Markers

Various nuclear markers used for DNA fingerprinting include random amplified polymorphic DNA (RAPD), amplified fragment length polymorphism (AFLP), and microsatellites. All these markers are described below in detail under different sections.

However, the most commonly used markers in animals are mitochondrial DNA markers particularly cytochrome b (Parson et al. 2000) and cytochrome oxidase subunit I (COI) (Hebert et al. 2003a, b), as their mutation rates roughly coincide with the rate of species evolution. Moreover, in comparison to nuclear DNA, where in diploid species a single nucleus carries just two copies of each marker, mitochondrial DNA is present in multiple copies within a single cell, and so the mitochondrial DNA markers are present in multiple copies. Thus, the chances of success increase significantly when analyzing trace evidence samples that typically contain relatively little cellular material (Butler and Levin 1998; Budowle et al. 2003). However, it is recommended that multiple genes, preferably both mitochondrial and nuclear genes, are used for species delimitation (Brower et al. 1996; Maddison 1997; Page 2000; Sites et al. 1996).

Various techniques used in species identification are described below:

20.4.1.2.1 DNA Nucleotide Sequencing

It is a primary method of species identification and usually involves sequencing around 500 bases of DNA to provide a species-specific sequence. Sequencing identifies each nucleotide (base) within a specific target region of DNA (the genetic marker). DNA nucleotide sequencing is particularly advan-tageous because it enables the designing of universal PCR primers that can be used for the amplification of DNA from a wide range of species without any prior information regarding the sample (Verma and Singh 2003). It also enables the designing of species-specific primers so that more than one species can be identified in a single multiplex PCR reaction. Furthermore, sequencing provides data for developing genetic markers such as single nucleotide polymorphisms (SNPs) and microsatellites, which describe specific areas of sequence variation.

Various evidence samples are processed, and DNA is isolated as already described above. After determining the nucleotide sequence of this unknown DNA, its identification is done through the use of a reference database search whereby the unknown sequence is compared to those of known samples: a measure of the similarity between sequences is calculated, and the most similar species is attributed to the sample (Parson et al. 2000; Branicki et al. 2003). Therefore, it is necessary to have a well representation of that species in the public databases. Closely related species may have sequence similarities of 90–95 % or higher. Moreover, since the level of similarity also depends on the total length of the sequence which is being matched, it is ultimately left to the experience and judgment of the forensic scientist to evaluate the strength of evidence when undertaking a sequence similarity match.

We can also construct a phylogenetic tree and determine the evolutionary relationships between the test sample and reference sequences from each possible species. The position of the test sample in the tree allows the closest reference species to be identified as the likely source (Avise 1994; Roman and Bowen 2000; Verma et al. 2003).

20.4.1.2.2 SNP Typing or Genotyping

We can go for either full DNA sequencing or sequence only certain regions of the genome that contain various genetic markers which are species specific. The differences observed among species at a genetic marker are largely due to single base pair changes in the DNA sequence, known as single nucleotide polymorphisms (SNPs). Therefore, SNP typing, or genotyping, allows specific variable sites in a DNA sequence to be investigated enabling species differentiation through faster and cheaper tests that do not require such long fragments of high-quality DNA as is required in DNA sequencing methods. However, in comparison to DNA sequencing, lesser information is gained in SNP typing.

SNP typing requires shorter DNA sequences and thus minimizes the size of the genetic marker targeted. This is often necessary to obtain results from samples that either are degraded or have been highly processed, fragmenting the DNA (Butler et al. 2003; Hajibabaei et al. 2006). However, there is also an increased risk of sample misidentification associated with SNP typing. This must be considered when designing assays and interpreting forensic data.

There are multiple methods for typing SNP markers, some of which are described below.

PCR-RFLP

In this technique the DNA segment of interest is amplified using PCR and is then subjected to digestion by restriction enzymes. Restriction enzymes are a special category of enzymes that recognize specific base pair sequence motifs (that are often mirror images) and cut the amplified fragment at these sites. Species that differ in nucleotide composition at the restriction enzyme recognition sites will differ in whether or not the enzyme cuts the DNA or in the position at which the enzymes cut the sequences. This generates DNA fragments of differing lengths (i.e., polymorphic fragments), in which the number and size of the fragments depend on the number of cutting sites in the DNA fragment of interest (Upholt 1977). Electrophoresis of the digested DNA samples thus reveals characteristic banding patterns in different taxa. Such a banding pattern becomes a DNA fingerprint of that taxa or species and is termed as RFLP profile. Similarly, in cases where restriction enzyme sites coincide with an SNP marker, a different RFLP profile is generated in different species. However, selection of restriction enzymes for PCR-RFLP analysis must ensure that the variability between species is appropriately represented so that all species tested can be accurately discriminated from each other by their unique RFLP profile.

Using primers 5′GTGCTACGAAAGCAGG3′ and 5′GGCGCGGATACTTGCATGTG3′, specific amplification of mitochondrial DNA D-loop

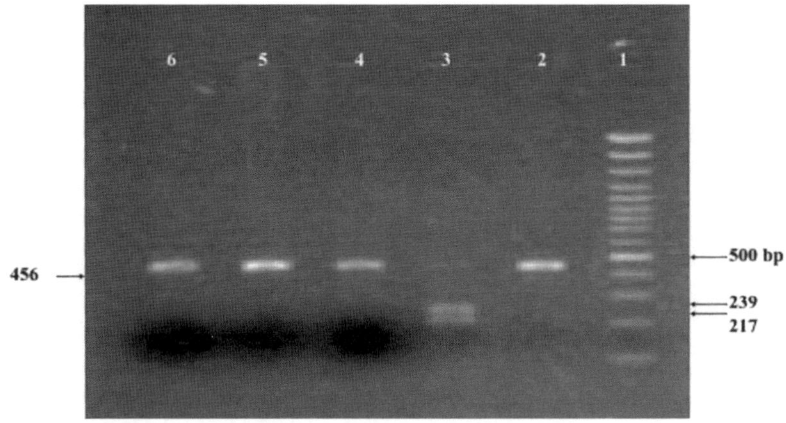

Fig. 20.2 Species specific gene amplification of different animal species; 100-bp ladder (lane 1), cattle (lane 2), buffalo (lane 3), goat (lane 4), sheep (lane 5), and nilgai (lane 6)

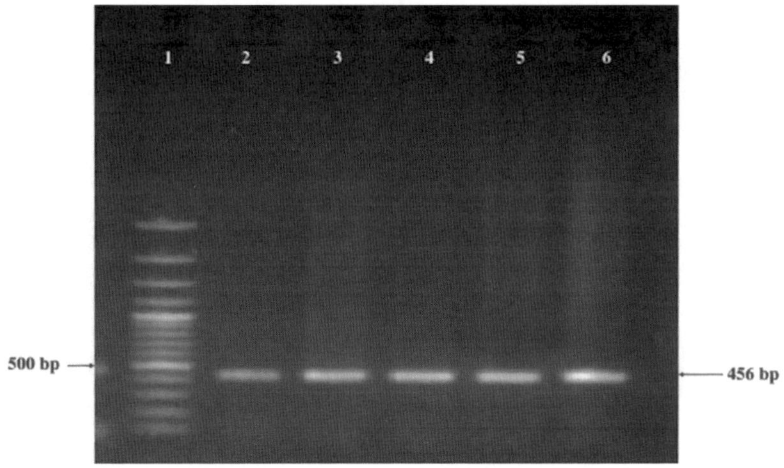

Fig. 20.3 Restriction enzyme digestion of species-specific PCR products using *Hha*1; 100-bp ladder (lane 1), cattle (lane 2), buffalo (lane 3), sheep (lane 4), goat (lane 5), and nilgai (lane 6)

region was carried out followed by restriction analysis of amplicons with *Dra*I endonuclease, and a unique RFLP profile was generated for wild boar and domestic swine, offering a useful tool to reveal fraud in meat substitutions as well as in legal cases to verify wild boar meat (Samaraweera et al. 2011). In order to differentiate between five different species (cattle, buffalo, sheep, goat, and nilgai), cytochrome b gene-specific primer pair (primer pair-D) was designed which successfully amplified a 456-bp fragment of cytochrome b gene in all the five species (Fig. 20.2). RFLP analysis of this fragment of cytochrome b gene was done using restriction enzymes Hha1, Alu1, and Apo1, and a unique RFLP profile was generated for each of the different species (Figs. 20.3 and 20.4) (Prasad et al. 2008; unpublished data). Moreover, the primer pair was shown to be very sensitive and could successfully amplify DNA extracted from boiled and processed tissue samples (Fig. 20.5). A patent was awarded to the authors for this study.

Allele-Specific PCR

Primers used in PCR for amplifying genetic markers can be designed for regions where DNA

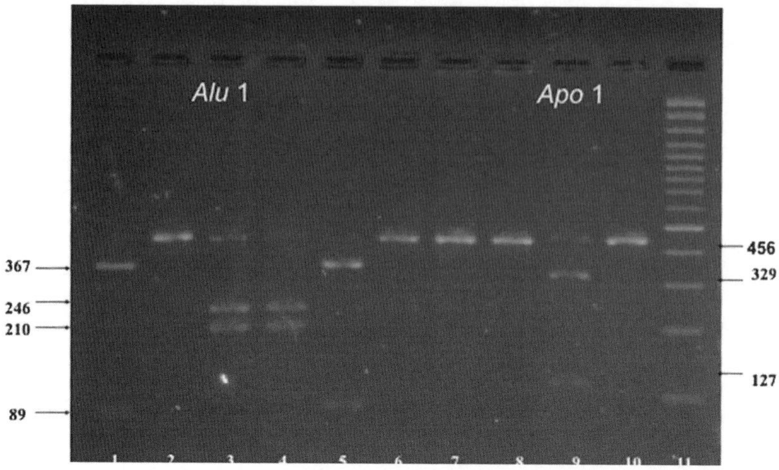

Fig. 20.4 Restriction digestion of species-specific PCR products using Alu 1 and Apo 1; cattle (lane 1, 6), buffalo (lane 2, 7), goat (lane 3, 8), sheep (lane 4, 9), nilgai (lane 5, 10), and 100-bp ladder (lane 11)

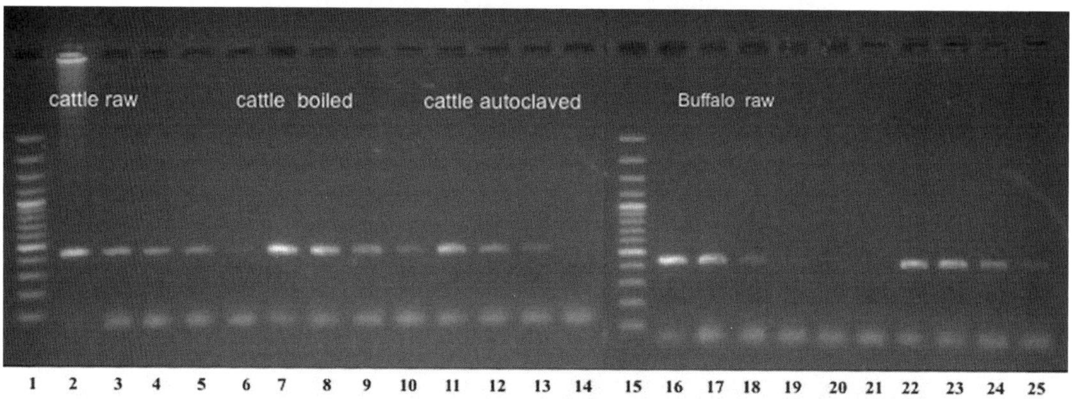

Fig. 20.5 Species-specific gene amplification of raw and processed tissue samples of cattle and buffalo; cattle raw DNA 0.3 ng (lane 6), cattle boiled DNA 0.5 ng (lane 10), cattle autoclaved DNA 2 ng (lane 13), buffalo raw DNA 3.4 ng (lane 18), and buffalo autoclaved DNA 10 ng (lane 24) were amplified

varies between species and populations. Such primers are known as allele-specific primers. SNP is responsible for generation of different alleles of a genetic marker in a population or species. Allele-specific primers are designed so that nucleotide sequence of the primers includes the SNP site, and that PCR only works when DNA from the target sequence is present in a sample. Thus, when such primers are used in PCR for amplifying unknown DNA samples, amplification of only that allele of a genetic marker occurs which has a sequence complementary to the allele-specific primers. However, for the development of species-specific primers, it is necessary to consider sequence data from all species likely to be encountered, for the design of putative primers. For example, species-specific primers that amplify the nuclear ITS2 region and the mitochondrial cytochrome b have been used to develop assays for the identification of various shark species from dried fins or meat (Moore et al. 2003; Clarke et al. 2006; Magnussen et al. 2007; Pank et al. 2001; Shivji et al. 2005).

Species-specific primer pairs based on mitochondrial D-loop and 12S ribosomal ribonucleic acid (rRNA) gene were designed that successfully amplified 629- and 322-bp DNA fragments, respectively, from the DNA sample extracted from pig meat. Therefore, a highly specific single-step polymerase chain reaction (PCR) was optimized which provides a valuable tool for identification of pig meat and to avoid its fraudulent substitution and adulteration (Arun et al. 2012).

20.4.1.2.3 Multiplex PCR

Species-specific primers have an advantage that they can be used in a multiplex PCR which is a more robust technique for species identification. Here several primers are added to a PCR to simultaneously amplify different DNA regions of the target species in a single PCR reaction. However, universal primers that amplify across all potential taxa should be included as a control in the multiplex assay. This reduces the incidence of false negatives. If the universal primer amplifies but the species-specific primers fail to amplify, it can be confirmed that target species is absent in the sample. On the other hand, if both the universal and species-specific primers fail to amplify, then there is a problem with PCR reaction, and so the result is inconclusive. If more than one species-specific primers are used in a multiplex PCR reaction, several different species can be identified in a single assay. For example, in a study, Shivji et al. (2005) used six different species-specific primers and two universal shark primers (for positive controls) to identify six species of sharks commonly encountered in North Atlantic fisheries (Shivji et al. 2005).

Allele-Specific Probes

In this case instead of allele-specific primers, universal primers are used in combination with a specific probe. Universal primers are the primers that are designed for conserved regions of DNA and are used in PCR for amplifying genetic markers. Thus, when such primers are used in PCR for amplifying unknown DNA samples, amplification of all the alleles of a genetic marker occurs. However, subsequent use of different probes that are designed to attach to different DNA sequence variants allows the base present at a SNP site to be detected.

Even though, SNP typing methods are applied in the context of species detection, rather than species identification, they have the advantage of allowing the analysis of samples where multiple species are present. A more powerful method of species identification can be designed by combining the power of DNA sequencing and SNP typing. By using species-specific PCR primers, a single DNA sequence for the target species is generated from a sample containing multiple species DNA. This sequence can then be identified to species level to categorically demonstrate its presence in the sample.

20.4.1.2.4 Random Amplified Polymorphic DNA (RAPD)

It is a simple and least expensive technique. In this technique arbitrary primers which are usually ten bases long are used to amplify random segments of DNA. No prior information about the DNA sequence for the targeted gene is required, as the primers will bind somewhere in the sequence, but it is not certain exactly where. If the sample DNA has in it sequences complimentary to the primer oligonucleotide, PCR products will be detected, but if the sample DNA does not have the complimentary sequences, no product will be detected. Hence, random primers will or will not amplify a segment of DNA, depending on positions that are complementary to the nucleotide sequence of the primers. For example, no amplification occurs if primers anneal too far apart or 3′ ends of the primers are not facing each other. Therefore, in this technique we have to keep on using different random primers until we hit a pair of random primers that give a unique DNA profile for each DNA sample. Since the technique relies on a large, intact DNA template sequence, it cannot be used in case of degraded DNA samples.

Alternatively, bands which are specific to the target species are identified from the DNA fingerprint, extracted, and sequenced. These sequences are then used to design primers that will specifically amplify the species-specific region, termed a sequence characterized amplified region (*SCAR*). Subsequently, these primers can be used for rapid identification of species. SCAR method is a highly reproducible technique.

20.4.1.2.5 Amplified Fragment Length Polymorphism (AFLP)

This technique combines some features of both RFLP and RAPD analyses and has greater differentiation power than RAPD. It is a PCR-based tool and uses restriction-enzyme-digested fragment as template for PCR amplification using primers that are complementary to the adaptor sequence, the restriction site sequence, and a few nucleotides inside the restriction site fragments. The amplified fragments are then visualized on denaturing polyacrylamide gels.

In practice, suppose we have two DNA samples. Digest each genomic DNA with a combination of two restriction enzymes, commonly *EcoR*1 and *Mse*1 for animals. Double-stranded *EcoR*1- and *Mse*1-specific linkers are then ligated to the fragment ends in both the samples. A pre-selective amplification of *EcoR*1/*Mse*1 templates is carried out using primers that anneal to the linker ends of the fragments and have one selection nucleotide at their 3′ end. A final selective PCR amplification step is carried out in which additional selection nucleotides are added at the end of the *EcoR*1- and *Mse*1-selective primers, and the primers are fluorescently labeled. Fragments are separated by migration through a polyacrylamide gel and detected by a laser as the fragments migrate through the gel. In this way a DNA profile is created for both the samples which can then be analyzed. If the two samples belong to the same species, they will have a similar DNA profile. AFLPs arise due to a difference in restriction sites in the samples, mutations beyond the restriction sites that affect complementarity with the selection nucleotides of the AFLP primers, or due to deletions and insertions within the amplified restriction fragments. Like RAPD, this technique has an advantage that no prior sequence information is needed.

20.4.1.2.6 Microsatellite Genotyping

This technique is more robust than SNP genotyping and RAPD and AFLP analyses because of hypervariability of microsatellite DNA markers. For the purpose of species identification, those microsatellite markers are chosen that show a considerable variation among species but are generally conserved within species. If on the other hand the markers show considerable variation within species or populations, the technique can be effectively used for individual identification. This technique is discussed in detail in the section describing individual identification.

20.4.2 Identification of Geographic Origin

Identification of geographic origin has applications almost only in wildlife forensics rather than in human forensics. In order to overcome a mismatch between wildlife legislation, which usually operates within political boundaries, and species distributions, which are governed by biological and environmental factors, wildlife crime investigations often seek for answers to questions concerning the geographic origin of a sample. For example, to enforce CITES regulations, the need to determine the geographic source of a specific sample, in addition to identifying the species, cannot be ruled out. Similarly, the effective management of marine-protected areas requires that illegally harvested stocks be distinguished from those taken legally from elsewhere. Returning lost or captured animals to their native geographic area after genotypically assigning them to that population reduces the risk of outbreeding depression. The possibility of corruption of the evolutionary processes leading to divergence among geographic isolates can also be minimized by such targeted releases.

When it comes to forensic science, identifying the geographic origin means identifying its reproductive population of origin. Since populations are most often delineated by geographic rather than reproductive barriers, they are often capable of sharing genetic material. Therefore, in comparison to species identification, DNA markers are less likely to show discrete differences among different populations. Geographic origin identification thus requires the source population to be sufficiently genetically distinct from other candidate populations and heavily relies on the existence of population data from multiple areas. However, when populations are so isolated from one another that there is effectively no exchange of genetic material between them, genetic differences gradually accumulate over evolutionary time to a point where members of an isolated region share the same types of genetic marker (alleles) within their population but exhibit different alleles to that of any other population. Such markers can be effectively used for identifying populations or assigning geographic origin to a sample.

Geographic origins of an individual can be identified if there is known genetic structure within the region of interest using the following methods:

20.4.2.1 Phylogeography

This method assesses the geographic distribution of genealogical lineages where specific mtDNA haplotypes are associated with broad geographic regions (Avise et al. 1987). For example, hypervariable mtDNA control region, or D-loop, is often used as a marker in geographic origin identification with individual control region sequence types (haplotypes) corresponding to specific populations.

20.4.2.2 Population Assignment Methods

When sufficient mitochondrial DNA variation is absent, various genetic markers from the nuclear genome that show variability among regions can be used for the identification of geographic origin. These allelic differences at hypervariable nuclear DNA genetic markers between groups of individuals (populations) form the basis of population assignment methods. The hypervariable markers most often used in such methods are AFLPs (amplified fragment length polymorphisms) and microsatellites (also called short tandem repeats or STRs). Although these markers do show discrete differences, individual alleles are often distributed across populations. Therefore, differentiation can only be achieved on the basis of differing allele frequencies. The frequency of the alleles observed in a population can be used to characterize its genetic structure and to assess the probability of a sample originating from that area. Therefore, population assignment methods heavily rely on the development of large genetic databases that provide representative allele frequencies for all of the potential source populations and also on the use of statistical analysis to provide quantitative probabilities of assignment for an unknown sample to each of those populations.

20.4.3 Individual Identification, Sexing, and Parentage

20.4.3.1 Individual Identification

The use of DNA profiling for the individual identification of genetic evidence is of particular importance in human forensics and is considerably lesser relevant to animal and wildlife forensics, where its main applications include identification of stolen animals and the authentication of legally traded wildlife products or in cases of poaching, where it may be necessary to demonstrate that a horn, tusk, bone, or skin has originated from a specific individual. It has over the past 20 years revolutionized human forensic analysis.

Those genetic markers that are highly variable within species are likely to show differences among individuals. By targeting such markers through DNA profiling studies, a unique DNA profile for each specific individual can be arrived at. Such a DNA profile becomes the DNA fingerprint of that individual and can be used for individual identification.

Various methods of individual identification, sexing, and parentage are described below.

20.4.3.1.1 Microsatellite Genotyping

This technique is best suited for parentage analysis. Microsatellites are hypervariable markers, also known as short tandem repeats (STRs) or simple sequence repeats (SSRs). Microsatellites are short sequence motifs consisting of two, three, or four nucleotides and can be repeated 3–100 times. Microsatellites have a high mutation rate predominantly due to slippage of the polymerase during DNA replication. The longer the loci, the more will be the alleles due to the greater potential for slippage. Consequently, the greater the number of markers used, the lesser are the chances that another individual has the same series of alleles (same profile). One common example of a microsatellite is a $(CA)_n$ repeat, where n varies between alleles. These repeats are frequent in human and other genomes and are present every few thousand base pairs.

Microsatellites are codominant markers with bi-allelic or multi-allelic presentation in an individual or a population, respectively. Gene variants or alleles are inherited from both parents. These alleles are amplified in a PCR reaction and visualized on a polyacrylamide gel. Homozygote individuals will have the same-sized STR repeats, whereas heterozygote individuals will have different-sized repeats. Since at a particular microsatellite locus there are often many alleles present, it becomes possible to identify the progenitor of a particular allele within pedigrees. Therefore, microsatellites are also ideal for determining paternity.

For individual identification, DNA profiles are produced for different samples and thereby analyzed. If two samples produce different DNA profiles, it means that the samples did not originate from the same individual. On the other hand, if the samples share the same profile, there is a significant possibility that the samples are from the same individual. However, it becomes necessary to calculate the probability that the two individuals have the same profile by chance. Various factors which affect this probability include the number and variability of markers in the profile, the frequency of the alleles in the species, and the degree of relationship between the individuals in the population from where the samples were taken. Therefore, it becomes necessary to have representative sample of DNA profiles from the population. The genetic profiles can be generated using information from the mitochondrial control region, sex determination markers, or microsatellite loci.

In case of wildlife forensics, there are situations where animals are sold as meat or highly processed products. Identification of an individual based on their unique genetic profile has a particular application in such cases where it is used to monitor the number of animals entering commercial markets. Similarly, in order to distinguish between legally and illegally obtained specimens, individual identification is needed. However, it is necessary to maintain a DNA register where each legal specimen is DNA profiled in a certified laboratory and the profiles are lodged in a database.

20.4.3.1.2 DNA Nucleotide Sequencing, SNP Typing, RAPD, and AFLP

As with species identification or identification of geographic origin, individual identification and paternity testing can also be done by SNP typing, RAPD, AFLP, and DNA nucleotide sequencing, using both nuclear and mitochondrial DNA markers. However, here it is important that the genetic markers have a considerable variability within species or populations so that a unique DNA profile can be assigned to each individual.

20.4.3.2 Sexing

There are situations where it becomes necessary to determine the sex of an individual. For example, in case of immature animals in which secondary sexual characters have not developed, sexing on the basis of morphological characters becomes very difficult. In case of Asian elephants where tusks are only present in males, drastic declines in the numbers of males from hunting for their ivory can result in unbalanced sex ratios. Similarly, monitoring the trade in some countries where qualified hunters are restricted to hunting only one sex requires determination of the sex of the animals hunted.

In animals that have a heteromorphic sex chromosome system, molecular sexing relies on either the PCR amplification of fragments specific to the Y or W chromosomes or the amplification of homologous fragments from both sex chromosomes. The former approach can be used in placental mammals and marsupials, in which amplification of the Y-specific SRY locus is expected only in males. In the latter approach, amplification of homologous fragments from both sex chromosomes such as ZFX and ZFY genes is carried out using the same pair of PCR primers and later differentiated on the basis of polymorphism between the two fragments using various techniques like PCR-RFLP.

Sex-specific molecular markers, however, are not universal and need to be developed independently for different classes of organisms. For example, in lizards *Calotes versicolor*, even though SRY gene is present in males, it is also present in 50 % of females.

Various sex-specific genes that have been extensively used in sex determination include:

20.4.3.2.1 SRY Gene

Also known as sex-determining region Y gene, it is a sex-determining gene present on the Y chromosome in placental mammals and marsupials and initiates male sex determination. It can be detected in males by nucleic acid hybridization or by PCR amplification using specific primers. However, it is important to include an appropriate control, usually another nuclear gene such as actin, in the PCR reaction to avoid generation of false females in case failure of amplification is because of technical reasons.

In a study Gowans et al. (2000) used primers developed for the sperm whale SRY gene and amplified a 147-bp fragment of the SRY gene from tissue samples taken by harpoon biopsy from live northern bottlenose whales and provided an accurate sexing procedure for these animals (Gowans et al. 2000). An inexpensive test was developed based on the SRY gene on the Y chromosome for identifying male Asian elephants from poached carcasses (Gupta et al. 2006).

20.4.3.2.2 ZFY and ZFX Genes

ZFY gene encodes a zinc finger-containing protein known as zinc finger Y-chromosomal protein that functions as a transcription factor, whereas ZFX gene encodes zinc finger X-chromosomal protein. Therefore, ZFY is a Y-chromosome-linked gene, and ZFX is an X-chromosome-linked gene.

In a study Fontanesi et al. (2008) developed a sexing method for three leporid species (*Oryctolagus cuniculus*, *Lepus europaeus,* and *Lepus timidus*) based on the analysis by PCR-RFLP of point mutations that differentiate the ZFX and ZFY gene sequences. A 432-bp fragment of exon 11 of the ZFX and ZFY gene was amplified using a same set of primers, and restriction enzyme digestion was carried out using different enzymes which generated a separate RFLP profile for male and female DNA samples (Fontanesi et al. 2008). In another study, Malisa et al. (2005) successfully identified the sex of 13 animal species, 9 of which were wild (wildebeest, zebra, Thomson's gazelle, impala, reedbuck, kongoni, oryx, warthog, and hippopotamus) and 4 were domestic (bovine, caprine, ovine, and porcine), following successful amplification of gender-specific, SRY and ZFY/ZFX, chromosomal domains using sex-specific primers (Malisa et al. 2005).

20.4.3.2.3 Chromobox-Helicase-DNA-Binding Gene (CHD1)

In case of avians, sex-specific markers, one on the Z chromosome and one on the W chromosome, can be used successfully for gender determination because avian males are homozygotes (ZZ), whereas females are heterozygotes (ZW). For example, in birds other than some flightless species (ratites), chromobox-helicase-DNA-binding gene is localized on the W (female-specific) chromosome (CHD1W), although it has a homologue (CHD1Z) on the Z chromosome. PCR primers were designed to amplify the intron of the CHD1 gene, and fragments were generated which consistently differ in size between the W and Z versions (Griffiths et al. 1998).

20.4.3.3 Parentage

Genetic markers are inherited from one generation to the next. Therefore, DNA profiles can be used to verify parent-offspring relationships. Various genetic markers, for example, microsatellites, that are used for individual identification can also be used for parentage verification analysis. In order to confirm a particular parent-offspring relationship, the alleles present in the DNA profile of an individual must also be present in its putative parents, one allele per marker in each parent. In case alleles observed in an individual do not correspond to those in the putative parental profiles, then the possibility of the individual being their offspring can be excluded.

20.5 Conclusion

With the discovery of newer genetic markers, production of more reference databases, validation of laboratory techniques, development of more efficient protocols for sample recovery and DNA extraction, and the ability to type samples of low quantity and quality, DNA analysis is going to revolutionize animal DNA forensics. Genome sequencing is becoming faster and less expensive, paving the way for the development of novel markers for forensic identification and the subsequent production of reference data. The increase in genomic data generated through high-throughput sequencing technologies will enable the discovery and application of markers associated with adaptive traits, thereby allowing greater resolution of geographic population identification. It will also provide universal nuclear primers that can amplify informative regions over a broad range of taxa.

So long as the emerging technologies will keep on adding to the forensic genetic toolbox, current genetic technologies, with the help of various nuclear and mitochondrial DNA markers, are capable of addressing most forensic questions. The choice of a genetic marker depends on the forensic question to be addressed and the ecology and genetic knowledge of the species. However, in order to validate their routine use for forensic application and admissibility as evidence, further research needs to be done. Furthermore, it is necessary to establish a network of accredited animal and wildlife DNA forensic laboratories.

References

Arun K, Rajiv RK, Brahma DS, Sanjod KM, Deepak S, Palanisamy G (2012) Species specific polymerase chain reaction (PCR) assay for identification of pig (Sus domesticus) meat. Afr J Biotechnol 11:15590–15595

Avise JC (1994) Molecular markers, natural history and evolution. Chapman & Hall, New York

Avise JC, Arnold J, Martin Ball R, Bermingham E, Lamb T, Neigel JE, Reeb CA, Saunders NC (1987) Intraspecific phylogeography: the mitochondrial DNA bridge between population genetics and systematics. Annu Rev Ecol Syst 18:489–522

Baker LE, McCormick WF, Matteson KJ (2001) A silica-based mitochondrial DNA extraction method applied to forensic hair shafts and teeth. J Forensic Sci 46:126–130

Branicki W, Kupiec T, Pawlowski R (2003) Validation of cytochrome b sequence analysis as a method of species identification. J Forensic Sci 48:83–87

Brower AVZ, DeSalle R, Vogler A (1996) Gene trees, species trees, and systematics: a cladistic perspective. Annu Rev Ecol Syst 27:423–450

Budowle B, Allard MW, Wilson MR, Chakraborty R (2003) Forensics and mitochondrial DNA: applications, debates, and foundations. Annu Rev Genomics Hum Genet 4:119–141

Butler JM, Levin BC (1998) Forensic applications of mitochondrial DNA. Trends Biotechnol 16:158–162

Butler JM, Shen Y, McCord BR (2003) The development of reduced size STR amplicons as tools for analysis of degraded DNA. J Forensic Sci 48:1054–1064

Chaisomchit S, Wichajarn R, Chowpreecha S, Chareonsiriwatana W (2003) A simple method for extraction and purification of genomic DNA from dried blood spots on filter paper. Southeast Asian J Trop Med Public Health 34:641–645

Chapman DD, Abercrombie DL, Douady CJ, Pikitch EK, Stanhope MJ, Shivji MS (2003) A streamlined, bi-organelle, multiplex PCR approach to species identification: application to global conservation and trade monitoring of the great white shark, *Carcharodon carcharias*. Conserv Genet 4:415–425

Clarke SC, Magnussen JE, Abercrombie DL, McAllister MK, Shivji MS (2006) Identification of shark species composition and proportion in the Hong Kong shark fin market based on molecular genetics and trade records. Conserv Biol 20:201–211

Cline RE, Laurent NM, Foran DR (2003) The fingernails of Mary Sullivan: developing reliable methods for selectively isolating endogenous and exogenous DNA from evidence. J Forensic Sci 48:1–6

Courchamp F, Angulo E, Rivalan P, Hall RJ, Signoret L, Bull L, Meinard Y (2006) Rarity value and species extinction: the anthropogenic allee effect. PLoS Biol 4:2405–2410

Deguilloux MF, Pemonge MH, Petit RJ (2002) Novel perspectives in wood certification and forensics: dry wood as a source of chloroplast, mitochondrial and nuclear DNA. Proc R Soc Lond B Biol Sci 269:1039–1046

Fontanesi L, Davoli R, Nanni Costa L, Beretti F, Scotti E, Tazzoli M, Tassone F, Colombo M, Buttazzoni L, Russo V (2008) Investigation of candidate genes for glycolytic potential of porcine skeletal muscle: association with meat quality and production traits in Italian Large White pigs. Meat Sci 80:780–787

Gerber AS, Loggins R, Kumar S, Dowling TE (2001) Does nonneutral evolution shape observed patterns of DNA variation in animal mitochondrial genomes? Annu Rev Genet 35:539–566

Gill P, Jeffreys AJ, Werret DJ (1985) Forensic applications of DNA fingerprints. Nature 318:577–579

Giovambattista G, Ripoli MV, Liron JP, Villegas Castagnasso EE, Peral-Garcia P, Lojo MM (2001) DNA typing in a cattle stealing case. J Forensic Sci 46:1484–1486

Gowans S, Dalebout ML, Hooker SK, Whitehead H (2000) Reliability of photographic and molecular techniques for sexing bottlenose whales, Hyperoodon ampullatus. Can J Zool 78:1224–1229

Greenberg BD, Newbold JE, Sugino A (1983) Intraspecific nucleotide sequence variability surrounding the origin of replication in human mitochondrial DNA. Gene 21:33–49

Griffiths R, Double MC, Orr K, Dawson RJG (1998) A DNA test to sex most birds. Mol Ecol 7:1071–1075

Gupta SK, Verma SK, Singh L (2005) Molecular insight into a wildlife crime: the case of a peafowl slaughter. Forensic Sci Int 154:214–217

Gupta SK, Thangaraj K, Singh L (2006) A simple and inexpensive molecular method for sexing and identification of the forensic samples of elephant origin. J For Sci 51:805–807

Hajibabaei M, Smith MA, Janzen DH, Rodriguez JJ, Whitfield JB, Hebert PDN (2006) A minimalist barcode can identify a specimen whose DNA is degraded. Mol Ecol Notes 6:959–964

Hebert PDN, Cywinska A, Ball SL, deWaard JR (2003a) Biological identifications through DNA barcodes. Proc R Soc Lond B Biol Sci 270:313–321

Hebert PDN, Ratnasingham S, deWaard JR (2003b) Barcoding animal life: cytochrome c oxidase subunit 1 divergences among closely related species. Proc R Soc Lond B Biol Sci 270:96–99

Higuchi RG, von Beroldigen CH, Sensabaugh GF, Ehrlich HA (1988) DNA typing from single hairs. Nature 332:543–545

Hoss M, Paabo S (1993) DNA extraction from Pleistocene bones by a silica- based purification method. Nucleic Acids Res 21:3913–3914

Hsieh HM, Huang LH, Tsai LC, Kuo YC, Meng HH, Linacre A et al (2003) Species identification of rhinoceros horns using the cytochrome b gene. Forensic Sci Int 136:1–11

Johnson WE, O'Brien SJ (1997) Phylogenetic reconstruction of the Felidae using 16S rRNA and NADH–5 mitochondrial genes. J Mol Evol 44:S98–S116

Keller RP, Lodge DM (2007) Species invasions from commerce in live aquatic organisms: problems and possible solutions. Bioscience 57:428–436

Lips KR, Brem F, Brenes R, Reeve JD, Alford RA, Voyles J, Carey C, Livo L, Pessier AP, Collins JP (2006) Emerging infectious disease and the loss of biodiversity in a neotropical amphibian community. Proc Natl Acad Sci U S A 103:3165–3170

Loreille OM, Diegoli TM, Irwin JA, Coble MD, Parsons TJ (2007) High efficiency DNA extraction from bone by total demineralization. Forensic Sci Int Genet 1:191–195

Maddison WP (1997) Gene trees in species trees. Syst Biol 46:523–536

Magnussen JE, Pikitch EK, Clarke SC, Nicholson C, Hoelzel AR, Shivji MS (2007) Genetic tracking of basking shark products in international trade. Anim Conserv 10:199–207

Malisa A, Paul G, Sakurani B, Sam W, Benezeth M (2005) Species and gender differentiation between and among domestic and wild animals using mitochondrial and sex-linked DNA markers. Afr J Biotechnol 4:1269–1274

Mei JV, Alexander JR, Adam BW, Hannon WH (2001) Use of filter paper for the collection and analysis of human whole blood specimens. J Nutr 131: 1631S–1636S

Moore MK, Bemiss JA, Rice SM, Quattro JM, Woodley CM (2003) Use of restriction fragment length polymorphisms to identify sea turtle eggs and cooked meats to species. Conserv Genet 4:95–103

Nguyen TH, Nguyen DHC, Phan TP, Nguyen THL, Nguyen DTG (2012) Extraction of human genomic DNA from dried blood spots and hair roots. Int J Biosci Biochem Bioinform 2:21–26

Normile D (2004) Invasive species—expanding trade with China creates ecological backlash. Science 306:968–969

Ojeda GN, Amavet PS, Rueda EC, Siroski PA (2012) DNA extraction from skins of wild (Hydrochoerus hydrochaeris and Pecari tajacu) and domestic (Sus scrofa domestica) species using a novel protocol. Genet Mol Res 11:672–678

Page RDM (2000) Extracting species trees from complex gene trees: reconciled trees and vertebrate phylogeny. Mol Phylogenet Evol 14:89–106

Pank M, Stanhope M, Natanson L, Kohler N, Shivji M (2001) Rapid and simultaneous identification of body parts from the morphologically similar sharks Carcharhinus obscurus and Carcharhinus plumbeus (Carcharhinidae) using multiplex PCR. Marine Biotechnol 3:231–240

Parson W, Pegoraro K, Niederstatter H, Foger M, Steinlechner M (2000) Species identification by means of the cytochrome b gene. Int J Legal Med 114:23–28

Pedersen AB, Jones KE, Nunn CL, Altizer S (2007) Infectious diseases and extinction risk in wild mammals. Conserv Biol 21:1269–1279

Peppin L, McEwing R, Carvalho GR, Ogden R (2008) A DNA based approach for the forensic identification of Asiatic black bear (Ursus thibetanus) in a traditional Asian medicine. J Forensic Sci 53:1358–1362

Pfeiffer H, Huhne J, Ortmann C, Waterkamp K, Brinkmann B (1999) Mitochondrial DNA typing from human axillary, pubic and head hair shafts—success rates and sequence comparisons. Int J Leg Med 112:287–290

Reed RN (2005) An ecological risk assessment of non-native boas and pythons as potentially invasive species in the United States. Risk Anal 25:753–766

Roman J, Bowen BW (2000) The mock turtle syndrome: genetic identification of turtle meat purchased in the south-eastern United States of America. Anim Conserv 3:61–65

Samaraweera M, Himali SMC, Zeng SC, Jianlin H, Silva P (2011) Development of molecular tools to differentiate Sri Lankan wild boar (Sus scrofa affinis) meat from exotic and village pig (Sus scrofa domestica) meat. Trop Agric Res 23:11–20

Sambrook J, Fritsch EF, Maniatis T (1989) Molecular cloning: a laboratory manual, 2nd edn. Cold Spring Harbor Laboratory Press, Cold Spring Harbor

Shivji M, Clarke S, Pank M, Natanson L, Kohler N, Stanhope M (2002) Genetic identification of pelagic shark body parts for conservation and trade monitoring. Conserv Biol 16:1036–1047

Shivji MS, Chapman DD, Pikitch EK, Raymond PW (2005) Genetic profiling reveals illegal international trade in fins of the great white shark, Carcharodon carcharias. Conserv Genet 6:1035–1039

Sites JW, Davis SK, Guerra T, Iverson JB, Snell HL (1996) Character congruence and phylogenetic signal in molecular and morphological data sets: a case study in the living iguanas (Squamata, Iguanidae). Mol Biol Evol 13:1087–1105

Skerratt LF, Berger L, Speare R, Cashins S, McDonald KR, Phillott AD, Hines HB, Kenyon N (2007) Spread of chytridiomycosis has caused the rapid global decline and extinction of frogs. Ecohealth 4:125–134

Smith KF, Sax DF, Lafferty KD (2006) Evidence for the role of infectious disease in species extinction and endangerment. Conserv Biol 20:1349–1357

Spinks PQ, Shaffer HB (2007) Conservation phylogenetics of the Asian box turtles (Geoemydidae, Cuora): mitochondrial introgression, numts, and inferences from multiple nuclear loci. Conserv Genet 8:641–657

Suenaga E, Nakamura H (2005) Evaluation of three methods for effective extraction of DNA from human hair. J Chromatogr B 820:137–141

Sullivan KM, Hopgood R, Gill P (1992) Identification of human remains by amplification and automated sequencing of mitochondrial DNA. Int J Legal Med 105:83–86

Turan C (2008) Molecular systematic analyses of Mediterranean skates (Rajiformes). Turk J Zool 32:1–6

Upholt WB (1977) Estimation of DNA-sequence divergence from comparison of restriction endonuclease digests. Nucleic Acids Res 4:1257–1265

Verma SK, Singh L (2003) Novel universal primers establish identity of an enormous number of animal species for forensic application. Mol Ecol Notes 3:28–31

Verma SK, Prasad K, Nagesh N, Sultana M, Singh L (2003) Was elusive carnivore a panther? DNA typing of faeces reveals the mystery. Forensic Sci Int 137: 16–20

Vigilant L (1999) An evaluation of techniques for the extraction and amplification of DNA from naturally shed hairs. Biol Chem 380:1329–1331

Volo SBD, Reynolds RT, Douglas MR, Antolin MF (2008) An improved extraction method to increase DNA yield from molted feathers. Condor 110:762–767

Walsh DJ, Corey AC, Cotton RW, Forman L, Herrin GL Jr, Word CJ, Garner DD (1992) Isolation of deoxyribonucleic acid (DNA) from saliva and forensic science samples containing saliva. J Forensic Sci 37:387–395

Weigle SM, Smith LD, Carlton JT, Pederson J (2005) Assessing the risk of introducing exotic species via the live marine species trade. Conserv Biol 19:213–223

Wetton JH, Tsang CFS, Roney CA, Spriggs AC (2004) An extremely sensitive species-specific ARMs PCR test for the presence of tiger bone DNA. Forensic Sci Int 140:139–145

Wilson MR, Polanskey D, Butler J, DiZinno JA, Replogle J, Budowle B (1995) Extraction, PCR amplification and sequencing of mitochondrial DNA from human hair shafts. BioTechniques 18:662–669

Yoshii T, Tamura K, Ishiyama I (1992) Presence of a PCR inhibitor in hairs (Japanese). Nippon Hoigaku Zasshi 46:313–316

Yoshii T, Tamura K, Taniguchi T, Akiyama K, Ishiyama I (1993) Water soluble eumelanin as a PCR inhibitor and a simple methods for its removal (Japanese). Nippon Hoigaku Zasshi 47:323–329

Zhang F, Jiang Z (2006) Mitochondrial phylogeography and genetic diversity of Tibetan gazelle (Procapra picticaudata): implications for conservation. Mol Phylogenet Evol 41:313–321

21 Rapid Detection of Viruses Using Loop-Mediated Isothermal Amplification (LAMP): A Review

P. Saharan, P. Khatri, S. Dingolia, J.S. Duhan, and S.K. Gahlawat

Abstract

Most of the diseases caused by viral infection are found to be fatal, and the diagnosis is difficult due to confusion with other causative agents. So, a highly efficient molecular-based advance detection technique, i.e., loop-mediated isothermal amplification (LAMP) method, is developed for diagnosis of viral infections by various workers. It is based on amplification of DNA at very low level under isothermal conditions, using a set of four specifically designed primers and a DNA polymerase with strand displacement activity. This technique is found to be superior than most of the molecular techniques like PCR, RT-PCR, and real-time PCR due to its high specificity, sensitivity, and rapidity. Major advantage of LAMP method is its cost-effectiveness as it can be done simply by using water bath or dry bath. Here, in this review information regarding almost all the effective LAMP techniques which is developed so far for diagnosis of numerous viral pathogens is presented.

21.1 Introduction

Viruses exist in living system since living cells first evolved, so found wherever there is life exist. Viruses do not form fossils so the mode of their investigation/detection is only by using molecular techniques (Iyer et al. 2006). Most of the viruses are harmful to animals (http://en.wikipedia.org/wiki/Introduction_to_viruses) and cause a number of infectious diseases. Common human diseases caused by viruses include the common cold, influenza, chickenpox, and cold sores. Ebola, AIDS, avian influenza, and SARS are various serious diseases caused by viruses (Emiliani 1993). The relative ability of viruses to cause disease is known as virulence of virus. Viral diseases sometimes cause confusion with any other causative agent such as the possible connection between human herpesvirus six (HHV6) and neurological diseases such as multiple sclerosis and chronic fatigue syndrome (Thomas 2008). There is controversy over whether the bornavirus, previously thought to cause neurological diseases in horses, could be responsible

P. Saharan • P. Khatri • S. Dingolia
J.S. Duhan • S.K. Gahlawat (✉)
Department of Biotechnology, Chaudhary Devi Lal University, Sirsa 125055, India
e-mail: skgcdlu@gmail.com

R.K. Salar et al. (eds.), *Biotechnology: Prospects and Applications*,
DOI 10.1007/978-81-322-1683-4_21, © Springer India 2013

for psychiatric illnesses in human (Mandal 2013). According to Shors (2008) viruses transmit by two modes, i.e., vertical (from mother to child) and horizontal (from person to person). Vertical transmission include hepatitis B virus and HIV where the baby is born already infected with the virus. *Varicella-zoster* virus is another but rarest example of vertical transmission. Although this virus causes relatively mild infections in humans, it can be fatal to the fetus and newly born baby (Mandal 2013). Generally living populations are affected by horizontal transmission of viral infection where viral transmission occurs via exchange of blood and by sexual activity (HIV, hepatitis B, and hepatitis C), through mouth by exchange of saliva (Epstein-Barr virus), through contaminated food or water (norovirus), by breathing viruses in the form of aerosols (influenza virus) and by insect vectors such as mosquitoes (dengue) (Collier et al. 1998). Currently, 21 families of viruses are known to cause disease in humans (http://en.wikipedia.org/wiki/virus classification). So due to broad spectrum of harmful infections/diseases caused by viruses, a rapid and accurate diagnosis is essential for timely therapeutic interventions (Barenfanger et al. 2000).

Several methods such as viral antigen, genomic sequences, and/or antibodies are used for the detection of viral infection. Viral isolation and serological assays are standard methods of viral diagnosis. Both viral isolation and serological testing require substantial time to obtain accurate final results. On the other hand, use of histological and biochemical methods for virus detection may sometimes fail to detect the virus antigen during the early stages of infection. Among various conventional methods, viral culture and antigenic detection (immunological detection) methods of viral detection are often costly, time-consuming, and less sensitive. Parida (2008) suggested that nucleic acid amplification is one of the most valuable tools in virtually all life science fields, including application-oriented fields such as clinical medicine in which diagnosis of infectious diseases, genetic disorders, and genetic traits is based on this novel method. Several amplification methods, namely, nucleic acid sequence-based amplification (NASBA), self-sustained sequence replication (3SR), strand displacement amplification (SDA), as well as polymerase chain reaction (PCR), have been invented in past days (Chan and Fox 1999). Nonquantitative PCR and immunological detection methods are usually only qualitative tests with limited sensitivity and cannot distinguish infective viruses (Enteric Viruses: Rapid Detection and Identification Methods. http://www.rapidmicrobiology.com/test-methods/Enteric-viruses.php). PCR is one of the conventional methods which are affordable, but it has some intrinsic limitations such as the requirement of rapid thermal cycling, insufficient specificity, and low-amplification efficiency (Mori et al. 2001). Nested PCR and multiplex PCR have been developed to address the need for rapid identification of viruses to serotype level with more accuracy (Ratcliff et al. 2007). Another novel technique, i.e., real-time PCR, is developed which in turn brought true quantization of target nucleic acids out of the pure research laboratory and into the diagnostic laboratory (Espy et al. 2006). Mackay et al. (2002) reported quantitative PCR (QPCR) assays as most established assay for the detection of viral load and therapy monitoring. Nucleic acid-based assays or real-time quantitative assay might eventually replace virus isolation and conventional RT-PCR as the new gold standard for the rapid diagnosis of virus infection in the acute-phase samples (Ratcliff et al. 2007). Besides, all these nucleic acid amplification methods have several intrinsic disadvantages of requiring either a high-precision instrument for amplification or a complicated method for detection of amplified products. The high cost of instruments required for performing the real-time assays restricted its use to laboratories with good financial resources (Parida et al. 2008).

To overcome all above limitations in viral diagnosis, a loop-mediated isothermal amplification (LAMP) method was developed as an alternative method for viral detection. LAMP is a powerful and novel nucleic acid amplification method that amplifies a few copies of target DNA with high specificity, efficiency, and rapidity under isothermal conditions, using a set of four specially designed primers and a DNA polymerase with strand displacement activity (Notomi et al. 2000; Parida et al. 2008; Tomita et al. 2008). When detecting the RNA genome of a pathogen such as an RNA virus, LAMP has been merged with reverse transcription (RT) into RT-LAMP to allow nucleic acid amplification (Soliman et al. 2009). This technique requires only simple and cost-effective equipment amenable to use in clinical laboratories (Enomoto et al. 2005). Compared to PCR and real-time PCR, the LAMP has the advantages of reaction simplicity and detection sensitivity. It has been claimed that the LAMP method can detect as few as 100 copies of DNA template in blood samples, equal roughly to 5 parasites/μl (Han et al. 2007). This sensitivity is notably higher than any known immunochromatography-based RDTs, which are recommended by WHO as part of the global malaria control strategy (Long 2009). LAMP has potential applications for clinical diagnosis as well as surveillance of infectious diseases in developing countries without requiring sophisticated equipment or skilled personnel (Mori and Notomi 2009; Parida et al. 2008).

21.2 Background of LAMP Technique

Although the inception of LAMP refers back to 1998, the popularity of LAMP starts only after 2003 following emergence of West Nile and SARS viruses. This technique was first described and initially evaluated for detection of hepatitis B virus DNA by Notomi et al. (2000). First of all, LAMP has been applied to many kinds of pathogens causing food-borne diseases (Lukinmaa et al. 2004). LAMP kits for detecting *Salmo-nella*, *Legionella*, *Listeria*, verotoxin-producing *Escherichia coli*, and *Campylobacter* have been commercialized. Due to its advantages, later on, this method is successfully employed for detection of variety of bacteria, viruses, parasites, and fungus.

21.3 Advantages of LAMP Method over Conventional Methods

1. Amplifies a few copies of target DNA.
2. Easy to handle.
3. Higher sensitivity and specificity is attributed to continuous amplification under isothermal condition employing four primers recognizing six distinct regions of the target (Parida 2008).
4. High efficiency.
5. Rapid detection.
6. Cost-effective as it can be carried out with simplest equipment, i.e., dry bath/water bath (Enomoto et al. 2005).
7. Vast application area in detection of different types of microorganisms (detection of variety of bacteria, viruses, parasites, and fungus) (Endo et al. 2004; Okafuji et al. 2005).
8. Can detect RNA genome of a pathogen by reverse transcription using reverse transcriptase enzyme (Soliman et al. 2009).
9. LAMP product can be easily detected visually either by using SYBR Green I or by using turbidometer.
10. Free online software Primer-Explorer IV software program (http://primerexplorer.jp/e/) is available for LAMP primer design.

21.4 Detection of Viruses by LAMP Method

LAMP assay is also helpful in detection of RNA template by combo use of reverse transcriptase along with DNA polymerase (Notomi et al. 2000; Whiting and Champoux 1998). Due to this, LAMP has already found wide application in RNA virus detection, such as foot-and-mouth disease (Dukes et al. 2006), swine vesicular disease (Blomstrom et al. 2008), Taura syndrome (Kiatpathomchai et al. 2007), severe acute respiratory syndrome coronavirus (Hong et al. 2004), norovirus (Fukuda et al. 2006), human papilloma virus (Hagiwara et al. 2007), cytomegalovirus (Reddy et al. 2010), human immunodeficiency virus (Hosaka et al. 2009), and H5N1 avian influenza virus (Jayawardena et al. 2007).

21.4.1 Human Viruses

LAMP method has been developed for detection of human viruses including chickenpox virus (Okamoto et al. 2004); mumps virus (Yoshida et al. 2008); Colorado tick fever virus Coltivirus; respiratory syncytial virus (Ushio et al. 2005); flavivirus causing West Nile fever (Parida et al. 2004); enterovirus 71 (Shi et al. 2011); human immunodeficiency virus (HIV) causing acquired immune deficiency syndrome (AIDS) (Hosaka et al. 2009); pandemic (H1N1) 2009 virus (Lee et al. 2010); cytomegalovirus causing cytomegalovirus inclusion disease (Reddy et al. 2010); chikungunya virus (Parida et al. 2007); Japanese encephalitis virus (Parida et al. 2006); herpes simplex virus type 2 causing genital herpes (Enomoto et al. 2005); adult T-cell leukemia and hairy cell leukemia viruses; human T-cell lymphotropic-1 and human T-cell lymphotropic-2 retroviruses (Komiyama et al. 2009); foot-and-mouth disease virus (Chen et al. 2011); Epstein-Barr virus (EBV) (Iwata et al. 2006), influenza A and B viruses causing influenza (Ito et al. 2006); Ebola virus (Kurosaki et al. 2007); human papilloma viruses type 6, 11, 16 and 18 (Hagiwara et al. 2007); and HIV-1 virus (Curtis et al. 2009).

Mumps virus was circulating throughout the world, but in the United States the acceptance of MMR combined vaccine against measles, mumps, and rubella has reduced the number of patients suffering from mumps (Centers for Disease Control and Prevention 1998; Peltola et al. 1994). Mumps patients having high vaccine coverage are prone to suffering from secondary vaccine failure (SVF) (Pugh et al. 2002). Due to this critical problem, rapid detection of mumps virus by a fast diagnostic technique was necessary. Therefore, Okafuji et al. (2005) developed a fast LAMP assay for detection of mumps virus genome with a detection of 0.1 PFU along with the same sensitivity as RT-nested PCR within 60 min only. Similarly, Yoshida et al. (2008) developed RT-LAMP assay for mumps virus with a detection limit of 0.12 PFU/μl. According to a WHO report, approximately two billion people are infected with hepatitis B virus (HBV) and more than 350 million have a chronic HBV infection. Molecular diagnostic assays have provided insight into the pathogenesis and natural history of HBV infection and have facilitated detection of drug resistance (Lok et al. 2001; Corden et al. 2003). Hepatitis B virus (HBV) is responsible for hepatocellular carcinoma (a liver disease) in approximately 350–400 million people all over the world (Dienstag 2008). Cai et al. (2008) detected hepatitis B virus by RtF-LAMP and determined the lower detection limit of LAMP by Probit analysis at the 95 % detection level (210 copies/ml). So, RtF-LAMP was more precise in detecting low titers of the virus which makes it more useful than real-time PCR in clinical applications. Moslemi et al. (2009) also detected hepatitis B virus by LAMP method and observed that PCR sensitivity was up to 40 particles, while the LAMP sensitivity test was verified up to four particles.

Human respiratory syncytial virus (RSV) is a major causative agent of lower respiratory tract infections in children and adults worldwide. Children born preterm or with underlying cardiopulmonary disorders are at specially high risk of developing severe and lethal RSV respiratory tract infections (American Academy of Pediatrics 2003; Girard et al. 2005; Saijo et al. 1993, 1994). A reverse transcription loop-mediated isothermal amplification (RT-LAMP) assay was developed by Shirato et al. (2007) to amplify the genome of RSV subgroups A and B, in order to improve current diagnostic methods for RSV infection. For RSV detection, primers were designed according to the highly conserved nucleotide sequence located in the matrix protein region of the nucleoprotein region. RT-LAMP can detect viral RNA in 10^{-1} to 10^{-2} PFU of RSV.

Generally, smallpox and AIDS viruses attack only humans, but in 1995 the first report of a chimpanzee being infected with HIV was reported. Reverse transcription loop-mediated isothermal amplification assay was developed for rapid detection of HIV type 1 (group M) (Hosaka et al. 2009). The 100 % detection limit of HIV-1 RT-LAMP was determined using a standard strain (WHO HIV-1 [97/656]) in octuplicate and found to be 120 copies/ml.

Hand-foot-and-mouth disease (HFMD) is a common infectious disease in young children and infants and is characterized by fever, ulcers in the mouth, and vesicles on the hands and feet. Two enteroviruses, i.e., human enterovirus 71 (EV71) and coxsackievirus A16 (CVA16), are responsible for this disease. Schmidt et al. (1974) reported that EV71 was first isolated from an infant with encephalitis in California in 1969. HFMD epidemics cause serious public health threats among infants throughout the world (Lee et al. 2009; McMinn 2002). Jiang et al. (2011) developed an RT-LAMP assay targeting the VP3 gene of EV71 and further evaluated with 40 clinical stool samples collected during the epidemics of HFMD in Guangxi province of China in 2009. This assay was found to be more sensitive and accurate, so it may become a useful alternative in resource-limited hospital laboratories and rural clinics for diagnosis of EV71. The sensitivity of the EV71-specific RT-LAMP is 0.01 PFU per reaction mixture, which is similar to the rRT-PCR assay. Based on 40 clinical samples from suspected HFMD cases, 92.9 % diagnostic sensitivity was achieved with the RT-LAMP assay.

Out of 15 hemagglutinin subtypes and 9 neuraminidase subtypes, only 3 hemagglutinin (H1 to H3) and 2 NA (N1 and N2) subtypes have established stable lineages in the human population since the last century (Webby and Webster 2003). Highly contagious nature of influenza A viruses (Fraser et al. 2004) leads the disease rapidly spread in close communities. Poon et al. (2005) developed LAMP assay for detection of influenza A virus with detection limit of ten copies per reaction.

Novel H1N1 (letters refer to haemagglutinin and neuraminidase proteins on the surface of the virus; the numbers refer to slight variations in the forms of each protein) 2009 pandemic virus affected individuals worldwide and caused over 18,138 deaths since its initial outbreak in March 2009. Comparative study of sequences available into public databases suggests that since April 2009 a range of genetic variation has arisen in H1N1 viruses (Zhang et al. 2009). To eradicate the probability of false-negative results due to mutations in these genes, the RT-LAMP method was designed by Hiromoto et al. (2000) and Welman and Arora (2000) in order to detect the gene encoding the matrix protein, a type-specific antigen of influenza virus (highly conserved and undergoes less evolutionary change). Out of 57 pandemic (H1N1) 2009 virus-positive samples identified by virus isolation only one false-negative result was obtained when the clinical samples were tested with M-specific RT-LAMP. There are two possible explanations for the failure of the amplification. It was worthy to note that, as the M segment of pandemic (H1N1) 2009 virus is phylogenetically close to the avian-like swine lineage, one might be concerned about cross-reactivity when using the M-specific RT-LAMP with other avian-origin influenza viruses. However, unlike swine influenza, avian influenza A viruses do not commonly circulate in humans, and therefore the chances of this potential misinterpretation are low. On the other hand, learning from previous experiences during H5N1 outbreaks, any avian-origin influenza virus detected in a human specimen indicates a cross-species infection, which means that special precautions need to be taken. Hence, any possible lack of specificity related to the M-based nucleic acid assay should not pose a particular problem for its routine use, and any human sample with a risk of being infected by avian influenza should also be tested using an H5N1-specific assay to obtain a definite diagnosis (Carr et al. 2009). The sensitivities of the pandemic (H1N1) 2009 virus HA-specific RT-LAMP assay and a TaqMan real-time reverse transcription PCR assay performed with the SW H1 primer/probe set (WHO 2009b) were compared by testing ten-fold serial dilutions of in vitro-transcribed target RNA (from 100 to 104 copies) in quadruplicate. Both assays were equally sensitive, and the detection limit was ten copies of target RNA per reaction volume. All the positive amplifications by the pandemic (H1N1) 2009 virus HA-specific RT-LAMP assay were achieved in less than 40 min. The pandemic (H1N1) 2009 virus HA-specific RT-LAMP assay targets only the HA gene, so this RT-LAMP method has difficulty with the detection of pandemic (H1N1) 2009 viruses which have HA genes that are difficult to amplify even by rRT-PCR. This may be a limitation of the pandemic (H1N1) 2009 virus HA-specific RT-LAMP method. This limitation could be overcome by combining the pandemic (H1N1) 2009 virus HA-specific RT-LAMP assay with conventional RT-PCR for the matrix gene of influenza A virus (WHO 2009a) or by designing influenza A virus consensus RT-LAMP assay primers and performing the two RT-LAMP assays together. Based on above facts and data, there is an urgent need for the development of an easy, accurate, and simple method for the diagnosis of this novel pandemic virus (Lee et al. 2010). Lee et al. (2010) developed M-specific reverse transcription loop-mediated isothermal amplification for detection of pandemic (H1N1) 2009 virus. Based on validation by virus isolation, the specificity and sensitivity of this M-specific RT-LAMP assay were 100 % and 98.25 %, respectively. A newly emerged pandemic human influenza virus 2009 which is a triple reassortant including segments from swine, avian, and human influenza viruses (Garten et al. 2009; Shinde et al. 2009; Smith et al. 2009) has spread throughout the world. Kubo et al. (2010) developed an HA-specific LAMP assay for the detection of pandemic (H1N1) 2009 virus with 100 % specificity and 97.8 % sensitivity. This method was also considered as a novel molecular method for diagnosis of pandemic influenza in resource-limited settings.

In the early stages of ocular manifestations, clinical differentiation between patients of retinitis associated with CMV and other herpesvirus infections is often difficult (Knox et al. 1998). Viral retinitis is commonly caused by herpes simplex virus type 1 (HSV-1), HSV-2, varicella-zoster virus (VZV), cytomegalovirus (CMV), and occasionally by Epstein-Barr virus (EBV) (Madhavan et al. 2004). It is very important to differentiate CMV retinitis from HSV and VZV retinitis early in the course of infection because the therapeutic agent to be used for treatment differs from virus to virus (Madhavan et al. 2004). So LAMP assay was developed for their differentiation. Reddy et al. (2010) developed a simple and cost-effective LAMP assay for the rapid detection of CMV DNA in patients infected with viral retinitis. It is highly sensitive as its lower detection limit is ten copies of CMV DNA per microliter. Also it is a time-saving technique; the time required for amplification and detection of the product is about 75 min by the LAMP assay, whereas that for PCR is 4 h.

In recent years, emergence or reemergence of severe arboviral hemorrhagic fevers caused by mosquito-borne viruses such as dengue virus and chikungunya (CHIK) virus has been frequently reported in the Indian subcontinent. Parida et al. (2007) developed a rapid and real-time detection of chikungunya virus by RT-LAMP assay. RT-LAMP assay correctly identified 21 additional positive borderline cases from acute-phase patient serum samples that were missed by conventional RT-PCR ($P < 0.0001$) with a detection limit of 20 copies. Humans infected with CHIK virus produce an illness often characterized by sudden onset of fever, headache, fatigue, nausea, vomiting, rash, myalgia, severe arthralgia, and polyarthralgia (typical clinical sign of the disease is very painful). Self-limiting symptoms last from 1 to 10 days.

The most common cause of childhood viral encephalitis is Japanese encephalitis virus (JEV) in the world, causing an estimated 50,000 infections and 10,000 deaths annually. Several serological tests, such as the hemagglutination inhibition test, enzyme-linked immunosorbent assay (ELISA), serum neutralization techniques, and dot enzyme immunoassay, had been used for the detection of antibody for JEV infection (Burke et al. 1982; Soloman et al. 1998). For many years, the hemagglutination inhibition test had been employed, but this has various practical limitations. Besides being cumbersome, it requires paired serum samples and therefore cannot give an early diagnosis. In 2006, Toriniwa and Komiya standardized RT-LAMP for the rapid detection and quantification of JEV by targeting the E gene, but still the application of this novel gene amplification system for the clinical diagnosis of JE patients during epidemic situations needs to be established through evaluation with a large number of clinical samples (Toriniwa and Komiya 2006). Further, Parida et al. (2006) also developed LAMP assay for detection of JEV, but this time particular importance is given to the substantial reduction in time required for the confirmation of results by RT-LAMP assay (in 30 min compared to 3–4 h in the case of RT-PCR). The RT-LAMP assay demonstrated exceptionally higher sensitivity compared to that of RT-PCR, with a detection limit of 0.1 PFU.

Dengue virus is a mosquito-borne flavivirus and the most widely prevalent arbovirus in tropical and subtropical regions of Asia, Africa, and Central and South America (Gubler 1997). Four distinct serotypes, i.e., DEN-1, DEN-2, DEN-3, and DEN-4, produce a spectrum of illness ranging from in apparent infection to moderate febrile illness and severe, fatal hemorrhagic disease. The three basic methods routinely practiced by most laboratories are virus isolation, characterization, detection of dengue virus-specific antibodies, and detection of genomic sequences by nucleic acid amplification techniques (Guzman and Kouri 1996; Innis et al. 1989; WHO 1997). Both virus isolation and PRNT assays are time-consuming and tedious and require more than a week for completion. Several investigators have reported on fully automatic real-time PCR assays for the detection of dengue virus in acute-phase serum samples (Callahan et al. 2001; Houng et al. 2001; Laue et al. 1999; Seah et al. 1995). During the past decade, various nucleic acid amplification techniques such as RT-PCR, nested PCR, Taqman real-time RT-PCR, SYBR Green I real-time RT-PCR, and NASBAs have been developed for the rapid identification of dengue virus to the serotype level with more accuracy (Kuno 1998; Lanciotti et al. 1992; Morita et al. 1991; Seah et al. 1995; Sudiro et al. 1997). Magnitude of amplification can be obtained easily by these PCR-based methods, but they require either high-precision instruments for the amplification or elaborate methods for detection of the amplified products. In addition, these methods are often cumbersome to adapt to routine clinical use, especially in peripheral health care settings and private clinics. Parida et al. (2005) developed RT-LAMP assay for reducing time required for the confirmation of the detection of dengue virus. LAMP assay detects dengue virus within 30 min only (less than 1 h), whereas RT-PCR requires 3–4 h for same detection. Also it was found to be 10- to 100-fold more sensitive than RT-PCR, with a detection limit of 0.1–1.0 PFU of virus.

Varicella (chickenpox) is a common and highly contagious disease in childhood that is caused by primary infection with varicella-zoster virus (VZV). The virus infects the dorsal root ganglia after primary infection, and its reactivation leads to shingles (zoster) in older individuals. In 1995, universal varicella immunization was started in the United States with VZV Oka (vOka strain) vaccine. Molecular analysis of the VZV genome (obtained from vesicular regions) used to discriminate the vOka strains and wild-type strains was based on restriction fragment length polymorphism analysis of the viral genomes from isolated viruses (La Russa et al. 1992). But this method had limited general clinical use because the virus had to be isolated from the vesicular region. PCR methods were used to overcome this problem, PCR which can detect single nucleotide polymorphism (SNP) mutations in open reading frame 38 (ORF38), ORF54, and ORF62 had been developed to distinguish the vOka and wild-type strains (Argaw et al. 2000; Quinlivan et al. 2005). Higashimoto et al. (2008) developed LAMP method to distinguish between the varicella-zoster virus (VZV) vaccine (vOka) strain and wild-type strains. Two single nucleotide polymorphisms (SNPs) (nucleotide [nt] 105705 for VR-1 VZV LAMP and nt 106262 for VR-2 VZV LAMP) located in the open reading frame 62 gene were selected as LAMP targets. The detection limits of both LAMP methods were 100 copies per reaction, which is higher than that of the previously reported VZV LAMP method (Okamoto et al. 2004), and also this method can detect viral DNA directly from swab samples without need of DNA extraction.

Commercial kits have been developed by using LAMP method, and some of them have been adopted as the officially recommended methods for the routine identification and surveillance of pathogens in Japan (Mori and Notomi 2009).

21.4.2 Plant Viruses

LAMP had been developed for many plant viruses such as tobacco mosaic virus (Liu et al. 2010), banana streak virus (Peng et al. 2012), cauliflower mosaic virus (Fukuta et al. 2004), tomato spotted wilt virus (Fukuta et al. 2004), yellow mosaic virus (Fukuta et al. 2003), potato virus Y (Nie 2005), potato spindle tuber viroid (Tsutsumi et al. 2010), peach latent mosaic viroid (Tsutsumi et al. 2010), wheat yellow mosaic virus (Zhang et al. 2011), necrotic spots virus (Gunimaladevi et al. 2005), and chrysanthemum stunt virus (Fukuta et al. 2005).

The most commonly known peach viroid is peach latent mosaic viroid (PLMVd) which causes a vast majority of natural infections (except leaf symptoms) but long time is required for the onset of symptoms. So a sensitive, accessible, reliable, cost-effective, and fast diagnostic method is in demand to control PLMVd spread and the production of healthy and of high-quality propagation material. Boubourakas et al. (2009) developed an RT-LAMP protocol for the detection of PLMVd with a detection limit of 10^{-5}. RT-LAMP products were confirmed for the presence of corresponding sequences of the selected target by subjecting a portion of the amplified products to the restriction enzyme analysis (Kubota et al. 2008).

Tsutsumi et al. (2010) developed RT-LAMP for the detection of potato spindle tuber viroid (PSTVd-S). Potato spindle tuber viroid (PSTVd) is a species of the genus *Pospiviroid* in the family Pospiviroidae that occurs in China (He et al. 1987), Europe, and so on. PSTVd infects potatoes (*Solanum tuberosum*), tomatoes (*Lycopersicon esculentum*), and other *Solanum* spp. (Singh 1973). PSTVd causes stunting, smaller and elongated tubers, and yield losses of potatoes (Pfannenstiel and Slack 1980). The occurrence of PSTVd has not been recorded in Japan, except a temporary occurrence in a greenhouse in Fukushima Prefecture in 2008. As a result of investigation of its cause, it could be thought to originate from imported tomato seeds. Sensitivity for PSTVd-S was determined from extracted DNA of infected potato leaf and serially diluted up to 10^{-4} times with sterilized water. It was possible to detect PSTVd from the extracted total nucleic acid diluted 10^{-3} in RT-LAMP method and diluted to 10^{-2} times in RT-PCR method. RT-LAMP method was ten times higher sensitive than RT-PCR method.

Nie (2005) developed reverse transcription loop-mediated isothermal amplification of DNA for the detection of potato virus *Y.* Coat protein (CP) gene was selected for designing the primers in such a way that a loop could be formed and elongated during DNA amplification. LAMP assay was found to be 10-fold more sensitive than conventional PCR.

Wheat yellow mosaic as one of the most devastating soilborne disease of winter wheat (*Triticum aestivum* L.) was first reported in Japan in the 1920s and China in the 1960s (Sawada 1927; Kusume et al. 1997) which spread continually in Japan and China (Wang et al. 1980; Han et al. 2000). A soilborne pathogen, i.e., wheat yellow mosaic virus (WYMV), was the causal agent responsible behind this disease, and it is transmitted by the fungus-like organism polymyxa graminis (Inouye 1969). Several common methods were used to detect WYMV, and ELISA was found as a reliable and suitable method for detecting high-throughput WYMV (Hariri et al. 1996a, b; Geng et al. 2003), but the sensitivity of ELISA might not be sufficiently high to detect low concentrations of WYMV; plus virus-specific antiserum is required for ELISA assay. Zhang et al. (2011) developed RT-LAMP as a rapid detection method for wheat yellow mosaic virus. RT-LAMP could detect WYMV from total RNA diluted up to 10^{-5}, while RT-PCR detected only up to 10^{-3}. So, RT-LAMP was 100 times more sensitive than conventional reverse transcriptase-polymerase chain reaction (RT-PCR).

Banana streak virus (BSV) is a significant constraint to banana production and genetic improvement. Therefore, Peng et al. (2012) developed an assay for rapid detection of banana streak virus by loop-mediated isothermal amplification assay in South China. LAMP assay could detect as low as 1 pg/μl template DNA. Test results of all field samples collected from different regions of South China showed that LAMP is more sensitive than PCR.

21.4.3 Other Animal Viruses

LAMP had been developed for the following viruses. Foot-and-mouth disease virus (Dukes et al. 2006), viral conjunctivitis (Wakabayashi et al. 2004), porcine reproductive and respiratory syndrome virus and porcine circovirus (Rovira et al. 2009), duck virus enteritis (Ji et al. 2009), monkeypox virus (Iizuka et al. 2009), infectious hematopoietic necrosis virus (IHNV) (Gunimaladevi et al. 2004), Newcastle disease virus (Pham et al. 2005), and pestiviruses cause diseases in animals such as classical swine fever (CSF) and bovine viral diarrhea/mucosal disease (BVD/MD). Equine arteritis virus (EAV), porcine reproductive and respiratory syndrome virus (PRRSV), lactate dehydrogenase elevating virus (LDV) of mice, and simian hemorrhagic fever virus (SHFV) all belong to the Arteriviridae family of viruses.

Coronaviruses are positive-strand, enveloped RNA viruses. Viruses in this group are important pathogens of mammals and birds. They cause enteric or respiratory tract infections in a variety of animals including humans, livestock, and pets. Severe acute respiratory syndrome (SARS) is caused by a coronavirus called SARS-CoV. CoV-based vectors have potential in vaccine development and for gene therapy (Theil 2007). Severe acute respiratory syndrome (SARS) first spread in Guangdong Province, China in November 2002 (Zhong et al. 2003). SARS is diagnosed by two major approaches: first, detection of antibodies against SARS-CoV was a sensitive and specific diagnostic approach, but seroconversion can be detected only around after day 10 of illness (Peiris et al. 2003a). In contrast, PCR-based tests have been shown to be useful for early SARS diagnosis (Berger et al. 2004). Quantitative PCR approaches are a powerful tool for identifying SARS-CoV early after disease onset (Drosten et al. 2003; Poon et al. 2003; Ng et al. 2003; Grant et al. 2003). Because of the requirement of sophisticated instrumentation and expensive reagents, these rapid molecular tests might not be the method of choice in field situations. The detection limit of the assay was 10 copies/reaction, and positive signals were consistently observed in reactions containing ~50 copies of the target sequence. Quantitative PCR method was used for SARS diagnosis (Poon et al. 2003). Thai et al. (2004) developed a one-step, single-tube, accelerated, real-time, quantitative RT-LAMP assay for the early and rapid diagnosis of SARS-CoV. The sensitivity and specificity of RT-LAMP assay for detecting viral RNA in clinical specimens with regard to RT-PCR were 100 and 87 %, respectively. The RT-LAMP assay was found to be 100-fold more sensitive than RT-PCR, with a detection limit of 0.01 PFU in clinical samples. In addition, by using a real-time RT-LAMP assay, the quantitation of virus concentration in the clinical sample is possible, which will indicate the early stage of the virus infection as well as potential source transmitters. However, the detection rates for SARS in the LAMP assay are similar to those with our conventional PCR-based assays (Chan et al. 2004). Rapid detection of the severe acute respiratory syndrome (SARS) coronavirus by a loop-mediated isothermal amplification assay was developed by Poon et al. (2004). The detection limit of the assay was 10 copies/reaction reactions containing ~50 copies of the target sequence. Viral load peaks in the second week of infection in SARS patients due to which its detection is difficult. LAMP assay is an inexpensive and accurate alternative method for detection of virus at the end of first week of infection (Peiris et al. 2003b).

Foot-and-mouth disease (FMD) causes severe economic losses and affected both national and international trade within the livestock and animal products (King et al. 2006). So, it was quite urgent and necessary to control causative virus responsible behind this veterinary infection. Chen et al. (2011) successfully detected foot-and-mouth disease viral RNA by reverse transcription loop-mediated isothermal amplification method with the detection limit of 10 copies, whereas RT-PCR detected 100 copies per reaction.

African swine fever virus causes a viral hemorrhagic fever with high mortality rates in pigs but persistently infects its natural hosts, warthogs, bushpigs, and soft ticks, with no disease signs (Viruses of Animals, George Durrell http://EzineArticles.com/?Expert=George_Durrell). Pseudorabies virus causes Aujeszky's disease in pigs and is extensively studied as a model for basic processes during lytic herpesvirus infection. Bovine herpesvirus-1 causes bovine infectious rhinotracheitis and pustular vulvovaginitis. The avian infectious laryngotracheitis virus is phylogenetically different from these two viruses. A large variety of influenza A viruses infect wild aquatic birds. Major economic losses in poultry production are due to highly pathogenic avian influenza virus (HPAIV) which also threatens human health. For rapid AIV detection, new assays would be designed that can be carried out with a minimum of specific equipment as compared to conventional methods. Although RT-LAMP was markedly more vulnerable to reduce the detection limits because of strain-specific sequence variation than subtype-specific real-time RT-PCR. Because of simplicity, LAMP proves to be a promising technology for AIV diagnosis (Notomi et al. 2000; Parida et al. 2008). Real-time RT-PCR also represents the gold standard for AIV RNA detection as it is considerably more sensitive than conventional RT-PCR (Pasick 2008). All RNA samples tested specifically positive in the real-time RT-LAMP were also assayed for sensitivity and compared with real-time RT-PCR. All AIV-targeted RT-LAMP assays developed so far have been compared with conventional RT-PCR, and sensitivity was calculated on the basis of infectivity, not taking into account the amount of RNA copies (Chen et al. 2008; Imai et al. 2006, 2007; Jayawardena et al. 2007). Chen et al. (2010) diagnosed H9 avian influenza virus by using reverse transcription loop-mediated isothermal amplification with a detection limit of ten copies per reaction, and no cross-reactivity was observed from the samples of other related viruses including H5N1, H3N2 subtype of AIV, and Newcastle disease virus. Again in 2011, Nagatani et al. detected RNA of influenza virus by reverse transcription LAMP using a USB-powered portable potentiostat.

Xie et al. (2010) developed an assay for rapid detection of infectious laryngotracheitis virus isolates. LAMP assay was found ten-fold more sensitive than the routine PCR assay, with a detection limit of 46 copies per reaction.

Gunimaladevi et al. (2004) developed LAMP method for detection of infectious hematopoietic necrosis virus (IHNV) in rainbow trout (*Oncorhynchus mykiss*) with a ten-fold higher sensitivity compared with conventional nested PCR.

Porcine circoviruses (PCV) are the smallest autonomously replicating viruses in eukaryotic cells. Among known two types of PCV, porcine circovirus type 1 (PCV1) does not cause disease, but porcine circovirus type 2 (PCV2) causes a new emerging and multifactorial disease in swine known as postweaning multisystemic wasting syndrome (PMWS) (Bolin et al. 2001) and porcine dermatitis and nephropathy syndrome (PDNS). Porcine circovirus type 2 (PCV2) syndrome is responsible for great losses to the pig industry. PCV2 is a non-enveloped, circular, single-stranded DNA virus. Presence of PCR inhibitors (including organic and inorganic substances such as detergents, antibiotics, phenolic compounds, enzymes, polysaccharides, fats, proteins, and salts) in the analysis of real biological samples limits the usefulness of PCR. These inhibitors reduce the amplification efficiency (Darwich et al. 2002; Guo et al. 2010). PCV2 was detected by reverse transcription LAMP using membrane protein M gene for primers with a detection limit ten fold greater than conventional PCR (Chen et al. 2009). Chen et al. (2008) detected PCV2 by LAMP by using only four primers, while Zhao et al. (2011) detected same virus by efficient LAMP assay using six primers including two loop primers for detection of target ORF2 gene in PCV2 (which encodes Rep protein that is involved in virus replication) with a detection limit of 10 copies, whereas the limit by conventional PCR was 1,000 copies. LAMP assay had advantages in specificity, selectivity, and rapidity over other nucleic acid amplification methods (Mori et al. 2006). LAMP had been further advanced by using forward loop primers (Nagamine et al. 2002). The LAMP assay developed had a detection limit of ten copies for PCV2, which was 100-fold more sensitive than conventional PCR.

Rovira et al. (2009) developed reverse transcription LAMP for porcine reproductive and respiratory syndrome virus detection. The limit of detection ranged between 10^{-2} and 10^{-4} (50 % tissue culture infective) dose/ml.

Swine transmissible gastroenteritis coronavirus (TGEV) is a kind of single-stranded RNA virus, which produces villous atrophy and enteritis, leading to the serious financial loss to the whole pig industry. The traditional detection methods including virus isolation, virus immunodiagnostic assays, and PCR tests have shortcomings such as precise instruments requirement, demand of elaborate result analysis, high cost, and long detection time, which prevent these methods from being widely used (Reynolds and Garwes 1979; Rodak et al. 2005; Denac et al. 1997; Paton et al. 1997). TGEV can be detected by using loop-mediated isothermal amplification with the detection limit of about 10 pg RNA which is ten times more sensitive than that of PCR and having no cross-reaction with other viruses (Chen et al. 2010) and found that the most conserved fragment of 187 bp was found in the nucleocapsid protein gene which showed highly homology among different TGEV strains/isolates (more than 97 %) and low homology among other similar viruses (less than 52.5 %). LAMP could not only qualitatively detect the TGEV but also quantitatively analyze the virus. It was concluded that real-time fluorescence LAMP for quantitatively detection of TGEV was established by adding 1X SYBR Green I in the LAMP reaction.

Duck virus enteritis is a serious disease among farmed and free-living ducks (Anatidae) and a constant threat to the commercial duck industry in China. Ji et al. (2009) developed LAMP method by using set of four specific primers designed to recognize six distinct genomic sequences of UL6 protein from duck plague virus (DPV). The optimum reaction temperature and time were verified to be 61.5 °C and 60 min, respectively. Marek's disease viral genome in chicken feathers can be easily detected by LAMP with detection limit of ten copies of the MEQ gene in the MD viral genome along with ten times higher sensitivity than the traditional PCR methods (Angamuthu et al. 2012).

By using 12 dogs and 2 cats, Boldbaatar et al. (2009) developed reverse transcription LAMP for detection of rabies virus with detection limit of 10^3 copies of viral RNA (corresponding to approximately 5 fg of RNA).

Nimitphak et al. (2010) developed a rapid and sensitive detection of *Penaeus monodon nucleopolyhedrovirus (PemoNPV)* by loop-mediated isothermal amplification combined with a lateral-flow dipstick. With a DNA template extracted from PemoNPV-infected shrimp, the LAMP-LFD detection limit was 0.1 pg, whereas one-step PCR and nested PCR followed with gel electrophoresis was 1 pg. Fang et al. (2011) developed for predicting viruses accurately by a multiplex microfluidic loop-mediated isothermal amplification chip. Multiplex gene assay is a valuable molecular tool not only in academic science but also in clinical diagnostics. An octopus-like multiplex microfluidic loop-mediated isothermal amplification (mμ LAMP) assay was proved a robust approach for predicting viruses for the rapid analysis of multiple genes in the point-of-care format. It has ability of analyzing multiple genes qualitatively and quantitatively as it is highly specific, operationally simple, and cost-/time-effective with the detection limit of less than 10 copies/μl in 2 μl quantities of sample within 0.5 h. A multiplex microfluidic LAMP chip for differentiating three human influenza A substrains and identifying eight important swine viruses had been successfully developed.

In 2008, Blomstrom et al. developed single-step reverse transcriptase LAMP assay for simple and rapid detection of swine vesicular disease virus with a detection limit of 50 viral RNA copies per μl.

Rift Valley fever virus (RVFV) is a zoonotic mosquito-borne virus with tripartite negative-strand RNA genome composed of a large segment (L), encoding the viral transcriptase, a medium segment (M), coding for the two external glycoproteins (GN and GC), and an S segment, which codes for the nucleocapsid protein (N) and a nonstructural protein (NSs) (Schmaljohn 1996). The virus causes explosive outbreaks in animals and humans (Durand et al. 2003) and has been observed in Egypt, in Mauritania, and more recently in the Arabian Peninsula (Bird et al. 2007). Domestic animals are sensitive to RVFV infection and amplify the virus to high titers (Zeller and Bouloy 2000). Among cattle, sheep, goats, pigs, and camels, infection causes fever and anorexia. Peyrefitte et al. (2008) also developed a real-time reverse transcription LAMP for rapid detection of Rift Valley fever virus. The assay is highly sensitive and comparable to real-time RT-PCR, with a detection limit of ~10 RNA copies per assay.

Newcastle disease is a highly contagious viral infection of poultry caused by a paramyxovirus called avian paramyxovirus type 1 (APMV-1), one of the nine serotypes of the virus identified (Alexander and Manvell 1997). Pham et al. (2005) developed loop-mediated isothermal amplification for rapid detection of Newcastle disease virus. The detection limit of both methods, i.e., PCR as well as LAMP, was 0.5 pg or 9×10^4 copies/reaction determined by using a serial ten-fold dilution of the plasmids.

Parida et al. (2004) developed real-time reverse transcription LAMP method for rapid detection of West Nile virus by using envelope gene of West Nile (WN) virus. Suebsing et al. (2011) developed reverse transcription LAMP method for detection of infectious hematopoietic necrosis virus in *Oncorhynchus keta*. The limit of detection was 0.01 fg of RNA extracted from IHNV-infected CHSE-214 cells, compared with 1.0 fg for nested RT-PCR. Ren and Li (2011) developed reverse transcription loop-mediated isothermal amplification for rapid detection of porcine epidemic diarrhea virus. Six primers were designed to amplify the nucleocapsid (N) gene of PEDV. The optimal reaction condition for RT-LAMP amplifying PEDV N gene was achieved at 63 °C for 50 min. It was found that RT-LAMP assay was more sensitive than gel-based RT-PCR and enzyme-linked immunosorbent assay. It was capable of detecting PEDV from clinical samples and differentiating PEDV from porcine transmissible gastroenteritis virus, porcine rotavirus, porcine pseudorabies virus, porcine reproductive and respiratory syndrome virus, and avian infectious bronchitis virus. Taura syndrome virus (TSV) was first discovered in Ecuador in 1992 (Jimenez 1992). Kiatpathomchai et al. (2007) developed a rapid and sensitive detection of Taura syndrome virus by reverse transcription loop-mediated isothermal amplification. The detection of TSV using RT-LAMP was ten times more sensitive than the RT-PCR, but less sensitive than nested RT-PCR. Sappat et al. (2011) detected the shrimp Taura syndrome virus by loop-mediated isothermal amplification using a designed portable multichannel turbidimeter. When using the same TSV-RNA template sets, the results revealed the same detection limit of 10^{-6} dilution for RT-LAMP when performed using a commercial heating block (Major Science, Taiwan) and the designed turbidimeter. When RT-LAMP reactions were carried out at 63 °C for 30 min using 2 µl of the ten-fold serially diluted RNA template extracted from TSV-infected shrimp, the turbidity measurement was able to detect 100 fg of total RNA. This detection limit was equivalent to that of RT-LAMP-AGE and nested RT-PCR. When the RNA *in vitro* transcript was used as the template, the detection limit by both of turbidity measurement and AGE was ten copies. A prior study indicated that the LAMP method was capable of amplification starting with only six copies of template (Nagamine et al. 2002).

Xie et al. (2011) rapid detection of group I avian adenoviruses by a loop-mediated isothermal amplification. In total, 72 out of 184 cloacal swab samples from poultry were identified as positive by LAMP, whereas 45 out of 184 were identified as positive by conventional PCR test. Liu (2011) reverse transcription loop-mediated isothermal amplification for the rapid detection of infectious bronchitis virus. Infectious bronchitis virus (IBV) is a major cause of disease in domestic fowl and causes an acute, highly contagious disease of the respiratory tracts and sometimes urogential tracts (King and Cavanagh 1991). The result indicated detection limit of IBV RT-LAMP was 10 copies/

tube. In addition, the reaction time of RT-LAMP method is 45 min, which is more rapid than conventional RT-PCR, and the reaction only needs a laboratory water bath. Kiatpathomchai et al. (2007) detected RT-TSV RNA in *P. vannamei* (collected from shrimp farms) by LAMP assay, and the sensitivity of RT-LAMP appears to be ten times more than RT-PCR. Wang et al. (2011) detected infectious bursal disease virus (IBDV) in one simple step by reverse transcription loop-mediated isothermal amplification (RT-LAMP) and further identified the very virulent strain from non-vvIBDVs with a simply post-amplification restriction enzyme analysis. A set of two inner, two outer, and two loop primers were designed on the basis of sequence analysis to target the VP5 gene and showed great specificity with no cross-reaction to the other common avian pathogens. Detection limit was determined by both color change inspection and agarose gel electrophoresis was 28 copies viral RNA, which was almost as sensitive as a real-time RT-PCR previously developed.

Last year in 2012, Foord et al. developed hendra virus detection using loop-mediated isothermal amplification. Hendra virus (HeV) is a zoonotic paramyxovirus endemic in Australian Pteropus bats (fruit bats or flying foxes). Although bats appear to be unaffected by the virus, HeV can spread from fruit bats to horses, causing severe disease. Human infection results from close contact with the blood, body fluids, and tissues of infected horses. HeV is a biosecurity level 4 (BSL-4) pathogens, with high fatality rate in humans and horses. Current assays for HeV detection require complex instrumentation and are generally time-consuming. Analytical sensitivity and specificity of the HeV-LAMP assay was equal to a TaqMan assay developed previously.

The viruses for which LAMP has not been developed till now are as follows: equine arteritis virus, simian hemorrhagic fever virus, tick-borne encephalitis virus, bornavirus, infectious bovine rhinotracheitis virus, feline herpesvirus, bovine respiratory syncytial virus, bovine rhinotracheitis virus, equine anemia virus, cowpox virus, elephant endotheliotropic herpesvirus, turkeypox virus, myxomatosis virus, menangle virus, rabbit hemorrhagic disease virus, and rabbit hemorrhagic virus.

21.5 Conclusions

Compared with RT-PCR and real-time PCR, LAMP has the advantages of reaction simplicity and detection sensitivity. The higher sensitivity and specificity of the LAMP reaction is attributed to continuous amplification under isothermal conditions employing six primers recognizing eight distinct regions of the target. Additionally, the higher amplification efficiency of the LAMP reaction yields large amounts of by-product (pyrophosphate ion) leading to a white precipitate of magnesium pyrophosphate in the reaction mixture. Since the increase in turbidity of the reaction mixture—according to the production of precipitate—correlates with the amount of DNA synthesized, real-time monitoring of the LAMP reaction can be achieved by real-time measurement of turbidity. The LAMP assay has emerged as a powerful gene amplification tool for the rapid identification of microbial infections and is being increasingly used by various investigators for the rapid detection and typing of emerging viruses, including WNV, severe acute respiratory syndrome (SARS), dengue, and JEV.

21.6 Future Prospects

Some other viruses for which LAMP has not been developed till now are as follows: tomato bushy stunt virus, peanut stunt virus, peony ring spot virus, psoriasis A virus, bean common mosaic virus, lettuce mosaic virus, fanleaf disease virus, wheat streak mosaic virus, bud blight virus, rose mosaic virus, zucchini yellow mosaic virus, cankers virus, alfalfa mosaic virus, bean yellow mosaic virus, tobacco etch virus, wound tumor virus, pea enation mosaic virus, quirking virus (on datura), sugarcane mosaic virus, etc.

References

Alexander DJ, Manvell RJ (1997) Antigenic diversity and similarities detected in Avian paramyxovirus type 1 (Newcastle disease virus) isolates using monoclonal antibodies. Avian Pathol 26:399–418

American Academy of Pediatrics Committee on Infectious Diseases and Committee on Fetus Newborn (2003) Revised indications for the use of palivizumab and respiratory syncytial virus immune globulin intravenous for the prevention of respiratory syncytial virus infections. Pediatrics 112:1442–1446

Angamuthu R, Baskaran S, Gopal DR, Devarajan J, Kathaperumal K (2012) Rapid detection of the Marek's disease viral genome in chicken feathers by loop-mediated isothermal amplification. J Clin Microbiol 50(3):961–965

Argaw T, Cohen JI, Klutch M, Lekstrom K, Yoshikawa T, Asano Y, Krause PR (2000) Nucleotide sequences that distinguish Oka vaccine from parental Oka and other varicella zoster virus isolates. J Infect Dis 181:1153–1157

Barenfanger J, Drake C, Leon N, Mueller T, Troutt T (2000) Clinical and financial benefits of rapid detection of respiratory viruses: an outcomes study. J Clin Microbiol 38:2824–2828

Berger A, Drosten C, Doerr HW, Sturmer M, Preiser W (2004) Severe acute respiratory syndrome (SARS)-paradigm of an emerging viral infection. J Clin Virol 29:13–22

Bird BH, Khristova ML, Rollin PE, Ksiazek TG, Nichol ST (2007) Complete genome analysis of 33 ecologically and biologically diverse Rift Valley fever virus strains reveals widespread virus movement and low genetic diversity due to recent common ancestry. J Virol 81:2805–2816

Blomstrom AL, Hakhverdyan M, Reid SM, Dukes JP, King DP, Belak S, Berg MA (2008) One-step reverse transcriptase loop-mediated isothermal amplification assay for simple and rapid detection of swine vesicular disease virus. J Virol Methods 147:188–193

Boldbaatar B, Inoue S, Sugiura N, Noguchl A, Orbina JRC, Demetria C, Miranda ME, Yamada A (2009) Rapid detection of rabies virus by reverse transcription loop-mediated isothermal amplification. Jpn J Infect Dis 62(3):187–191

Bolin SR, Stoffregen WC, Nayar GP, Hamel AL (2001) Postweaning multisystemic wasting syndrome induced after experimental inoculation of cesarean-derived, colostrum-deprived piglets with type 2 porcine circovirus. J Vet Diagn Invest 13:185–194

Boubourakas IN, Fukuta S, Kyriakopoulou P (2009) Sensitive and rapid detection of peach latent mosaic viroid by the reverse transcription loop-mediated isothermal amplification. J Virol Methods 160(1–2): 63–68

Burke DS, Nisalak A, Ussery MA (1982) Antibody capture immunoassay detection of Japanese encephalitis virus immunoglobulin M and G antibodies in cerebrospinal fluid. J Clin Microbiol 16:1034–1042

Cai T, Lou GQ, Yang J, Xu D, Meng ZH (2008) Development and evaluation of real time loop-mediated isothermal amplification for hepatitis B virus DNA quantification: a new tool for HBV management. J Clin Virol 41:270–276

Callahan JD, Wu SJ, Dion-Schultz A, Mangold BE, Peruski LF, Watts DM, Porter KR, Murpgy GR, Suharyono W, King CC, Hayes CG, Temenak JJ (2001) Development and evaluation of serotype and group-specific fluorogenic reverse transcriptase PCR (TaqMan) assays for dengue virus. J Clin Microbiol 39:4119–4124

Carr MJ, Gunson R, Maclean A, Coughlan S, Fitzgerald M, Scully M, O'Herlihy B, Ryan J, O'Flanagan D, Connell J, Carman WF, Hall WW (2009) Development of a real-time RT PCR for the detection of swinelineage influenza A (H1N1) virus infections. J Clin Virol 45:196–199

Centers for Disease Control and Prevention (1998) Measles, mumps, rubella vaccine use and strategies for elimination of measles, rubella, and congenital rubella syndrome and control of mumps: recommendations of the Advisory Committee on Immunization Practices (ACIP). Morb Mortal Wkly Rep 47(RR-8):1–57

Chan AB, Fox JD (1999) NASBA and other transcription-based amplification methods for research and diagnostic microbiology. Rev Med Microbiol 10:185–196

Chan KH, Poon LLM, Cheng VCC, Guan Y, Hung IFN, Kong J, Yam LY, Seto WH, Yuen KY, Peiris JS (2004) Detection of SARS coronavirus in patients with suspected SARS. Emerg Infect Dis 10(2):294–299

Chen HT, Zhang J, Sun DH, Ma LN, Liu XT, Cai XP, Liu YS (2008) Development of reverse transcription loop mediated isothermal amplification for rapid detection of *H9* Avian influenza virus. J Virol Methods 151:200–203

Chen W, Seifert K, Levesque CA (2009) A high density COX1 barcode oligonucleotide array for identification and detection of species of *Penicillium* subgenus *Penicillium*. Mol Ecol Resour 9:114–129

Chen HT, Zhang J, Ma YP, Ma LN, Ding YZ, Liu XT, Cai XP, Ma LQ, Zhang YG, Liu YS (2010) Reverse transcription loop-mediated isothermal amplification for the rapid detection of infectious bronchitis virus in infected chicken tissues. Mol Cell Probes 24(2): 104–106

Chen HT, Zhang J, Liu YS, Liu X (2011) Detection of foot-and-mouth disease virus by reverse transcription loop-mediated isothermal amplification. Virol J 8:510

Collier L, Balows A, Sussman M (1998) Topley and Wilson's microbiology and microbial infections, vol 1, 9th edn, Virology. Mahy B, Collier L (vol ed). Arnold, London. ISBN 0-340-66316-2

Corden S, Ballard AL, Ijaz S, Barbara JA, Gilbert N, Gilson RJ, Boxall EH, Tedder RS (2003) HBV DNA levels and transmission of hepatitis B by health care workers. J Clin Virol 27(1):52–58

Curtis K, Rudolph D, Owen MS (2009) Sequence-specific detection method for reverse transcription, loop-mediated isothermal amplification of HIV-1. J Med Virol 81:966–972

Darwich L, Segales J, Domingo M, Mateu E (2002) Changes in CD4+, CD8+, CD4+ CD8+ and immunoglobulin M-positive peripheral blood mononuclear cells of postweaning multisystemic wasting syndrome-affected pigs and age-matched uninfected wasted and healthy pigs correlate with lesions and porcine circovirus type 2 load in lymphoid tissues. Clin Diagn Lab Immunol 9:236–242

Denac H, Moser C, Tratschin JD, Hofmann MA (1997) An indirect ELISA for the detection of antibodies against porcine reproductive and respiratory syndrome virus using recombinant nucleocapsid protein as antigen. J Virol Methods 65:169–181

Dienstag JL (2008) Hepatitis B virus infection. N Engl J Med 359:1486–1500

Drosten C, Gunther S, Preiser W, van der Werf S, Brodt HR, Becker S, Rabenau H, Panning M, Kolesnikova L, Fouchier RA, Berger A, Burguiere AM, Cinatl J, Eickmann M, Escriou N, Grywna K, Kramme S, Manuguerra JC, Muller S, Rickerts V, Sturmer M, Vieth S, Klenk HD, Osterhaus AD, Schmitz H, Doerr HW (2003) Identification of a novel coronavirus in patients with severe acute respiratory syndrome. N Engl J Med 348:1967–1976

Dukes JP, King DP, Alexandersen S (2006) Novel reverse transcription loop-mediated isothermal amplification for rapid detection of foot-and-mouth disease virus. Arch Virol 151:1093–1106

Durand JP, Bouloy M, Richecoeur L, Peyrefitte CN, Tolou H (2003) Rift valley fever virus infection among French troops in Chad. Emerg Infect Dis 9:751–752

Emiliani C (1993) Extinction and viruses. Biosystems 31:155–159

Endo S, Komori T, Ricci G, Sano A, Yokoyama K, Ohori A, Kamei K, Franco M, Miyaji M, Nishimura K (2004) Detection of gp43 of *Paracoccidioides brasiliensis* by the loop mediated isothermal amplification (LAMP) method. FEMS Microbiol Lett 234:93–97

Enomoto Y, Yoshikawa T, Ihira M, Akimoto S, Miyake F, Usui C, Suga S, Suzuki K (2005) Rapid diagnosis of herpes simplex virus infection by a loop-mediated isothermal amplification method. J Clin Microbiol 43(2):951–955

Espy MJ, Uhl JR, Sloan LM, Buckwalter SP, Jones MF, Vetter EA, Yao JD, Wengenack NL, Rosenblatt JE, Cockerill FR, Smith TF (2006) Real-time PCR in clinical microbiology: applications for routine laboratory testing. Clin Microbiol Rev 19:165–256

Fang X, Chen H, Yu S, Jiang X, Kong J (2011) Predicting viruses accurately by a multiplex microfluidic loop-mediated isothermal amplification chip. Anal Chem 83(3):690–695

Foord AJ, Middleton D, Heine HG (2012) Hendra virus detection using loop-mediated isothermal amplification. J Virol Methods 181(1):93–96

Fraser C, Riley S, Anderson RM, Ferguson NM (2004) Factors that make an infectious disease outbreak controllable. Proc Natl Acad Sci U S A 101:6146–6151

Fukuda S, Takao S, Kuwayama M, Shimazu Y, Miyazaki K (2006) Rapid detection of norovirus from fecal specimens by real-time reverse transcription-loop-mediated isothermal amplification assay. J Clin Microbiol 44:1376–1381

Fukuta S, Kato S, Yoshida K, Mizukami Y, Ishida A, Ueda J (2003) Detection of tomato yellow leaf curl virus by loop-mediated isothermal amplification reaction. J Virol Methods 112:35–40

Fukuta S, Ohishi K, Yoshida K, Mizukami Y, A I, Kanbe M (2004) Development of immunocapture reverse transcription loop-mediated isothermal amplification for the detection of tomato spotted wilt virus from *Chrysanthemum*. J Virol Methods 121:49–55

Fukuta S, Nimi Y, Oishi K, Yoshimura Y, Anai N, Hotta M, Fukaya M, Kato T, Oya T, Kambe M (2005) Development of reverse transcription loop-mediated isothermal amplification (RT-LAMP) method for detection of two viruses and chrysanthemum stunt viroid. Ann Rep Kansai Plant Prot Soc 47:31–36

Garten RJ, Davis CT, Russell CA, Shu B, Lindstrom S, Balish A, Sessions WM, Xu X, Skepner E, Deyde V, Okomo-Adhiambo M, Gubareva L, Barnes J, Smith CB, Emery SL, Hillman MJ, Rivailler P, Smagala J, de Graaf M, Burke DF, Fouchier RA, Pappas C, Alpuche-Aranda CM, Lopez-Gatell H, Olivera H, Lopez I, Myers CA, Faix D, Blair PJ, Yu C, Keene KM, Dotson PD Jr, Boxrud D, Sambol AR, Abid SH, St George K, Bannerman T, Moore AL, Stringer DJ, Blevins P, Demmler-Harrison GJ, Ginsberg M, Kriner P, Waterman S, Smole S, Guevara HF, Belongia EA, Clark PA, Beatrice ST, Donis R, Katz J, Finelli L, Bridges CB, Shaw M, Jernigan DB, Uyeki TM, Smith DJ, Klimov AI, Cox NJ (2009) Antigenic and genetic characteristics of swine-origin 2009 A (H1N1) influenza viruses circulating in humans. Science 325:197–201

Geng B, Han C, Zhai Y, Wang H, Yu J, Liu Y (2003) Detection of wheat yellow mosaic virus by heterogeneous animal double-antibody sandwich ELISA. Virol Sin 18(1):76–78

Girard MP, Cherian T, Pervikov Y, Kieny MP (2005) A review of vaccine research and development: human acute respiratory infections. Vaccine 23:5708–5724

Grant PR, Garson JA, Tedder RS, Chan PK, Tam JS, Sung JJ (2003) Detection of SARS coronavirus in plasma by real-time RT-PCR. N Engl J Med 349: 2468–2469

Gubler DJ (1997) Dengue and dengue hemorrhagic fever: its history and resurgence as a global public health problem. In: Gubler DJ, Kuno G (eds) Dengue and dengue hemorrhagic fever. CAB International, New York, pp 1–22

Gunimaladevi I, Kono T, Venugopal MN, Sakai M (2004) Detection of koi herpesvirus in common carp, *Cyprinus carpio* L., by loop-mediated isothermal amplification. J Fish Dis 27(10):583–589

Gunimaladevi I, Kono T, Lapatra SE, Sakai M (2005) A loop mediated isothermal amplification (LAMP) method for detection of infectious hematopoietic necrosis virus (IHNV) in rainbow trout (*Oncorhynchus mykiss*). Arch Virol 150(5):899–909

Guo LJ, Lu YH, Wei YW, Huang LP, Liu CM (2010) Porcine circovirus type 2(PCV2): genetic variation and newly emerging genotypes in China. Virol J 7:273

Guzman MG, Kouri G (1996) Advances in dengue diagnosis. Clin Diagn Lab Immunol 3:621–627

Hagiwara M, Sasaki H, Matsuo K, Honda M, Kawase M, Nakagawa H (2007) Loop mediated isothermal amplification method for detection of human papillomavirus type 6, 11, 16 and 18. J Med Virol 79:605–615

Han C, Li D, Xing Y, Zhu K, Tian Z, Cai Z, Yu J, Liu Y (2000) Wheat yellow mosaic virus widely occurring in wheat (*Triticum aestivum*) in China. Plant Dis 84(6): 627–630

Han ET, Watanabe R, Sattabongkot J, Khuntirat B, Sirichaisinthop J, Iriko H, Jin L, Takeo S, Tsuboi T (2007) Detection of four plasmodium species by genus- and species-specific loop-mediated isothermal amplification for clinical diagnosis. J Clin Microbiol 45(2521):2528

Hariri D, Delaunay T, Gomes L, Filleur S, Plovie C, Lapierre H (1996a) Comparison and differentiation of wheat yellow mosaic virus (WYMV), wheat spindle streak mosaic virus (WSSMV) and barley yellow mosaic virus (BaYMV) isolates using WYMV monoclonal antibodies. Eur J Plant Pathol 102(3):283–292

Hariri D, Lapierre H, Filleur S, Plovie C, Delaunay T (1996b) Production and characterization of monoclonal antibodies to barley yellow mosaic virus and their use in detection of four bymoviruses. J Phytopathol 144(6):331–336

He LY, Zhang HL, Huang H (1987) Potato diseases in Asia: resent and expected developments. Acta Hortic 213:129–142

Higashimoto Y, Ihira M, Ohta A, Inoue S, Usui C, Asano Y, Yoshikawa T (2008) Discriminating between Varicella-Zoster virus vaccine and wild-type strains by loop mediated isothermal amplification. J Clin Microbiol 46(8):2665–2670

Hiromoto Y, Yamazaki Y, Fukushima T, Saito T, Lindstrom SE, Omoe K et al (2000) Evolutionary characterization of the six internal genes of H5N1 human influenza A virus. J Gen Virol 81:1293–1303

Hong TC, Mai QL, Cuong DV, Parida M, Minekawa H, Notomi T, Hasebe F, Morita K (2004) Development and evaluation of a novel loop mediated isothermal amplification method for rapid detection of severe acute respiratory syndrome coronavirus. J Clin Microbiol 42:1956–1961

Hosaka N, Ndembi N, Ishizaki A, Kageyama S, Numazaki K, Ichimura H (2009) Rapid detection of human immunodeficiency virus type 1 group M by a reverse transcription-loop mediated isothermal amplification assay. J Virol Methods 157:195–199

Houng HH, Chen RC, Vaughn MDW, Kanesa-thasan K (2001) Development of a fluorogenic RT-PCR system for quantitative detection of dengue virus serotypes 1–4 using conserved and serotype specific 3-noncoding sequence. J Virol Methods 95:19–32

http://en.wikipedia.org/wiki/Introduction_to_viruses

http://en.wikipedia.org/wiki/Virus_classification

http://primerexplorer.jp/e/

Enteric viruses: rapid detection and identification methods. http://www.rapidmicrobiology.com/test-methods/Enteric-viruses.php

Human diseases caused by viruses by Dr. Ananya Mandal, MD (Pharmacology), Assistant Professor (ed). Published online in News Medical, 15 Nov 2013. http://www.news-medical.net/health/Human-Diseases-Caused-by-Viruses.aspx

Iizuka I, Saijo M, Shiota T, Ami Y, Suzaki Y, Nagata N, Hasegawa H (2009) Loop- mediated isothermal amplification-based diagnostic assay for monkeypox virus infections. J Med Virol 81:1102–1108

Imai M, Ninomiya A, Minekawa H, Notomi T, Ishizaki T, Tashiro M, Odagiri T (2006) Development of H5-RT-LAMP (loop-mediated isothermal amplification) system for rapid diagnosis of H5 avian influenza virus infection. Vaccine 24:6679–6682

Imai M, Ninomiya A, Minekawa H, Notomi T, Ishizaki T, Van Tu P, Tien NT, Tashiro M, Odagiri T (2007) Rapid diagnosis of H5N1 avian influenza virus infection by newly developed influenza H5 hemagglutinin gene-specific loop-mediated isothermal amplification method. J Virol Methods 141:173–180

Innis BL, Nisalak A, Nimmannitya S, Kusalerdchariya S, Chongswasdi V, Suntayakorn S, Puttisri P, Hoke CH (1989) An enzyme-linked immunosorbent assay to characterize dengue infections where dengue and Japanese encephalitis co-circulate. Am J Trop Med Hyg 40:418–427

Inouye T (1969) Filamentous particles as the causal agent of yellow mosaic disease of wheat. Nogaku Kenkyu 53:61–68

Ito M, Watanabe M, Nakagawa N, Ihara T, Okuno Y (2006) Rapid detection and typing of influenza A and B by loop-mediated isothermal amplification: comparison with immunochromatography and virus isolation. J Virol Methods 135:272–275

Iwata S, Shibata Y, Kawada J, Hara S, Nishiyama Y, Morishima T, Ihira M (2006) Rapid detection of Epstein–Barr virus DNA by loop-mediated isothermal amplification method. J Clin Virol 37:128–133

Iyer LM, Balaji S, Koonin EV, Aravind L (2006) Evolutionary genomics of nucleo cytoplasmic large DNA viruses. Virus Res 117(1):156–184

Jayawardena S, Cheung CY, Barr I, Chan KH, Chen H, Guan Y, Peiris JSM, Poon LLM (2007) Loop-mediated isothermal amplification for influenza A (H5N1) virus. Emerg Infect Dis 13(6):899–901

Ji J, Du LQ, Xie QM, Cao YC, Zuo KJ, Xue CY, Ma JY, Chen F, Bee YZ (2009) Rapid diagnosis of duck plagues virus infection by loop-mediated isothermal amplification. Res Vet Sci 87(1):53–58

Jiang T, Liu J, Deng YQ, Xu LJ, Li XF, Han JF, Cao RY, Qin ED, Qin CF (2011) Development and evaluation of a reverse transcription-loop-mediated Isothermal amplification assay for rapid detection of enterovirus 71. J Clin Microbiol 49(3):870–874

Jimenez R (1992) Sindrome de Taura (Resumen). Acuacultura del Ecuador. In: Jiminez R (ed) Revista especializada de la camara nacional de acuacultura, vol 1. Guayaquil, pp 1–16

Kiatpathomchai W, Jareonram W, Jitrapakdee S, Flegel TW (2007) Rapid and sensitive detection of Taura syndrome virus by reverse transcription loop-mediated isothermal amplification. J Virol Methods 146:125–128

King DJ, Cavanagh D (1991) Infectious bronchitis. In: Calnek BW, Barnes HJ, Beard CW, Reid WM, Yoder HW Jr (eds) Disease of poultry, 9th edn. Iowa State University Press, Ames, pp 471–484

King DP, Ferris NP, Shaw AE, Reid SM, Hutchings GH, Giuffre AC, Robida JM, Callahan JD, Nelson WM, Beckham TR (2006) Detection of foot-and-mouth disease virus: comparative diagnostic sensitivity of two independent real-time reverse transcription polymerase chain reaction assays. J Vet Diagn Invest 18:93–97

Knox CM, Chandler D, Short GA, Margolis TP (1998) Polymerase chain reaction-based assays of vitreous samples for the diagnosis of viral retinitis. Ophthalmology 105:37–45

Komiyama C, Suzuki K, Miura Y, Sentsui H (2009) Development of loop-mediated isothermal amplification method for diagnosis of bovine leukemia virus infection. J Virol Methods 157(2):175–179

Kubo T, Agoh M, le Mai Q, Fukushima K, Nishimura H, Yamaguchi A, Hirano M, Yoshikawa A, Hasebe F, Kohno S, Morita K (2010) Development of a reverse transcription-loop-mediated isothermal amplification assay for detection of pandemic (H1N1) 2009 virus as a novel molecular method for diagnosis of pandemic influenza in resource-limited settings. J Clin Microbiol 48:728–735

Kubota R, Vine BG, Alvarez AM, Jenkins DM (2008) Detection of *Ralstonia solanacearum* by loop-mediated isothermal amplification. Phytopathology 98:1045–1051

Kuno G (1998) Universal diagnostic RT-PCR protocol for arboviruses. J Virol Methods 72:27–41

Kurosaki Y, Takada A, Ebihara H, Grolla A, Kamo N, Feldmann H, Kawaoka Y, Yasuda J (2007) Rapid and simple detection of ebola virus by reverse transcription-loop-mediated isothermal amplification. J Virol Methods 141:78–83

Kusume T, Tamada T, Hattori H, Tsuchiya T, Kubo K, Abe H, Namba S, Tsuchizaki T, Kishi K, Kashiwazaki S (1997) Identification of a new wheat yellow mosaic virus strain with specific pathogenicity towards major wheat cultivars grown in Hokkaido. Ann Phytopathol Soc Jpn 63(2):107–109

La Russa P, Lungu O, Hardy I, Gershon A, Steinberg SP, Silverstein S (1992) Restriction fragment length polymorphism of polymerase chain reaction products from vaccine and wild type varicella-zoster virus isolates. J Virol 66:1016–1020

Lanciotti RS, Calisher CH, Gubler DJ, Chang GJ, Vorndam AV (1992) Rapid detection and typing of dengue viruses from clinical samples by using reverse transcriptase-polymerase chain reaction. J Clin Microbiol 30:545–551

Laue T, Emmerich P, Schmitz H (1999) Detection of dengue virus RNA in patients after primary or secondary dengue infection by using the TaqMan automated amplification system. J Clin Microbiol 37:2543–2547

Lee MF, Chen YH, Peng CF (2009) Evaluation of reverse transcription loop mediated isothermal amplification in conjunction with ELISA hybridization assay for molecular detection of Mycobacterium tuberculosis. J Microbiol Methods 76:174–180

Lee MS, Shih HC, Lu JJ, Su MC, Deng MC, Wu CC, Lin FY, Lin KH, Chen PY, Hsu WL (2010) M-specific reverse transcription loop-mediated isothermal amplification for detection of pandemic (H1N1) 2009 virus. Eur J Clin Invest 41(4):434–441

Liu YS (2011) Reverse transcription loop-mediated isothermal amplification for the rapid detection of infectious bronchitis virus. In: Gaca M (ed) Approaches to bronchitis. InTech. ISBN: 978-953-307-770-3, doi:10.5772/18586. Available from: http://www.intechopen.com/books/approaches-to-bronchitis/reverse-transcription-loop-mediated-isothermal-amplification-for-the-rapid-detection-of-infectious-b

Liu Y, Wang Z, Qian Y, Mu J, Shen L, Wang F, Yang J (2010) Rapid detection of tobacco mosaic virus using the reverse transcription loop-mediated isothermal amplification method. Arch Virol 155(10):1681–1685

Lok AS, Heathcote EJ, Hoofnagle JH (2001) Management of hepatitis B: 2000 summary of a workshop. Gastroenterology 120(7):1828–1853

Long EG (2009) Requirements for diagnosis of malaria at different levels of the laboratory network in Africa. Am J Clin Pathol 131:858–860

Lukinmaa S, Nakari UM, Eklund M, Siitonen A (2004) Application of molecular genetic methods in diagnostics and epidemiology of food-borne bacterial pathogens. APMIS 112:908–929

Mackay IM, Arden KE, Nitsche A (2002) Real-time PCR in virology. Nucleic Acids Res 30:1292–1305

Madhavan HN, Priya K, Biswas J (2004) Current perspectives of herpes viral retinitis and choroiditis. Indian J Pathol Microbiol 47:453–468

McMinn PC (2002) An overview of the evolution of enterovirus 71 and its clinical and public health significance. FEMS Microbiol Rev 26:91–107

Mori Y, Notomi T (2009) Loop-mediated isothermal amplification (LAMP): a rapid, accurate and cost-effective diagnostic method for infectious diseases. J Infect Chemother 15:62–69

Mori Y, Nagamine K, Tomita N, Notomi T (2001) Detection of loop mediated isothermal amplification reaction by turbidity derived from magnesium pyrophosphate formation. Biochem Biophys Res Commun 289:150–154

Mori Y, Hirano T, Notomi T (2006) Sequence specific visual detection of LAMP reactions by addition of cationic polymers. BMC Biotechnol 3:6

Morita K, Tanaka M, Igarashi A (1991) Rapid identification of dengue virus serotypes by using polymerase chain reaction. J Clin Microbiol 29:2107–2110

Moslemi E, Mohammad Shahhosseiny H, Javadi G, Praivar K (2009) Loop mediated isothermal amplification (LAMP) for rapid detection of HBV in Iran. Afr J Microbiol Res 3(8):439–445

Nagamine K, Hase T, Notomi T (2002) Accelerated reaction by loop-mediated isothermal amplification using loop primers. Mol Cell Probes 16:223–229

Nagatani N, Yamanaka K, Saito M, Koketsu R, Sasaki T, Ikuta K, Miyahara T, Tamiya E (2011) Semi-real time electrochemical monitoring for influenza virus RNA by reverse transcription loop-mediated isothermal amplification using a USB powered portable potentiostat. Analyst 136:5143–5150

Ng EK, Ng PC, Hon KL, Cheng WT, Hung EC, Chan KC, Chiu RW, Li AM, Poon LL, Hui DS, Tam JS, Fok TF, Lo YM (2003) Serial analysis of the plasma concentration of SARS coronavirus RNA in pediatric patients with severe acute respiratory syndrome. Clin Chem 49:2085–2088

Nie X (2005) Reverse transcription loop-mediated isothermal amplification of DNA for detection of Potato virus Y. Plant Dis 89(6):605–610

Nimitphak T, Meemetta W, Arunrut N, Senapin S, Kiatpathomchai W (2010) Rapid and sensitive detection of penaeus monodon nucleopolyhedrovirus (PemoNPV) by loop-mediated isothermal amplification combined with a lateral-flow dipstick. Mol Cell Probes 24(1):1–5

Notomi T, Okayama H, Masubuchi H, Yonekawa T, Watanabe K, Amino N, Hase T (2000) Loop-mediated isothermal amplification of DNA. Nucleic Acids Res 28(12):E63

Okafuji T, Yoshida N, Fujino M, Motegi Y, Ihara T, Ota Y, Notomi T, Nakayama T (2005) Rapid diagnostic method for the detection of mumps virus genome by loop-mediated isothermal amplification. J Clin Microbiol 43:1625–1631

Okamoto S, Yoshikawa T, Ihira M, Suzuki K, Shimokata K, Nishiyama Y, Asano Y (2004) Rapid detection of *Varicella-zoster* virus infection by a loop-mediated isothermal amplification method. J Med Virol 74:677–682

Parida MM (2008a) Rapid and real-time detection technologies for emerging viruses of biomedical importance. J Biosci 33:617–628

Parida M, Posadas G, Inoue S, Hasebe F, Morita K (2004) Real-time reverse transcription loop-mediated isothermal amplification for rapid detection of West Nile virus. J Clin Microbiol 42(1):257

Parida MM, Horioke K, Ishida H, Dash PK, Saxena P, Jana AM, Islam MA, Inoue S, Hosaka N, Morita K (2005) Rapid detection and differentiation of dengue virus serotypes by real-time reverse transcription loop-mediated isothermal amplification assay. J Clin Microbiol 43:2895–2903

Parida MM, Santhosh SR, Dash PK, Tripathi NK, Saxena P, Ambuj S, Sahni AK, Lakshmana Rao PV, Morita K (2006) Development and evaluation of reverse transcription-loop-mediated isothermal amplification assay for rapid and real-time detection of Japanese encephalitis virus. J Clin Microbiol 44:4172–4178

Parida MM, Santhosh SR, Dash PK, Tripathi NK, Lakshmi V, Mamidi N, Shrivastva A, Gupts N, Saxena P, Pardeep Babu J, Lakshmana Rao PV, Morita K (2007) Rapid and real-time detection of Chikungunya virus by reverse transcription loop-mediated isothermal amplification assay. J Clin Microbiol 45:351–357

Parida M, Sannarangaiah S, Dash PK, Rao PV, Morita K (2008) Loop mediated isothermal amplification (LAMP): a new generation of innovative gene amplification technique; perspectives in clinical diagnosis of infectious diseases. Rev Med Virol 18:407–421

Pasick J (2008) Advances in the molecular based techniques for the diagnosis and characterization of avian influenza virus infections. Transbound Emerg Dis 55:329–338

Paton D, Ibata G, Sands J, Mc Goldrick A (1997) Detection of transmissible gastroenteritis virus by RT-PCR and differentiation from porcine respiratory coronavirus. J Virol Methods 66:303–309

Peiris JSM, Lai S, Poon LLM, Guan Y, Yam L, Lim W, Nicholls J, Yee W, Yan W, Cheung M, Chen V, Chan KH, Tsang D, Yung R, Ng T, Yuen KY (2003a) Coronavirus as a possible cause of severe acute respiratory syndrome. Lancet 361:1319–1325

Peiris JSM, Chu CM, Cheng VC, Chan KS, Hung IF, Poon LL, Law KI, Tang BS, Hon TY, Chan CS, Chan KH, Ng JS, Zheng BJ, Ng WL, Lai RW, Guan Y, Yuen KY, HKU/UCH SARS Study Group (2003b) Clinical progression and viral load in a community outbreak of coronavirus-associated SARS pneumonia: a prospective study. Lancet 361:1767–1772

Peltola H, Heinonen OP, Valle M, Paunio M, Virtanen M, Karanko V, Cantell K (1994) The elimination of indigenous measles, mumps, and rubella from Finland by a 12-year, two dose vaccination program. N Engl J Med 331:1397–1402

Peng J, Fan Z, Huang J (2012) Rapid detection of banana streak virus by loop mediated isothermal amplification assay in South China. J Phytopathol 160:248–250

Peyrefitte CN, Boubis L, Coudrier D, Bouloy M, Grandadam M, Tolou HJ, Plumet S (2008) Real-time reverse-transcription loop-mediated isothermal amplification for rapid detection of Rift valley fever virus. J Clin Microbiol 46(11):3653–3659

Pfannenstiel MA, Slack SA (1980) Response of potato cultivars to infection by the potato spindle tuber viroid. Phytopathology 70:922–926

Pham HM, Nakajima C, Oaci K, Onuma M (2005) Loop mediated isothermal amplification for rapid detection of Newcastle disease virus. J Clin Microbiol 43(5): 1646–1650

Poon LL, Wong OK, Chan KH, Luk W, Yuen KY, Peiris JS, Guan Y (2003) Rapid diagnosis of a coronavirus associated with severe acute respiratory syndrome (SARS). Clin Chem 49:953–955

Poon LLM, Leung CSW, Tashiro M, Chan KH, Wong BWY, Yuen KY, Guan Y, Peiris JSM (2004) Rapid detection of the Severe Acute Respiratory Syndrome (SARS) Coronavirus by a loop-mediated isothermal amplification assay. Clin Chem 50(6):1050–1052

Poon LLM, Leung CSW, Chan KH, Lee JHC, Yuen KY, Guan Y, Peiris JS (2005) Detection of human influenza

A viruses by loop-mediated isothermal amplification. J Clin Microbiol 43:427–430

Pugh RN, Akinosi B, Pooransingh S, Kumar J, Grant S, Livesley E, Linnane J, Ramaiah S (2002) An outbreak of mumps in the metropolitan area of Walsall, United Kingdom. Int J Infect Dis 6:283–287

Quinlivan M, Gershon AA, Steinberg SP, Breuer J (2005) An evaluation of single nucleotide polymorphisms used to differentiate vaccine and wild type strains of *Varicella zoster* virus. J Med Virol 75:174–180

Ratcliff RM, Chang G, Kok T, Sloots TP (2007) Molecular diagnosis of medical viruses. Curr Issues Mol Biol 9:87–102

Reddy AK, Balne PK, Reddy RK, Mathai A, Kaur I (2010) Loop-mediated isothermal amplification assay for the diagnosis of retinitis caused by herpes simplex virus-1. Clin Microbiol Infect 17:210–213

Ren X, Li P (2011) Development of reverse transcription loop-mediated isothermal amplification for rapid detection of porcine epidemic diarrhea virus. Virus Genes 42(2):229–235

Reynolds D, Garwes D (1979) Virus isolation and serum antibody-responses after infection of cats with transmissible gastroenteritis virus-brief report. Arch Virol 60:161–166

Rodak L, Nevorankova Z, Valicek L, Smitalova R (2005) Use of monoclonal antibodies in blocking ELISA detection of transmissible gastroenteritis virus in feces of piglets. J Vet Med B Infect Dis Vet Public Health 52:105–111

Rovira A, Abrahante J, Murtaugh M, Claudia MZ (2009) Reverse transcription lamp for the detection of porcine reproductive and respiratory syndrome virus. J Vet Diagn Invest 21(3):350–354

Saijo M, Ishii T, Kokubo M, Takimoto M, Takahashi Y (1993) Respiratory syncytial virus infection in lower respiratory tract and asthma attack in hospitalized children in North Hokkaido, Japan. Acta Paediatr Jpn 35:233–237

Saijo M, Takahashi S, Kokubo M, Saino T, In-Yaku F, Ishii T, Takimoto M, Takahashi Y (1994) The role of respiratory syncytial virus in acute bronchiolitis in small children in northern Japan. Acta Paediatr Jpn 36:371–374

Sappat A, Jaroenram W, Puthawibool T, Lomas T, Tuantranont A, Kiatpathomchai W (2011) Detection of shrimp Taura syndrome virus by loop-mediated isothermal amplification using a designed portable multichannel turbidimeter. J Virol Methods 175:141–148

Sawada E (1927) Control of wheat yellow mosaic virus. J Plant Prot 14:444–449

Schmaljohn C (1996) Bunyaviridae: the viruses and their replication. In: Fields BN, Knipe DM, Howley PM (eds) Fields virology, 3rd edn. Lippincott Raven Publishers, Raven Press, New York, pp 1447–1471

Schmidt NJ, Lennette EH, Ho HH (1974) An apparently new enterovirus isolated from patients with disease of the central nervous system. J Infect Dis 129: 304–309

Seah CLL, Chow VTK, Tan HC, Chan YC (1995) Rapid single-step RT-PCR typing of dengue viruses using NS3 gene primers. J Virol Methods 51:193–200

Shi W, Li K, Ji Y, Jiang Q, Shi M, Mi J (2011) Development and evaluation of reverse transcription-loop-mediated isothermal amplification assay for rapid detection of enterovirus 71. BMC Infect Dis 11:197

Shinde V, Bridges CB, Uyeki TM, Shu B, Balish A, Xu X, Lindstrom S, Gubareva LV, Deyde V, Garten RJ, Harris M, Gerber S, Vagasky S, Smith F, Pascoe N, Martin K, Dufficy D, Ritger K, Conover C, Quinlisk P, Klimov A, Bresee JS, Finelli L (2009) Triple reassortant swine influenza A (H1) in humans in the United States, 2005–2009. N Engl J Med 360:2616–2625

Shirato K, Nishimura H, Saijo M, Okamoto M, Noda M, Tashiro M, Taguchi M (2007) Diagnosis of human respiratory syncytial virus infection using reverse transcription loop mediated isothermal amplification. J Virol Methods 139:78–84

Shors T (2008) Understanding viruses. Jones and Bartlett Publishers, Sudbury. ISBN 978-0-7637-2932-9

Singh RP (1973) Experimental host range of the potato spindle tuber "virus". Am Potato J 50:111–123

Smith GJ, Vijaykrishna D, Bahl J, Lycett SJ, Worobey M, Pybus OG, Ma SK, Cheung CL, Raghwani J, Bhatt S, Peiris JS, Guan Y, Rambaut A (2009) Origins and evolutionary genomics of the 2009 swine-origin H1N1 influenza A epidemic. Nature 459:1122–1125

Soliman H, Midtlyng PJ, El-Matbouli M (2009) Sensitive and rapid detection of infectious pancreatic necrosis virus by reverse transcription loop mediated isothermal amplification. J Virol Methods 158:77–83

Soloman T, Thao LTT, Dung NM, Kneen R, Hung NT, Nisalak A, Vaughn DW, Farrar J, Hien TT, White JN, Cardosa MJ (1998) Rapid diagnosis of Japanese encephalitis by using an immunoglobulin M dot enzyme immunoassay. Clin Microbiol 36:2030–2034

Sudiro TM, Ishiko H, Green S, Vaughn DW, Nisalak A, Kalayanarooj S, Rothman AL, Raengsakulrach B, Janus J, Kurane I, Ennis FA (1997) Rapid diagnosis of dengue viremia by reverse transcriptase polymerase chain reaction using 3-noncoding region universal primers. Am J Trop Med Hyg 56:424–429

Suebsing R, Jeon CH, Oh MJ, Kim JH (2011) Reverse transcriptase loop-mediated isothermal amplification assay for infectious hematopoietic necrosis virus in *Oncorhynchus keta*. Dis Aquat Organ 94:1–8

Thai H, Le M, Vuong C, Parida M, Minekawa H, Notomi T, Hasebe F, Morita K (2004) Development and evaluation of a novel loop-mediated isothermal amplification method for rapid detection of severe acute respiratory syndrome coronavirus. J Clin Microbiol 42(5):1956

Theil V (2007) Coronaviruses: molecular and cellular biology. Caister Academic Press, Norfolk. ISBN 78-1-904455-16-5

Thomas C (2008) Animal viruses: molecular biology. Caister Academic Press, Norfolk. ISBN 978-1-904455-22-6

Tomita N, Mori Y, Kanda H, Notomi T (2008) Loop-mediated isothermal amplification (LAMP) of gene sequences and simple visual detection of products. Nat Protoc 3:877–882

Toriniwa H, Komiya T (2006) Rapid detection and quantification of Japanese encephalitis virus by real-time reverse transcription loop-mediated isothermal amplification. Microbiol Immunol 50:379–387

Tsutsumi N, Yanagisawa H, Fujiwara Y, Ohara T (2010) Detection of potato spindle tuber viroid by reverse transcription loop mediated isothermal amplification. Res Bull Plant Prot Serv Jpn 46:61–67

Ushio M, Yui I, Yoshida N, Fujino M, Yonekawa T, Ota Y, Notomi T, Nakayama T (2005) Detection of respiratory syncytial virus genome by subgroups-A, B specific reverse transcription loop-mediated isothermal amplification (RT-LAMP). J Med Virol 77:121–127

Viruses of animals by George Durrell. Published online 13 Oct 2008. http://ezinearticles.com/?Viruses-of-Animals&id=1558098

Wakabayashi T, Yamashita R, Kakita T, Kakita M, Oshika T (2004) Rapid and sensitive diagnosis of adenoviral keratoconjunctivitis by loop-mediated isothermal amplification (LAMP) method. Curr Eye Res 29(4–5):219–224

Wang M, Liu G, Lu X (1980) A primary confirmation on wheat spindle streak mosaic virus occurring in China. Agric Sci Technol Sichuan 1:34–35

Wang Y, Kang Z, Gao H, Gao Y, Qin L, Lin H, Yu F, Qi X, Wang X (2011) A one-step reverse transcription loop-mediated isothermal amplification for detection and discrimination of infectious bursal disease virus. Virol J 8:108

Webby RJ, Webster RG (2003) Are we ready for pandemic influenza? Science 302:1519–1522

Welman M, Arora DJ (2000) Genomic analysis of matrix gene and antigenic studies of its gene product (M1) of a swine influenza virus (H1N1) causing chronic respiratory disease in pigs. Virus Genes 21:157–165

Whiting SH, Champoux JJ (1998) Properties of strand displacement synthesis by Moloney murine leukemia virus reverse transcriptase: mechanistic implications. J Mol Biol 8(278):559–577

WHO (2009a) WHO information for laboratory diagnosis of new influenza A (H1N1) virus in humans. Posting date 21 May 2009. WHO, Geneva. http://www.who.int/csr/resources/publications/swineflu/WHO_Diagnostic_RecommendationsH1N1_20090521.pdf

WHO (2009b) CDC protocol of realtime RTPCR for influenza A (H1N1). Posting date 30 April 2009. WHO, Geneva. http://www.who.int/csr/resources/publications/swineflu/CDCRealtimeRTPCR_SwineH1Assay-2009_20090430.pdf

World Health Organization (1997) Dengue haemorrhagic fever: diagnosis, treatment and control, 2nd edn. World Health Organization, Geneva

Xie QM, Ji J, Pickens TT, Du LQ, Cao YC (2010) Rapid detection of infectious laryngotracheitis virus isolates by loop-mediated isothermal amplification. J Virol Methods 165(1):71–75

Xie Z, Tang Y, Fan Q, Liu J, Pang Y, Deng X, Xie Z, Peng Y, Xie L, Khan MI (2011) Rapid Detection of Group I avian adenoviruses by a loop-mediated isothermal amplification. Avian Dis 55(4):575–579

Yoshida N, Fujino M, Miyata A, Nagai T, Kamada M, Sakiyama H, Ihara T, Kumagai T, Okafuji T, Okafuji T, Nakayama T (2008) Mumps virus reinfection is not a rare event confirmed by reverse transcription loop-mediated isothermal amplification. J Med Virol 80:517–523

Zeller H, Bouloy M (2000) Infections by viruses of the families Bunyaviridae and Filoviridae. Rev Sci Tech 19:79–91

Zhang H, Thekisoe O, Aboge G, Kyan H, Yamagishi J, Inoue N, Nishikawa Y (2009) Toxoplasma gondii: sensitive and rapid detection of infection by loop-mediated isothermal amplification (LAMP) method. Exp Parasitol 122:47–50

Zhang ZY, Liu XJ, Li DW, Yu JL, Han CG (2011) Rapid detection of wheat yellow mosaic virus by reverse transcription loop-mediated isothermal amplification. Virol J 8:550

Zhao K, Shi W, Han F, Xu Y, Zhu L, Zou Y, Wu X, Zhu H, Tan F, Tao S, Tang X (2011) Specific, simple and rapid detection of porcine circovirus type 2 using the loop mediated isothermal amplification method. Virol J 8:126

Zhong NS, Zheng BJ, Li YM, Poon LLM, Xie ZH, Chan KH, Li PH, Tan SY, Chang Q, Xie JP, Liu XQ, Xu J, Li DX, Yuen KY, Peiris GY (2003) Epidemiology and cause of severe acute respiratory syndrome (SARS) in Guangdong, People's Republic of China, in February, 2003. Lancet 362:1353–1358

About the Editors

R.K. Salar, Ph.D., is an Associate Professor and Chairperson of the Department of Biotechnology, Chaudhary Devi Lal University, Sirsa, India. He received his postdoctoral fellowship from the Ministry of Education, Slovak Republic, to pursue his postdoctorate from Department of Biochemical Technology, Slovak University of Technology in Bratislava, Slovakia, and visited several other countries including Japan, Norway, Austria, and Hungary. His current interests are in fermentation technology and harnessing medicinal values of plants and microbes. Dr. Salar also received several research and development projects from UGC, New Delhi; HSCST, Chandigarh, and international travel grant from DST, New Delhi. He is a Reviewer of *Journal of Basic Microbiology*, *Annals of Applied Biology* (Wiley-VCH, Germany), and *Frontiers in Life Sciences* (Taylor & Francis, UK).

Dr. S.K. Gahlawat, Ph.D., is a Professor at the Department of Biotechnology and Dean of the Faculty of Life Sciences, Chaudhary Devi Lal University, Sirsa, India. He has been awarded Boyscast fellowship and DBT Overseas Associateship for doing postdoctoral research in FRS Marine Laboratory, Aberdeen, UK. He has received several R&D projects from UGC, ICAR, and Haryana Govt. His research interests include development of molecular diagnostic methods for bacterial and viral diseases. He is a Referee of *International Journal of Aquaculture*, Netherlands, and *Journal of Applied Microbiology*.

Dr. P. Siwach, Ph.D., is an Associate Professor at the Department of Biotechnology, Chaudhary Devi Lal University, Sirsa, India. She has received R&D projects from UGC, New Delhi. Her current research areas include molecular marker analysis for improvement of quality traits, micropropagation, and biochemical characterization of medicinal plants.

Dr. J.S. Duhan, Ph.D., is an Assistant Professor at the Department of Biotechnology, Chaudhary Devi Lal University, Sirsa, India. He has received R&D projects from UGC and CSIR, New Delhi. His specialization includes microbial biotechnology and biological nitrogen fixation.

R.K. Salar et al. (eds.), *Biotechnology: Prospects and Applications*,
DOI 10.1007/978-81-322-1683-4, © Springer India 2013

Index

R.K. Salar et al. (eds.), *Biotechnology: Prospects and Applications*,
DOI 10.1007/978-81-322-1683-4, © Springer India 2013

Printed by Printforce, the Netherlands